Microneedle Technology

Microneedle Technology

Advances and Biomedical Applications

Edited by

Executive Editors

Chuanbin Wu
Jinan University, China

Xin Pan
Sun Yat-sen University, China

Associate Editors

Tingting Peng
Jinan University, China

Guilan Quan
Jinan University, China

ELSEVIER

Elsevier
Radarweg 29, PO Box 211, 1000 AE Amsterdam, Netherlands
125 London Wall, London EC2Y 5AS, United Kingdom
50 Hampshire Street, 5th Floor, Cambridge, MA 02139, United States

Notices

Knowledge and best practice in this field are constantly changing. As new research and experience broaden our understanding, changes in research methods, professional practices, or medical treatment may become necessary.

Practitioners and researchers must always rely on their own experience and knowledge in evaluating and using any information, methods, compounds, or experiments described herein. In using such information or methods they should be mindful of their own safety and the safety of others, including parties for whom they have a professional responsibility.

To the fullest extent of the law, neither the Publisher nor the authors, contributors, or editors, assume any liability for any injury and/or damage to persons or property as a matter of products liability, negligence or otherwise, or from any use or operation of any methods, products, instructions, or ideas contained in the material herein.

ISBN: 978-0-443-45544-5

Publisher: Matthew Deans
Acquisitions Editor: Glyn Jones
Editorial Project Manager: Akansha Rathi
Production Project Manager: Paul Prasad Chandramohan
Cover Designer: Mark Rogers

Typeset by TNQ Tech

Contents

7. Microneedle technology for diabetes therapy

Wenhao Wang and Chuanbin Wu

8. Microneedle technology for antitumor therapy

Siyuan Peng, Guilan Quan and Xin Pan

11. Microneedle technology for treating skin bacterial infections

Tingting Peng, Weishi Ye and Xin Pan

12. Microneedle technology for cosmetic applications

Yanping Fu and Guilan Quan

13. Microneedle technology for biological monitoring and diagnosis

Ying Chen and Guilan Quan

Contributors

Tianxiang Chen, School of Mechatronic Engineering, Guangdong Polytechnic Normal University, Guangzhou, China

Ying Chen, State Key Laboratory of Bioactive Molecules and Druggability Assessment/International Cooperative Laboratory of Traditional Chinese Medicine Modernization and Innovative Drug Development of Ministry of Education (MOE) of China/College of Pharmacy, Jinan University, Guangzhou, China

Yanping Fu, State Key Laboratory of Bioactive Molecules and Druggability Assessment/International Cooperative Laboratory of Traditional Chinese Medicine Modernization and Innovative Drug Development of Ministry of Education (MOE) of China/College of Pharmacy, Jinan University, Guangzhou, China

Mingyu Gong, State Key Laboratory of Bioactive Molecules and Druggability Assessment/International Cooperative Laboratory of Traditional Chinese Medicine Modernization and Innovative Drug Development of Ministry of Education (MOE) of China/College of Pharmacy, Jinan University, Guangzhou, China

Wanshan Hu, State Key Laboratory of Bioactive Molecules and Druggability Assessment/International Cooperative Laboratory of Traditional Chinese Medicine Modernization and Innovative Drug Development of Ministry of Education (MOE) of China/College of Pharmacy, Jinan University, Guangzhou, China

Fan Jia, School of Pharmaceutical Science, Sun Yat-Sen University, Guangzhou, China

Zeshi Jiang, School of Pharmaceutical Science, Sun Yat-Sen University, Guangzhou, China

Jubo Jian, State Key Laboratory of Bioactive Molecules and Druggability Assessment/International Cooperative Laboratory of Traditional Chinese Medicine Modernization and Innovative Drug Development of Ministry of Education (MOE) of China/College of Pharmacy, Jinan University, Guangzhou, China

Xin Pan, School of Pharmaceutical Science, Sun Yat-Sen University, Guangzhou, China

Huanhuan Pan, State Key Laboratory of Bioactive Molecules and Druggability Assessment/International Cooperative Laboratory of Traditional Chinese Medicine Modernization and Innovative Drug Development of Ministry of Education (MOE) of China/College of Pharmacy, Jinan University, Guangzhou, China

Tingting Peng, State Key Laboratory of Bioactive Molecules and Druggability Assessment/International Cooperative Laboratory of Traditional Chinese Medicine Modernization and Innovative Drug Development of Ministry of Education (MOE) of China/College of Pharmacy, Jinan University, Guangzhou, China

Siyuan Peng, School of Pharmaceutical Science, Sun Yat-Sen University, Guangzhou, China

Guilan Quan, State Key Laboratory of Bioactive Molecules and Druggability Assessment/International Cooperative Laboratory of Traditional Chinese Medicine Modernization and Innovative Drug Development of Ministry of Education (MOE) of China/College of Pharmacy, Jinan University, Guangzhou, China

Tao Ren, State Key Laboratory of Bioactive Molecules and Druggability Assessment/International Cooperative Laboratory of Traditional Chinese Medicine Modernization and Innovative Drug Development of Ministry of Education (MOE) of China/College of Pharmacy, Jinan University, Guangzhou, China

Wenhao Wang, School of Pharmaceutical Science, Sun Yat-Sen University, Guangzhou, China

Chuanbin Wu, State Key Laboratory of Bioactive Molecules and Druggability Assessment/International Cooperative Laboratory of Traditional Chinese Medicine Modernization and Innovative Drug Development of Ministry of Education (MOE) of China/College of Pharmacy, Jinan University, Guangzhou, China

Yuhuan Wu, School of Pharmaceutical Science, Sun Yat-Sen University, Guangzhou, China

Wentao Wu, School of Pharmaceutical Science, Sun Yat-Sen University, Guangzhou, China

Weishi Ye, School of Pharmaceutical Science, Sun Yat-Sen University, Guangzhou, China

Meihong Zhang, State Key Laboratory of Bioactive Molecules and Druggability Assessment/International Cooperative Laboratory of Traditional Chinese Medicine Modernization and Innovative Drug Development of Ministry of Education (MOE) of China/College of Pharmacy, Jinan University, Guangzhou, China

Xinyu Zhang, State Key Laboratory of Bioactive Molecules and Druggability Assessment/International Cooperative Laboratory of Traditional Chinese Medicine Modernization and Innovative Drug Development of Ministry of Education (MOE) of China/College of Pharmacy, Jinan University, Guangzhou, China

Chunxian Zhou, State Key Laboratory of Bioactive Molecules and Druggability Assessment/International Cooperative Laboratory of Traditional Chinese Medicine Modernization and Innovative Drug Development of Ministry of Education (MOE) of China/College of Pharmacy, Jinan University, Guangzhou, China

Preface

Transdermal Drug Delivery Systems (TDDS) occupy a pivotal position in the global pharmaceutical market owing to their high safety, sustained-controlled drug release, and exceptional patient compliance. Research and Markets projects the global TDDS market to surpass $90 billion by 2025, growing at a compound annual growth rate (CAGR) of 11.6%. Among emerging physical enhancement technologies, microneedles (MNs) have attracted significant attention in recent years. Comprising micron-scale needle arrays, microneedle patches penetrate the stratum corneum to create microchannels, effectively bridging the non-invasiveness of transdermal delivery with the high efficiency of injections. The key advantages of microneedles include: (1) Breaking traditional barriers: Unlike conventional transdermal methods, microneedles enable precise delivery of diverse therapeutics (*e.g.*, small molecules, proteins, drug-loaded particles, and cells), regardless of molecular weight, hydrophilicity, or size limitations; (2) Improved patient experience: Their bandage-like design simplifies administration and reduces pain by avoiding subcutaneous nerve stimulation. Additionally, microneedles offer solid-state storage, enhancing biologic stability while lowering cold-chain logistics costs. Integrated with point-of-care testing (POCT), they further support biofluid sampling for early disease diagnosis. These transformative benefits earned microneedle technology the top spot in *Scientific American* and the World Economic Forum's *Top 10 Emerging Technologies of 2020.*

The evolution of microneedle technology traces back to Gerstel and Place's conceptual proposal (ALZA Corporation, 1976), though microfabrication constraints delayed practical implementation until Hashi's team developed silicon-based microneedle patches in 1995. A landmark study by Henry's team (Georgia Institute of Technology, 1998) demonstrated their efficacy, showing a >1000-fold enhancement in calcein permeation. Subsequent milestones included Miyano's dissolving microneedles (Ritsumeikan University, 2005) and Donnelly's hydrogel microneedles (Queen's University Belfast, 2012). Recent advances have diversified microneedle designs beyond solid/coated/hollow types to innovative architectures like cryomicroneedles, core-shell structures, and layered systems. Coupled with micro/nanotechnology and smart biomaterials, responsive microneedles now achieve precise spatiotemporal control over drug loading and release.

Microneedle technology has evolved beyond transdermal drug delivery to include mucosal administration (*e.g.*, oral and ocular applications), demonstrating considerable clinical promise in treating infectious diseases, cancer, diabetes, and autoimmune disorders. To overcome critical challenges such as mechanical performance and drug delivery efficiency, researchers have developed systematic quality evaluation frameworks and proposed innovative strategies to enhance mechanical strength and therapeutic efficacy. Despite exponential growth in global microneedle research, industrialization progress remains sluggish. Of the 100+ microneedle-related clinical trials worldwide, most are in Phase I/II, and no microneedle-based drug has yet reached the market. The key industrialization bottlenecks of microneedle products include: (1) Drug-loading constraints: Coated microneedles face limited capacity, while polymeric variants struggle to meet clinical dosage requirements due to the microliter-scale volume of needle tips; (2) Manufacturing complexity: Micron-level precision and stringent sterile production requirements present significant technical barriers; (3) Regulatory gaps: Unified quality standards, safety evaluation protocols, and end-to-end regulatory systems are urgently needed to accelerate translation.

This book provides a comprehensive overview of microneedle technology, covering its fundamental principles, characterization methods, fabrication materials/techniques, and constructive strategies to address key challenges. It also explores applications in disease treatment and biosensing-based diagnostics, offering readers a clear and efficient understanding of the field. By deepening knowledge of microneedle systems, this work aims to empower researchers to design higher-performance products, ultimately bridging the gap between laboratory innovation and clinical adoption.

Chapter 1

Classification and development history of microneedles

Wanshan Hu, Tingting Peng and Chuanbin Wu
State Key Laboratory of Bioactive Molecules and Druggability Assessment/International Cooperative Laboratory of Traditional Chinese Medicine Modernization and Innovative Drug Development of Ministry of Education (MOE) of China/College of Pharmacy, Jinan University, Guangzhou, China

1. Introduction

Microneedles (MNs) are composed of hundreds of tiny needle tips, each with a length ranging from 25 to 2000 μm, integrated into a base in an array formation. They can directly penetrate the stratum corneum barrier, creating hundreds of microscopic channels for drug delivery without reaching subcutaneous pain nerves and blood vessels, thereby reducing pain.[1,2] As a result, MNs combine the dual advantages of transdermal and injectable preparations, overcoming the limitations of traditional transdermal preparations on molecular weight (<500 Da) and lipophilicity ($1 < \log P < 3$).[3,4] In 2020, MNs were ranked first among the top 10 emerging technologies with the potential to change the world by Scientific American and the World Economic Forum. As a revolutionized transdermal delivery system (TDDS), MNs have attracted significant attention in the biomedical field. In this chapter, we primarily describe the development, advantages, and classification of MNs.

2. The development of microneedle technology

Since ancient times, humans have sought minimally invasive methods to alleviate pain during medical treatments. The origins of microneedles (MNs) can be traced back several centuries to the acupuncture practices of ancient China. However, the modern understanding of MNs has developed in tandem with advancements in human anatomy and medical science (Fig. 1.1). The term "microneedle" was first introduced in 1921 when Robert Chambers used

Microneedle Technology. https://doi.org/10.1016/B978-0-443-45544-5.00012-9

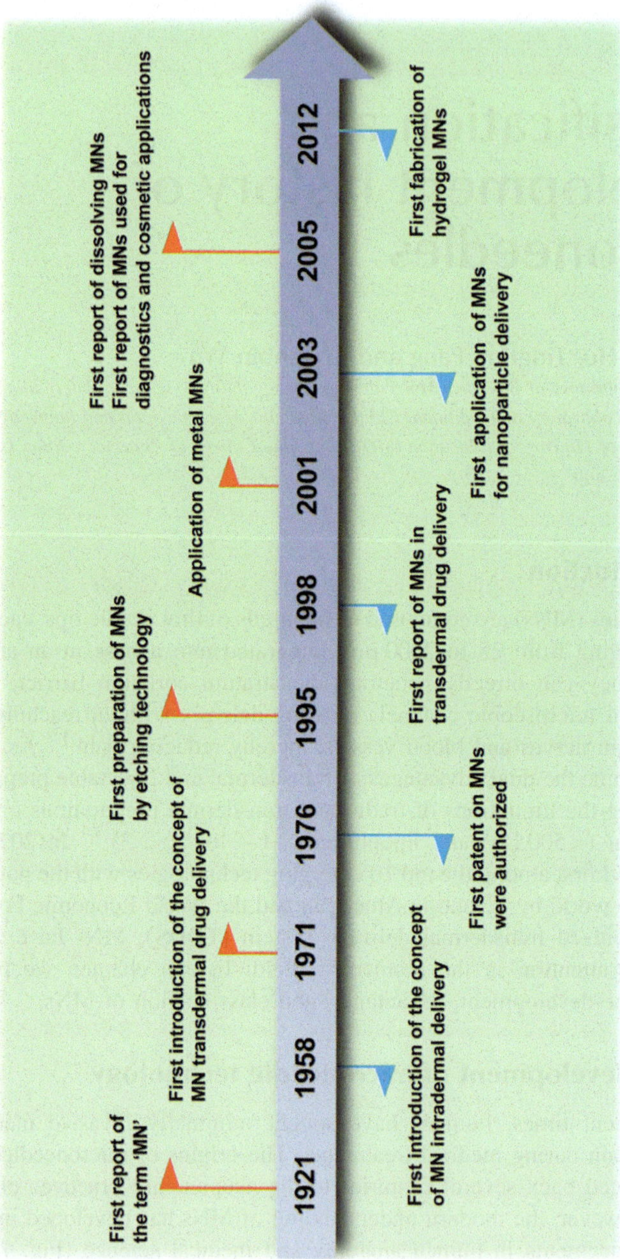

FIGURE 1.1 Schematic illustrating the major events in the development of MNs.

a needle to inject into the nucleus of an echinoderm egg for microdissection purposes.[5] Later, in 1958, Alan Richard Wagner proposed the concept of intradermal injection using MNs and filed a patent. However, limitations in fabrication technology at the time prevented practical MN production, and no significant breakthroughs were achieved before the patent expired in 1979.

The exploration of MNs as a novel TDDS began in 1976, when Martin S. Gerstel and Virgil A. Place of Alza Corporation filed the first patent for MNs.[6,7] In this patent, MNs are defined as a drug delivery device comprising multiple projections extending from a drug reservoir, designed to penetrate the skin for local or systemic drug delivery. This patent laid the foundation for the application of MNs in TDDS. With the advancement of microelectronics and microfabrication techniques, significant progress in MN technology occurred in the 1990s. These developments enabled the production of MNs with diverse geometric parameters and materials, solidifying their role in the pharmaceutical field.[7] In 1998, Henry et al.[8] demonstrated the first practical application of MN technology for the transdermal delivery of calcein. This study showed that pretreatment of skin with MNs enhanced skin permeability by up to four orders of magnitude compared to traditional topical creams.

In 2001, Alza Corporation published a research report on the transdermal delivery of antisense oligonucleotides using solid metal MNs. Two years later, McAllister et al.[9] conducted the first study exploring the feasibility of MN-assisted transdermal delivery of macromolecules and nanoparticles. Using both solid and hollow MNs, they successfully delivered insulin, albumin, and 100 nm latex beads into human cadaver skin. In 2005, Miyano et al.[10] introduced the first case of dissolving MNs. They developed maltose MNs loaded with ascorbic acid-2-glucoside as a model drug and conducted trials on healthy human volunteers. The results showed that the MNs were well-tolerated, dissolved in the skin, and released ascorbic acid into the epidermis and dermis. In the same year, two significant advancements were reported: the use of hollow glass MNs for extracting skin interstitial fluid (ISF) for glucose monitoring[11,12] and their application in cosmetic procedures, specifically collagen induction therapy.[19]

In 2012, Donnelly et al.[13] pioneered the fabrication of hydrogel MNs made from crosslinked polymers, significantly enhancing the transdermal delivery efficiency of macromolecules and expanding the range of possible cargoes. Since then, MNs have found applications beyond cosmetics and drug delivery, including biological fluid sampling, allergy testing, vaccination, and photodynamic therapy. In recent years, the rapid advancement of MN technology has led to a substantial increase in the number of research publications and their citation rates, reflecting the growing interest and innovation in this field.

3. Advantages of microneedles

In the past decades, microneedles (MNs) have been increasingly exploited to treat various diseases, owing to their notable advantages (Fig. 1.2). The advantages of MNs are detailed below.

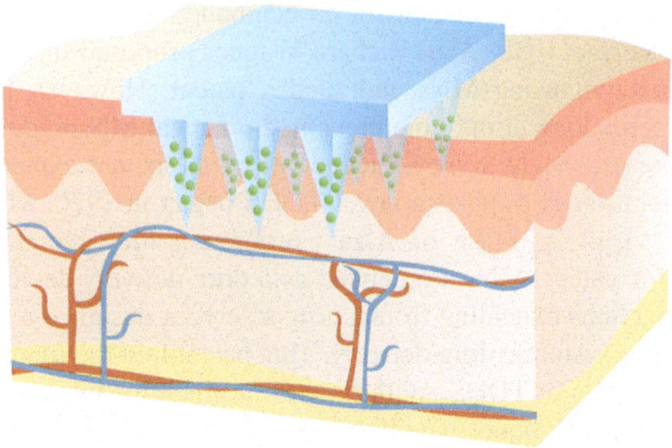

FIGURE 1.2 The schematic illustration of MN-mediated transdermal delivery.

(1) **Improving drug delivery efficiency**: MNs can effectively penetrate the stratum corneum barrier, demonstrating significantly enhanced drug delivery efficiency compared to conventional TDDS.[8] Furthermore, MNs exhibit remarkable versatility in therapeutic delivery, capable of administering bioactive agents, including chemical compounds, proteins, nanoparticles, microparticles, and even living cells and bacteria.[14]
(2) **Reducing side effects**: MNs can deliver drugs directly to skin tissue, preventing drugs from entering the systemic circulation and reducing side effects.[15] Therefore, MNs are especially suitable for the treatment of skin diseases.
(3) **Enhancing patient compliance**: Unlike traditional hypodermic needles, optimally sized MNs penetrate the skin without reaching dermal pain receptors or blood vessels, providing a pain-free administration method that significantly improves patient compliance, especially among individuals with needle phobia.[16] Moreover, their user-friendly design for self-administration further enhances treatment adherence.
(4) **Offering a user-friendly and cost-effective administration route**: MNs, as compact solid formulations, offer exceptional portability and storage convenience. Their application mode resembles that of transdermal patches, enabling patient self-administration and reducing reliance on medical professionals. This user-centric design significantly decreases the necessity for repeated clinical visits, thereby alleviating the burden on healthcare systems. Moreover, by enhancing immunogenicity through precise delivery to immunologically active skin layers, MNs demonstrate particular efficacy in mass vaccination campaigns against epidemic infectious diseases, effectively minimizing risks associated with large-scale public gatherings.[17]
(5) **Showing higher ecological benefit**: MNs composed of biodegradable materials, including polymers, polysaccharides, and proteins, can

dissolve within the skin without leaving residual needle fragments. In contrast to conventional injections, biodegradable MNs eliminate the need for medical waste disposal and reduce the risk of needle-related infections, thereby showing greater ecological benefits.[18]

(6) **Offering physiotherapy**: Beyond drug delivery, the mechanical stimulation of skin tissues induced by microneedle puncture can activate cell signaling pathways, converting mechanical forces into biochemical signals within cells, thereby generating therapeutic effects. The physical therapeutic benefits of MNs have been demonstrated in various tissue regeneration applications, including promoting hair regeneration,[19] accelerating wound healing,[20] and enhancing collagen synthesis.[21]

4. Classification of microneedles

Since the advent of microneedles (MNs), researchers have developed various types of MNs to cater to the requirements of diverse applications (Fig. 1.3). The evolution and utilization of MNs have become increasingly convenient and environmentally sustainable. The distinct categories of MNs are outlined as follows.

4.1 Three-step administration of microneedles

4.1.1 Solid microneedles

Solid microneedles (MNs) represent the first and second generations of MNs. Solid MNs usually have enough mechanical strength to pierce the cuticular barrier as their constituent materials are rigid.[22] During drug administration, solid MNs are used to pierce the skin to create microchannels, and then, they are removed from the insertion sites and replaced with drug-loaded patches, creams, solutions, or a backing layer (Fig. 1.3a). As the created microchannels are in a dynamic recovery process and might be closed during administration, the permeation of drugs from solid MNs is usually passive, resulting in poor efficiency of delivery and low accuracy of the dose.[23] Additionally, silicon or metal-based solid MNs are biologically incompatible, and needle breakage in the skin might be harmful to the health of the patient.

4.1.2 Hollow microneedles

Hollow microneedles (MNs) are microsyringes, that is, the length and diameter of their needles are at the micrometer level. The needles used in hollow MNs have an empty cavity to hold a drug solution and for dispersion. The drug delivery through hollow MNs usually includes the following three steps (Fig. 1.3b): The hollow MNs first pierce the skin, and then, the drugs are injected into the skin through the empty cavity under pressure or by electric driving.[24] After administration, the hollow MNs are pulled out from the skin

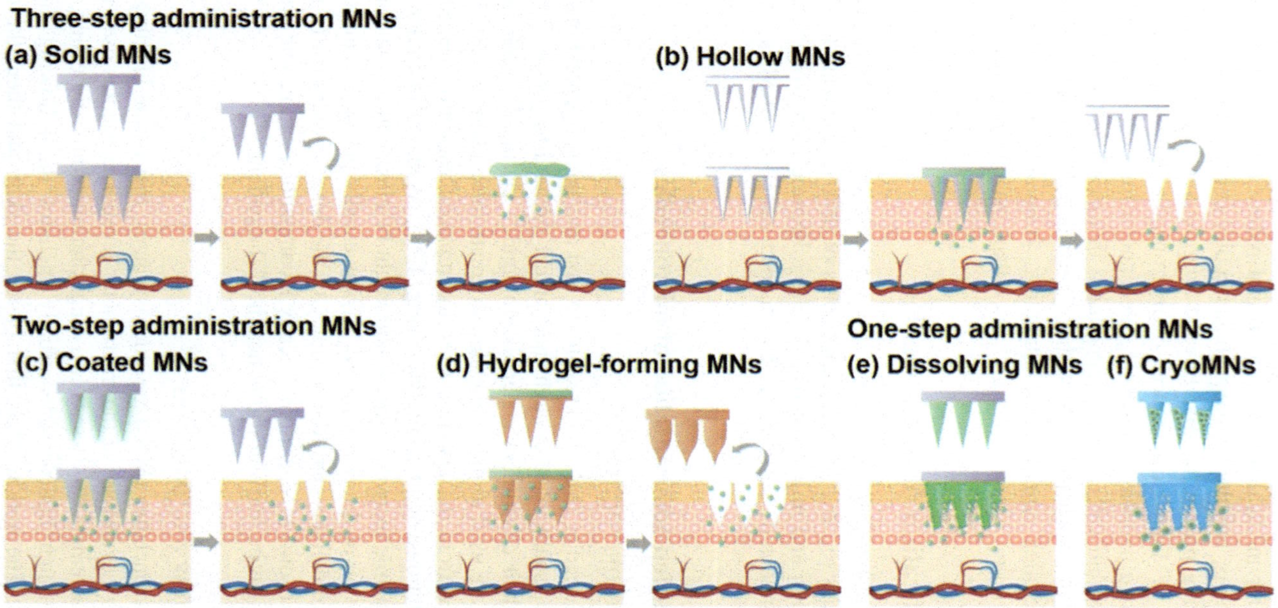

FIGURE 1.3 Classification of MNs. *Reprinted from Ref. 40 with permission.*

for recovery. Compared to other types of MNs, hollow MNs offer a flexible flow rate for accurately delivering high doses of drugs and are primarily used for assisting the delivery of biomacromolecules, including proteins,[25] genes,[26] and vaccines.[27] However, hollow MNs are difficult to manufacture, and they also have the risk of needle fracture and lumen blockage during administration.[28]

4.2 Two-step administration of microneedles

4.2.1 Coated microneedles

In coated microneedles (MNs), the drug formulation is coated onto the surface of solid MNs via dip coating,[29] inkjet deposition,[30] inkjet printing,[31] spray drying,[32] and micromolding.[33] When coated MNs are inserted into the skin, the coated cargo first dissolves in the cutaneous tissues, and the remaining MNs can be removed from the skin for recycling (Fig. 1.3c). The major issues associated with the preparation and administration of coated MNs include the low drug-loading capacity, as the capacity is restricted by the coating thickness and needle size, and low efficiency of delivery due to nonhomogeneous coating and premature drug loss.

4.2.2 Hydrogel-forming microneedles

Hydrogel-forming microneedles (MNs) consist of swellable polymeric needles and a drug-loaded backing layer. After the MNs are inserted into the skin, the needles absorb interstitial fluid from tissues and expand to create microchannels for continuous drug delivery (Fig. 1.3d).[34] After administration, the MNs can be removed from the insertion site. Compared to other MNs, hydrogel-forming MNs can resist the closure of skin orifices to a greater extent. They can achieve higher delivery efficiency and longer drug release by tuning the swellability of polymers.[35] They also have certain advantages over dissolving MNs in drug-loading, as the backing layer has a larger space to load drugs.[36]

4.3 One-step administration of microneedles

4.3.1 Dissolving microneedles

Dissolving microneedles (MNs) consist of drug-loaded needles attached to a backing layer. Drug administration by dissolving MNs is simplified into one step, where the MNs pierce the skin to release cargoes without additional manipulation (Fig. 1.3e).[15] Thus, a good selection of needle materials with enough mechanical strength and, hence, good skin penetration ability, is critical for preparing dissolving MNs. The mechanical properties of dissolving MNs might be susceptible to high humidity, high temperature, and physical stress. Therefore, these MNs need to be protected from fracture during their preparation, application, and storage. Overall, dissolving MNs have excellent biocompatibility and are suitable for patients who need long-term therapy.[37]

4.3.2 Cryomicroneedles

Cryomicroneedles (CryoMNs) are the latest type of MNs that are specifically designed to deliver living cells[38] or bacteria[39] for treating melanoma or eye infection, which not only keep the cargoes active but also enable MNs with sufficient mechanical strength to perform skin insertion. When CryoMNs are inserted into the skin, the needles made from biodegradable polymers dissolve to release the cargo, which further proliferates and promotes therapeutic effects (Fig. 1.3f). CryoMNs enhance the stability and bioactivity of living cargoes during preparation and long-term storage, but their application is still restricted by the requirements of aseptic manufacturing, storage, and transportation at ultralow temperatures, and limited cell volume.[38] However, CryoMNs can be used for producing MNs loaded with biopharmaceutics (e.g., exosomes, vaccines, genes, and biomacromolecules).

References

1. Waghule, T.; Singhvi, G.; Dubey, S. K.; Pandey, M. M.; Dua, K. Microneedles: A Smart Approach and Increasing Potential for Transdermal Drug Delivery System. *Biomed. Pharmacother.* **2019,** *109,* 1249–1258.
2. Sartawi, Z.; Blackshields, C.; Faisal, W. Dissolving Microneedles: Applications and Growing Therapeutic Potential. *J. Control Release* **2022,** *348,* 186–205.
3. Chowdhury, M. R.; Moshikur, R. M.; Wakabayashi, R.; Moniruzzaman, M.; Goto, M. Biocompatible Ionic Liquids Assisted Transdermal Co-delivery of Antigenic Protein and Adjuvant for Cancer Immunotherapy. *Int. J. Pharm.* **2021,** *601,* 120582.
4. Li, D.; Hu, D.; Xu, H.; Patra, H. K.; Liu, X.; Zhou, Z.; Tang, J.; Slater, N.; Shen, Y. Progress and Perspective of Microneedle System for Anti-cancer Drug Delivery. *Biomaterials* **2021,** *264,* 120410.
5. Chambers, R. Microdissection Studies, III. Some Problems in the Maturation and Fertilization of the Echinoderm Egg. *Biol. Bull.* **1921,** *41* (6), 318–350.
6. Iapichino, M.; Maibach, H.; Stoeber, B. Quantification Methods Comparing In Vitro and In Vivo Percutaneous Permeation by Microneedles and Passive Diffusion. *Int. J. Pharm.* **2023,** *638,* 122885.
7. Sabri, A. H.; Ogilvie, J.; Abdulhamid, K.; Shpadaruk, V.; McKenna, J.; Segal, J.; Scurr, D. J.; Marlow, M. Expanding the Applications of Microneedles in Dermatology. *Eur. J. Pharm. Biopharm.* **2019,** *140,* 121–140.
8. Henry, S.; McAllister, D. V.; Allen, M. G.; Prausnitz, M. R. Microfabricated Microneedles: A Novel Approach to Transdermal Drug Delivery. *J. Pharm. Sci.* **1998,** *87* (8), 922–925.
9. McAllister, D. V.; Wang, P. M.; Davis, S. P.; Park, J. H.; Canatella, P. J.; Allen, M. G.; Prausnitz, M. R. Microfabricated Needles for Transdermal Delivery of Macromolecules and Nanoparticles: Fabrication Methods and Transport Studies. *Proc. Natl. Acad. Sci. U. S. A.* **2003,** *100* (24), 13755–13760.
10. Miyano, T.; Tobinaga, Y.; Kanno, T.; Matsuzaki, Y.; Takeda, H.; Wakui, M.; Hanada, K. Sugar Micro Needles as Transdermic Drug Delivery System. *Biomed. Microdevices* **2005,** *7* (3), 185–188.

11. Wang, P. M.; Cornwell, M.; Prausnitz, M. R. Minimally Invasive Extraction of Dermal Interstitial Fluid for Glucose Monitoring Using Microneedles. *Diabetes Technol. Ther.* **2005,** *7* (1), 131–141.
12. Fernandes, D. Minimally Invasive Percutaneous Collagen Induction. *Oral Maxillofac. Surg. Clin. North Am.* **2005,** *17* (1), 51–63.
13. Donnelly, R. F.; Singh, T. R. R.; Garland, M. J.; Migalska, K.; Majithiya, R.; McCrudden, C. M.; Kole, P. L.; Mahmood, T. M. T.; McCarthy, H. O.; Woolfson, A. D. Hydrogel-forming Microneedle Arrays for Enhanced Transdermal Drug Delivery. *Adv. Funct. Mater.* **2012,** *22* (23), 4879–4890.
14. Zhang, J.; Li, H.; Albakr, L.; Zhang, Y.; Lu, A.; Chen, W.; Shao, T.; Zhu, L.; Yuan, H.; Yang, G.; et al. Microneedle-enabled Therapeutics Delivery and Biosensing in Clinical Trials. *J. Control Release* **2023,** *360,* 687–704.
15. Vora, L. K.; Sabri, A. H.; Naser, Y.; Himawan, A.; Hutton, A. R. J.; Anjani, Q. K.; Volpe-Zanutto, F.; Mishra, D.; Li, M.; Rodgers, A. M.; et al. Long-acting Microneedle Formulations. *Adv. Drug Deliv. Rev.* **2023,** *201,* 115055.
16. Al-Japairai, K. A. S.; Mahmood, S.; Almurisi, S. H.; Venugopal, J. R.; Hilles, A. R.; Azmana, M.; Raman, S. Current Trends in Polymer Microneedle for Transdermal Drug Delivery. *Int. J. Pharm.* **2020,** *587,* 119673.
17. Li, M.; Yang, L.; Wang, C.; Cui, M.; Wen, Z.; Liao, Z.; Han, Z.; Zhao, Y.; Lang, B.; Chen, H.; et al. Rapid Induction of Long-Lasting Systemic and Mucosal Immunity via Thermostable Microneedle-Mediated Chitosan Oligosaccharide-Encapsulated DNA Nanoparticles. *ACS Nano* **2023,** *17* (23), 24200–24217.
18. Yilmaz, E. G.; Ece, E.; Erdem, Ö.; Es, I.; Inci, F. A Sustainable Solution to Skin Diseases: Ecofriendly Transdermal Patches. *Pharmaceutics* **2023,** *15* (2), 579.
19. Younis, N.; Puigmal, N.; Kurdi, A. E.; Badaoui, A.; Zhang, D.; Morales-Garay, C.; Saad, A.; Cruz, D.; Rahy, N. A.; Daccache, A.; et al. Microneedle-Mediated Delivery of Immunomodulators Restores Immune Privilege in Hair Follicles and Reverses Immune-Mediated Alopecia. *Adv. Mater.* **2024,** *36* (31), e2312088.
20. Lyu, S.; Dong, Z.; Xu, X.; Bei, H. P.; Yuen, H. Y.; Cheung, C. W. J.; Wong, M. S.; He, Y.; Zhao, X. Going Below and Beyond the Surface: Microneedle Structure, Materials, Drugs, Fabrication, and Applications for Wound Healing and Tissue Regeneration. *Bioact. Mater.* **2023,** *27,* 303–326.
21. Ocampo-Garza, S. S.; Fabbrocini, G.; Ocampo-Candiani, J.; Cinelli, E.; Villani, A. Micro Needling: A Novel Therapeutic Approach for Androgenetic Alopecia, a Review of Literature. *Dermatol. Ther.* **2020,** *33* (6), e14267.
22. Babu, M. R.; Vishwas, S.; Khursheed, R.; Harish, V.; Sravani, A. B.; Khan, F.; Alotaibi, B.; Binshaya, A.; Disouza, J.; Kumbhar, P. S.; et al. Unravelling the Role of Microneedles in Drug Delivery: Principle, Perspectives, and Practices. *Drug. Deliv. Transl. Res.* **2024,** *14* (6), 1393–1431.
23. Avcil, M.; Çelik, A. Microneedles in Drug Delivery: Progress and Challenges. *Micromachines* **2021,** *12* (11), 1321.
24. Nguyen, T. T.; Nguyen, T. T. D.; Tran, N. M.; Vo, G. V. Advances of Microneedles in Hormone Delivery. *Biomed. Pharmacother.* **2022,** *145,* 112393.
25. Han, X.; Li, H.; Zhou, D.; Chen, Z.; Gu, Z. Local and Targeted Delivery of Immune Checkpoint Blockade Therapeutics. *Acc. Chem. Res.* **2020,** *53* (11), 2521–2533.
26. Dul, M.; Stefanidou, M.; Porta, P.; Serve, J.; O'Mahony, C.; Malissen, B.; Henri, S.; Levin, Y.; Kochba, E.; Wong, F. S.; et al. Hydrodynamic Gene Delivery in Human Skin Using a Hollow Microneedle Device. *J. Control Release* **2017,** *265,* 120–131.

27. Niu, L.; Chu, L. Y.; Burton, S. A.; Hansen, K. J.; Panyam, J. Intradermal Delivery of Vaccine Nanoparticles Using Hollow Microneedle Array Generates Enhanced and Balanced Immune Response. *J. Control Release* **2019,** *294,* 268–278.
28. Waghule, T.; Singhvi, G.; Dubey, S. K.; Pandey, M. M.; Gupta, G.; Singh, M.; Dua, K. Microneedles: A Smart Approach and Increasing Potential for Transdermal Drug Delivery System. *Biomed. Pharmacother.* **2019,** *109,* 1249–1258.
29. Caudill, C. L.; Perry, J. L.; Tian, S.; Luft, J. C.; DeSimone, J. M. Spatially Controlled Coating of Continuous Liquid Interface Production Microneedles for Transdermal Protein Delivery. *J. Control Release* **2018,** *284,* 122–132.
30. Boehm, R. D.; Jaipan, P.; Skoog, S. A.; Stafslien, S.; VanderWal, L.; Narayan, R. J. Inkjet Deposition of Itraconazole onto Poly(glycolic Acid) Microneedle Arrays. *Biointerphases* **2016,** *11* (1), 011008.
31. Uddin, M. J.; Scoutaris, N.; Klepetsanis, P.; Chowdhry, B.; Prausnitz, M. R.; Douroumis, D. Inkjet Printing of Transdermal Microneedles for the Delivery of Anticancer Agents. *Int. J. Pharm.* **2015,** *494* (2), 593–602.
32. McGrath, M. G.; Vrdoljak, A.; O'Mahony, C.; Oliveira, J. C.; Moore, A. C.; Crean, A. M. Determination of Parameters for Successful Spray Coating of Silicon Microneedle Arrays. *Int. J. Pharm.* **2011,** *415* (1–2), 140–149.
33. Chen, B. Z.; He, M. C.; Zhang, X. P.; Fei, W. M.; Cui, Y.; Guo, X. D. A Novel Method for Fabrication of Coated Microneedles with Homogeneous and Controllable Drug Dosage for Transdermal Drug Delivery. *Drug Deliv. Transl. Res.* **2022,** *12* (11), 2730–2739.
34. Turner, J. G.; White, L. R.; Estrela, P.; Leese, H. S. Hydrogel-Forming Microneedles: Current Advancements and Future Trends. *Macromol. Biosci.* **2021,** *21* (2), e2000307.
35. Li, X.; Zhao, Z.; Zhang, M.; Ling, G.; Zhang, P. Research Progress of Microneedles in the Treatment of Melanoma. *J. Control Release* **2022,** *348,* 631–647.
36. Migdadi, E. M.; Courtenay, A. J.; Tekko, I. A.; McCrudden, M. T. C.; Kearney, M. C.; McAlister, E.; McCarthy, H. O.; Donnelly, R. F. Hydrogel-Forming Microneedles Enhance Transdermal Delivery of Metformin Hydrochloride. *J. Control Release* **2018,** *285,* 142–151.
37. Lee, K. J.; Jeong, S. S.; Roh, D. H.; Kim, D. Y.; Choi, H. K.; Lee, E. H. A Practical Guide to the Development of Microneedle Systems—In Clinical Trials or on the Market. *Int. J. Pharm.* **2020,** *573,* 118778.
38. Chang, H.; Chew, S. W. T.; Zheng, M.; Lio, D. C. S.; Wiraja, C.; Mei, Y.; Ning, X.; Cui, M.; Than, A.; Shi, P.; et al. Cryomicroneedles for Transdermal Cell Delivery. *Nat. Biomed. Eng.* **2021,** *5* (9), 1008–1018.
39. Cui, M.; Zheng, M.; Wiraja, C.; Chew, S. W. T.; Mishra, A.; Mayandi, V.; Lakshminarayanan, R.; Xu, C. Ocular Delivery of Predatory Bacteria with Cryomicroneedles against Eye Infection. *Adv. Sci.* **2021,** *8* (21), e2102327.
40. Peng, T.; Chen, Y.; Hu, W.; Huang, Y.; Zhang, M.; Lu, C.; Pan, X.; Wu, C. Microneedles for Enhanced Topical Treatment of Skin Disorders: Applications, Challenges, and Prospects. *Engineering* **2023,** *30,* 170–189.

Chapter 2

Materials and fabrication methods of microneedles

Yuhuan Wu, Fan Jia, Weishi Ye and Xin Pan
School of Pharmaceutical Science, Sun Yat-Sen University, Guangzhou, China

1. Introduction

Microneedles (MNs) have emerged as a revolutionary transdermal drug delivery platform, offering advantages such as painless administration, enhanced patient compliance, and improved therapeutic efficacy.[1] The choice of materials and fabrication methods for MNs is paramount, as they critically influence the properties and applications of the microneedles.[2]

With the advancement of material science and manufacturing technologies, an increasing number of materials, ranging from traditional silicon and metals to modern biodegradable carbohydrates and polymers, are being explored for MN fabrication. Understanding the unique properties of these materials, such as their biocompatibility, biodegradability, and mechanical strength, is essential for tailoring MNs to meet specific therapeutic requirements. For instance, silicon, with its exceptional mechanical properties, enables the creation of MNs capable of reliable skin penetration,[3] whereas biodegradable carbohydrates and polymers offer flexibility and biocompatibility suitable for MN applications.[4] In parallel, microfabrication techniques have undergone substantial evolution, enabling the precise and scalable production of MNs with complex geometries and functional capabilities. Techniques such as photolithography, micromolding, and advanced 3D-printing methods not only expand the range of potential applications for MNs but also enhance their effectiveness in delivering drugs.[1]

This chapter mainly summarizes the materials and fabrication methods utilized in MN technology, aiming to guide future research directions and facilitate the development of innovative MNs for diverse therapeutic applications.

Microneedle Technology. https://doi.org/10.1016/B978-0-443-45544-5.00004-X

2. Materials used to fabricate microneedles

The choice of materials for MN fabrication significantly impacts their formability, mechanical properties, and application. The materials used for MN fabrication can be broadly classified into silicon, metals, ceramics, carbohydrates, and polymers, and each of them possesses unique characteristics that render them suitable for the preparation of different types of MNs. Silicon, exhibiting outstanding mechanical properties, enables the fabrication of MNs with reliable skin-penetration capability. Metals, known for their high mechanical strength and excellent electrical conductivity, are particularly well-suited for MN designs requiring precise control and durability. Ceramics, characterized by their high hardness and wear resistance, are especially appropriate for MNs intended for deep skin treatment or cutting applications. Carbohydrates, featuring excellent biocompatibility, tunable biodegradability, and natural abundance, are increasingly employed for creating dissolving MNs (DMNs) with controlled drug release profiles. Polymers, with their versatile physicochemical properties along with ease of processing and molding, are ideal for creating disposable or biodegradable MNs.

The ongoing advancement of novel materials, including biodegradable polymers, smart responsive materials, and surfaces with engineered properties, has significantly expanded the potential applications of MNs. These innovations not only enhance MNs' utility in traditional medical domains (e.g., drug delivery and vaccination) but also propel their adoption in emerging fields such as medical aesthetics, biosensing, and tissue engineering, thereby accelerating the evolution of MN technology.

2.1 Silicon

Silicon was among the earliest materials employed in microcrystalline components for drug delivery systems, as the capability to fabricate micron or submicron structures only became feasible in the 1990s with the development of industrial-grade high-precision microelectronic tools.[5] Silicon has demonstrated exceptional utility in microfabrication and microelectromechanical systems (MEMS) due to several inherent advantages.[6] Its foremost merit lies in the remarkable flexibility during the molding process, enabling facile production of microstructures with precisely controlled geometries and dimensions. Both monocrystalline and polycrystalline silicon variants offer customizable solutions adaptable to diverse application requirements. Furthermore, silicon possesses numerous advantageous physical properties that enhance its versatility as a functional material. Crucially, established silicon substrate manufacturing techniques provide unparalleled precision while supporting cost-effective mass production.[7]

Silicon is an anisotropic crystalline material whose mechanical properties vary significantly with crystallographic orientation, exhibiting elastic moduli ranging from 50 to 180 GPa.[8] Notably, these values are comparable to those of metallic alloys used in orthopedic implants. Silicon MNs demonstrate sufficient mechanical strength for reliable skin penetration, making them effective for transdermal drug delivery applications.[3] Owing to its manufacturability, silicon has been extensively employed to produce MNs with diverse geometries, including variations in tip sharpness, aspect ratios, and array densities.[9] Silicon can be used to prepare solid MNs, hollow MNs, and coated MNs.[5] However, silicon-based MN fabrication presents several challenges: (1) high material costs, (2) complex manufacturing process, (3) long manufacturing time, and (4) multistep etching procedures that increase yield variability.[10] Furthermore, the inherent brittleness of silicon may lead to MN fracture during insertion, posing risks of dermal retention and inflammatory responses.[11]

2.2 Metals

Metals have been extensively utilized in medical applications for decades, with stainless steel (e.g., hypodermic needles) and titanium (e.g., implants and prosthetics) being representative examples. Given the well-established safety profile of these materials in medical device manufacturing, their application in MN production may avoid additional safety concerns, thereby facilitating regulatory approval. The primary metals employed in MN fabrication include stainless steel, titanium, palladium, palladium-cobalt alloys, and nickel.[12] These materials exhibit excellent mechanical properties. For instance, medical-grade stainless steel (SUS316L) demonstrates a Young's modulus of approximately 180 GPa, while pure titanium and its alloys typically possess lower values. The widely used Ti-6Al-4V ELI alloy, common in implantable devices, has a Young's modulus of around 110 GPa.[13] Notably, the elastic modulus of metals is comparable to the upper range of silicon-based materials (up to 180 GPa). Moreover, metals surpass silicon in fracture toughness while maintaining comparable yield strength. These characteristics suggest that metals may offer superior suitability over silicon for MN fabrication.

2.2.1 Stainless steel

Stainless steel was the first metallic material used in MN production. As a special steel alloy, it exhibits excellent mechanical properties and corrosion resistance, making it the most common material for surgical implants, particularly AISI 316L and AISI 317L ultra-low carbon stainless steel variants. Consequently, stainless steel remains the most widely adopted material for metallic MNs. Stainless steel MNs can be fabricated through various methods, including excimer laser processing, infrared laser processing, femtosecond laser ablation, electrical discharge machining, and electrohydraulic atomization.[14]

2.2.2 *Titanium*

Titanium is an excellent alternative to stainless steel, especially in medical applications. As a silvery-white transition metal, titanium offers numerous advantages, including light weight, high strength, chemical stability, exceptional temperature resistance, acid and alkali corrosion resistance, and low density.[15] Although its mechanical properties are less robust than those of stainless steel, titanium remains sufficiently strong for biomedical uses. Notably, titanium MN has been developed primarily for applications such as biosensors and transdermal drug delivery systems.

2.3 Ceramics

Another material employed in MN fabrication is ceramics. Ceramic MNs are typically produced by casting ceramic slurries into micromolds using micromolding technology. This approach offers the advantage of cost-effective process development, as the technology can be readily scaled up for mass production.[5]

The primary ceramic material used for MN fabrication is alumina (Al_2O_3),[16] which offers several advantages, most notably its exceptional chemical resistance. The stability of Al_2O_3 arises from high-energy ionic and covalent bonds between aluminum and oxygen atoms, rendering the material highly resistant to corrosion and harsh environmental conditions. While alumina demonstrates strong compressive strength, it exhibits brittleness under tensile stress,[17] with notably lower tensile strength compared to metallic materials. This mechanical limitation was further corroborated by Bystrova et al.,[18] who observed fracture events during artificial skin insertion of Al_2O_3 MNs. Alumina-based MNs are typically employed as drug-coated solid arrays, utilizing a 'coat-and-poke' delivery mechanism. Interestingly, the inherent porosity of alumina ceramics enables these MNs to retain active substances within their open pore networks, enabling controlled drug release.[16] Beyond alumina, other biocompatible ceramics such as calcium sulfate dihydrate [Gypsum ($CaSO_4 \cdot 2H_2O$)] and calcium phosphate dihydrate [Brushite ($CaHPO_4 \cdot 2H_2O$)] have been utilized for MN fabrication.[19] These materials, previously established as drug-eluting bone cements, offer favorable mechanical properties and drug loading capacity.[20] Recent advancements have introduced organo-ceramic hybrid materials like Ormocer for MN production.[21] Comprising organically modified silicon alkoxides and organic monomers that form a 3D network,[22] these materials enable property customization through compositional adjustments and synthesis parameter optimization to meet specific application requirements.

2.4 Silica glass

Silica glass is an alternative to the above materials. Glass MNs can be quickly fabricated into various geometries for small-scale laboratory use. This material is physiologically inert, allows for easy observation of fluid flow, and can be fabricated in dimensions similar to those of microfabricated MNs.[23] Glass MNs described in the literature are hollow and are commonly used to inject drugs through the cavity in the needles.[24] Borosilicate glass, with a lower elastic modulus (64 GPa), exhibits greater resilience. However, silica glass is inherently brittle, and both silicon and silica glass share similar fracture toughness values. A key limitation of glass MNs is their reliance on time-consuming manual fabrication processes.[25] As a result, while glass MNs remain useful for experimental research, their commercial application in drug delivery remains impractical.

2.5 Carbohydrates

MNs can be easily fabricated by molding hot melts or slurries of carbohydrate materials using silicon or metallic MNs as the primary template.[26] The drug is incorporated into the mixture before the formulation is poured into the mold. These MNs are designed to dissolve upon skin insertion, thereby releasing the drug payload.[27] Carbohydrates serve as excellent alternatives to other materials due to their low cost and favorable safety profile for human health.[28]

2.5.1 Hyaluronic acid

Hyaluronic acid (HA), a linear non-sulfated glycosaminoglycan that is a major component of the skin's extracellular matrix,[29] has emerged as one of the most promising materials for DMN fabrication due to its exceptional biocompatibility, biodegradability, non-immunogenicity, and high water solubility.[30] As a naturally occurring polysaccharide that plays crucial roles in dermal metabolism and tissue hydration, HA offers distinct advantages for transdermal drug delivery systems, including rapid dissolution in tissue fluids, excellent skin penetration capability, and complete clearance through interaction with ubiquitously expressed CD44 receptors.[31] While HA-based DMNs demonstrate sufficient mechanical strength for effective skin insertion, studies have shown that their mechanical properties can be further enhanced through formulation optimization, such as incorporating hydroxypropyl β-cyclodextrin, which not only improves structural integrity but also enhances the solubility of poorly water-soluble drugs.[32] However, certain limitations must be considered when using HA as a matrix material, particularly for ocular applications where potential migration to periocular tissues may lead to complications such as infection or inflammation; these risks can be mitigated through enzymatic degradation using hyaluronidase.[33] The versatility of HA as a DMN material continues to drive innovation in transdermal delivery systems, with ongoing research focused on optimizing its mechanical properties and expanding its therapeutic applications.

2.5.2 Chitosan

Chitosan (CS), derived from the N-deacetylation of chitin, is the sole natural cationic polysaccharide containing free amino groups, making it particularly valuable for biomedical applications.[34] This unique biopolymer offers multiple advantages, including excellent biocompatibility, biodegradability, and inherent immunostimulatory properties that can enhance both humoral and cellular immune responses.[35] For DMN systems, CS demonstrates strong adsorption capacity and remarkable capabilities in encapsulating and protecting various active substances (e.g., antigens, antibodies, and enzymes), enabling sustained drug release upon dissolution in the skin.[34,36] Studies have also demonstrated that CS maintains its mechanical integrity in MN formulations without significantly affecting drug loading or drug release behavior.[37] However, due to its inherently moderate mechanical strength, CS-based MNs often require supportive arrays to enhance structural stability.

2.5.3 Other carbohydrates

In addition to HA and chitosan, other carbohydrates such as maltose, fructose, sucrose, and trehalose are widely used in the fabrication of DMNs (Table 2.1).[26,38] These carbohydrates offer several advantages, including high water solubility, biocompatibility, biodegradability, low cost, wide availability, ease of manufacturing, and sustainability for drug delivery applications.[39,40] Their rapid dissolution in the skin's interstitial fluid enables efficient drug release, making them particularly suitable for transdermal delivery systems.[41] However, a key limitation of these carbohydrates is their relatively low mechanical strength, which may compromise microneedle penetration efficiency when used alone.[42] To address this, they are often combined with polymers to form composite MNs, enhancing structural integrity while retaining the benefits of fast dissolution.[43,44]

Beyond their role as structural materials, these carbohydrates play a crucial part in stabilizing biomacromolecules, including peptides, proteins, and vaccines.[45] Their hydroxyl-rich structures interact with these biologics via hydrogen bonding, preventing denaturation and maintaining bioactivity during microneedle fabrication and storage. Additionally, their high water solubility facilitates rapid MN separation from the backing layer upon skin insertion, enabling the design of fast-separable DMNs that enhance transdermal drug delivery efficiency.[46]

TABLE 2.1 Representative monosaccharides and oligosaccharides are used to prepare MNs.

Name	Type (monosaccharide/oligosaccharide)	Molecular formula
Maltose	Oligosaccharide (disaccharide)	$C_{12}H_{22}O_{11}$
Fructose	Monosaccharide	$C_6H_{12}O_6$
Sucrose	Oligosaccharide (disaccharide)	$C_{12}H_{22}O_{11}$
Trehalose	Oligosaccharide (disaccharide)	$C_{12}H_{22}O_{11}$

2.6 Protein

Protein is a pivotal material for MN fabrication due to its inherent biocompatibility, biodegradability, and alignment with sustainable green technologies.[47] Natural proteins like silk fibroin, gelatin, collagen, zein, fish-scale, and mussel adhesive proteins are key protein-based materials used in MN fabrication.[48] These proteins exhibit unique structural and functional properties like strength, self-adhesion, and self-assembly, making them ideal for specific applications such as targeted drug delivery to special tissues (e.g., cardiovascular tissues).[49] Compared to silicon, metals, and ceramics, protein-based MN generally exhibit lower mechanical strength. Strategies such as polymer blending or cross-linking can enhance their rigidity to achieve sufficient skin penetration capability.

2.6.1 Silk fibroin

Silk fibroin is a versatile biomaterial for MN fabrication, valued for its high water solubility, tunable mechanical properties, and mild processing conditions compatible with sensitive biomolecules (e.g., insulin,[50] vaccines[51]). Its rapid dissolution in bodily fluids enables immediate drug release. Structural adjustments such as methanol treatment or light-activated cross-linking to form β-sheets can further enhance its mechanical stability and controlled degradation.[52] Silk fibroin can be molded at room temperature without organic solvents, facilitating the production of pure-silk fibroin, composite-silk fibroin, or hydrogel-forming MN. Composite formulations address limitations in mechanical strength and aqueous stability, such as blending with proline to regulate crystallinity and drug release rates.[53] Despite its advantages, silk fibroin requires purification to minimize contamination risks.

2.6.2 Gelatin

Gelatin is a protein obtained through the partial hydrolysis of skin collagen, which has been commercialized and available since 1685.[54] Its biocompatibility, biodegradability, and thermo-responsive hydrogel-forming capability make it a prime candidate for MN fabrication. Gelatin-based MNs leverage rapid dissolution in skin fluids for immediate drug release, while adjustable cross-linking enables sustained release.[55] The blind spot of gelatin is its insufficient mechanical strength, which can be compensated for by adding other materials (e.g., synthetic polymers, bio-ceramics, ceramics, and polysaccharides) or cross-linking. For instance, photo-polymerized gelatin methacryloyl (GelMA) can balance the degradation kinetics and mechanical strength of MN.[56]

2.6.3 Zein

Zein is a plant-derived prolamin protein first isolated from corn in 1821. Its high mechanical strength, thermal stability, and resistance to humidity/abrasion make it ideal for MN fabrication.[57] Zein-based MN form tough, glossy structures with excellent skin penetration capability and low water vapor permeability, enhancing drug formulation shelf-life. Zein exhibits low immunogenicity and exceptional biocompatibility, while its hydrophobic nature enables efficient loading of protein-based therapeutics through molecular interactions.[58] Despite brittleness in pure zein-based MN, incorporating plasticizers or composite strategies mitigates handling-induced cracking.[59]

2.6.4 Fish scale

Fish scales are an abundant natural source of type I collagen derived from species like tilapia and cod, offering a sustainable and cost-effective platform for MN fabrication.[60] Fish collagen-based MN exhibit intrinsic self-adhesive properties, high porosity, and water absorption capacity, which enable efficient drug loading and skin adhesion. Unlike mammalian collagen, fish collagen minimizes ethical concerns while maintaining low immunogenicity.[61] By blending with nanocellulose, fish collagen-based MNs achieve tunable dissolution rates in the form of rapid-dissolving MN or hydrogel MN.[62] Pure fish collagen-based MNs composites with nanocellulose effectively delay dissolution and improve moisture resistance.

2.6.5 Other protein

Mussel adhesive proteins (MAPs) are extremely adhesive materials with low toxicity and immunogenicity. Exceptionally, MAPs are impervious to water and turbulent forces.[63] Based on that, MAPs-based MN can be widely explored as an ideal platform in wound healing and drug delivery to the target tissues, from vascular and gastrointestinal to skin.[64] Remarkably, protein-based MN have common advantages like high biocompatibility, good biodegradability, and renewable character. Composite proteins and cross-linking strategies have been employed to enhance mechanical strength; their hydrophilic nature still compromises both mechanical strength and pharmaceutical stability of MN in environmental moisture. The technical challenge of protein isolation is a barrier to large-scale production.[65] Moreover, protein is difficult to adopt traditional molding processes of MN because of its high crystallinity and inadequate thermal properties. Digital light processing (DLP) 3D printing can do direct 3D construction of molecules dissolved in aqueous solution without the use of molds, which may avoid thermal degradation risks and realize precise structural control of protein-based MN.[66]

2.7 Polymers

Compared to metals and inorganic materials, polymers exhibit superior biocompatibility, excellent biodegradability, and cost-effectiveness. To date, numerous polymers have been successfully utilized in the fabrication of MNs, such as poly(vinylpyrrolidone) (PVP),[67] poly(vinylalcohol) (PVA),[68] PLGA,[69] carboxymethyl cellulose (CMC),[70] poly(methyl methacrylate) (PMMA).[71] These materials are primarily employed to produce DMNs and hydrogel-forming MNs, though they can also be used to manufacture solid, hollow, and coated MNs in specific application[72] (Fig. 2.1).

2.7.1 Poly(vinylpyrrolidone)

Polyvinylpyrrolidone (PVP), a prominent member of the N-vinylamide polymer family, has emerged as a critical matrix material for DMNs due to its exceptional biocompatibility and solubility. This synthetic polymer achieves rapid aqueous dissolution through polar amide groups in its molecular chains while simultaneously conferring superior mechanical strength to the microneedles, enabling effective penetration of the stratum corneum barrier. However, the abundance of hydrophilic groups in PVP molecules also leads to significant hygroscopicity, making the microneedles susceptible to moisture absorption during storage and application, which may result in structural softening and deformation. To address this limitation, researchers have proposed several solutions: (1) optimizing the molecular weight specifications of PVP (e.g., K30 or K90) to modulate its hygroscopic properties[73]; (2) incorporating hydrophobic excipients such as cyclodextrins to effectively reduce moisture absorption.[74] These strategies have significantly improved the clinical reliability of PVP-based MNs.

For the delivery of macromolecular drugs, PVP exhibits unique advantages. Its mild aqueous processing avoids the use of organic solvents and high temperatures, thereby preserving the integrity of sensitive therapeutics such as proteins.[75,76] Studies by Chen et al.[74] demonstrated that cyclodextrin-modified PVP microneedles not only exhibited reduced hygroscopicity but also achieved improvement in mechanical strength and enhanced tip sharpness.

2.7.2 Poly(vinylalcohol)

Polyvinyl alcohol (PVA) is a water-soluble synthetic polymer that has gained widespread use as a matrix material for DMNs due to its excellent biocompatibility, thermal stability, and film-forming properties. As a hydrophilic polymer containing numerous hydroxyl groups, PVA demonstrates good water

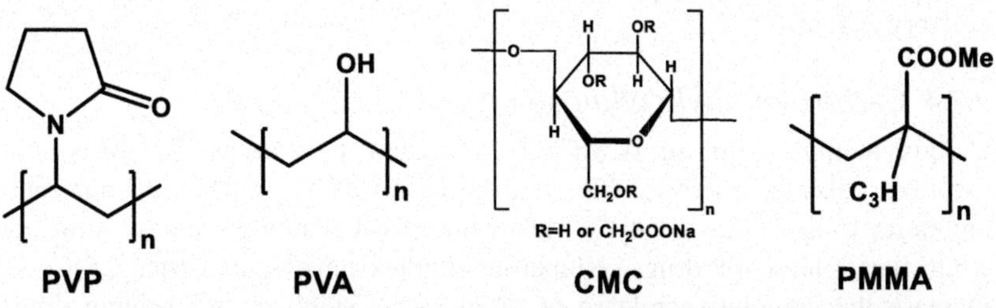

FIGURE 2.1 Chemical structure of the polymers used to prepare microneedles.

retention capacity and enables rapid dissolution of DMNs upon skin insertion, making it particularly suitable for transdermal drug delivery applications.[77] The polymer's ability to form hydrogen bonds with other materials allows for versatile modifications to optimize DMN performance.

While PVA offers several advantages for DMN fabrication, its application as a sole matrix material presents certain limitations. The mechanical strength of pure PVA is often insufficient for reliable skin penetration, and its dissolution rate may be too slow for some drug delivery applications. These challenges have been effectively addressed through the development of polymer blends, such as PVA/PVP composites that exhibit enhanced mechanical properties through intermolecular hydrogen bonding.[74] Other successful approaches include combining PVA with gelatin or PLGA to improve structural robustness[49] or incorporating trehalose to enhance vaccine stability.[78,79] Such composite systems have expanded PVA's utility in DMN technology, enabling applications ranging from conventional drug delivery to the intradermal administration of thermosensitive biologics and vaccines.[77] The adaptability of PVA through these material modifications has established its importance in advanced transdermal delivery systems, particularly for applications requiring both rapid dissolution and maintained drug stability.[79]

2.7.3 Poly(lactic-co-glycolic acid)

Poly(lactic-co-glycolic acid) (PLGA), a widely used polymeric material for fabricating DMNs, exhibits remarkable advantages in biomedical applications.[80] This biodegradable copolymer is synthesized through random copolymerization of lactic acid (LA) and glycolic acid (GA), demonstrating excellent biocompatibility and controllable degradation characteristics.[81] The degradation products (lactic acid and glycolic acid) can be completely metabolized by the human body, eliminating potential toxicity risks associated with long-term material retention.[82] The degradation rate of PLGA can be precisely modulated from days to months by adjusting the LA:GA ratio (e.g., 50:50 or 75:25), making it suitable for various drug delivery requirements.[83,84] Furthermore, the mechanical properties of PLGA can be optimized through molecular weight adjustment and blending modification, ensuring both the structural integrity of microneedles.[73] These tunable characteristics position PLGA as an exceptionally versatile material for advanced transdermal drug delivery systems.

2.7.4 Carboxymethyl cellulose

Carboxymethyl cellulose (CMC) is a natural polysaccharide. It is non-cytotoxic, elastic, water-soluble, and its dissolved by-products do not cause any harm to the human body.[85] Meanwhile, CMC can also achieve slow and controlled release of drugs. Numerous studies have shown that CMC can increase the cumulative release of drugs. As a stabilizer for protein drugs,

CMC-based DMN can be used to deliver vaccines, nucleic acids, and other bioactive drugs.[86] For example, Kim et al.[87] used CMC-based DMNs for vaccine delivery. However, it was shown that the activity of amino acids decreases as the concentration of the carboxymethylcellulose solution increases. Therefore, a low concentration of matrix solution should be used to deliver amino acids, as the increase in carboxymethylcellulose raises the pH of the matrix solution. Sodium CMC is also not recommended as a matrix for the delivery of certain acidic drugs. Lee et al.[88] used carboxymethylcellulose as a matrix to produce DMN for the delivery of lidocaine. The experimental results showed that the matrix material could provide the mechanical strength and drug content required for the clinical application of DMN. Moreover, the viscosity, biocompatibility, and biodegradability were met.

3. Microneedle fabrication methods

3.1 Microelectromechanical systems

Micromachining technology, also known as microelectromechanical systems (MEMS), represents one of the most promising approaches for the fabrication of MNs.[89] MEMS technology employs diverse tools and methodologies to create three-dimensional structures with dimensions ranging from sub-centimeter to sub-micron scales. This technology has found extensive applications in biomedical fields, including DNA sequencing devices, drug delivery systems, chemical analysis platforms, and biosensors.[90] Three main steps are usually involved in the MEMS fabrication process to produce functional devices: (1) deposition of a thin material film on a substrate, (2) adding a patterned mask to the top of the film by photolithographic imaging, and (3) selective etching of the film to make a mask.[91]

3.2 Photolithography

The development of microelectronics and micromachining has its roots in lithography. Various lithography techniques exist, including ion beam lithography, photolithography, X-ray lithography, and electron beam lithography. Among these, photolithography is the most widely used method.[92] Photolithography can be used to fabricate dissolving, hydrogel, solid, and hollow MNs. Additionally, this technique can produce silicon or polymer MNs by creating a countermould based on the MN structure.[93]

Photolithography involves transferring a master pattern onto the surface of a substrate material, typically a silicon wafer. In this process, the silicon wafer is first heated to 900–1150 °C in the presence of steam or humidified oxygen to grow a thin oxide layer on its surface. A UV-sensitive photoresist organic polymer is then applied to the oxide-coated wafer via spin-coating (1500–8000 rpm) to achieve a uniform thickness. Subsequently, the photoresist is baked at 75–100.8 °C for 10 minutes to remove residual solvent. As the solvent evaporates, the photoresist layer adheres to the silicon wafer. After solvent removal, a photomask consisting of a transparent glass plate and an opaque chrome pattern is placed in contact with the photoresist-coated surface.[94] B The wafer is then exposed to UV radiation (150–500 nm), transferring the photomask pattern onto the photoresist-coated wafer. The UV exposure triggers chemical reactions in the irradiated regions of the photoresist, which can be classified as either negative or positive reactions, depending on the resist type. These reactions alter the solubility of the exposed photoresist. During the development process, the soluble regions (either exposed or unexposed, depending on the resist polarity) are removed using a rinsing solution, either wet (solvent-based) or dry (plasma or vapor-phase) methods.[94] This process leaves a patterned photoresist layer on the wafer surface, which serves as a protective mask for subsequent etching or oxidation steps, resulting in well-defined oxide structures. After the advancement process, residual photoresist is removed through an oxygen-plasma treatment known as descumming.[95] The resulting oxide pattern (positive or negative) serves as the photomask, which is then used to mask subsequent processing steps.[15] In MEMS, oxides act as sacrificial masks to create new layers or are etched further to form deep 3D cavities. These structures ultimately evolve into complete 3D devices.[96]

3.3 Etching

Etching is a widely used technique in MN fabrication, typically performed after photolithography. This process employs chemical or physical methods to selectively remove the unprotected material from the substrate, thereby transferring the desired pattern onto its surface.[91] The etching process plays a critical role in defining key structural features of MNs, including their tapered tip geometry, base dimensions, interneedle spacing, needle length, and shape. Based on the mechanism of material removal, etching methods are broadly categorized into wet etching and dry etching.[97]

Wet etching serves as an effective technique for fabricating various silicon MN structures, including hollow silicon MNs,[98] ultrashort silicon MNs,[99] and porous silicon MNs.[100] This process of wet etching involves immersing silicon wafers, metals, or other substrates in a chemical etchant solution. To define specific patterns, a protective mask resistant to the etchant is applied, ensuring the desired structure is preserved after etching. Wet etching offers notable advantages such as low equipment costs, high throughput, and excellent etching uniformity. Olivia Howells et al.[101] developed a simplified wet etching method to produce in-plane sharp silicon MNs suitable for both solid and hollow designs. Their study characterized the performance of MNs with varying needle sizes through skin insertion tests and mechanical evaluations, further validating the efficacy of in-plane MNs for transdermal delivery of insulin and hyaluronan. However, the application of wet etching is limited to producing MNs that have a low aspect ratio.[102]

Dry etching is a thin-film etching technique that utilizes plasma. This process involves the use of inert or reactive gases at very low pressures, where gas ions are excited by high-frequency energy to accelerate toward the substrate, enabling physical or chemical etching of the target material.[91] Compared to wet etching, dry etching offers superior control over MN density and geometry, avoids crystallographic effects, and can produce cylindrically symmetrical MNs.[102] A prominent dry etching method is silicon deep reactive ion etching (DRIE), which operates at low pressure and generates reactive ion plasma on the material surface. This technique is increasingly prevalent and can fabricate hollow MNs with diverse shapes or intricate architectures. For instance, Bolton et al.[103] developed hollow beveled silicon MNs using a three-step DRIE process. In another study, researchers employed DRIE to fabricate MNs by first etching channels from the back of a silicon wafer, followed by alternating isotropic and anisotropic etching on the front side.[104] This approach yielded spiked MNs with heights ranging from 310 to 400 μm.

3.4 Micromolding

The micromolding method, also known as micromould casting, is one of the most commonly used techniques for MN preparation in laboratory research. This method primarily employs external forces, such as vacuum or centrifugation, to make the matrix materials of MN fully fill the mold cavities. The molds are then dried and demolded to obtain the final products. The fabrication process of MNs by micromolding consists of three main steps: (1) Fabricating a positive master mold to serve as a template for replication; (2) Creating a male mold by casting a curable liquid material onto the master mold, allowing it to cure, and then peeling it off. To date, the majority of male molds used for MN production have been made from polydimethylsiloxane (PDMS). The primary reasons for choosing this material are its flexibility and its ability to accurately reproduce the structure of the master mold[105]; and (3) producing the final MNs by casting another curable material onto the negative mold, curing it, and peeling it off to obtain MNs on a flexible substrate.[39]

Many polymers have been used for the preparation of MNs. However, carbohydrates were initially considered as the main candidates for potential MN materials.

3.5 Droplet-born air blowing

The droplet-born air blowing (DAB) method is a relatively gentle and rapid technique for preparing DMNs. In this method, the drug content in the MNs is controlled by adjusting the air pressure and blowing time. The double-layer structure of the MNs ensures complete drug delivery without wastage.[106]

Additionally, using air blowing to solidify the droplets and enhance the mechanical properties of the MNs avoids the need for light or heat exposure, thereby preserving drug stability and activity.[107] The specific steps of the DAB method are as follows: (1) Prepare the drug solution and drug-polymer solution; (2) Disperse the drug-matrix mixture on a flat surface; (3) Lower the upper plate to contact the droplets; (4) Raise the upper plate to a predetermined distance (equal to the combined length of the two dissolving MNs); (5) Apply air blowing to solidify the MNs; (6) After complete solidification, separate the upper and lower plates, yielding two sets of dissolving MN patches simultaneously[1] (Fig. 2.2).

3.6 3D printing

Conventional MN preparation methods are limited in producing structurally diverse MNs. In contrast, 3D printing offers a novel approach by layering polymer materials using 3D printers. This technology is versatile, eco-friendly, cost-effective, and capable of fabricating MNs with precise architectures.

3D printing for MN fabrication can be broadly categorized into extrusion/nozzle-based methods and vat photopolymerization techniques.[108] Among extrusion/nozzle-based methods, inkjet (IJ) printing and fused deposition modeling (FDM) are commonly employed. Vat photopolymerization techniques include digital light processing (DLP), laser stereolithography (laser SLA), two-photon polymerization (TPP), and direct metal laser sintering (DMLS).

3.6.1 Extrusion or nozzle 3D printing

Inkjet printing fabricates three-dimensional structures through the sequential deposition of molten polymers, polymer solutions, or ink binder droplets onto a substrate.[109] These droplets solidify via cooling, chemical curing, or solvent evaporation, depending on the material system. Inkjet technologies are broadly classified into two modes: continuous inkjet (CIJ) and drop-on-demand (DoD). In CIJ systems, a pressurized ink stream passes through a nozzle and is fragmented into droplets by piezoelectric actuation. An electrostatic field generated between the nozzle and deflection electrodes then precisely guides the droplets to predetermined substrate positions. Due to its precision, inkjet printing is widely employed for the surface modification of coated MNs. For instance, Boehm et al.[110] successfully modified the surfaces of solid MNs through the inkjet printing technique and fabricated biodegradable acid anhydride copolymer microneedles containing quantum dots by strategies of visible light dynamic mask microstereolithography–micromolding and inkjet printing. Their study presents a novel method for fabricating polymeric microneedle arrays, enabling transdermal delivery of diverse pharmacological agents.

Fused deposition modeling (FDM) fabricates three-dimensional objects by extruding molten thermoplastic polymer filaments through a heated nozzle.

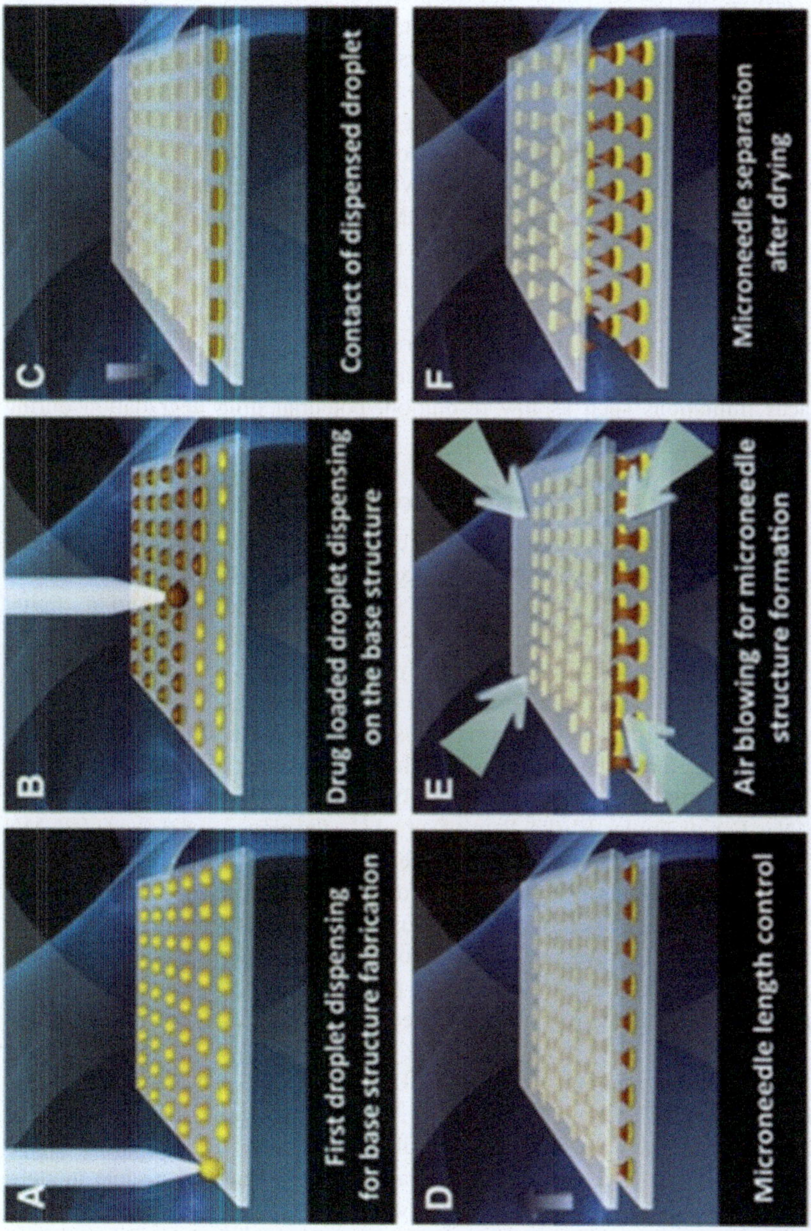

FIGURE 2.2 The operating steps of the droplet-born air-blowing method. *Reproduced from Ref. 1 with permission. Copyright 2021, Elsevier.*

The nozzle deposits the material in a predefined two-dimensional pattern on a platform through robotic motion, and successive layers are stacked by adjusting the vertical position of the platform or nozzle to achieve the desired geometry.[108] FDM offers several advantages, including low cost, compatibility with a wide range of materials, and user-friendly operation. To enhance tip sharpness, postprocessing techniques such as mechanical stretching or chemical etching can be employed. For instance, in the fabrication of insulin-loaded nanoparticles, FDM was used to print cylindrical MNs from a bioink containing sodium alginate and hydroxyapatite, followed by axial stretching to sharpen the tips.[111] Additionally, FDM printers can utilize biodegradable bioplastics like polylactic acid (PLA), mitigating environmental concerns. A notable example is the work by Khosraviboroujeni et al., who pioneered the use of FDM 3D printing combined with injection molding to fabricate biodegradable polymer microneedle arrays (BPMNAs) for painless, sustained transdermal delivery of estradiol valerate.[112] Despite these benefits, FDM faces limitations such as low resolution, potential drug degradation during processing, and nozzle clogging issues.

3.6.2 Reduced light polymerization 3D printing

DLP is a versatile technique for fabricating both porous MNs[113] and well-structured hollow MNs.[114] Its high flexibility also enables the production of personalized MNs.[115,116] In DLP, a visible light projector and a digital micromirror device (DMD) illuminate a liquid photopolymer from below, curing one layer at a time. After each layer is solidified, it is separated from the vat, allowing fresh photopolymer to fill the gap. The build platform then rises, and the process repeats, irradiating and solidifying subsequent layers, until the final 3D structure is complete.[108] DLP has a wide range of applications due to its high printing accuracy and flexibility to change the shape and size of the MNs. For example, Eliška Petrová et al.[117] developed a DLP-printed MN patch containing imiquimod (IMQ), offering a novel delivery method for drugs insoluble in printing resin but suitable for topical application. Donghyeok Shin et al. utilized DLP to fabricate MNs using filipin proteins and riboflavin as photoinitiators, followed by antiserration and dehydration. The resulting silk fibroin MNs were flexible, capable of penetrating pig skin to deliver rhodamine B into the epidermis, and exhibited robust compressive strength without structural failure.[118]

SLA utilizes a computer-controlled UV laser to cure photosensitive resin in a point-by-point manner (Fig. 2.3A). After solidification of each layer, the process is repeated layer-by-layer until the complete 3D structure is fabricated.[119,120] As the only technology capable of printing large-scale models, SLA has become one of the most widely adopted 3D printing techniques.[121] A notable application was demonstrated by Uddin et al.,[122] who combined SLA 3D printing with inkjet coating to fabricate cross-MNs for cisplatin delivery to

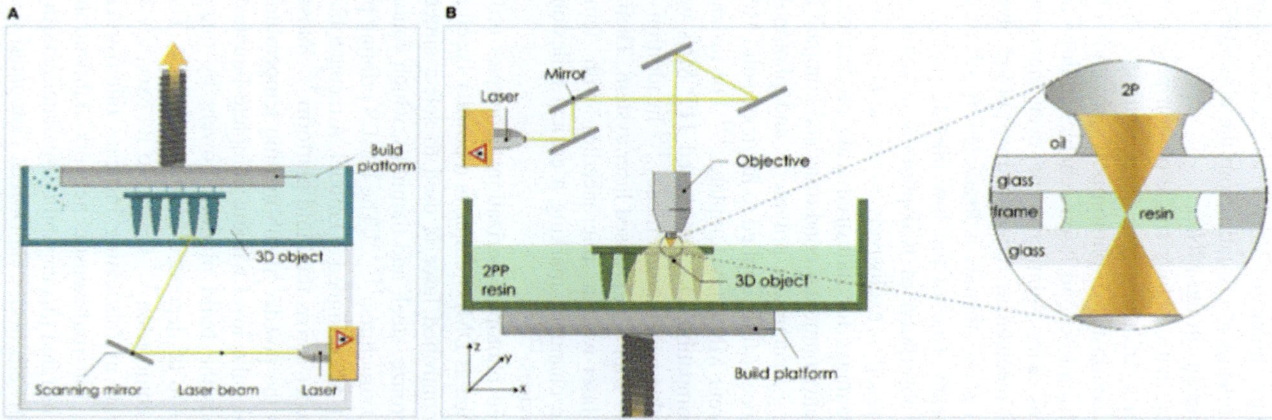

FIGURE 2.3 Schematic illustration of reduced light polymerization 3D-printing technology: (A) Laser stereolithography (laser SLA); (B) Two-photon polymerization (TPP). *Reproduced from Ref. 133 with permission; Copyright 2022, Elsevier.*

A-431 epidermoid skin tumors. Porcine skin penetration tests revealed that these printed MNs required significantly lower penetration force compared to metallic MNs, indicating enhanced safety. Franz cell diffusion studies further showed nearly 100% release of cisplatin from the cross-MNs, even at higher drug loading concentrations in the MNs.

SLA is a technique that employs single-photon polymerization (SPP) to print 3D micro/nanostructures; however, the resolution of SLA is low due to optical diffraction. In contrast, two-photon polymerization (TPP) technology, utilizing near-infrared femtosecond lasers, enables high-precision fabrication of 3D microstructures with submicron or even nanoscale resolution[123] (Fig. 2.3B). The TPP mechanism relies on two-photon absorption: A resin droplet deposited on a glass substrate is irradiated by a focused ultrafast (e.g., femtosecond) laser beam. Due to the low photon energy, resin molecules require simultaneous absorption of two photons to initiate polymerization, confining the curing reaction exclusively to the laser focal volume.[124] TPP offers significant advantages for industrial-scale production, including high tolerance to standard environmental conditions and compatibility with diverse low-cost materials (e.g., ceramics, polymers, and photopolymers).[125] This technology has been successfully applied to fabricate solid MNs,[126–128] hollow MNs,[129,130] and different MN array templates.[131] For instance, Doraiswamy et al., pioneered biocompatible Ormocer-based TPP MNs with superior mechanical strength and skin penetration efficacy.[132] In addition, Trautmann et al. fabricated hollow MNs with a TPP-based laser system. Their work combined femtosecond direct laser-written microneedles with femtosecond laser-generated microfluidic channels, marking a critical milestone toward multifunctional medical devices.[129]

The aforementioned photopolymerization techniques utilize liquid materials such as resin, whereas DMLS employs a powder-based approach. In this process, a computer-controlled laser is used to select sinters or melt metal alloy powders (e.g., bronze, steel, and titanium), which contain no binders or fluxes, to construct 3D structures.[125,134] The DMLS printing process begins by exporting data from a CAD-designed 3D model. A recoater blade or roller then evenly distributes powder across the build platform.[135] Subsequently, a high-precision laser sinters the powder at sub-melting temperatures, enabling particle fusion at the molecular level without complete liquefaction.[135,136] After each layer is sintered, the build platform lowers by one layer thickness, allowing fresh powder to be deposited for the next cycle. This layer-by-layer process repeats until completion.[135] Owing to its exceptional resolution and accuracy, DMLS excels at fabricating intricate microstructures, making it particularly valuable for multisensor network fabrication. For instance, Krieger et al.[137] demonstrated this capability by manufacturing sharp medical-grade stainless steel MN electrodes via DMLS. Comparative studies with non-MN 3D-printed stainless steel electrodes and conventional Ag/AgCl wet electrodes revealed superior performance: the MN electrodes exhibited

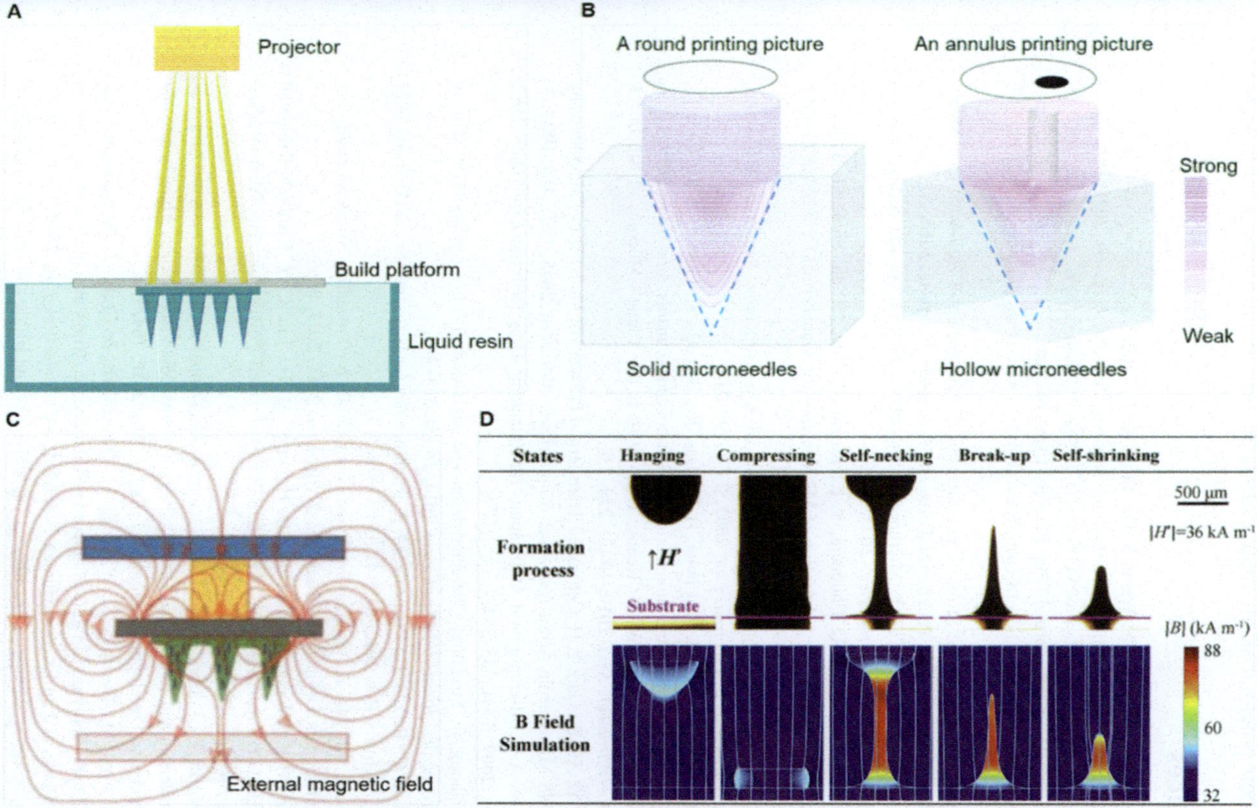

FIGURE 2.4 Schematic illustration of 3D-printing technologies with an emerging layer-by-layer process for fabricating microneedles (MNs). (A) Schematic diagram of manufacturing MNs by single-photon optical lithography (SPOL). (B) Spatial distribution of light intensity of the MN formation process by SOPL. (C) Schematic diagram of manufacturing MNs by magnetic-driven lithography (MDL). (D) MN formation process with the magnetic-rotational-driven lithography (MRDL) method under an external magnetic field. *Reproduced from Ref. 138. Copyright 2021, Elsevier. (D) Reproduced from Ref. 139 with permission. Copyright 2018, Elsevier.*

significantly lower electrode-skin impedance in healthy volunteers during bicep EMG recordings, while maintaining wearer comfort for up to 6 h, suggesting their potential as alternatives to standard wet electrodes[137] (Fig. 2.4).

References

1. Zhang, L. J.; Guo, R. R.; Wang, S. Q.; Yang, X. T.; Ling, G. X.; Zhang, P. Fabrication, Evaluation and Applications of Dissolving Microneedles. *Int. J. Pharm.* **2021,** *604.*
2. Sabri, A. H.; Kim, Y.; Marlow, M.; Scurr, D. J.; Segal, J.; Banga, A. K.; Kagan, L.; Lee, J. B. Intradermal and Transdermal Drug Delivery Using Microneedles-Fabrication, Performance Evaluation and Application to Lymphatic Delivery. *Adv. Drug Delivery Rev.* **2020,** *153,* 195–215.
3. Khanna, P.; Luongo, K.; Strom, J. A.; Bhansali, S. Sharpening of Hollow Silicon Microneedles to Reduce Skin Penetration Force. *J. Micromech. Microeng.* **2010,** *20* (4).
4. Luo, X. J.; Yang, L.; Cui, Y. Microneedles: Materials, Fabrication, and Biomedical Applications. *Biomed. Microdevices* **2023,** *25* (3).
5. Indermun, S.; Luttge, R.; Choonara, Y. E.; Kumar, P.; Toit, LC du; Modi, G.; Pillay, V. Current Advances in the Fabrication of Microneedles for Transdermal Delivery. *J. Control. Release* **2014,** *185,* 130–138.
6. Paul, O.; Gaspar, J.; Ruther, P. Advanced Silicon Microstructures, Sensors, and Systems. *IEEJ Trans. Electr. Electron. Eng.* **2007,** *2* (3), 199–215.
7. Chorsi, M. T.; Curry, E. J.; Chorsi, H. T.; Das, R.; Baroody, J.; Purohit, P. K.; Ilies, H.; Nguyen, T. D. Piezoelectric Biomaterials for Sensors and Actuators. *Adv. Mater.* **2019,** *31* (1).
8. Hopcroft, M. A.; Nix, W. D.; Kenny, T. W. What Is the Young's Modulus of Silicon? *J. Microelectromech. Syst.* **2010,** *19* (2), 229–238.
9. Xie, Y.; Xu, B.; Gao, Y. H. Controlled Transdermal Delivery of Model Drug Compounds by MEMS Microneedle Array. *Nanomed. Nanotechnol. Biol. Med.* **2005,** *1* (2), 184–190.
10. Banga, A. K. Microporation Applications for Enhancing Drug Delivery. *Expet Opin. Drug Deliv.* **2009,** *6* (4), 343–354.
11. O'Mahony, C. Structural Characterization and *In-Vivo* Reliability Evaluation of Silicon Microneedles. *Biomed. Microdevices* **2014,** *16* (3), 333–343.
12. Kim, Y. C.; Park, J. H.; Prausnitz, M. R. Microneedles for Drug and Vaccine Delivery. *Adv. Drug Delivery Rev.* **2012,** *64* (14), 1547–1568.
13. Niinomi, M.; Nakai, M. Titanium-Based Biomaterials for Preventing Stress Shielding between Implant Devices and Bone. *Int. J. Biomater.* **2011,** *2011,* 836587.
14. Chen, M. C.; Wang, K. W.; Chen, D. H.; Ling, M. H.; Liu, C. Y. Remotely Triggered Release of Small Molecules from $LaB_6@SiO_2$-Loaded Polycaprolactone Microneedles. *Acta Biomater.* **2015,** *13,* 344–353.
15. Larrañeta, E.; Lutton, R. E. M.; Woolfson, A. D.; Donnelly, R. F. Microneedle Arrays as Transdermal and Intradermal Drug Delivery Systems: Materials Science, Manufacture and Commercial Development. *Mater. Sci. Eng. R Rep.* **2016,** *104,* 1–32.
16. Verhoeven, M.; Bystrova, S.; Winnubst, L.; Qureshi, H.; de Gruijl, T. D.; Scheper, R. J.; Luttge, R. Applying Ceramic Nanoporous Microneedle Arrays as a Transport Interface in Egg Plants and an Ex-Vivo Human Skin Model. *Microelectron. Eng.* **2012,** *98,* 659–662.
17. Simons, E. C.; Weerheijm, J.; Sluys, L. J. Simulating the Brittle and Ductile Responses of Alumina Ceramics Under Dynamic Loading. *Eng. Fract. Mech.* **2019**, 216.

18. Bystrova, S.; Luttge, R. Micromolding for Ceramic Microneedle Arrays. *Microelectron. Eng.* **2011,** *88* (8), 1681–1684.
19. Cai, B.; Xia, W.; Bredenberg, S.; Engqvist, H. Self-setting Bioceramic Microscopic Protrusions for Transdermal Drug Delivery. *J. Mater. Chem. B* **2014,** *2* (36), 5992–5998.
20. Ginebra, M. P.; Traykova, T.; Planell, J. A. Calcium Phosphate Cements as Bone Drug Delivery Systems: A Review. *J. Control. Release* **2006,** *113* (2), 102–110.
21. Doraiswamy, A.; Ovsianikov, A.; Gittard, S. D.; Monteiro-Riviere, N. A.; Crombez, R.; Montalvo, E.; Shen, W. D.; Chichkov, B. N.; Narayan, R. J. Fabrication of Microneedles Using Two Photon Polymerization for Transdermal Delivery of Nanomaterials. *J. Nanosci. Nanotechnol.* **2010,** *10* (10), 6305–6312.
22. Gittard, S. D.; Narayan, R. J.; Jin, C.; Ovsianikov, A.; Chichkov, B. N.; Monteiro-Riviere, N. A.; Stafslien, S.; Chisholm, B. Pulsed Laser Deposition of Antimicrobial Silver Coating on Ormocer® Microneedles. *Biofabrication* **2009,** *1* (4).
23. Wang, P. M.; Cornwell, M.; Hill, J.; Prausnitz, M. R. Precise Microinjection into Skin Using Hollow Microneedles. *J. Invest. Dermatol.* **2006,** *126* (5), 1080–1087.
24. Martanto, W.; Moore, J. S.; Kashlan, O.; Kamath, R.; Wang, P. M.; O'Neal, J. M.; Prausnitz, M. R. Microinfusion Using Hollow Microneedles. *Pharm. Res.* **2006,** *23* (1), 104–113.
25. McAllister, D. V.; Wang, P. M.; Davis, S. P.; Park, J. H.; Canatella, P. J.; Allen, M. G.; Prausnitz, M. R. Microfabricated Needles for Transdermal Delivery of Macromolecules and Nanoparticles: Fabrication Methods and Transport Studies. *Proc. Natl. Acad. Sci. U. S. A.* **2003,** *100* (24), 13755–13760.
26. Lee, K.; Lee, C. Y.; Jung, H. Dissolving Microneedles for Transdermal Drug Administration Prepared by Stepwise Controlled Drawing of Maltose. *Biomaterials* **2011,** *32* (11), 3134–3140.
27. Donnelly, R. F.; Morrow, D. I. J.; Singh, T. R. R.; Migalska, K.; McCarron, P. A.; O'Mahony, C.; Woolfson, A. D. Processing Difficulties and Instability of Carbohydrate Microneedle Arrays. *Drug Dev. Ind. Pharm.* **2009,** *35* (10), 1242–1254.
28. Miyano, T.; Tobinaga, Y.; Kanno, T.; Matsuzaki, Y.; Takeda, H.; Wakui, M.; Hanada, K. Sugar Micro Needles as Transdermic Drug Delivery System. *Biomed. Microdevices* **2005,** *7* (3), 185–188.
29. Saha, I.; Rai, V. K. Hyaluronic Acid Based Microneedle Array: Recent Applications in Drug Delivery and Cosmetology. *Carbohydr. Polym.* **2021,** *267.*
30. Kim, D. S.; Choi, J. T.; Kim, C. B.; Shin, Y. R.; Park, P. G.; Kim, H.; Lee, J. M.; Park, J. H. Microneedle Array Patch (MAP) Consisting of Crosslinked Hyaluronic Acid Nanoparticles for Processability and Sustained Release. *Pharm. Res.* **2020,** *37* (3).
31. Cui, Z.; Liao, J.; Cheong, N.; Longoria, C.; Cao, G. Y.; DeLisser, H. M.; Savani, R. C. The Receptor for Hyaluronan-Mediated Motility (CD168) Promotes Inflammation and Fibrosis after Acute Lung Injury. *Matrix Biol.* **2019,** *78–79,* 255–271.
32. Lin, S. Q.; Quan, G. L.; Hou, A. L.; Yang, P. P.; Peng, T. T.; Gu, Y. K.; Qin, W. B.; Liu, R. B.; Ma, X. Y.; Pan, X.; et al. Strategy for Hypertrophic Scar Therapy: Improved Delivery of Triamcinolone Acetonide Using Mechanically Robust Tip-Concentrated Dissolving Microneedle Array. *J. Control. Release* **2019,** *306,* 69–82.
33. Fallacara, A.; Baldini, E.; Manfredini, S.; Vertuani, S. Hyaluronic Acid in the Third Millennium. *Polymers* **2018,** *10* (7).
34. Chen, M. C.; Lai, K. Y.; Ling, M. H.; Lin, C. W. Enhancing Immunogenicity of Antigens through Sustained Intradermal Delivery Using Chitosan Microneedles with a Patch-Dissolvable Design. *Acta Biomater.* **2018,** *65,* 66–75.

35. Jamaledin, R.; Makvandi, P.; Yiu, C. K. Y.; Agarwal, T.; Vecchione, R.; Sun, W. J.; Maiti, T. K.; Tay, F. R.; Netti, P. A. Engineered Microneedle Patches for Controlled Release of Active Compounds: Recent Advances in Release Profile Tuning. *Adv. Therapeut.* **2020,** *3* (12).
36. Vasiliev, Y. M. Chitosan-Based Vaccine Adjuvants: Incomplete Characterization Complicates Preclinical and Clinical Evaluation. *Expet Rev. Vaccine* **2015,** *14* (1), 37–53.
37. Gopan, G.; Jose, J.; Khot, K. B.; Bandiwadekar, A. The Use of Cellulose, Chitosan and Hyaluronic Acid in Transdermal Therapeutic Management of Obesity: A Review. *Int. J. Biol. Macromol.* **2023,** *244.*
38. Martin, C. J.; Allender, C. J.; Brain, K. R.; Morrissey, A.; Birchall, J. C. Low Temperature Fabrication of Biodegradable Sugar Glass Microneedles for Transdermal Drug Delivery Applications. *J. Control. Release* **2012,** *158* (1), 93–101.
39. Qi, X. L.; Tong, X. Q.; Pan, W. H.; Zeng, Q. K.; You, S. Y.; Shen, J. L. Recent Advances in Polysaccharide-Based Adsorbents for Wastewater Treatment. *J. Cleaner Prod.* **2021,** *315.*
40. Onesto, V.; Di Natale, C.; Profeta, M.; Netti, P. A.; Vecchione, R. Engineered PLGA-PVP/VA Based Formulations to Produce Electro-Drawn Fast Biodegradable Microneedles for Labile Biomolecule Delivery. *Prog. Biomater.* **2020,** *9* (4), 203–217.
41. Nguyen, H. X.; Banga, A. K. Fabrication, Characterization and Application of Sugar Microneedles for Transdermal Drug Delivery. *Ther. Deliv.* **2017,** *8* (5), 249–264.
42. Raphael, A. P.; Crichton, M. L.; Falconer, R. J.; Meliga, S.; Chen, X. F.; Fernando, G. J. P.; Huang, H.; Kendall, M. A. F. Formulations for Microprojection/microneedle Vaccine Delivery: Structure, Strength and Release Profiles. *J. Control. Release* **2016,** *225,* 40–52.
43. Dillon, C.; Hughes, H.; O'Reilly, N. J.; McLoughlin, P. Formulation and Characterisation of Dissolving Microneedles for the Transdermal Delivery of Therapeutic Peptides. *Int. J. Pharm.* **2017,** *526* (1–2), 125–136.
44. Lin, S. Q.; Cai, B. Z.; Quan, G. L.; Peng, T. T.; Yao, G. T.; Zhu, C. N.; Wu, Q. L.; Ran, H.; Pan, X.; Wu, C. B. Novel Strategy for Immunomodulation: Dissolving Microneedle Array Encapsulating Thymopentin Fabricated by Modified Two-step Molding Technology. *Eur. J. Pharm. Biopharm.* **2018,** *122,* 104–112.
45. Alimardani, V.; Abolmaali, S. S.; Yousefi, G.; Rahiminezhad, Z.; Abedi, M.; Tamaddon, A.; Ahadian, S. Microneedle Arrays Combined with Nanomedicine Approaches for Transdermal Delivery of Therapeutics. *J. Clin. Med.* **2021,** *10* (2).
46. Hou, A. L.; Quan, G. L.; Yang, B. B.; Lu, C.; Chen, M. L.; Yang, D.; Wang, L. L.; Liu, H.; Pan, X.; Wu, C. B. Rational Design of Rapidly Separating Dissolving Microneedles for Precise Drug Delivery by Balancing the Mechanical Performance and Disintegration Rate. *Adv. Healthcare Mater.* **2019,** *8* (21).
47. Fonseca, D. F.; Vilela, C.; Silvestre, A. J.; Freire, C. S. J. A Compendium of Current Developments on Polysaccharide and Protein-Based Microneedles. *Int. J. Biol. Macromol.* **2019,** *136,* 704–728.
48. Bhatnagar, S.; Kulkarni, R. B.; Manimaran, R.; Venuganti, V. V. K. Protein-Based Microneedles for Drug and Vaccine Delivery. In *Engineered Biomaterials: Progress and Prospects;* World Scientific, 2022; pp 261–312.
49. Lim, S.; Park, T. Y.; Jeon, E. Y.; Joo, K. I.; Cha, H. J. Double-Layered Adhesive Microneedle Bandage Based on Biofunctionalized Mussel Protein for Cardiac Tissue Regeneration. *Biomaterials* **2021,** *278,* 121171.
50. Zhu, M.; Liu, Y.; Jiang, F.; Cao, J.; Kundu, S. C.; Lu, S. Combined Silk Fibroin Microneedles for Insulin Delivery. *ACS Biomater. Sci. Eng.* **2020,** *6* (6), 3422–3429.

51. Stinson, J. A.; Raja, W. K.; Lee, S.; Kim, H. B.; Diwan, I.; Tutunjian, S.; Panilaitis, B.; Omenetto, F. G.; Tzipori, S.; Kaplan, D. L.; et al. Silk Fibroin Microneedles for Transdermal Vaccine Delivery. *ACS Biomater. Sci. Eng.* **2017,** *3* (3), 360–369.
52. Grabska-Zielińska, S.; Sionkowska, A. How to Improve Physico-Chemical Properties of Silk Fibroin Materials for Biomedical Applications?—Blending and Cross-Linking of Silk Fibroin—A review. *Materials* **2021,** *14* (6), 1510.
53. Wang, S.; Zhu, M.; Zhao, L.; Kuang, D.; Kundu, S. C.; Lu, S. Insulin-loaded Silk Fibroin Microneedles as Sustained Release System. *ACS Biomater. Sci. Eng.* **2019,** *5* (4), 1887–1894.
54. Foox, M.; Zilberman, M. Drug Delivery from Gelatin-Based Systems. *Expet Opin. Drug Deliv.* **2015,** *12* (9), 1547–1563.
55. Kirchmajer, D. M.; Watson, C. A.; Ranson, M. Gelapin, a Degradable Genipin Cross-Linked Gelatin Hydrogel. *RSC Adv.* **2013,** *3* (4), 1073–1081.
56. Chen, K.; Li, B.; Xu, H.; Wu, J.; Li, J.; Sun, W.; Fang, M.; Wang, W.; Wang, S.; Zhai, X. Zeolitic Imidazole Framework-8 Loaded Gelatin Methacryloyl Microneedles: A Transdural and Controlled-Release Drug Delivery System Attenuates Neuroinflammation after Spinal Cord Injury. *Int. J. Biol. Macromol.* **2024,** *256,* 128388.
57. Mahanty, A.; Abbasi, Y. F.; Bera, H.; Chakraborty, M.; Maruf, MA Al. Zein-Based Nanomaterials in Drug Delivery and Biomedical Applications. In *Biopolymer-Based Nanomaterials in Drug Delivery and Biomedical Applications;* Elsevier, 2021; pp 497–518.
58. Bhatnagar, S.; Chawla, S. R.; Kulkarni, O. P.; Venuganti, V. V. K. Zein Microneedles for Transcutaneous Vaccine Delivery: Fabrication, Characterization, and In Vivo Evaluation Using Ovalbumin as the Model Antigen. *ACS Omega* **2017,** *2* (4), 1321–1332.
59. Zhang, X.; Gao, M.; Zhang, Y.; Dong, C.; Xu, M.; Hu, Y.; Luan, G. Effect of Plasticizer and Zein Subunit on Rheology and Texture of Zein Network. *Food Hydrocoll.* **2022,** *123,* 107140.
60. Olatunji, O.; Igwe, C. C.; Ahmed, A. S.; Alhassan, D. O.; Asieba, G. O.; Diganta, B. D. Microneedles from Fish Scale Biopolymer. *J. Appl. Polym. Sci.* **2014,** *131* (12).
61. Al-Nimry, S.; Dayah, A. A.; Hasan, I.; Daghmash, R. Cosmetic, Biomedical and Pharmaceutical Applications of Fish Gelatin/Hydrolysates. *Mar. Drugs* **2021,** *19* (3), 145.
62. Olatunji, O.; Olsson, R. T. Microneedles from Fishscale-Nanocellulose Blends Using Low Temperature Mechanical Press Method. *Pharmaceutics* **2015,** *7* (4), 363–378.
63. Kord Forooshani, P.; Lee, B. P. Recent Approaches in Designing Bioadhesive Materials Inspired by Mussel Adhesive Protein. *J. Polym. Sci. Polym. Chem.* **2017,** *55* (1), 9–33.
64. Jeon, E. Y.; Lee, J.; Kim, B. J.; Joo, K. I.; Kim, K. H.; Lim, G.; Cha, H. J. Bio-inspired Swellable Hydrogel-Forming Double-Layered Adhesive Microneedle Protein Patch for Regenerative Internal/External Surgical Closure. *Biomaterials* **2019,** *222,* 119439.
65. Barati, M.; Hashemi, S.; Sayed Tabatabaei, M.; Zarei Chamgordani, N.; Mortazavi, S. M.; Moghimi, H. R. Protein-Based Microneedles for Biomedical Applications: A Systematic Review. *Biomed. Microdevices* **2024,** *26* (2), 19.
66. Shin, D.; Hyun, J.; Chemistry, E. Silk Fibroin Microneedles Fabricated by Digital Light Processing 3D Printing. *J. Ind. Eng. Chem.* **2021,** *95,* 126–133.
67. Zare, M. R.; Khorram, M.; Barzegar, S.; Sarkari, B.; Asgari, Q.; Ahadian, S.; Zomorodian, K. Dissolvable Carboxymethyl Cellulose/Polyvinylpyrrolidone Microneedle Arrays for Transdermal Delivery of Amphotericin B to Treat Cutaneous Leishmaniasis. *Int. J. Biol. Macromol.* **2021,** *182,* 1310–1321.

68. Zhang, X. P.; Wang, B. B.; Li, W. X.; Fei, W. M.; Cui, Y.; Guo, X. D. In Vivo Safety Assessment, Biodistribution and Toxicology of Polyvinyl Alcohol Microneedles with 160-Day Uninterruptedly Applications in Mice. *Eur. J. Pharm. Biopharm.* **2021,** *160,* 1–8.
69. Pawley, D. C.; Goncalves, S.; Bas, E.; Dikici, E.; Deo, S. K.; Daunert, S.; Telischi, F. Dexamethasone (DXM)-Coated Poly(lactic-*co*-glycolic Acid) (PLGA) Microneedles as an Improved Drug Delivery System for Intracochlear Biodegradable Devices. *Adv. Therapeut.* **2021,** *4* (11).
70. Nguyen, T. T.; Nguyen, T. T. D.; Tran, N. M. A.; Vo, G. V. Advances of Microneedles in Hormone Delivery. *Biomed. Pharmacother.* **2022,** *145.*
71. Choi, S. O.; Kim, Y. C.; Park, J. H.; Hutcheson, J.; Gill, H. S.; Yoon, Y. K.; Prausnitz, M. R.; Allen, M. G. An Electrically Active Microneedle Array for Electroporation. *Biomed. Microdevices* **2010,** *12* (2), 263–273.
72. DeMuth, P. C.; Min, Y. J.; Huang, B.; Kramer, J. A.; Miller, A. D.; Barouch, D. H.; Hammond, P. T.; Irvine, D. J. Polymer Multilayer Tattooing for Enhanced DNA Vaccination. *Nat. Mater.* **2013,** *12* (4), 367–376.
73. Zhang, X. Y.; Zhou, C. X.; Chen, T. X.; Jiang, Z. S.; Lu, C.; Wu, C. B.; Pan, X.; Huang, Z. W.; Peng, T. T. State-of-the-Art Strategies to Enhance the Mechanical Properties of Microneedles. *Int. J. Pharm.* **2024,** *663.*
74. Chen, W.; Wang, C.; Yan, L.; Huang, L. B.; Zhu, X. Y.; Chen, B.; Sant, H. J.; Niu, X. R.; Zhu, G. Y.; Yu, K. N.; et al. Improved Polyvinylpyrrolidone Microneedle Arrays with Non-stoichiometric Cyclodextrin. *J. Mater. Chem. B* **2014,** *2* (12), 1699–1705.
75. Sun, W. C.; Araci, Z.; Inayathullah, M.; Manickam, S.; Zhang, X. X.; Bruce, M. A.; Marinkovich, M. P.; Lane, A. T.; Milla, C.; Rajadas, J.; et al. Polyvinylpyrrolidone Microneedles Enable Delivery of Intact Proteins for Diagnostic and Therapeutic Applications. *Acta Biomater.* **2013,** *9* (8), 7767–7774.
76. Sullivan, S. P.; Koutsonanos, D. G.; Martin, M. D.; Lee, J. W.; Zarnitsyn, V.; Choi, S. O.; Murthy, N.; Compans, R. W.; Skountzou, I.; Prausnitz, M. R. Dissolving Polymer Microneedle Patches for Influenza Vaccination. *Nat. Med.* **2010,** *16* (8), 915–920.
77. Feng, Y. H.; Liu, J. L.; Zhu, D. D.; Hao, Y. Y.; Guo, X. D. Multiscale Simulations of Drug Distributions in Polymer Dissolvable Microneedles. *Colloids Surf. B-Biointerfaces* **2020,** *189.*
78. Demir, Y. K.; Akan, Z.; Kerimoglu, O. Sodium Alginate Microneedle Arrays Mediate the Transdermal Delivery of Bovine Serum Albumin. *PLoS One* **2013,** *8* (5).
79. Liao, J. F.; Lee, J. C.; Lin, C. K.; Wei, K. C.; Chen, P. Y.; Yang, H. W. Self-Assembly DNA Polyplex Vaccine inside Dissolving Microneedles for High-Potency Intradermal Vaccination. *Theranostics* **2017,** *7* (10), 2593–2605.
80. Malek-Khatabi, A.; Razavi, M. S.; Abdollahi, A.; Rahimzadeghan, M.; Moammeri, F.; Sheikhi, M.; Tavakoli, M.; Rad-Malekshahi, M.; Rad, Z. F. Recent Progress in PLGA-Based Microneedle-Mediated Transdermal Drug and Vaccine Delivery. *Biomater. Sci.* **2023,** *11* (16), 5390–5409.
81. Mahar, R.; Chakraborty, A.; Nainwal, N.; Bahuguna, R.; Sajwan, M.; Jakhmola, V. Application of PLGA as a Biodegradable and Biocompatible Polymer for Pulmonary Delivery of Drugs. *AAPS PharmSciTech* **2023,** *24* (1).
82. Battisti, M.; Vecchione, R.; Casale, C.; Pennacchio, F. A.; Lettera, V.; Jamaledin, R.; Profeta, M.; Di Natale, C.; Imparato, G.; Urciuolo, F.; et al. Non-invasive Production of Multi-Compartmental Biodegradable Polymer Microneedles for Controlled Intradermal Drug Release of Labile Molecules. *Front. Bioeng. Biotechnol.* **2019,** *7.*
83. Narmani, A.; Jahedi, R.; Bakhshian-Dehkordi, E.; Ganji, S.; Nemati, M.; Ghahramani-Asl, R.; Moloudi, K.; Hosseini, S. M.; Bagheri, H.; Kesharwani, P.; et al. Biomedical

Applications of PLGA Nanoparticles in Nanomedicine: Advances in Drug Delivery Systems and Cancer Therapy. *Expet Opin. Drug Deliv.* **2023,** *20* (7), 937–954.

84. Harguindey, A.; Domaille, D. W.; Fairbanks, B. D.; Wagner, J.; Bowman, C. N.; Cha, J. N. Synthesis and Assembly of Click-Nucleic-Acid-Containing PEG-PLGA Nanoparticles for DNA Delivery. *Adv. Mater.* **2017,** *29* (24).
85. Nam, K.; Mahmood, T. M. T.; Ahmad, I.; Ramli, S. Improving Rate of Gelatin/Carboxymethylcellulose Dissolving Microneedle for Transdermal Drug Delivery. *Sains Malays.* **2020,** *49* (9), 2269–2279.
86. Pettignano, A.; Charlot, A.; Fleury, E. Carboxyl-Functionalized Derivatives of Carboxymethyl Cellulose: Towards Advanced Biomedical Applications. *Polym. Rev.* **2019,** *59* (3), 510–560.
87. Kim, Y. C.; Quan, F. S.; Compans, R. W.; Kang, S. M.; Prausnitz, M. R. Formulation of Microneedles Coated with Influenza Virus-Like Particle Vaccine. *AAPS PharmSciTech* **2010,** *11* (3), 1193–1201.
88. Lee, B. M.; Lee, C.; Lahiji, S. F.; Jung, U. W.; Chung, G.; Jung, H. Dissolving Microneedles for Rapid and Painless Local Anesthesia. *Pharmaceutics* **2020,** *12* (4).
89. Chircov, C.; Grumezescu, A. M. Microelectromechanical Systems (MEMS) for Biomedical Applications. *Micromachines* **2022,** *13* (2).
90. Ashraf, M. W.; Tayyaba, S.; Afzulpurkar, N. Micro Electromechanical Systems (MEMS) Based Microfluidic Devices for Biomedical Applications. *Int. J. Mol. Sci.* **2011,** *12* (6), 3648–3704.
91. Kulkarni, D.; Damiri, F.; Rojekar, S.; Zehravi, M.; Ramproshad, S.; Dhoke, D.; Musale, S.; Mulani, A. A.; Modak, P.; Paradhi, R.; et al. Recent Advancements in Microneedle Technology for Multifaceted Biomedical Applications. *Pharmaceutics* **2022,** *14* (5).
92. Tran, K. T. M.; Nguyen, T. D. Lithography-based Methods to Manufacture Biomaterials at Small Scales. *J. Sci. Adv. Mater. Dev.* **2017,** *2* (1), 1–14.
93. Waghule, T.; Singhvi, G.; Dubey, S. K.; Pandey, M. M.; Gupta, G.; Singh, M.; Dua, K. Microneedles: A Smart Approach and Increasing Potential for Transdermal Drug Delivery System. *Biomed. Pharmacother.* **2019,** *109,* 1249–1258.
94. Moreau, W. M. *Semiconductor Lithography: Principles, Practices, and Materials;* Springer US, 1988.
95. Charoensri, T.; Kulthong, K.; Pornrungruang, P.; Wongwanakul, R.; Kunyanee, C.; Aueviriyavit, S.; Rahong S. Enterocyte Culture on a Hybrid Transwell-Inserted Gut-on-a-Chip for Liquid-Liquid and Air-Liquid Interface Conditions. *ACS Omega* **2025**, *10* (45), 53865–53873.
96. Banks, D. *Microengineering, MEMS, and Interfacing: A Practical Guide;* CRC Press, 2006.
97. Donnelly, R. F.; Singh, T. R. R.; Woolfson, A. D. Microneedle-Based Drug Delivery Systems: Microfabrication, Drug Delivery, and Safety. *Drug Deliv.* **2010,** *17* (4), 187–207.
98. O'Mahony, C.; Sebastian, R.; Tjulkins, F.; Whelan, D.; Bocchino, A.; Hu, Y.; O'Brien, J.; Scully, J.; Hegarty, M.; Blake, A.; et al. Hollow Silicon Microneedles, Fabricated Using Combined Wet and Dry Etching Techniques, for Transdermal Delivery and Diagnostics. *Int. J. Pharm.* **2023,** *637,* 122888.
99. Li, W. Z.; Huo, M. R.; Zhou, J. P.; Zhou, Y. Q.; Hao, B. H.; Liu, T.; Zhang, Y. Super-Short Solid Silicon Microneedles for Transdermal Drug Delivery Applications. *Int. J. Pharm.* **2010,** *389* (1–2), 122–129.
100. Tabassum, N.; Alba, M.; Yan, L.; Voelcker, N. H. Porous Silicon Microneedles for Enhanced Transdermal Drug Delivery. *Adv. Therapeut.* **2022.**

101. Howells, O.; Blayney, G. J.; Gualeni, B.; Birchall, J. C.; Eng, P. F.; Ashraf, H.; Sharma, S.; Guy, O. J. Design, Fabrication, and Characterisation of a Silicon Microneedle Array for Transdermal Therapeutic Delivery Using a Single Step Wet Etch Process. *Eur. J. Pharm. Biopharm.* **2022,** *171,* 19–28.
102. Faraji Rad, Z.; Prewett, P. D.; Davies, G. J. An Overview of Microneedle Applications, Materials, and Fabrication Methods. *Beilstein J. Nanotechnol.* **2021,** *12,* 1034–1046.
103. Bolton, C. J. W.; Howells, O.; Blayney, G. J.; Eng, P. F.; Birchall, J. C.; Gualeni, B.; Roberts, K.; Ashraf, H.; Guy, O. J. Hollow Silicon Microneedle Fabrication Using Advanced Plasma Etch Technologies for Applications in Transdermal Drug Delivery. *Lab Chip* **2020,** *20* (15), 2788–2795.
104. Roxhed, N.; Gasser, T. C.; Griss, P.; Holzapfel, G. A.; Stemme, G. Penetration-Enhanced Ultrasharp Microneedles and Prediction on Skin Interaction for Efficient Transdermal Drug Delivery. *J. Microelectromech. Syst.* **2007,** *16* (6), 1429–1440.
105. Ruan, S. Y.; Zhang, Y. T.; Feng, N. P. Microneedle-Mediated Transdermal Nanodelivery Systems: A Review. *Biomater. Sci.* **2021,** *9* (24), 8065–8089.
106. Kim, J. D.; Kim, M.; Yang, H.; Lee, K.; Jung, H. Droplet-Born Air Blowing: Novel Dissolving Microneedle Fabrication. *J. Control. Release* **2013,** *170* (3), 430–436.
107. Huh, I.; Kim, S.; Yang, H.; Jang, M.; Kang, G.; Jung, H. Effects of Two Droplet-Based Dissolving Microneedle Manufacturing Methods on the Activity of Encapsulated Epidermal Growth Factor and Ascorbic Acid. *Eur. J. Pharm. Sci.* **2018,** *114,* 285–292.
108. Parhi, R. Recent Advances in 3D Printed Microneedles and Their Skin Delivery Application in the Treatment of Various Diseases. *J. Drug Deliv. Sci. Technol.* **2023,** *84,* 104395.
109. Economidou, S. N.; Lamprou, D. A.; Douroumis, D. 3D Printing Applications for Transdermal Drug Delivery. *Int J Pharm* **2018,** *544* (2), 415–424.
110. Boehm, R. D.; Miller, P. R.; Hayes, S. L.; Monteiro-Riviere, N. A.; Narayan, R. J. Modification of Microneedles Using Inkjet Printing. *AIP Adv.* **2011,** *1* (2).
111. Wu, M.; Zhang, Y.; Huang, H.; Li, J.; Liu, H.; Guo, Z.; Xue, L.; Liu, S.; Lei, Y. Assisted 3D Printing of Microneedle Patches for Minimally Invasive Glucose Control in Diabetes. *Mater. Sci. Eng. C* **2020,** *117,* 111299.
112. Khosraviboroujeni, A.; Mirdamadian, S. Z.; Minaiyan, M.; Taheri, A. Preparation and Characterization of 3D Printed PLA Microneedle Arrays for Prolonged Transdermal Drug Delivery of Estradiol Valerate. *Drug Delivery and Translational Research* **2022,** *12* (5), 1195–1208.
113. Lim, S. H.; Kathuria, H.; Amir, M. H. B.; Zhang, X.; Duong, H. T. T.; Ho, P. C.; Kang, L. High Resolution Photopolymer for 3D Printing of Personalised Microneedle for Transdermal Delivery of Anti-wrinkle Small Peptide. *J. Control Release* **2021,** *329,* 907–918.
114. Mathew, E.; Pitzanti, G.; Gomes Dos Santos, A. L.; Lamprou, D. A. Optimization of Printing Parameters for Digital Light Processing 3D Printing of Hollow Microneedle Arrays. *Pharmaceutics* **2021,** *13* (11).
115. Choi, C. K.; Kim, J. B.; Jang, E. H.; Youn, Y. N.; Ryu, W. H. Curved Biodegradable Microneedles for Vascular Drug Delivery. *Small* **2012,** *8* (16), 2483–2488.
116. Kim, M.; Kim, T.; Kim, D. S.; Chung, W. K. Curved Microneedle Array-Based sEMG Electrode for Robust Long-Term Measurements and High Selectivity. *Sensors (Basel)* **2015,** *15* (7), 16265–16280.
117. Petrová, E.; Chvíla, S.; Štěpánek, F.; Zbytovská, J.; Lamprou, D. A. Imiquimod Nanocrystal-Loaded Dissolving Microneedles Prepared by DLP Printing. *Drug Deliv. Transl. Res.* **2024.**

118. Shin, D.; Hyun, J. Silk Fibroin Microneedles Fabricated by Digital Light Processing 3D Printing. *J. Ind. Eng. Chem.* **2020**.
119. Oliaei, S. N. B.; Nasseri, B. Stereolithography and Its Applications. *Additive and Subtractive Manufacturing: Emergent Technologies* **2020,** 229–249.
120. Wang, J.; Zhang, Y.; Aghda, N. H.; Pillai, A. R.; Thakkar, R.; Nokhodchi, A.; Maniruzzaman, M. Emerging 3D Printing Technologies for Drug Delivery Devices: Current Status and Future Perspective. *Adv. Drug Deliv. Rev.* **2021,** *174,* 294–316.
121. Quan, H.; Zhang, T.; Xu, H.; Luo, S.; Nie, J.; Zhu, X. Photo-Curing 3D Printing Technique and Its Challenges. *Bioact. Mater.* **2020,** *5* (1), 110–115.
122. Uddin, M. J.; Scoutaris, N.; Economidou, S. N.; Giraud, C.; Chowdhry, B. Z.; Donnelly, R. F.; Douroumis, D. 3D Printed Microneedles for Anticancer Therapy of Skin Tumours. *Mater. Sci. Eng. C Mater. Biol. Appl.* **2020,** *107,* 110248.
123. Song, X.; Xing, J. 3D Printing Technology Based on Two-Photon Polymerization. *CIESC Journal* **2015,** *66* (9), 3324–3332.
124. Economidou, S. N.; Pere, C. P. P.; Reid, A.; Uddin, M. J.; Windmill, J. F. C.; Lamprou, D. A.; Douroumis, D. 3D Printed Microneedle Patches Using Stereolithography (SLA) for Intradermal Insulin Delivery. *Mater. Sci. Eng. C Mater. Biol. Appl.* **2019,** *102,* 743–755.
125. Sirbubalo, M.; Tucak, A.; Muhamedagic, K.; Hindija, L.; Rahić, O.; Hadžiabdić, J.; Cekic, A.; Begic-Hajdarevic, D.; Cohodar Husic, M.; Dervišević, A.; et al. 3D Printing-A "Touch-Button" Approach to Manufacture Microneedles for Transdermal Drug Delivery. *Pharmaceutics* **2021,** *13* (7).
126. Gittard, S. D.; Ovsianikov, A.; Chichkov, B. N.; Doraiswamy, A.; Narayan, R. J. Two-Photon Polymerization of Microneedles for Transdermal Drug Delivery. *Expert Opin. Drug Deliv.* **2010,** *7* (4), 513–533.
127. Aksit, A.; Arteaga, D. N.; Arriaga, M.; Wang, X.; Watanabe, H.; Kasza, K. E.; Lalwani, A. K.; Kysar, J. W. In-vitro Perforation of the Round Window Membrane via Direct 3-D Printed Microneedles. *Biomed. Microdevices* **2018,** *20* (2), 47.
128. Plamadeala, C.; Gosain, S. R.; Hischen, F.; Buchroithner, B.; Puthukodan, S.; Jacak, J.; Bocchino, A.; Whelan, D.; O'Mahony, C.; Baumgartner, W.; et al. Bio-inspired Microneedle Design for Efficient Drug/Vaccine Coating. *Biomed. Microdevices* **2019,** *22* (1), 8.
129. Trautmann, A.; Roth, G. L.; Nujiqi, B.; Walther, T.; Hellmann, R. Towards a Versatile Point-of-Care System Combining Femtosecond Laser Generated Microfluidic Channels and Direct Laser Written Microneedle Arrays. *Microsyst. Nanoeng.* **2019,** *5,* 6.
130. Liao, C.; Anderson, W.; Antaw, F.; Trau, M. Two-Photon Nanolithography of Tailored Hollow Three-Dimensional Microdevices for Biosystems. *ACS Omega* **2019,** *4* (1), 1401–1409.
131. Cordeiro, A. S.; Tekko, I. A.; Jomaa, M. H.; Vora, L.; McAlister, E.; Volpe-Zanutto, F.; Nethery, M.; Baine, P. T.; Mitchell, N.; McNeill, D. W.; et al. Two-Photon Polymerisation 3D Printing of Microneedle Array Templates with Versatile Designs: Application in the Development of Polymeric Drug Delivery Systems. *Pharm. Res.* **2020,** *37* (9), 174.
132. Doraiswamy, A.; Jin, C.; Narayan, R. J.; Mageswaran, P.; Mente, P.; Modi, R.; Auyeung, R.; Chrisey, D. B.; Ovsianikov, A.; Chichkov, B. Two Photon Induced Polymerization of Organic-Inorganic Hybrid Biomaterials for Microstructured Medical Devices. *Acta Biomater.* **2006,** *2* (3), 267–275.
133. Sirbubalo, M.; Tucak, A.; Muhamedagic, K.; Hindija, L.; Rahic, O.; Hadziabdic, J.; Cekic, A.; Begic-Hajdarevic, D.; Husic, M. C.; Dervisevic, A.; et al. 3D Printing-A "Touch-Button" Approach to Manufacture Microneedles for Transdermal Drug Delivery. *Pharmaceutics* **2021,** *13* (7).

134. Beg, S.; Almalki, W. H.; Malik, A.; Farhan, M.; Aatif, M.; Rahman, Z.; Alruwaili, N. K.; Alrobaian, M.; Tarique, M.; Rahman, M. 3D Printing for Drug Delivery and Biomedical Applications. *Drug Discov. Today* **2020,** *25* (9), 1668–1681.
135. Charoo, N. A.; Ali, S. F. B.; Mohamed, E. M.; Kuttolamadom, M. A.; Ozkan, T.; Khan, M. A.; Rahman, Z. Selective Laser Sintering 3D Printing—An Overview of the Technology and Pharmaceutical Applications. *Drug Dev. Ind. Pharm.* **2020,** *46* (6), 869–877.
136. Jamróz, W.; Szafraniec, J.; Kurek, M.; Jachowicz, R. 3D Printing in Pharmaceutical and Medical Applications—Recent Achievements and Challenges. *Pharm. Res.* **2018,** *35* (9), 176.
137. Krieger, K. J.; Liegey, J.; Cahill, E. M.; Bertollo, N.; Lowery, M. M.; O'Cearbhaill, E. D. Development and Evaluation of 3D-Printed Dry Microneedle Electrodes for Surface Electromyography. *Adv. Mater. Technol.* **2020,** *5* (10), 2000518.
138. Chen, Z. P.; Ren, L.; Li, J. Y.; Yao, L. B.; Chen, Y.; Liu, B.; Jiang, L. L. Rapid Fabrication of Microneedles Using Magnetorheological Drawing Lithography. *Acta Biomater.* **2018,** *65,* 283–291.
139. Elahpour, N.; Pahlevanzadeh, F.; Kharaziha, M.; Bakhsheshi-Rad, H. R.; Ramakrishna, S.; Berto, F. 3D Printed Microneedles for Transdermal Drug Delivery: A Brief Review of Two Decades. *Int. J. Pharm.* **2021,** *597.*

Chapter 3

Technologies for characterization of microneedles

Huanhuan Pan[1], Tao Ren[2] and Tingting Peng[1]
[1]*State Key Laboratory of Bioactive Molecules and Druggability Assessment/International Cooperative Laboratory of Traditional Chinese Medicine Modernization and Innovative Drug Development of Ministry of Education (MOE) of China/College of Pharmacy, Jinan University, Guangzhou, China;* [2]*School of Pharmaceutical Science, Sun Yat-Sen University, Guangzhou, China*

1. Introduction

Microneedles (MNs) have been widely applied in biomedical fields such as drug delivery, disease diagnosis, and tissue repair and regeneration, owing to their advantages of being painless and minimally invasive, and having high drug delivery efficiency. However, MN technology faces several challenges that need urgent resolution: (1) insufficient mechanical strength may lead to incomplete skin penetration, causing drug retention on the skin surface and compromising dosing accuracy; (2) limited drug-loading capacity restricts therapeutic efficacy; (3) some MN materials are poorly biocompatible, potentially causing skin irritation (e.g., erythema, edema, eschar) or even systemic adverse effects such as cytotoxicity, hematotoxicity, and organ toxicity. To address these challenges, it is essential to establish a comprehensive evaluation system that assesses the physicochemical properties and biological performance of MNs through multiparameter analyses, ensuring that they meet clinical standards. In this chapter, we systematically summarize current strategies for evaluating MN performance.

2. Approaches to evaluating the morphology and geometrics of microneedles

Maintaining the structural integrity of MNs is crucial for ensuring effective skin penetration. Additionally, MN geometrics, including the needle length,

Microneedle Technology. https://doi.org/10.1016/B978-0-443-45544-5.00010-5

density, base diameter, and tip-to-tip spacing, are key factors affecting their mechanical performance. The surface topography of MNs further affects drug-loading capacity and release behavior. Consequently, a comprehensive characterization of MN morphology, geometry, and surface topography can guide the optimization of formulation compositions, geometric parameters, and fabrication design. The relevant characterization methods are described in detail subsequently.

Optical microscopy offers a simple and rapid approach for assessing MN morphology and overall structure, allowing visual detection of defects such as fractures or bending. It also enables the quantitative measurement of key geometric parameters, including needle length, density, base diameter, and tip-to-tip spacing. To ensure consistent puncture depth and drug delivery accuracy, critical MN dimensions (length, diameter, and tip angle) must be controlled within a ±10% tolerance. For instance, Zhang et al.[1] developed microparticle-loaded gas-propelled MNs to enhance the drug-loading capacity and transdermal delivery efficiency simultaneously. Vacuum microperfusion and centrifugal microperfusion are two prevalent methods for fabricating dissolving MNs, in which highly viscous needle solutions are infused into the microcavities of female mold through pressure gradients and centrifugal forces, respectively. To optimize the fabrication parameters, the authors investigated the effects of varying centrifugal speeds and negative pressures on the demolding needle percentage (DNP) and morphological characteristics of MNs (Fig. 3.1A,B). Optical microscopy analysis revealed that MNs prepared by centrifugal microperfusion exhibited significant needle loss and fracture defects. In contrast, those fabricated via vacuum microperfusion under −0.08 to −0.09 MPa demonstrated intact needle structures with minimal fractures (Fig. 3.1C,D), indicating superior morphological integrity. The observed differences can be attributed to the distinct forming mechanism of MNs: centrifugal molding may induce phase separation between the polymer matrix and microparticles owing to their divergent mechanical properties. Asynchronous centrifugation further leads to nonuniform component distribution and structural fractures. Conversely, vacuum-assisted molding, driven by uniform pressure gradients, facilitates synchronous sedimentation of all components into the microcavities through a gentler process, significantly improving the structural integrity of MNs. Stereo-fluorescence microscopy, such as optical microscopy, can assess MN morphology and geometry. It also enables the visualization of fluorescence signal distribution in MNs loaded with fluorescent agents, which is a common method for inferring drug distribution within MNs. However, both techniques are constrained by limitations of resolution and cannot resolve microstructural details or fine surface features.

Compared with optical microscopy and stereo-fluorescence microscopy, scanning electron microscopy (SEM) provides higher resolution, enabling direct, high-magnification topographic imaging of surface morphology via

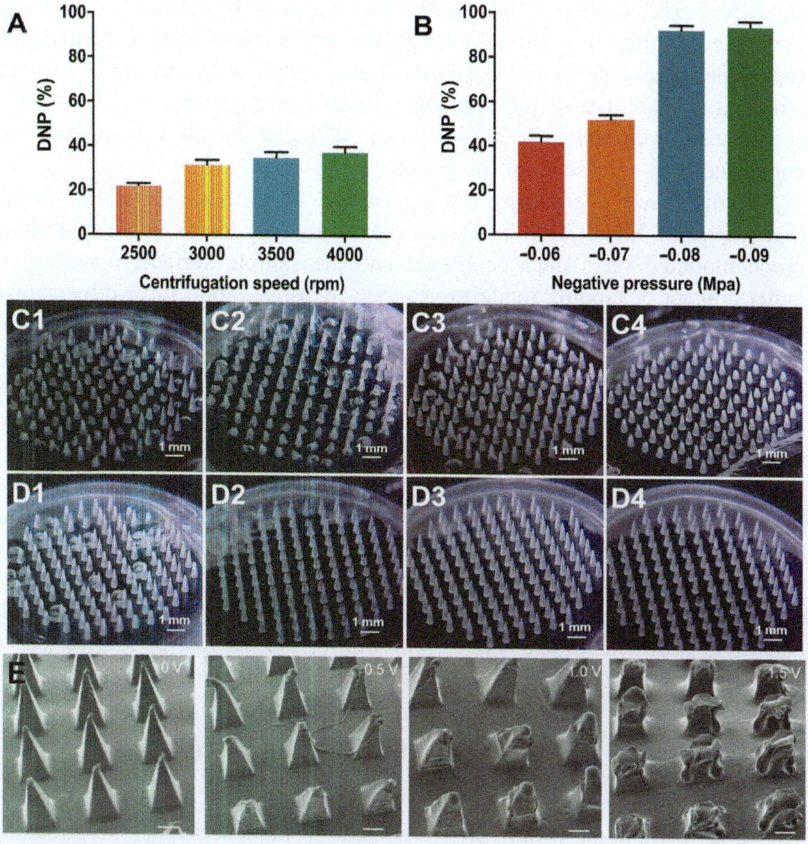

FIGURE 3.1 The influence of (A) centrifugation speed and (B) negative pressure on the demolding needle percentage (DNP) of particle-embedded gas-propelled microneedle (MNs) (n = 3). The morphology of MNs were prepared at centrifugation speeds of (C1) 2500 rpm, (C2) 3000 rpm, (C3) 3500 rpm, and (C4) 4000 rpm, and negative pressures of (D1) −0.06 MPa, (D2) −0.07 MPa, (D3) −0.08 MPa, (D4) and −0.09 MPa. (E) Scanning electron microscopy images to reveal the dissolution performance of MNs within 5 min under different voltages. *Scale bar* = 250 μm. *(A-D) were reprinted from Zhang et al.*[1] *with permission. (E) Reprinted from Jin et al.*[2] *with permission.*

secondary electron signals. Prior to SEM analysis, MNs are typically mounted on a V-shaped copper sheet using conductive tape and sputter-coated with a gold layer. For instance, Jin et al.[2] developed a flexible bioelectronic MN patch (FBMP) for actively controlled transdermal drug delivery. To investigate the release behavior of the FBMP, the MNs were inserted into porcine skin tissue, and different voltages were applied to the integrated heater for 5 min. After retrieval, the morphological changes and degradation behavior of the MNs were evaluated using SEM. As shown in Fig. 3.1E, the double-layered MNs

(DLMNs) exhibited no or only partial dissolution at 0, 0.5, and 1.0 V, while complete dissolution occurred at 1.5 V. This phenomenon can be attributed to the temperature-dependent degradation kinetics of DLMNs, as the final heater temperature varied with the applied voltage. Consequently, higher voltages accelerated drug release, while lower voltages enabled sustained and controlled drug release from the FBMP.

Atomic force microscopy (AFM) can be used to evaluate the surface topography of MNs. AFM measures the interaction forces between a sharp probe mounted on a microcantilever and the sample surface, operating in either contact or tapping mode to generate high-resolution three-dimensional topographic data. Unlike SEM, AFM not only visualizes the macroscopic morphology of MNs but also quantitatively analyzes microscopic parameters such as the surface roughness and tip curvature radius. These parameters are critical for assessing MN performance, including penetration efficiency and drug-loading capacity. Studies indicate that smoother MN surfaces facilitate more uniform drug coating and release, while a smaller tip curvature radius improves puncture efficiency.[3] For instance, Yang et al.[4] employed tetrahedral DNA nanostructures (TDNs) as nanocarriers to construct molecular recognition complexes on the surface of graphene-modified MN patches. AFM characterization revealed uniformly distributed nanoparticles with heights consistent with theoretical values, confirming the successful fabrication of TDNs.

3. Approaches used to evaluate the drug distribution of microneedles

The drug distribution within MNs significantly influences their drug-loading capacity and delivery efficiency. Optimizing this distribution has been shown to enhance drug-loading capacity while minimizing drug loss.[5] To analyze drug distribution, a fluorescent dye (e.g., Rhodamine B (RhB), fluorescein isothiocyanate [FITC], IR780) can be incorporated as a drug surrogate during MN fabrication. These dyes emit light upon excitation, enabling visualization of drug distribution within the MNs using confocal laser scanning microscopy (CLSM). For instance, Hu et al.[6] employed RhB as a fluorescent drug analog to investigate the spatial distribution of pharmaceuticals within MNs. Through CLSM analysis, a significant fluorescence signal was found to be enriched at the needle tips (Fig. 3.2A,B), demonstrating preferential drug accumulation in these regions. Alternatively, the drug itself can be chemically labeled with a fluorescent dye for direct tracking via CLSM. For instance, Wang et al.[7] developed glucose-responsive core-shell MNs featuring solid insulin powder cores and glucose-responsive shells to treat diabetes. To visualize the spatial distribution of insulin, insulin was labeled with FITC while the shell components were tagged with Cyanine 5. Optical and fluorescence microscopy imaging revealed homogeneous insulin distribution throughout the MNs (Fig. 3.2C). Cross-sectional analysis

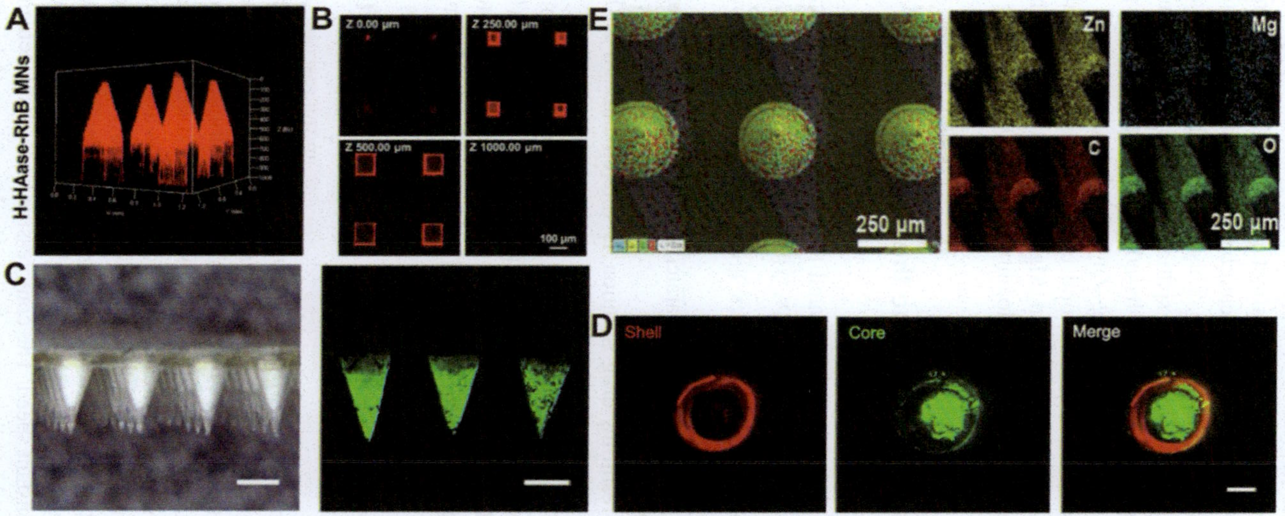

FIGURE 3.2 (A) 3D-reconstructed and (B) representative images at different depths of H-hyaluronidase (HAase)-Rhodamine B (RhB) microneedles (MNs) captured by confocal laser scanning microscopy (CLSM) to visualize drug distribution. (C) Microscopic image of solid MNs loaded with insulin powder (*left*) and fluorescence image of solid MNs loaded with fluorescein isothiocyanate (FITC)-labeled insulin powder (*right*). *Scale bars*, 500 μm. (D) CLSM images of glucose-responsive MNs made of a Cyanine 5-labeled shell and a fluorescein isothiocyanate-labeled insulin-loaded core from a cross-section view. *Scale bar*, 100 μm. (E) Top-down scanning electron microscopy image and energy-dispersive spectroscopy elemental mapping of the LHSQ-loaded MN patch. *(A and B) were reprinted from Hu et al.[6] with permission. (C and D) were reprinted from Wang et al.[7] with permission. (E) Reprinted from Liu et al.[8] with permission.*

via CLSM further confirmed the well-defined core-shell architecture (Fig. 3.2D), demonstrating successful fabrication of the dual-functional delivery system.

When the drug contains elements distinct from the MN matrix material, energy-dispersive spectroscopy (EDS) can be employed to determine the elemental composition of the MN surface. This approach allows for qualitative and quantitative assessment of the drug's spatial distribution. For instance, Liu et al.[8] developed γ-polyglutamic acid–based multifunctional MN patches integrating quercetin (Qu)-modified Mg–Zn-layered hydroxide salt (LHS) nanosheets (LHSQ) to combat diabetic oral ulcers. EDS elemental mapping revealed homogeneous distributions of Zn, Mg, O, and C throughout the MNs (Fig. 3.2E), confirming the successful loading of LHSQ within the needle matrix.

4. Approaches used to determine the drug loading of microneedles

The drug loading of MNs is closely related to their therapeutic efficacy. To determine the drug-loading capacity, steps typically performed are:

(1) Separation of needles: The needles are carefully detached from the baseplate using a scalpel blade;
(2) Dissolution and collection: The separated needles are collected and fully dissolved in a specific solvent;
(3) Filtration: The resulting drug solution is filtered through a 0.22-μm membrane;
(4) Quantification: The drug content in the filtrate is analyzed using techniques such as high-performance liquid chromatography (HPLC), ultraviolet-visible spectrophotometry, or fluorescence spectrophotometry, and the drug-loading capacity of the MNs is calculated accordingly.

5. Approaches used to evaluate the mechanical properties of microneedles

The mechanical properties of MNs are key parameters affecting the skin puncture of MNs. Texture analyzers, universal testers, micromechanical testers, and nanoindentation testers are commonly used instruments for directly determining the mechanical properties of MNs. Taking texture analyzers as an example, the mechanical properties of MNs can be measured by this instrument in two modes. In one mode depicted in Fig. 3.3A, MNs are placed on a stainless platform with the needles facing upward before the test. Then, the pressure probe connected to a load cell begins to move straight down and compress MNs at a constant speed. The compressive force–displacement curve can be further recorded and plotted to calculate the Young's modulus (Eq. 3.1) and Poisson's ratio (ν) (Eq. 3.2) of MNs, which represent the stiffness and

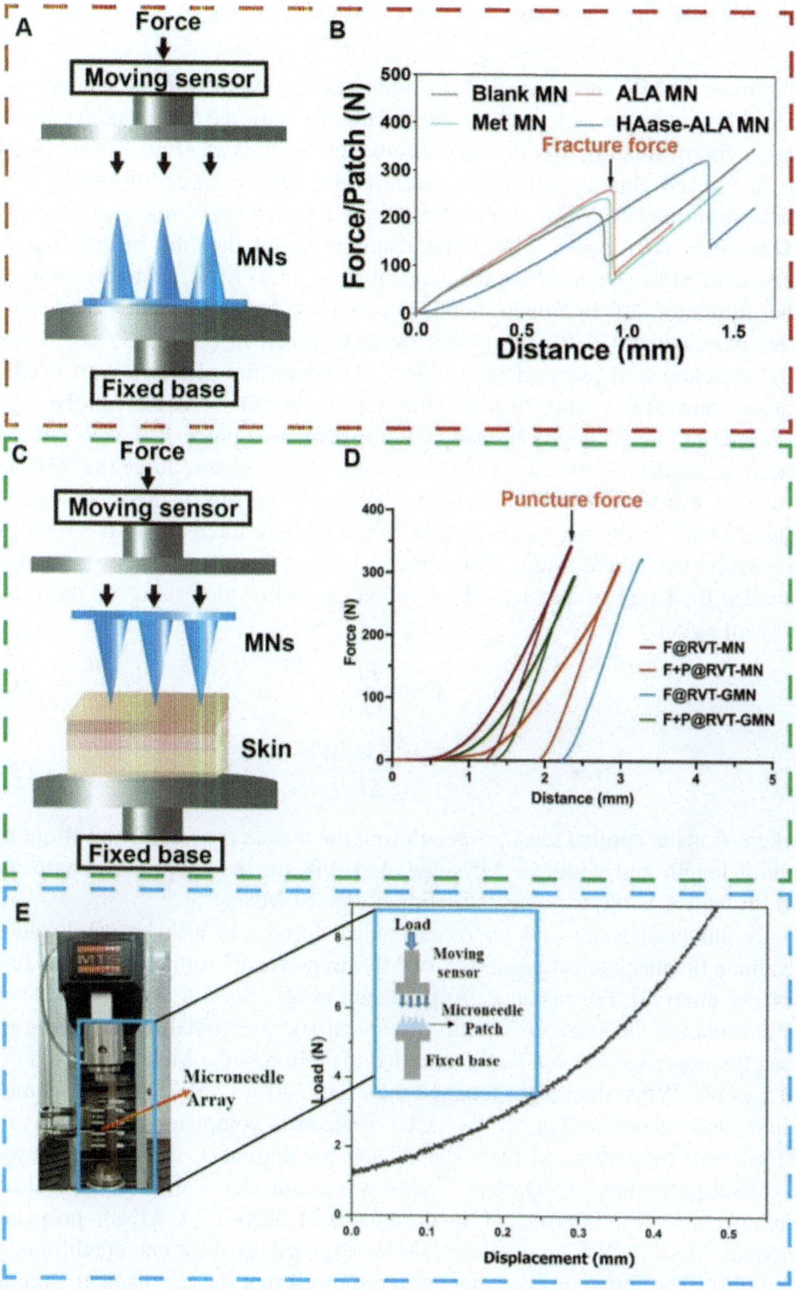

FIGURE 3.3 (A) Schematic diagram of texture analyzer used to measure fracture force of microneedles (MNs). (B) Force–displacement curves obtained from texture analyzer. (C) Schematic diagram of texture analyzer used to measure maximum puncture force that the MNs can withstand. (D) Force–displacement curves of MNs and gas-propelled MNs (GMNs) piercing skin tissues. (E) Fracture force–displacement curves obtained from universal testing machine. *ALA*, atomic force microscopy; *HAase*, hyaluronidase; *RVT*, rivastigmine. *(B) Reprinted from Huang et al.*[10] *with permission. (D) Reprinted from Zhang et al.*[11] *with permission. (E) Reprinted from Anbazhagan et al.*[12] *with permission.*

hardness of MN, respectively.[9] In addition, the point with a sudden drop occurring in the curve represents the fracture force of the MNs, and it can be a value for evaluating whether the needle tip is broken (Fig. 3.3B). Huang et al.[10] used this approach to measure the failure force of blank MNs, metformin-loaded MNs, 5-aminolevulinic acid (ALA), and hyaluronidase (HAase)/ALA-coloaded MNs. The result revealed that the introduction of HAase to MNs enhanced the mechanical strength of MNs, probably owing to the formation of hydrogen bond interactions between HAase and polyvinylpyrrolidone (PVP). In another mode depicted in Fig. 3.3C, the MNs is first attached to a piece of skin tissue, placed on the platform with needles facing downward, and further compressed by the pressure probe at a descending rate. This process can simulate the actual process of MNs piercing the skin. As displayed in Fig. 3.3D, the maximum puncture force that MN can resist is available from the compression force versus displacement curve. Zhang et al.[11] used this approach to detect the failure force of passive MNs and gas-propelled active MNs. The failure force of active MNs was slightly smaller than that of passive MNs, probably owing to loading of the effervescent agent:

$$Et = \frac{FL}{A\Delta L} \tag{3.1}$$

$$v = -\frac{\Delta W/W}{\Delta L/L} \tag{3.2}$$

where F is the applied tensile force during the tensile process, L and W are the initial length and width of MNs, ΔL and ΔW are the change in length and width, and A is the axial cross-sectional area of MNs.

A universal tester and micromechanical tester can also be employed to evaluate the mechanical properties of MNs in protocols similar that those for a texture analyzer. For instance, Anbazhagan et al.[12] used a universal tester to determine the mechanical strength of conical and pyramidal MNs. During the test, the upper workbench was descended to compress the MNs arrays at a rate of 1 mm/s. When the sensor touched the needle arrays, the load–displacement curve was recorded (Fig. 3.3E). After force was completely applied to the MNs, both MNs retained their shapes and tip sharpness, showing good mechanical properties. Wang et al.[13] used a micromechanical tester to evaluate the mechanical properties of hyaluronic acid MNs (HA MNs), polyvinyl alcohol MNs (MNs), and layered MNs exposed to different conditions of humidity. The force–displacement cures showed that the mechanical strength of all MNs decreased with increasing humidity. Fortunately, the layered MNs made of a PVA core and HA shell could still pierce the skin after exposure in an environment with no more than 60% humidity. The most commonly used function of a nanoindentation tester is to determine the hardness and elastic modulus.[14] The major components of nanoindentation testers include sensors, the actuators used to apply mechanical load and measure the indenter displacement, and the indenter tip. Unlike the instruments mentioned earlier, the pressure probe of the nanoindentation tester does not have a flat surface but rather a sharp indenter. During the nanoindentation test, the force–displacement curves are recorded when the indenter tip presses into the surface of materials under a preset load.[14] Dong et al.[15] utilized a nanoindenter with a sphere silicon indenter to determine the mechanical properties of MNs. The loading of a gold nanocage (AuNC) into MNs increased the resistance of MNs to indentation in a dose-dependent manner. When the loading amount of AuNC was increased from 0 to 4 μg/patch, the MNs showed a dramatic increase in the effective Young's moduli from 68.9 to 224.9 MPa. In contrast, the further loading of doxorubicin caused a decrease in the Young's modulus of the MNs from 68.9 to 38.4 MPa.

6. Approaches used to evaluate the skin puncture ability of microneedles

Effective penetration of MNs into skin tissue is essential for successful drug delivery. The insertion capability of MNs, including skin puncture efficiency and depth, can be assessed using methods such as dye staining, histological sectioning, and imaging techniques.

Skin puncture efficiency is a widely used metric that evaluates the penetration capability of MNs. The MNs are first inserted into the skin manually or with an applicator. For trypan blue (or methylene blue) staining, the dye solution is applied to the puncture site and left for several minutes. The skin is then rinsed with saline to remove excess dye, and the resulting microholes are imaged. The presence of blue punctures indicates a successful insertion, and the puncture rate is calculated as the ratio of blue holes to the total number of needles in the array (Fig. 3.4A). For example, Yang et al.[16] used trypan blue staining to evaluate the penetration efficiency of HAase MNs, achieving a rate exceeding 90%. Similarly, Haq et al.[17] used methylene blue to stain the skin surface for 10 min. The stained pinholes were then imaged by a dermatoscope, and the result showed an average penetration rate of 96% for MNs with a length of 280 μm.

The puncture depth is another key indicator of MN skin penetration performance. An appropriate depth ensures effective drug delivery to the target skin layer, such as the epidermis or dermis. The MN penetration depth can be measured using histological methods, optical coherence tomography (OCT), and other techniques. Histological analysis typically involves sectioning and staining skin samples, followed by microscopic examination of the puncture sites. For instance, Wang et al.[18] utilized hematoxylin-eosin (H&E) staining to evaluate the ability of skin punctures to dissolve MNs prepared from highly water-soluble cyclodextrin derivatives. The H&E staining images revealed that the MNs created distinct microscopic skin pores (Fig. 3.4B), with penetration depths of up to 200 μm. In contrast, OCT is a noninvasive imaging technique

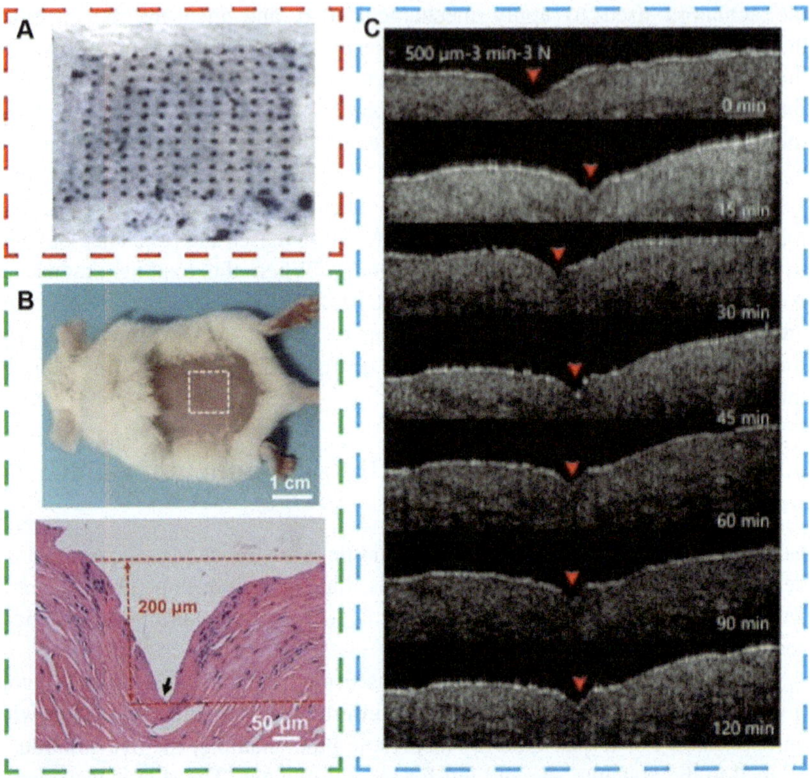

FIGURE 3.4 (A) Trypan blue staining of skin tissues used to evaluate the skin insertion ability of microneedles (MNs). (B) Optical microscopy image of the hematoxylin-eosin–stained porcine skin section. (C) Optical coherence tomography image illustrating changes in the pore size of microchannels created by MNs within 2 h. *(A) Reprinted from Yang et al.*[16] *with permission. (B) Reprinted from Wang et al.*[18] *with permission. (C) Reprinted from Huang et al.*[20] *with permission.*

that enables the real-time visualization of MN penetration depth and distribution in the skin, allowing dynamic monitoring of the puncture process.[19] Unlike staining methods, OCT provides high-resolution in situ imaging of the stratum corneum, epidermis, and upper dermis without requiring skin pretreatment. This technique facilitates quantification of the MN insertion depth to assess effectiveness. For instance, Huang et al.[20] used OCT to monitor microwell formation in real time, optimizing puncture conditions (Fig. 3.4C). Their results demonstrated that MNs with a length of 500 μm achieved optimal efficacy when applied for 3 min under 3 N pressure.

7. Approaches used to evaluate the dissolution and swelling behavior of microneedles

The dissolution or swelling properties of MNs directly affect the drug release kinetics. Dissolving MNs dissolve after penetrating the skin tissue, releasing the loaded drug into the skin. Since the dissolution process of MNs is accompanied by morphological changes, their postinsertion morphological evolution can be observed under an optical microscope. Specifically, the MNs are applied to ex vivo skin tissue, and an optical microscope is used to record their morphological changes before insertion and at different time points. By measuring variations in the length of MNs over time, a dissolution–time curve is plotted and their dissolution rate is subsequently calculated. For instance, Zhang et al.[11] developed a biomimetic MN system dually driven by pneumatic and capillary forces for rivastigmine (RVT) delivery, enabling the early intervention and long-term management of Alzheimer's disease. The researchers evaluated the skin dissolution behavior of MNs by monitoring their morphological changes in ex vivo skin tissue and quantifying residual needle height over time to generate dissolution profiles. The results demonstrated that the pneumatic MNs exhibited complete dissolution within 10 min, whereas passive MNs took 90 min to dissolve completely (Fig. 3.5A). Quantitative measurements at 10-min after administration demonstrated that the remaining height percentage of F@RVT-MN, F@RVT-GMN, F + P@RVT-MN, and F + P@RVT-GMN was 43.92% ± 3.54%, 37.80% ± 2.62%, 37.93%±3.47%, and 31.33% ± 2.63%, respectively (Fig. 3.5B). These findings indicate that pneumatic actuation significantly enhances the dissolution rates of MNs, particularly for the F + P@RVT-GMN formulation. This acceleration mechanism operates through two synergistic pathways: microbubble generation facilitates rapid matrix disintegration, and the guava-like porous architecture enhances interstitial fluid uptake via capillary action, promoting polymer swelling and dissolution.

Hydrogel-forming MNs enhance drug delivery by absorbing skin tissue fluid, swelling to form water channels within the needle matrix. This process facilitates the diffusion of drugs from the backing layer into the skin tissue. The swelling performance of hydrogel-forming MNs is a critical determinant of transdermal delivery efficiency. To evaluate the swelling properties of the MNs, this procedure can be employed: (1) immerse MNs in a buffer solution (32°C, pH 7.0) until swelling equilibrium is achieved (until their weight stabilizes); (2) gently remove surface moisture using filter paper; and (3) accurately weigh the MNs and calculate the swelling ratio at designated time intervals using Eq. (3.3). For instance, Li et al.[21] developed an ultra-swelling MN device (USMD) for the facile drug loading and transdermal delivery of hydrophilic drugs with a broad range of molecular weight. The USMD was composed of ultraswelling needle arrays made from crosslinked methacrylated hyaluronic acid (MeHA) and a resin substrate. The swelling capacity of USMDs can be flexibly adjusted by varying the crosslinking (CL) time of MeHA. In general, the shorter crosslinking time yielded larger pore sizes and higher swelling ratios in the USMDs (Fig. 3.5C,D):

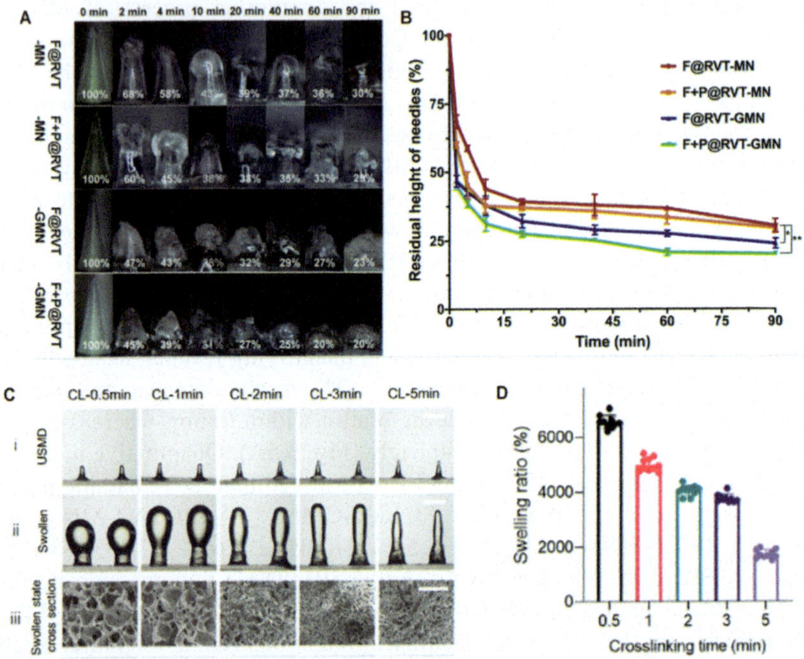

FIGURE 3.5 (A) Morphological changes in microneedles (MNs) and gas-propelled MNs (GMNs) after insertion into the skin for different times to evaluate their skin dissolvability. (B) Dissolution profiles of different MNs (n = 3, *$P < .05$ and **$P < .01$). (C) Simulation of drug-loading process for ultraswelling MN device (USMD) in phosphate-buffered saline (PBS). (i) pristine USMD, (ii) after 1 min swelling in PBS, (iii) cross-sectional scanning electron microscopy (SEM) of MNs in swollen state. (D) Swelling ratio of USMD with different UV crosslinking (CL) times (n = 10). *RVT*, rivastigmine. *(A and B) Reprinted from Zhang et al.*[11] *with permission. (C and D) Reprinted from Li et al.*[21] *with permission.*

$$\text{Swelling ratio } (\%) = \left[\frac{W_t - W_0}{W_0}\right] \tag{3.3}$$

where W_t is the weight of the MNs at time point t, and W_0 is the initial weight of the MNs.

8. Approaches used to evaluate the drug release, permeation, and retention behavior of microneedles

The drug release behavior of MNs is a key factor influencing the duration of drug action and therapeutic efficacy. In vitro drug release can be evaluated using a dialysis method. Briefly, the needle tips of the MNs are cut off and placed inside a dialysis bag, which is then immersed in a tube containing buffer solution (the release medium) to simulate a sink condition. The tube is placed in a water bath shaker maintained at 32°C and 100 rpm. At predetermined intervals, samples of the release medium outside the dialysis membrane are collected and the drug content is quantified via HPLC or UV spectrophotometry to generate the drug release profile. For instance, Deng et al.[22] developed a personalized biphasic MN patch (MNP) incorporating drug-loaded nanocontainers (NCs) for smart drug administration. The contraceptive hormone levonorgestrel (LNG) was used as a model drug and formulated into three MNP formulations with equivalent LNG doses by modulating the drug-to-NC ratio: type A (100% free LNG), type B (50% free LNG + 50% LNG-loaded NCs), and type C (100% LNG-loaded NCs). The authors used dialysis membrane assays to evaluate the in vitro drug release profile of MNP (Fig. 3.6A). The results demonstrated distinct release kinetics (Fig. 3.6B): In the first 4 h, type A showed the fastest drug release rate and released about 40% of the LNG, while type C showed the slowest cumulative drug release rate of 3.7% and type B demonstrated a release trend between those of types A and C. Subsequently, type A released 60% of the drug over 2 days, reaching a plateau, whereas types B and C showed only an initial burst LNG release in the first 4 days. The initial burst in types B and C could be attributed to the rapid PVP dissolution and surface-adsorbed, nonencapsulated, or loosely bound LNG diffusion from the membrane surface of polycaprolactone during hydration of the NCs.

Compared to conventional transdermal preparations, MNs can pierce the stratum corneum, significantly enhancing transdermal drug permeability. The transdermal permeation behavior of MNs can be influenced by multiple factors such as the compositions of formulations, MN geometries, and the mode of administration. The Franz diffusion cell is a widely used apparatus for evaluating the transdermal permeability of MNs. In a typical procedure, MNs are applied vertically to excised rat skin, which is then trimmed to an appropriate size and secured between the donor and receptor chambers of the diffusion cell using a horseshoe clamp. Prewarmed phosphate-buffered saline (PBS) (pH 7.0) is added to the receptor chamber as the dissolution medium, with the water bath maintained at 32°C and the stirring speed set at 100 rpm. At predetermined intervals, 1 mL of receptor fluid is sampled and replaced with an equal volume of fresh medium to maintain sink conditions. After the experiment, the skin tissue is removed, and any residual drug on the skin surface is gently wiped away with alcohol-soaked cotton. The skin is then homogenized, and the resulting homogenate is centrifuged. The supernatant is filtered through a 0.22-μm membrane, and the drug content in both the receptor fluid and skin homogenate is quantified via HPLC or UV spectrophotometry. This allows for the calculation of the transdermal permeation rate and amount of skin retention, respectively. For instance, Lin et al.[23] developed dissolving MN arrays loaded with triamcinolone acetonide (TA-DMNA) for the treatment of hypertrophic scar. The authors studied the Franz permeation behavior of TA-DMNA, commercial cream, and intralesional injections. The results

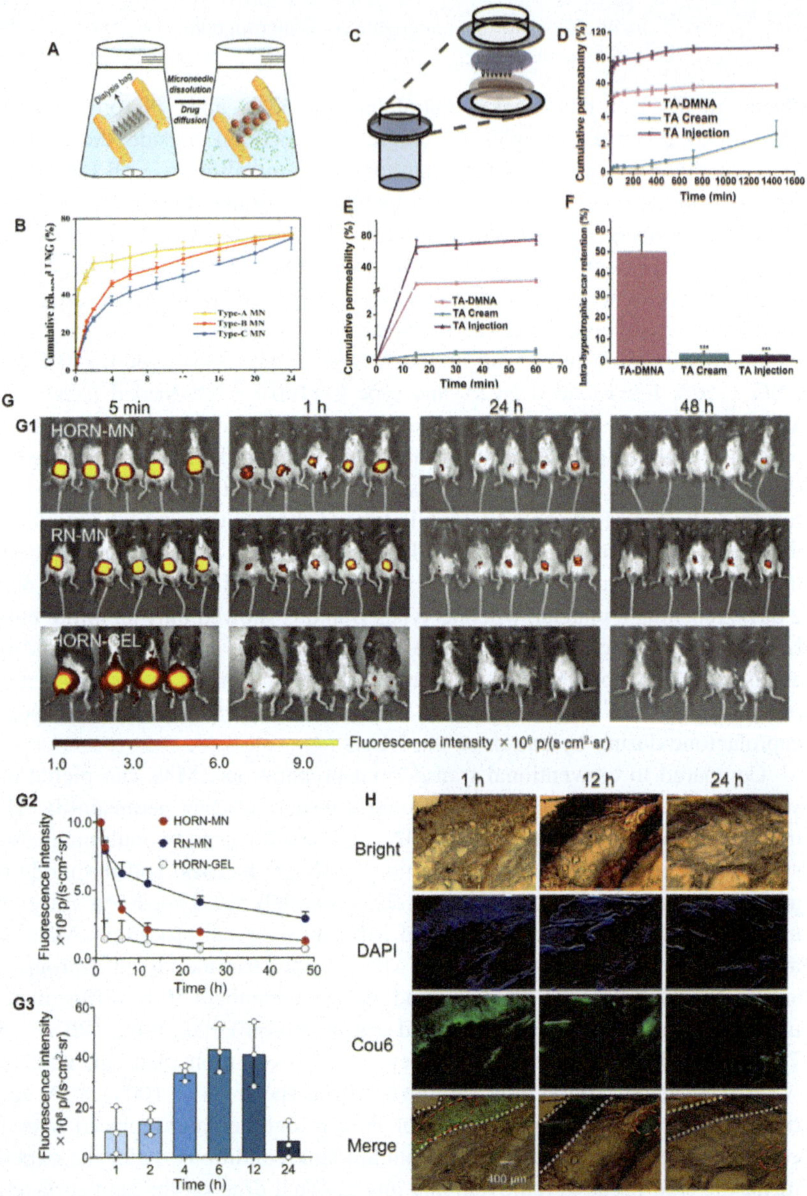

FIGURE 3.6 (A) Schematic diagram of in vitro drug release from microneedle (MN) patches (MNPs) using dialysis bags. (B) Drug release profiles of three types of MNPs (n = 3). Schematic illustration of ex-vivo transdermal permeation experiment. (C, D) Cumulative transdermal permeability profiles of MN arrays loaded with triamcinolone acetonide (TA-DMNA), TA cream, and TA injection or at (D) 24 h and (E)1 h (n = 3). (F) Drug retention amount of TA-DMNA, TA cream, and TA injection in hypertrophic scar tissue (n = 3). ***$P < .001$ versus TA-DMNA. (G) In vivo pharmacodynamic studies. (G1) In vivo fluorescence images. (G2) Quantification of drug retention. (G3) Quantification of drug penetration of rosiglitazone-loaded oleanolic acid dimer (HOA)-based nanoparticle MN (HORN-MN). (H) Fluorescence imaging of subcutaneous drug distribution. DAPI, 4',6-diamidino-2-phenylindole; *HORN-GEL*, Rosi-loaded HOA-based nanoparticles hyaluronic acid hydrogel; *RN-MN*, Rosi-loaded nanoparticle MN. *(A and B) were reprinted from Deng et al.*[22] *with permission. (C–F) were reprinted from Lin et al.*[23] *with permission. (G and H) were reprinted from Chen et al.*[24] *with permission.*

demonstrated that the TA-DMNA significantly enhanced drug retention in scar tissue while reducing transdermal permeability, offering superior advantages over conventional cream and injection therapies (Fig. 3.6C–F). These findings suggest that MNs represent a promising targeted topical treatment for skin diseases, minimizing systemic side effects while improving therapeutic efficacy.

The in vivo imaging system (IVIS) allows the real-time monitoring of drug distribution in the skin and major organs. Fluorescently labeled MNs were applied to the dorsal skin of mice, and IVIS was used to acquire fluorescence images at designated time points to track skin retention and biodistribution. The fluorescence intensity measured by IVIS in the skin showed a positive correlation with the amount of drug retained. At the end of the experiment, the mice were euthanized, and major organs (heart, liver, spleen, lungs, and kidneys) were excised for IVIS imaging to evaluate drug biodistribution. To assess transdermal permeation efficiency further, skin tissues from the MN application site were collected at predetermined time points for sectioning. The distribution of fluorescent drugs within the skin layers was then visualized using CLSM. Additionally, Image J software was employed to quantify the diffusion depth and total fluorescence intensity of the drugs. For instance, Chen et al.[24] developed three distinct MN patches to treat obesity: (1) rosiglitazone (Rosi)-loaded nanoparticle MNs (RN-MNs), (2) Rosi-loaded oleanolic acid dimer (HOA)-based nanoparticle MNs (HORN-MNs), and (3) Rosi-loaded HOA-based nanoparticles hyaluronic acid hydrogel (HORN-GEL). The researchers used an IVIS imaging system to evaluate the transdermal delivery efficacy of Dir-labeled MN patches. Compared to HORN-GEL, both HORN-MN and RN-MN exhibited significantly enhanced transdermal delivery efficiency and sustained effects over 48 h (82.0% $\pm$ 3.0% at 2 h for HORN-MN, 80.0% $\pm$ 6.3% at 2 h for RN-MN, and 14.0% $\pm$ 1.1% at 2 h for HORN-GEL) (Fig. 3.6G). Cryosections of the skin treated with Cou6-labeled HORN-MN were further photographed by fluorescence imaging to visualize the distribution of drug in subcutaneous tissue. As shown in Fig. 3.6H, fluorescent signals located in subcutaneous adipose tissue could be detected within 12 h and were still detectable for up to 24 h. This phenomenon can be attributed to the MNs' ability to puncture the stratum corneum effortlessly and dissolve within the dermal layer of abdominal skin within 5 min, facilitating the release of Rosi-loaded nanoparticles, followed by: (1) nanoparticle internalization by macrophages and white adipocytes, (2) lysosomal degradation of HOA into active oleanolic acid, and (3) releasing Rosi. Consequently, HORN-MN demonstrated significant antiobesity efficacy in an obese mouse model.

9. Approaches used to evaluate the in vivo safety of microneedles

The development of novel drug formulations must prioritize formulation safety. Good biocompatibility is a prerequisite for clinical application. The biocompatibility of MNs can be systematically evaluated at the cellular, tissue, and organ levels. In detail:

(1) Cytotoxicity: HaCaT cells are seeded in a 96-well plate and incubated with a series concentration of needle matrix solution for 24 or 48 h. After incubation, the cell viability of HaCaT can be measured using the CCK-8 assay to evaluate the biocompatibility of the materials used to prepare MNs.

(2) Hemolytic toxicity: Fresh Sprague Dawley (SD) rat blood is centrifuged to remove fibrinogen, and the supernatant is diluted with saline to prepare a 2% red blood cell (RBC) suspension. A series concentration of needle matrix solution is added to the RBC suspension to evaluate the hemolysis rate. PBS is used as the negative control, and Tween 80 is used as the positive control to evaluate the hemolytic properties of the materials employed to prepare MNs.

(3) Skin irritation test: SD rats are used as experimental animals. The hair on their backs is shaved and MN patches are applied to the dorsal skin. After administering MNs for a period, the skin is examined for erythema, edema, scabbing, and other reactions. At the experimental end point, skin tissue from the administration site is collected for H&E staining. Pathological changes in the skin are examined under an optical microscope.

(4) Skin barrier function: The skin barrier function is a critical physiological feature that primarily determines the skin's permeability to physical, chemical, biological, and aqueous substances. Since this permeability is mainly regulated by the epidermis, particularly the stratum corneum, it is commonly referred to as the stratum corneum barrier or epidermal barrier function. When MNs penetrate the skin, the resulting micropores can lead to increased transepidermal water loss (TEWL). Therefore, the measurement of TEWL is widely used to assess the skin's water permeability barrier by quantifying the water vapor pressure gradient on the skin surface, which reflects water dissipation. After anaesthetizing and acclimatizing the animal to ambient conditions, TEWL values are measured before and after MN treatment. A probe is applied and held on the intact skin before MN insertion until a stable reading is obtained. This is considered a base value that indicates an undisrupted skin barrier. TEWL values are then recorded immediately after the removal of MNs that serve as control for open pores. Barrier impairment leads to increased TEWL values, whereas a decline in TEWL signals barrier recovery. Skin barrier damage and recovery following MN administration can also be assessed using a skin moisture tester. The near-complete restoration of barrier function demonstrates the safety of MNs as a minimally invasive transdermal drug delivery method.

(5) Systemic toxicity: After administering MNs for a period, the major organs (heart, liver, spleen, lungs, and kidneys) of SD rats are collected for H&E staining and histopathological analysis. Blood samples are obtained from rats via tail vein puncture. An automated hematology analyzer is used to measure the counts and proportions of red blood cells, white blood cells, platelets, hemoglobin, and other parameters before and after MN administration. Additionally, an automated biochemical analyzer is employed to assess liver and kidney function-related indicators in the blood. The measured values are then compared with established normal reference ranges. The stability of these biochemical indicators supports the safety profile of the MN intervention.

10. Approaches used to evaluate the stability of microneedles

The evaluation of MN stability is a critical step to ensure their physical and chemical properties remain stable during storage. This assessment typically involves testing the mechanical properties, drug content, and impurity levels of MNs stored under various temperature, humidity, and light conditions. In evaluating systematic stability, the optimal storage conditions and shelf life of MNs can be determined, guaranteeing their safety and efficacy in clinical applications.

To evaluate the in vivo safety of MNs, potential adverse reactions such as skin irritation, allergic responses, and systemic toxicity, should be thoroughly assessed. These evaluations help identify local skin reactions (e.g., erythema, granulomatous formations, or inflammation) and systemic toxicity, ensuring that MNs are safe for use.

References

1. Zhang, M.; Yang, B.; Luan, X.; Jiang, L.; Lu, C.; Wu, C.; Pan, X.; Peng, T. State of the Art in Constructing Gas-Propelled Dissolving Microneedles for Significantly Enhanced Drug-Loading and Delivery Efficiency. *Pharmaceutics* **2023,** *15* (4), 1059.
2. Jin, T.; Wang, H.; Ullah, I.; Xie, W.; Lin, T.; Tan, Q.; Pan, X.; Yuan, Y. A Wireless Operated Flexible Bioelectronic Microneedle Patch for Actively Controlled Transdermal Drug Delivery. *Adv. Mater.* **2025,** *37* (11), e2417136.
3. Matias, Z.; Lopes, C. S.; Santos, N. C.; Carvalho, F. A. Nanotechnology Meets Medicine: Applications of Atomic Force Microscopy in Disease. *Biophys. Rev.* **2025,** *17* (2), 359–384.
4. Yang, B.; Wang, H.; Kong, J.; Fang, X. Long-Term Monitoring of Ultratrace Nucleic Acids Using Tetrahedral Nanostructure-Based NgAgo on Wearable Microneedles. *Nat. Commun.* **2024,** *15* (1), 1936.

5. Nadda, R.; Das, D. B. Optimising Distribution of Hollow Microneedle in Arrays for Transdermal Drug Delivery Considering Effects of Tissue Compression on Drug Permeability. *Int. J. Pharm.* **2025,** *673,* 125386.
6. Hu, W.; Peng, T.; Huang, Y.; Ren, T.; Chen, H.; Chen, Y.; Feng, D.; Wu, C.; Pan, X. Hyaluronidase-powered Microneedles for Significantly Enhanced Transdermal Delivery Efficiency. *J. Control Release* **2023,** *353,* 380–390.
7. Wang, S.; Yang, C.; Zhang, W.; Zhao, S.; You, J.; Cai, R.; Wang, H.; Bao, Y.; Zhang, Y.; Zhang, J.; et al. Glucose-Responsive Microneedle Patch with High Insulin Loading Capacity for Prolonged Glycemic Control in Mice and Minipigs. *ACS Nano* **2024**.
8. Liu, H.; Sun, Z.; Yin, M.; Ru, J.; Liu, X.; Chen, Z.; Liu, Y.; Sun, W.; Zheng, L.; Zhao, X.; et al. Polyphenol-Modified Mg-Zn Layered Hydroxide-Contained Microneedle Patch Enhance Mucosal Repair by Remolding Diabetic Oral Microenvironment. *Adv. Healthc. Mater.* **2025,** *14* (9), e2403883.
9. Yan, Q.; Weng, J.; Shen, S.; Wang, Y.; Fang, M.; Zheng, G.; Yang, Q.; Yang, G. Finite Element Analysis for Biodegradable Dissolving Microneedle Materials on Skin Puncture and Mechanical Performance Evaluation. *Polymers* **2021,** *13* (18), 3043.
10. Huang, Y.; Peng, T.; Hu, W.; Gao, X.; Chen, Y.; Zhang, Q.; Wu, C.; Pan, X. Fully Armed Photodynamic Therapy with Spear and Shear for Topical Deep Hypertrophic Scar Treatment. *J. Control Release* **2022,** *343,* 408–419.
11. Zhang, M.; Yang, B.; Ren, T.; Wang, X.; Chen, H.; Lu, C.; Wu, C.; Pan, X.; Peng, T. Dual Engine-Driven Bionic Microneedles for Early Intervention and Prolonged Treatment of Alzheimer's Disease. *J. Control Release* **2024,** *367,* 184–196.
12. Anbazhagan, G.; Suseela, S. B.; Sankararajan, R. Design, Analysis and Fabrication of Solid Polymer Microneedle Patch Using CO(2) Laser and Polymer Molding. *Drug. Deliv. Transl. Res.* **2023,** *13* (6), 1813–1827.
13. Wang, Q. L.; Zhang, X. P.; Chen, B. Z.; Guo, X. D. Dissolvable Layered Microneedles with Core-Shell Structures for Transdermal Drug Delivery. *Mater. Sci. Eng. C Mater Biol. Appl.* **2018,** *83,* 143–147.
14. Schuh, C. A. Nanoindentation Studies of Materials. *Mater. Today* **2006,** *9* (5), 32–40.
15. Dong, L.; Li, Y.; Li, Z.; Xu, N.; Liu, P.; Du, H.; Zhang, Y.; Huang, Y.; Zhu, J.; Ren, G.; et al. Au Nanocage-Strengthened Dissolving Microneedles for Chemo-Photothermal Combined Therapy of Superficial Skin Tumors. *ACS Appl. Mater. Interfaces* **2018,** *10* (11), 9247–9256.
16. Yang, B.; Dong, Y.; Shen, Y.; Hou, A.; Quan, G.; Pan, X.; Wu, C. Bilayer Dissolving Microneedle Array Containing 5-Fluorouracil and Triamcinolone with Biphasic Release Profile for Hypertrophic Scar Therapy. *Bioact. Mater.* **2021,** *6* (8), 2400–2411.
17. Haq, M. I.; Smith, E.; John, D. N.; Kalavala, M.; Edwards, C.; Anstey, A.; Morrissey, A.; Birchall, J. C. Clinical Administration of Microneedles: Skin Puncture, Pain and Sensation. *Biomed. Microdevices* **2009,** *11* (1), 35–47.
18. Wang, H.; Fu, Y.; Mao, J.; Jiang, H.; Du, S.; Liu, P.; Tao, J.; Zhang, L.; Zhu, J. Strong and Tough Supramolecular Microneedle Patches with Ultrafast Dissolution and Rapid-Onset Capabilities. *Adv. Mater.* **2022,** *34* (51), e2207832.
19. Donnelly, R. F.; Majithiya, R.; Singh, T. R.; Morrow, D. I.; Garland, M. J.; Demir, Y. K.; Migalska, K.; Ryan, E.; Gillen, D.; Scott, C. J.; et al. Design, Optimization and Characterisation of Polymeric Microneedle Arrays Prepared by a Novel Laser-Based Micromoulding Technique. *Pharm. Res.* **2011,** *28* (1), 41–57.
20. Huang, J.; Cui, Y.; Yang, Y.; Li, H.; Zhang, Y.; Yang, H.; Du, S.; Bai, J. Optical Coherence Tomography and Microdialysis for Microneedle-Mediated Penetration Enhancement Study of Paeoniflorin-Loaded Ethosomes. *Skin Pharmacol. Physiol.* **2021,** *34* (4), 183–193.

21. Li, Z.; Zhao, P.; Ling, Z.; Zheng, Y.; Qu, F.; Chang, H. An Ultraswelling Microneedle Device for Facile and Efficient Drug Loading and Transdermal Delivery. *Adv. Healthc. Mater.* **2024,** *13* (2), e2302406.
22. Deng, S.; Shuai, Y.; Zhang, S.; Sun, C.; Chang, L.; Xu, J.; Tong, L.; Ji, Q.; Li, M.; Dai, J.; et al. Personalized Demand-Responsive Biphasic Microneedle Patch for Smart Drug Administration. *Biomater. Sci.* **2023,** *11* (16), 5605–5617.
23. Lin, S.; Quan, G.; Hou, A.; Yang, P.; Peng, T.; Gu, Y.; Qin, W.; Liu, R.; Ma, X.; Pan, X.; et al. Strategy for Hypertrophic Scar Therapy: Improved Delivery of Triamcinolone Acetonide Using Mechanically Robust Tip-Concentrated Dissolving Microneedle Array. *J. Control Release* **2019,** *306,* 69–82.
24. Chen, S.; Wang, J.; Sun, L.; Xia, F.; Li, W.; Yuan, L.; Liu, C.; Li, P.; Bao, C.; Wang, M.; et al. A Quick Paster Type of Soluble Nanoparticle Microneedle Patch for the Treatment of Obesity. *Biomaterials* **2024,** *311*, 122687.

Chapter 4

Drug loading and release properties of microneedles

Meihong Zhang and Guilan Quan
State Key Laboratory of Bioactive Molecules and Druggability Assessment/International Cooperative Laboratory of Traditional Chinese Medicine Modernization and Innovative Drug Development of Ministry of Education (MOE) of China/College of Pharmacy, Jinan University, Guangzhou, China

1. Types of drugs delivered by microneedles

1.1 Small molecular drug delivery

The higher diffusion coefficient exhibited by small molecule drugs facilitates their easy permeation into the skin. However, for those requiring high doses, it remains challenging to attain effective drug concentrations above the therapeutic window solely through passive diffusion. The formation of microchannels in the skin induced by microneedles (MNs) has been demonstrated to enhance the permeability of small molecular drugs, augmenting their therapeutic efficacy.[1]

For example, solid MNs are generally employed to pretreat the skin followed by topical application of drug formulation at the penetration site. Naguib et al.[2] applied 5-fluorouracil cream to a B16-F10 melanoma mouse model upon solid MN roller insertion. The results demonstrated that the amount of penetration of 5-fluorouracil across the skin in the MN-pretreated group was enhanced by up to 4.5 times compared to that in the control group in a B16-F10 melanoma model, and tumor growth was nearly completely suppressed.

The amount of loading of small molecular drugs is contingent upon the type of MN, typically ranging from 50 to 600 μg per patch. In the case of coated MNs, the drug-loading capacity is susceptible to manipulation through alterations in the shape and quantity of needles within the array. Conversely, for dissolving or hydrogel MNs, the drug-loading capacity is predominantly influenced by the characteristics of the polymeric materials employed. The dissolution of hydrophilic drugs in their preparation solution can be utilized to

Microneedle Technology. https://doi.org/10.1016/B978-0-443-45544-5.00013-0

achieve effective loading to dissolve MNs. For example, hyaluronic acid dissolving MNs developed by Zhao et al.[3] demonstrated the capacity to load 122 μg of 5-aminolevulinic acid per patch. It is noteworthy that even with a reduced dosage, the photodynamic therapy efficacy of 5-aminolevulinic acid–encapsulated MNs exhibited superiority over the injection group. In another study, a dissolving MN system based on a polyvinyl pyrrolidone and polyvinyl alcohol composite matrix demonstrated a high loading capacity for doxorubicin hydrochloride, with a maximum loading of 533 μg per patch.[4]

The employment of hydrophobic drugs has been extensively implemented in the therapeutic management of diverse pathological conditions. However, the preponderance of dissolving MN preparation materials composed of water-soluble polymers required the constrained solubility of hydrophobic drugs within the needle matrix solution, culminating in inadequate drug loading. The delivery efficiency of insoluble drugs has been significantly improved by modulating molecular interactions between the drug and the polymer materials, including electrostatic adsorption, hydrophobic interaction, and host–guest encapsulation. In a related study, Lin et al.[5] utilized hydroxypropyl-β-cyclodextrin to form an inclusion complex with hydrophobic triamcinolone, enhancing the drug loading efficiency nearly fivefold. Meanwhile, Yang et al.[6] developed a core-shell structured MN system that successfully constructed a supersaturated needled solution system by combining a hydrophilic core material rich in hydrogen bond donors such as polyvinyl alcohol, with 1-methyl-D,L-tryptophan (1-MT) to form a hydrogen-bonding complex. This ultimately enabled the efficient delivery of the hydrophobic drug 1-MT.

1.2 Macromolecular drug delivery

Rhere has been a markedly increased focus on biomolecule therapeutics derived from macromolecular drugs. These therapeutics, which include peptides, proteins, and nucleic acids, have garnered significant attention owing to their ability to target specific receptors with high precision, notable potency, and generally favorable therapeutic profiles. Nevertheless, macromolecular drugs are associated with several disadvantages, including but not limited to poor stability, ease of inactivation, and a short half-life. In clinical practice, these macromolecular drugs are often administered via subcutaneous injection. This method of administration requires the maintenance of a cold chain and frequent administration, which results in high costs and poor patient compliance. MNs have been demonstrated to possess considerable promise as a platform for the delivery of biomolecules. The utilization of MNs has been demonstrated to enhance the stability and percutaneous penetration of biomolecules, augmenting the therapeutic efficacy. Consequently, to circumvent interference with the functionality of biomolecules, it is customary to coat the surface of dissolving or biodegradable MN tips with macromolecular drugs or to integrate them into dissolving MNs as needle materials.

1.2.1 Peptide delivery

Thymopentin, a synthetic water-soluble pentapeptide, is commercially available as a lyophilized powder commonly administered by intramuscular or intravenous injection. However, thymopentin exhibits an extremely short plasma half-life of approximately 30 s, while the therapeutic cycle lasts from 7 days to 6 months. The contradiction between the pharmacokinetic properties of thymopentin and the requirements for clinical administration has led to a significant reduction in patient compliance with the treatment regimen, because it necessitates frequent injections.[7] To address this issue, Lin et al.[8] adopted a modified two-step molding technology to construct thymopentin-loaded dissolving MNs, with a drug loading of 170 μg per patch. This method ensured the maintenance of structural integrity and biological activity after drug loading. The in vivo pharmacodynamics experiment demonstrated that thymopentin-containing MNs exhibited a immunomodulatory effect comparable to that of an intravenous thymopentin injection.

Liraglutide is a human glucagon-like peptide-1 (GLP-1) analog that has been demonstrated to significantly reduce the risk of glucose-dependent hypoglycemia in comparison with insulin. Consequently, it is regarded as a promising therapeutic option for the management of blood glucose levels in patients with the diagnosis of type 2 diabetes mellitus (T2DM). However, the administration of liraglutide necessitates a subcutaneous injection, which is disadvantageous for patients with diabetes mellitus because of its requirement of once-daily administration. In their study, You et al developed an Egg MNs (EMNs) with a three-layer structure loaded with liraglutide. After 4 h of administration, blood glucose levels in the EMN group were comparable to those in the injection group, demonstrating that the effect of liraglutide administered in the form of MNs was not altered. EMNs have the potential to reduce discomfort associated with hypodermic needles and to enhance patient compliance to a considerable extent.

Exendin-4, another GLP-1 receptor agonist, has been approved by the Food and Drug Administration for use as a potent candidate for clinical therapy in the treatment of T2DM. Nevertheless, the short blood circulation cycle and the high frequency of injections requiring twice-daily subcutaneous injections have greatly impeded its application in the clinic. Consequently, Chen et al.[10] exhibited a smart Exendin-4 delivery device based on MNs. When the concentration of blood glucose is elevated to hyperglycemia, a significant decrease in pH is triggered, resulting in the rapid release of Exendin-4 to regulate blood glucose. This suggests a pathology-specific drug release. Intelligent MNs facilitate the long-acting on-demand release of Exendin-4 (Fig. 4.1A).

1.2.2 Protein delivery

It has also been reported that therapeutic proteins, including but not limited to cytokines, antibodies, and insulin, can be effectively delivered using MNs.

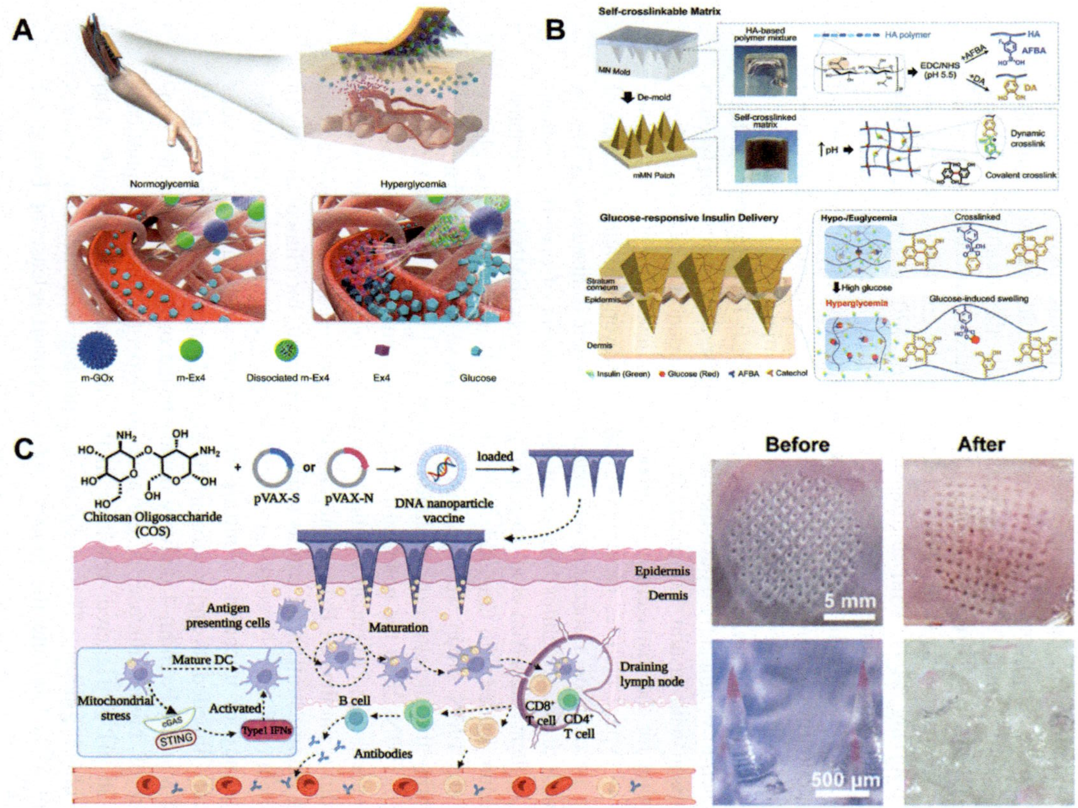

FIGURE 4.1 Microneedle (MN)-mediated macromolecular drug delivery. (A) Schematic diagram of a smart peptide Exendin-4 (Ex4) delivery system based on MNs. (B) Fabrication of self-cross-linkable polymer-based MNs for insulin delivery. (C) Schematic diagram illustrating MN-mediated intradermal administration of engineered DNA vaccine. *AFBA*, 4-amino-3-fluorophenylboronic acid; *cGAS,* cyclic GMP-AMP synthase; *DA*, dopamine; *DC*, dendritic cell;*EDC/NH*, N-hydroxysuccinimide; *HA,* hyaluronic acid; *IFNS,* interferons; *p-VAX,* pVAX1 plasmid vector; *GOx*, glucose oxidase. *(A) Reproduced from Chen et al.*[10] *with permission from the Nature Communications. (B) Reproduced from Liu et al.*[11] *with permission from the Advanced Science. (C) Reproduced from Li et al.*[12] *with permission from the ACS Nano.*

Le et al.[13] described the use of bioactive sucralfate-based MNs for delivering interleukin-4 (IL-4) to accelerate diabetic wound healing. The IL-4–loaded MN group showed the most rapid macroscopic wound closure rate on day 12, exhibiting a significantly diminished wound area (6.9%) in comparison to the liquid IL-4 solution (15.2%) group. The IL-4 group demonstrated no significant benefits in terms of wound healing, likely because of the poor tissue diffusion and limited bioactivity of liquid IL-4 solution on wounds. Therefore, the shortcoming of IL-4 delivery is remarkably ameliorated through MN administration.

Furthermore, imbalances in cytokines within the microenvironment of localized skin have been demonstrated to induce critical inflammatory dermatoses. These dermatoses are effectively mitigated by the topical administration of therapies such as anticytokine antibodies. From an anatomical and structural perspective, the cornea exhibits notable similarities to the skin, rendering MNs a viable and optimal choice for corneal drug delivery. Yu et al.[14] developed poly(vinyl alcohol) MNs for precise and sustained antitumor necrosis factor-α antibody (aT) delivery. In a murine model of corneal alkali burn, the application of aT-exosome-loaded MNs demonstrated enhanced therapeutic efficiency compared to conventional drug delivery approaches composed of topical administration and subconjunctival injection.

Insulin delivery represents a compelling illustration of the potential applications of MNs. The integration of MNs into the realm of insulin delivery holds considerable promise in facilitating the noninvasive modulation of glucose levels in patients with diabetes. This approach is characterized by its inherent simplicity and ease of use, further enhancing its appeal.[15] A novel self-cross-linkable and glucose-responsive polymer-based MN patch was designed by Liu et al.[11] to deliver insulin to treat hyperglycemia (Fig. 4.1B). The MN array was composed of a hyaluronic acid polymeric backbone, and the bioactivity of insulin is well maintained using biocompatible hyaluronic acid and protein-friendly cross-linking conditions. In comparison with insulin subcutaneous injections, insulin-loaded MNs have been shown to maintain tight control over blood glucose levels, remaining within the normal range for over 7 h. The proposed MN patch demonstrates a promising potential for the minimally invasive administration of insulin, as evidenced by in vivo experiments. This potential is attributed to its ability to deliver insulin with high efficiency during periods of hyperglycemia and its capacity to maintain the bioactivity of insulin.

1.2.3 Nucleic acid delivery

In the field of nucleic acid delivery, MN systems offer the benefit of transdermal penetration of nucleic acids owing to their minimally invasive characteristics. The clinical implementation of DNA and messenger RNA (mRNA) vaccines during the novel coronavirus disease (COVID-19) pandemic of 2019 has demonstrated the value of nucleic acid-based vaccines for emergency response.[16] In comparison with conventional delivery methods such as intramuscular or subcutaneous injections, MNs-mediated nucleic acid vaccination has been demonstrated to elicit more robust immune responses.[17] For instance, Li et al.[12] reported MNs for the delivery of DNA vaccines encoding the spike and nucleocapsid proteins of SARS-CoV-2 (Fig. 4.1C). In comparison with the intramuscular injection, the intradermal administration of the DNA vaccine via MNs elicited superior systemic and mucosal humoral immune responses with enhanced magnitude, polyfunctionality, and persistence. Notably, this MN patch demonstrates the capacity to be stored at ambient temperature for at least 1 month without a substantial decline in its immunogenicity.

In the domain of mRNA delivery, You and colleagues[12] pioneered a hyaluronic acid-based MNs platform for the transdermal delivery of mRNA encoding for extracellular matrix α1 type-I collagen. This approach resulted in a substantial enhancement in the efficacy of wrinkle treatment in a photoaged skin model. Furthermore, the realization of the three-dimensional structure of MNs has the potential to prolong collagen replacement time in photoaged skin by more than twice the duration of syringe needle delivery. These findings demonstrate that the MN platform successfully integrates spatially controlled nucleic acid delivery with sustained bioactivity, offering a highly promising translational solution for gene therapy.

1.3 Nanomedicine delivery

Despite their capacity for efficient transdermal drug delivery, the widespread application of MNs is hindered by their suboptimal therapeutic effects. The advent of nanomedicine has precipitated a paradigm shift in disease treatment, with nanoparticles (NPs) assuming a pivotal role in pharmaceutical studies aimed at augmenting the therapeutic efficacy of diverse pharmacological agents.[18] The combination of MNs and nanotechnology has been demonstrated to optimize the drug release behavior intelligently, broadening the scope of MN applications. The employment of NPs holds promise for ameliorating the multiple challenges associated with conventional pharmaceuticals, including suboptimal aqueous solubility, limited bioavailability, and nonspecific distribution within the body.[19] NPs modified with versatile moieties can respond to endogenous or exogenous stimuli, including pH, near-infrared light, ultrasound, temperature, and magnetic fields. This property facilitates the generation of customized drug release profiles in a spatiotemporally controllable manner. Furthermore, in comparison with small-molecule drugs, nanomedicines have the advantage of prolonged availability in the skin, enhancing the utilization of medications. However, accumulating evidence indicates that the penetration of NPs into the skin is limited after topical application.[20] MNs have been demonstrated to facilitate the efficient transdermal delivery of NPs.

Versatile NPs include inorganic NPs, organic NPs, organic-inorganic hybrid NPs, and nanoparticles/microdrug crystals.

1.3.1 Inorganic nanoparticle delivery

The integrity of MNs may be compromised during transdermal drug delivery owing to their inadequate mechanical strength. This deficiency can lead to a decline in the efficacy of drug delivery and the occurrence of adverse reactions such as skin irritation. It has been demonstrated that incorporating inorganic NPs can result in a substantial enhancement in the mechanical properties of MNs. Inorganic NPs have garnered significant attention owing to their exceptional mechanical stability, substantial specific surface area, and size properties that can be meticulously regulated, which are advantageous for biomedical applications. Inorganic NPs, which are extensively utilized in the domain of drug delivery, are principally composed of mesoporous silica NPs (MSNs), superparamagnetic iron oxide NPs, quantum dots, metal NPs, and carbon nanotubes.

Dong and his research group successfully developed a hyaluronic acid-based dissolving MN array loaded with both gold nanocages (AuNCs) and the chemotherapeutic drug doxorubicin[21] (Fig. 4.2A). The integration of AuNCs into MNs led to a substantial enhancement in their mechanical properties. The effective Young's modulus of the MNs exhibited a significant dose-dependent enhancement, with an increase from 68.9 to 224.9 MPa and an increase in AuNC loading from 0 to 4 μg/patch. A comparison of the MNs with AuNCs and the control group revealed that the former exhibited superior indentation resistance and higher mechanical strength. The enhancement may be attributed to the reinforcing effect of gold NPs on the polymer matrix. Specifically, MSN has been demonstrated to exhibit enhanced mechanical strength and skin adhesion when utilized in conjunction with MNs.[25,26] Besides, the unique physicochemical properties of inorganic NPs (e.g., magnetic, fluorescent, or photothermal effects) have the potential to enhance the diagnostic and therapeutic capabilities of MN systems, enabling their synergetic application in disease treatment and real-time monitoring.

1.3.2 Organic nanoparticle delivery

Organic NPs, which include polymer NPs, lipid NPs, and dendritic macromolecules, represent a class of highly promising drug delivery vehicles. These NPs have demonstrated significant potential for enhancing the delivery of insoluble drugs.[27] Liposome technology has been applied to enhance the water solubility and skin permeability of poorly soluble drugs, which has the potential to expand the scope of MN transdermal drug delivery. Hypericin, a potent lipid-soluble photosensitizer, has exhibited significant antitumor activity in photodynamic therapy. Nevertheless, its clinical application is hindered by significant challenges, including its reduced photoactivity, which is

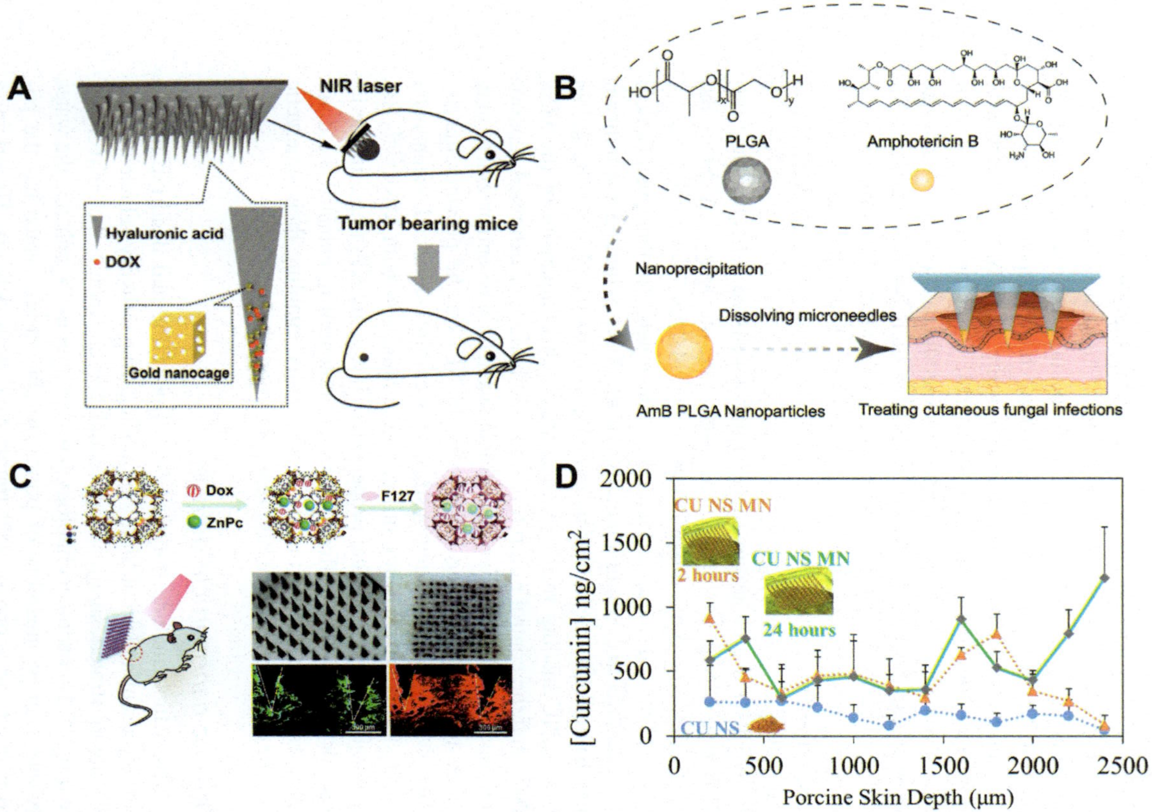

FIGURE 4.2 Microneedle (MN)-mediated nanomedicine delivery. (A) Schematic illustration showing doxorubicin (DOX)/gold nanocage-loaded dissolving MN system. (B) Schematic illustration of amphotericin B (AmB)-loaded PLGA nanoparticles integrated with dissolving MNs. (C) Schematic illustration of MN-mediated delivery of MIL-100(Fe) nanoparticles. (D) Schematic diagram of a polyvinyl alcohol-based MNs containing curcumin (CU) nanosuspension (NS). *NIR*, near infrared. *(A) Reproduced from Dong et al.*[21] *with permission from the ACS applied materials interfaces. (B) Reproduced from Peng et al.*[22] *with permission from the Journal of Drug Delivery Science Technology. (C) Reproduced from Luo et al.*[23] *with permission from the Biomaterials Science. (D) Reproduced from Sharif et al.*[24] *with permission from the Pharmaceutics.*

attributed to its high degree of hydrophobicity. To overcome these limitations, Heba et al.[28] reported a lipid nanocapsule (Hypericin-LNC) drug-carrying system, which achieved an encapsulation efficiency of up to 99.67%. The study further employed an array of hollow MNs to overcome the skin barrier problem successfully. This coadministration strategy ultimately achieved a significant nonmelanoma skin cancer inhibition rate of 85.84%, providing an effective solution for the clinical application of hypericin.

In another study, amphotericin B, a potent macrolide polyene antifungal drug, encountered significant delivery challenges owing to its extremely low aqueous solubility and permeability. The physicochemical properties of amphotericin B, with a molecular weight of up to 924 dalton (Da), contribute to its limited ability to penetrate the stratum corneum of the skin. The emergence of MN technology provides a breakthrough solution for the delivery of such difficult-to-deliver drugs, with the potential to achieve efficient intradermal delivery. Poly(lactic-co-glycolic acid) (PLGA) NPs offer significant advantages, including the ability to enhance the solubility of challenging drugs, along with excellent biodegradability and compatibility. Peng et al.[22] employed solvent deposition to load amphotericin B into PLGA NPs and subsequently integrated the NPs into dissolving MNs via a sequential loading process (Fig. 4.2B). The quantitative analysis showed that the MN tips were successfully enriched with amphotericin B PLGA NPs, exhibiting a drug loading capacity of 40.3 μg. Because the minimum inhibitory concentration of amphotericin B against most fungi is 0.5–1 μg/mL, the amount of drug loaded in this MN system fully meets the needs of antifungal therapy and demonstrates good potential for clinical application.

1.3.3 Organic-inorganic hybrid nanoparticle delivery

Metal-organic frameworks (MOFs) are a class of nanomaterials that are self-assembled from organic ligands and metal ions, exhibiting favorable biocompatibility and controllable drug release rates.[29] Compared with traditional drug delivery vehicles, such as micelles and lipid NPs, MOFs exhibit obvious advantages with regard to drug loading capacity and delivery efficiency. This is attributed to their tunable pore structure (ranging from microporous to macroporous) and high specific surface area.[30] This property compensates for the absence of drug loading resulting from the restricted physical space of MNs. The integration of MOFs into MNs has been demonstrated to enhance the drug-loading capacity per unit volume of MNs. Additionally, the pore structure and surface chemistry of MOFs can be regulated by design, conferring multiple stimulus-response characteristics.

Luo et al.[23] developed a multifunctional delivery system based on MOFs (Fig. 4.2C). This study utilized the unique high porosity and functional tunability of the MOF material (MIL-100[Fe]) to construct an integrated nanoplatform mediated by MNs for synergistic treatment with oxygen generation and chemophototherapy. The iron-based MOFs function as dual drug carriers for zinc phthalocyanine and doxorubicin hydrochloride. Additionally, they catalyzed the decomposition of H_2O_2 in the tumor microenvironment through a Fenton-like reaction, sustainably generating the oxygen necessary for treatment. The experimental studies demonstrated that the integrated system exhibited enhanced antitumor effects in a 4T1 tumor model. The drug contained within the system was capable of achieving controlled release from the MOF backbone under the conditions of an acidic tumor microenvironment.

Beyond their role in drug delivery, MOF materials exhibit a range of biomedical properties. For instance, Zn-MOF encapsulated MNs demonstrate both antimicrobial activity and degradability.[31] Cu-MOF and Mg-MOF have been demonstrated to be suitable for diabetic wound repair.[32] Consequently, the integration of MOF materials and MN technology can yield a novel system with a multitude of therapeutic effects.

1.3.4 Micro/nanodrug crystal delivery

The manipulation of particle size has been primarily used in the pharmaceutical industry to increase the dissolution rate of poorly water-soluble drugs. The production of drug microparticles was primarily accomplished through comminution techniques. However, the advent of innovative formulation methodologies has enabled the preparation of drug particles within the submicrometer range, specifically nanocrystals. Nanosuspensions encounter challenges in passive penetration of the stratum corneum owing to limitations in particle size. However, MNs have the capacity to form micron-sized channels within the stratum corneum through physical perforation, facilitating the precise delivery of nanosuspensions to the active epidermis. The intradermal administration of micro/nanodrug crystals using MNs has been reported in several publications. For instance, Vora et al.[33] prepared a cholecalciferol nanosuspension by sonoprecipitation. The drug penetration of MNs into the skin of neonatal porcine was significantly increased sevenfold over the course of 24 h. Similarly, when curcumin was embedded in polyvinyl alcohol-based MN arrays in the form of nanosuspensions, the MN arrays showed a significant enhancement in the deposition of curcumin in the deeper layers of the skin compared to the curcumin nanosuspension alone.[24] (Fig. 4.2D).

2. Drug release properties of microneedles

As the demand for precision drug delivery and personalized treatment increases, drug release mechanisms have become an increasingly valuable aspect of delivery system. The release kinetics can be modified through two distinct mechanisms: alteration of the drug's formation composition or exploitation of its intrinsic properties. By adjusting the composition of the

MNs or integrating it with nanotechnology, MN-based drug delivery systems can achieve controlled or responsive drug release profiles, meeting the varied demands for different treatments.

2.1 Rapid drug release

MNs with an immediate release effect were not previously favored by researchers because earlier studies indicated that the immediate release of a drug could lead to significant fluctuations in blood levels over a short period, potentially resulting in adverse outcomes. However, immediate-release MNs have been demonstrated to manifest a swift drug release profile. The efficacy of these MNs has been demonstrated in the treatment of certain diseases marked by rapid progression that necessitate prompt, effective medication to manage the condition (rapid hemostasis and analgesia) or that exert a local effect (Table 4.1). Typically, natural water-soluble polysaccharides and synthetic soluble polymers are commonly used to construct MNs with rapid release properties. These MNs dissolve in the skin in seconds or minutes and are consequently suitable for the rapid delivery of preloaded goods.

Natural carbohydrate materials that demonstrate optimal biocompatibility include maltose, dextran, and hyaluronic acid, among other carbohydrates. In this regard, hyaluronic acid is a prominent polymer utilized in the fabrication of rapid-release MNs, a choice that is largely influenced by its notable mechanical strength and the absence of a requirement for elevated temperatures during molding. Xu et al.[41] selected hyaluronic acid with a molecular weight of about 10 kD as the matrix to construct rapidly dissolving MNs for the coencapsulation of p53 DNA and IR820. The obtained MN patch demonstrated the capacity to penetrate the stratum corneum effectively and rapidly

TABLE 4.1 Rapid release system based on microneedles.

Therapeutic agent	Materials of microneedles	Release time (min)	Ref.
Bleomycin	Polylactic acid	15	34
Acyclovir	Gantrez S-97	15	35
Artemether	Hyaluronic acid	120	36
Exendin-4	Hyaluronic acid	5	37
Dihydroergotamine mesylate	Polyvinylpyrrolidone	2	38
Meloxicam	Polyvinylpyrrolidone	30	39
Sumatriptan	Dextran/hyaluronic acid	30	40

dissolve in the interstitial fluid within 5 min. In another study, Hou et al.[42] developed a rapidly separable dissolving MN system for immediate release of the antiinflammatory drug diclofenac sodium by a multistep casting technique. The system disintegrated rapidly in simulated skin within 30 s. Additionally, hyaluronic acid-based dissolving MNs have been demonstrated for the immediate release of different NPs such as PLGA NPs.[43]

In comparison, artificially synthesized polymers offer the benefit of a tunable structure and performance. Several polymers such as polyvinylpyrrolidone, hydroxypropyl methylcellulose, and polymethyl vinyl ether-alt-maleic anhydride (Gantrez) are utilized in rapid-release MN formulations.[44] For instance, Lee et al.[34] developed polylactic-acid MNs coated with bleomycin for the treatment of warts. In vitro studies demonstrated that more than 80% of the bleomycin was dissolved into the skin within 15 min. The result also indicated that tip-coated MNs were more effective than intralesional injections for distributing drugs into the subepidermal skin layer. Similarly, acyclovir is employed in the treatment of herpes labialis. However, its use is limited by its low skin permeability and suboptimal efficacy. Pamornapathomkul et al.[35] fabricated acyclovir MNs from Gantrez S-97 with sufficient mechanical strength. These MNs demonstrated rapid dissolution within 15 min, resulting in enhanced skin permeation compared to commercial creams. The permeation levels achieved were approximately 45 times higher than those of the commercial formulation. Moderate to severe acute migraine attacks can be treated with dihydroergotamine mesylate. To overcome the limitations of current formulations in subcutaneous injection and nasal sprays such as pain, adverse side effects, and poor bioavailability, Tas et al.[38] developed a dissolving MN patch composed of polyvinylpyrrolidone, which contains dihydroergotamine mesylate. This system was found to dissolve in porcine skin within 2 min. This remarkable property enables painless self-administration, quick onset of action, and high bioavailability.

2.2 Sustained drug release

The objective of sustained drug delivery is to administer a drug at a constant rate while maintaining a consistent concentration over a defined period, intending to minimize adverse effects. The MN-mediated sustained release system has been demonstrated to have several advantages over conventional delivery methods. It has been shown to reduce the frequency of administration, reducing the incidence of forgotten or missed doses. It has also been demonstrated to improve patient compliance and decrease the risk of infection at the site of administration. These benefits are especially notable compared to the intravenous or intramuscular routes. A salient design consideration for MNs equipped with sustained release mechanisms pertains to the necessity of maintaining the needle tips in the skin, independent of the patch backing. This imperative is pivotal in ensuring that the patch does not necessitate continuous wear. Consequently, the fabrication of transdermal MNs for the sustained release of drugs has been achieved by adopting biodegradable or swellable polymers as the needle matrix. The release of the drug can be tailored by selecting polymers with different molecular weights, degradability, and expandability (Table 4.2).

In the context of biodegradable MNs, the drug release profile is predominantly influenced by two primary factors: the hydrolysis and degradation of the polymeric matrix, and the rate of diffusion of the payload. Natural polymers (such as chitosan and silk fibroin) and synthetic polymers (e.g., polylactic acid, poly[lactic-co-glycolic acid]) are widely used as MN matrices for sustained drug release. Hormones and biomolecules (e.g., peptides and proteins) typically require sustained release over long periods. Li et al.[25] developed a novel MN patch with a core-shell structure to achieve sustained release of the contraceptive hormone levonorgestrel (LNG) for several months through a rate-control mechanism of the shell. In this approach, LNG was encapsulated in a poly(lactide-co-glycolide) core surrounded by a poly(L-lactide) shell and a poly(D,L-lactide) cap that were fabricated by sequential casting into MNs. This structure effectively limited the initial 24 h burst release of LNG to 5.8 ±0.5%, achieving approximately zero-order release for 6.2 ± 0.1 months in vitro. Additionally, Chen et al.[48] developed a chitosan-based dissolving MN patch

TABLE 4.2 Sustained drug delivery based on microneedles.

Therapeutic agent	Materials of microneedles	Release time (d)	Ref.
Doxorubicin and lipopolysaccharide	Sodium alginate and sulfobetaine methacrylate	7	45
Levonorgestrel	Poly(lactic acid)/poly(lactic-co-glycolic acid)	30	46
Etonogestrel	Polyvinyl alcohol/hydroxypropyl methylcellulose	7	47
Bovine serum albumin	Chitosan	8	48
Rhodamine B	Poly(lactic-co-glycolic acid)/poly-N-isopropyl acrylamide	3	49
Antigen ovalbumin	Chitosan	28	50
Bevacizumab	Polyvinyl alcohol	7	51
Insulin	Silk fibroin	3	52
Ovalbumin	Hyaluronic acid/chitosan	28	53
Horseradish peroxidase	Silk fibroin	2	54

for sustained transdermal delivery of hydrophilic macromolecular drugs. The release mechanism of the system is contingent upon the swelling and degradation processes of the chitosan matrix. In vitro release studies demonstrated the sustained release of bovine serum albumin over 8 days. This sustained release characteristic renders it particularly suitable for the delivery of proteins that require prolonged administration.

Hydrogel has been widely investigated in the field of transdermal drug delivery. The integration of MN technology with hydrogel facilitates the construction of a delivery system characterized by sustained release properties. Furthermore, this integration enables a substantial enhancement in the mechanical strength of MNs. From a structural perspective, the formation of hydrogels typically occurs through the process of chemical or physical cross-linking, which involves the formation of covalent or hydrogen bonds, electrostatic interactions, and other types of noncovalent interactions, resulting in a three-dimensional network structure. The primary benefit of this composite system is its ability to facilitate continuous, sustained release of the loaded drug. The sequential photoionic cross-linked hydrogel MN system of sodium alginate and sulfobetaine methacrylate, developed by Darge et al.,[45] achieved the efficient release of doxorubicin and lipopolysaccharide immediately within the first 24 h (release rates of 66.1 ± 7.4% and 59.4 ± 5.5%, respectively), followed by sustained release phase, in which the drug release cycle could be extended to about 7 days. Of particular significance is the polyvinyl alcohol-based hydrogel MN system developed by Pillai et al.[55] This system has been demonstrated to prolong the release cycle, achieving sustained release for up to 12 days using rhodamine as a model drug. This development provides a novel approach for the development of long-lasting transdermal drug delivery systems.

2.3 Stimuli-responsive drug release

Researchers have demonstrated heightened interest in stimuli-responsive MNs, which are predominantly formulated around polymer matrices. These matrices have been engineered to regulate and modulate the delivery of pharmaceutical agents. The stimulation response consists of physical signals that are externally stimulated (e.g., temperature, light, and electric field) or physiological signals that are internally stimulated (e.g., pH, glucose, and redox potential). The purpose of these designs is to achieve spatiotemporal modulation, enhancing therapeutic efficacy and mitigating the occurrence of potential side effects associated with off-target drug delivery (Fig. 4.3).[56]

2.3.1 External stimuli-responsive microneedles

Many efforts have been made to develop external stimuli-responsive MNs. The construction of responsive MNs may require the incorporation of domains that can be activated by external stimuli. Table 4.3 lists representative responsive MNs to external triggers and their potential in biomedical applications.

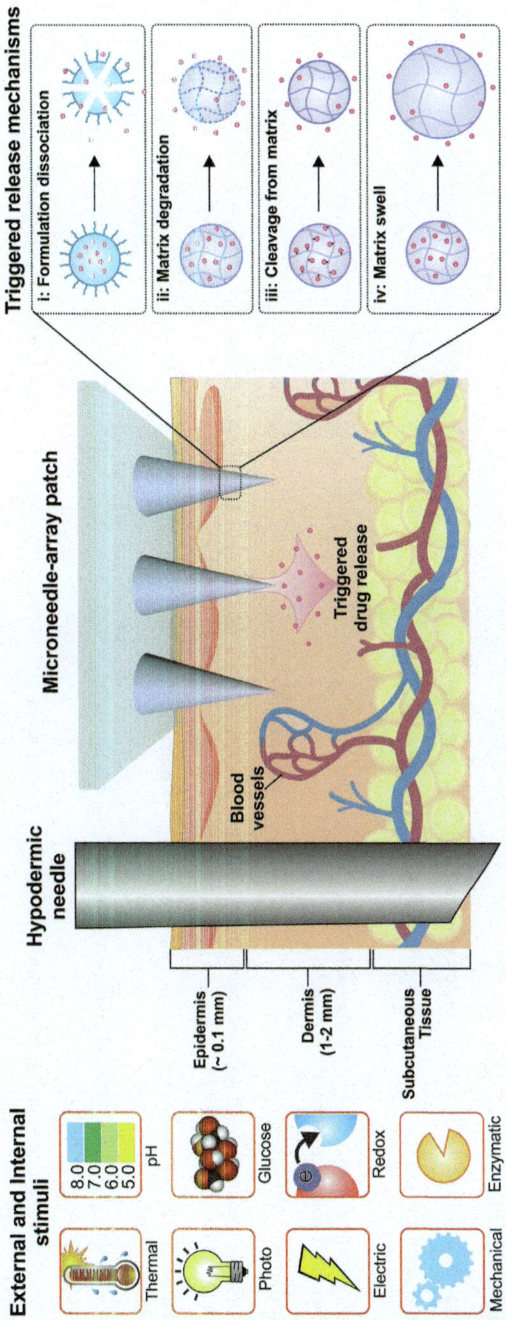

FIGURE 4.3 Different types of internal and external stimuli and trigger release mechanisms for stimuli-responsive microneedle systems. *Reproduced from Pooyan et al.*[56] *with permission from Materials Today.*

TABLE 4.3 Representative external stimuli-responsive microneedles and their biomedical applications.

Microneedle composition	Therapeutic agent	Stimulus	Condition	Ref.
Alginate hydrogel matrix coated with poly(lactide)	Peptide-nucleic acid	Ultraviolet light	miRNA sampling and detection	57
Poly(2-hydroxyethyl methacrylate-co-ethylene glycol dimethacrylate)	Ibuprofen	Ultraviolet light	Pain relief	58
Hyaluronic acid/Polyethylene glycolated gold nanorod	Doxorubicin	Near infrared light	Cancer therapy	59
Polycaprolactone/silica-coated lanthanum hexaboride (LaB_6@SiO_2)	Doxorubicin	Near infrared light	Cancer therapy	60
Poly(L-lactide) matrix coated with polyethylene glycolated gold nanorod and micelles	Docetaxel	Near infrared light	Cancer therapy	61
Hyaluronic acid with gold nanocage	Doxorubicin	Near infrared light	Cancer therapy	21
Polyvinylpyrrolidone with polydopamine/lauric acid-coated silica	Metformin	Near infrared light	Type 2 diabetes	62
Chitosan/graphene assembled in porous carbon nanocomposites	Cephalexin	Electric	Antimicrobial therapy	63
Polyethylene glycol-polyethylene oxide-polyethylene glycol triblock copolymer	Methotrexate	Thermo	Cancer therapy	64
Polyvinylpyrrolidone with graphene doped with gold and combined with a gold mesh	Metformin	Electric/thermo	Diabetes	65
Poly(caprolactone) as tip and polyvinyl alcohol/polyvinylpyrrolidone as base	Metformin	Mechanical/near infrared light	Diabetes	66
Hyaluronic acid as the tip and polycaprolactone as base	Vaccine antigens derived from canine influenza virus	Mechanical	Influenza vaccination	67

2.3.1.1 Thermo-responsive microneedles

Two primary sources of heat have been identified as contributors to the thermo-responsiveness of MNs. The initial heat source is derived from a thermal heater, while the subsequent heat source utilizes photothermal agents capable of converting light to localized heat, facilitating melting of the polymer. Polycaprolactone is characterized by its high biocompatibility and relatively low melting point, making it well-suited for utilization as a needle material in thermally triggered drug release applications. In the study by Zhang et al.,[68] metformin-loaded polycaprolactone MNs were fabricated, and it was demonstrated that these MNs exhibited excellent thermal ablation changes. When the MNs were exposed to an electric heating sheet, the polycaprolactone underwent a phase transition to a liquid state (approximately 50°C), which resulted in the liberation of the encapsulated metformin. Su et al.[69] incorporated a photothermal converter, IR780, into a dissolving polyvinylpyrrolidone MN patch and then coated it with a phase change material, 1-tetradecane. Under near infrared (NIR) light irradiation, the IR780 compound underwent photothermal conversion, leading to the melting of 1-tetradecane. This process subsequently induced the dissolution of the MNs, resulting in the release of the W379 peptide from the MNs into the surrounding regions.

2.3.1.2 Photo-responsive microneedles

In comparison with other triggers, light possesses advantageous characteristics, including its noninvasiveness, high spatial resolution, and temporal control. NIR-light-responsive materials have found widespread applications because NIR light is more effective than UV or visible light in tissue penetration, minimizing photodamage to the tissues involved.[70] Photo-responsive MNs are predominantly employed in combination with chemotherapy and photodynamic therapy. Specifically, photosensitizers are usually incorporated into the MNs in conjunction with drugs to exert therapeutic effects. For instance, the photosensitizer indocyanine green and a reactive oxygen species (ROS)-responsive doxorubicin prodrug were encapsulated stepwise into a polyvinylpyrrolidone MN platform. In the context of photodynamic therapy, the application of 808 nm NIR has been demonstrated to induce ROS generation through the photoexcitation of indocyanine green. This process subsequently activates doxorubicin prodrugs, resulting in the conversion of doxorubicin prodrugs into doxorubicin. This therapeutic approach has been observed to elicit a notable therapeutic response in melanoma.[71] Photosensitive MNs have been also applied in the treatment of diabetes, in which the release of the drug is regulated by an irradiation source. For example, Zhang et al.[62] presented polyvinylpyrrolidone-based MNs using an NIR irradiation system to control effective drug release behavior. The system contains polydopamine as a photothermal conversion agent and lauric acid as a phase change material. These two components were utilized to modify hollow mesoporous silica NPs loaded with metformin, an antidiabetic drug, to construct an NIR-responsive drug delivery system. Approximately 72.5% of metformin was released from MNs following five NIR laser on-off cycles. However, in the absence of NIR irradiation, only approximately 17.8% of the metformin could be released from MNs via a concentration diffusion pathway.

2.3.1.3 Electric-responsive microneedles

Various electrical stimulation-based therapies have been proposed for biomedical and therapeutic applications. MN patches represent a novel electrical stimulation device that has emerged in recent years.[72] Electro-responsive MNs are prepared through the incorporation of conducting polymers, such as polypyrrole, and nanostructures, including metallic materials. This integration enables the realization of spatiotemporal on-demand, and electrically triggered drug release.[73] Polypyrrole is a smart electro-responsive material that has been demonstrated to release drugs by transforming between oxidative and reductive states. In a study, polypyrrole-based MNs patches were designed to control the release of dexamethasone electrically to treat atopic dermatitis. The results demonstrated that MNs with electrical stimulation would release more dexamethasone into the skin and present a significant therapeutic effect in the atopic dermatitis model. However, the dexamethasone solution group and the group that did not receive electrical stimulation exhibited a reduced number of drugs entering the skin and demonstrated no discernible antiinflammatory effect.[74]

2.3.1.4 Magnetic-responsive microneedles

Magnetic-responsive MNs typically incorporate nanostructures with magnetic characteristics in a polymer matrix to manipulate the drug to target the lesion accurately by employing an external magnetic field. MNs that possess magnetic properties are commonly employed in bioimaging assays. For example, the fabrication of chitosan MNs patches was coupled with the incorporation of magnetic graphene quantum nanodots. This particular magnetic-responsive MN has demonstrated significant promise in the realm of targeted and tracked transdermal drug delivery.[75] A study by Fang and colleagues reported a magnetic-responsive MN patch that utilizes an externally applied magnetic field to regulate the release of medication, with the aim of treating androgenetic alopecia.[76] Mesoporous iron oxide NPs were prepared by encapsulating minoxidil (Mx-MIOs). Then, Mx-MIOs were loaded into polyvinyl alcohol-based MNs (Mx-MIOs@MNs). The in vitro study depicted a 20% release of Mx from the MNs at the end of 120 min without the application of a magnetic field. In contrast, a release of over 40% of Mx from MNs was observed within 120 min under magnetic field application for 1 min. Mx-MIOs@MNs without magnetic fields and Mx@MNs were also studied in 8-week-old depilated

C57BL/6 mice. The results showed an eightfold increase in hair growth density in mice treated with an Mx-MIOs@MNs + magnetic field. The on-off switching strategy has been effectively realized by the transdermal application of magnetic nanopatches, enabling on-demand drug delivery for the treatment of diverse diseases.

2.3.2 Internal stimuli-responsive microneedles

These MNs incorporate polymers that are sensitive to the emergence or variation of a biological signal (e.g., pH, ROS, glucose, or enzyme). In light of the underlying pathology, researchers may opt for diverse internal triggers. For example, they can exploit physiological signals characteristic of tumor niches (e.g., reduced conditions, lower pH, and overexpression of enzymes and compounds) to initiate drug release. Table 4.4 lists representative internal stimuli-responsive MNs and their potential biomedical applications.

2.3.2.1 pH-responsive microneedles

pH-responsive MNs are composed of polymers that undergo degradation, swelling, or collapse in response to changes in the pH of their surroundings. These materials are composed of hydrophilic and ionically functional polymer chains. For instance, pH-responsive polyelectrolyte multilayers were encapsulated on the surface of polycaprolactone MNs. The pH-responsive polymer underwent a charge conversion when exposed to the acidic tumor microenvironment, which could facilitate the release of genes for the treatment of subdermal cancer. The in vivo experiments demonstrated that the tumor inhibition rate of mice treated with pH-responsive MNs reached 90.1%, while the rates for mice treated with non–pH responsive MNs and those given an intravenous injection were only 46.4% and 30.5%, respectively.[94]

When a wound occurs, the skin's acidic environment is disrupted, leading to an alkaline shift in the underlying tissue. This elevated pH promotes bacterial proliferation on the skin surface and facilitates the formation of bacterial biofilms at the wound site. These biofilms compromise the skin's barrier function and are a major contributor to delayed wound healing. Ullah et al.[95] fabricated Eudragit S100-coated MNs capable of automatically releasing therapeutic drugs in response to specific alkaline conditions present within the wound. Eudragit S100 is a pH-sensitive polymer that exhibits dissolution at alkaline pH while maintaining structural stability in acidic environments. When the MNs were exposed to a normal skin environment, the release of the drug from the MNs was negligible. In contrast, significant release occurred within the wound alkaline environment. The development of these MNs in combination with antimicrobial and antiinflammatory agents demonstrated enhanced therapeutic efficacy against wound infections.

The use of a physiological pH-responsive MN patch was recognized as a critical strategy for the efficient delivery of DNA vaccines. Duong et al.[96]

TABLE 4.4 Representative internal stimuli-responsive microneedles and their biomedical applications.

Microneedle composition	Therapeutic agent	Stimulus	Condition	Ref.
Hyaluronic acid with hypoxia-responsive vehicles	Insulin	Glucose/ hypoxia	Diabetes	77
Photopolymerized *N*-vinylpyrrolidone, dimethylamino-ethyl acrylate, 3-acrylamido-phenylboronic acid, and ethylene glycol dimethacrylate	Insulin	Glucose	Diabetes	78
Polyvinylpyrrolidone loaded with mesoporous silica nanoparticles	Insulin	Glucose/ROS	Diabetes	79
Hyaluronic acid with H_2O_2-responsive vesicles	Insulin	Glucose/ROS	Diabetes	80
Hyaluronic acid with pH/H_2O_2-dual responsive polymersomes	Insulin	Glucose/ROS/ Hypoxia	Diabetes	81
Polyvinylpyrrolidone with charge-switchable particles	Insulin	Glucose	Diabetes	82
Reactive oxygen species–responsive poly(vinyl alcohol) hydrogel	Insulin	Glucose/ROS	Diabetes	83
Hydrogel containing phenylboronic domain and silk fibroin	Insulin	Glucose	Diabetes	84
Hyaluronic acid	Glucagon	Glucose	Insulin-induced hypoglycemia	85
Alginate-based microneedles with mineralized particles	Exendin-4	Glucose	Type 2 diabetes	10
Hyaluronic acid with dextran nanoparticles	Rosiglitazone	Glucose	Obesity	86

Poly(vinyl alcohol) with calcium phosphate nanoparticles	3,3′,5,5′-tetramethylbenzidine and horseradish peroxidase	Glucose/pH	Glucose sensing	87
Hyaluronic acid containing dextran nanoparticles	Antiprogrammed cell death protein 1 and indoleamine 2,3-dioxygenase	Glucose	Cancer immunotherapy	88
Dimethylmaleic anhydride-modified polylysine	p53-DNA	pH	Cancer therapy	89
Polycarbonate conjugated poly(β-amino ester urethane)	Adjuvants and DNA vaccines	pH	Cancer immunotherapy	90
Heparin conjugated hyaluronic acid	Heparin	Thrombin	Blood coagulation	91
Polyvinyl alcohol as the tip and hyaluronic acid/ diatomaceous earth as the base	Clindamycin	Reactive oxygen species	Acne vulgaris	92
Poly(vinyl alcohol)/polyvinylpyrrolidone hydrogels embedded with poly(caprolactone) nanoparticles	Carvacrol	Bacterial lipases	Wound repair	93

reported a novel charge reversal pH-responsive copolymer that functions as the release-layer of MNs. The copolymer contains cationic tertiary amine and anionic sulfonamide groups. These groups exhibit a positive charge at an acidic pH of 4.03 and a negative charge at a neutral pH of 7.4. The charge reversal characteristic of the copolymer has been demonstrated to trigger the release of the DNA vaccine effectively.

The incorporation of pH-responsive NPs in MNs is another potential candidate strategy for cancer treatment. For example, innovative self-degradable MN patches have been reported for the sustained delivery of antiprogrammed death-1 (aPD-1) antibodies in a physiologically controllable manner. The MN was composed of biocompatible hyaluronic acid integrated with pH-sensitive dextran NPs that encapsulate aPD-1 and glucose oxidase (GOx). The acidic environment promoted the self-dissociation of dextran NPs, which subsequently resulted in the substantial release of aPD-1. Compared to MN without a degradation trigger or intratumoral injection of free aPD-1 with an equivalent dose, the MN patch induced robust immune responses in the B16-F10 mouse melanoma model.[97]

2.3.2.2 Glucose-responsive microneedles

Upon insertion into the skin, MNs are exposed to cutaneous microcirculation, enabling real-time sensing of changes in the metabolic content of the biological environment. In the field of diabetes, glucose-responsive MNs have been developed to facilitate convenient and painless drug administration. To ensure optimal treatment, patients must engage in continuous self-monitoring of their blood glucose levels, a practice that facilitates the timely administration of antidiabetic medications.

Glucose-responsive MNs are administered via three methods. (1) Enzymes catalyze environmental changes: the method depends on GOx, which serves as the catalyst enzyme that converts glucose to gluconic acid. Consequently, the reaction leads to a decrease in pH, oxygen depletion, and H_2O_2 production. These changes are frequently emphasized as triggers of glucose-responsive MNs. (2) Glucose binding molecules: glucose-responsive devices have been prepared using alternative materials, such as lectins. The most prevalent lectin is the tetravalent binding protein concanavalin A. The sugar-binding site of concanavalin A can bind hydroxyl groups of α-D-mannopyranose, α-D-glucose pyranose, and other nonreducing sugars at C-3, C-4, and C-6. Therefore, the utilization of concanavalin A in glucose-responsive intelligent systems is a plausible prospect.[98] However, there is a series of problems in the practical application of the concanavalin A system, including leakage, stability and solubility in water, the length of response time, and biological toxicity. (3) Molecular recognition of diol-binding chemical molecules: alternatives to natural glucose-detecting compounds include a variety of synthetic molecules, such as phenylboronic acid, which has been demonstrated to bind to *cis*-1,2 or *cis*-1,3 diols in glucose and other similar diols.

Insulin plays a critical role in regulating blood glucose levels in the body. Inadequate secretion of insulin, or an inadequate response between cellular receptors and insulin, leads to abnormal glucose metabolism, which initiates diabetes. However, conventional insulin delivery systems have proven incapable of achieving the requisite precision in blood glucose level control, often resulting in fatal overinjections. Conversely, smart delivery systems have been demonstrated to regulate insulin release in a manner contingent upon fluctuations in blood glucose levels, a feature that renders them more conducive to diabetes management. The integration of MN technology has led to the development of an emerging method of responsive insulin delivery. It has been determined that glucose-responsive MNs comprise the majority of biologically responsive MNs.

Yang et al. synthesized a novel nanocomposite by loading insulin and GOx into cobalt-doped zeolitic imidazole framework-8. Subsequently, the nanocomposite was encapsulated into dissolving MNs.[99] In the presence of elevated blood glucose levels, GOx embedded within the system underwent a reaction that resulted in the conversion of glucose into gluconic acid. Subsequent to the local acidic environment, the nanocomposite underwent degradation, which resulted in the release of insulin. In another study, Yu and coworkers designed a glucose-responsive MN patch that contained glucose-responsive vesicles loaded with insulin and GOx.[100] The MNs were self-assembled from hypoxia-sensitive hyaluronic acid (HS-HA) conjugated with 2-nitroimidazole, a hydrophobic component that can be converted to hydrophilic 2-aminoimidazoles through bioreduction under hypoxic conditions. The local hypoxic microenvironment, induced by the enzymatic oxidation of glucose in the hyperglycemic state, has been shown to promote the reduction of HS-HA. This, in turn, has been demonstrated to trigger the dissociation of vesicles rapidly and the subsequent release of insulin. In type I diabetic mice, the administration of MNs led to the swift normalization of blood glucose levels, with the levels remaining within the normoglycemic range for a duration of 4 h.

Phenylboronic acid have been used for glucose-sensitive self-regulated insulin delivery, a methodology predicated on the principle of a reversible reaction between phenylboronic acid and *cis*-diol. Phenylboronic acid is a synthetic Lewis acid that reversibly binds to the hydroxyl group in glucose to form a cyclic ester via its phenyl borate moiety. The majority of phenylboronic acids exhibit a partial negative charge and hydrophilic property when the pH exceeds their pKa. However, once the pH of the solution falls below its pKa, the phenylboronic acid molecules become neutral and hydrophobic. The change in hydrophilicity of phenylboronic acid-functionalized materials induces the expansion or contraction of phenylboronic acid-based platforms. This, in turn, exerts an influence on the drug release profile. Consequently, several reports have indicated the potential of a glucose-responsive MN system based on phenylboronic acid for the management of diabetes mellitus. In the presence of blood glucose levels, glucose has been observed to bind to phenylboronic acid molecules. This has been demonstrated to result in an increase in the negative charge within the MN matrix and subsequent insulin release, facilitating the maintenance of blood glucose within the normal range for approximately 1 day.[78] Similarly, a glucose-sensitive MN patch was developed. This patch is based on phenylboronic acid-modified chitosan particles and was formulated to prevent the risk of hypoglycemia associated with subcutaneous injection.[101] The amino groups present in the particles improved the loading capacity of insulin through electrostatic interaction. A first-order response of glucose concentration was realized by changing the hydrophilic and hydrophobic properties of phenylboronic acid groups. Subsequently, the external polyvinyl alcohol gel was employed to envelop the broken particles, establishing bonds and structures with the phenylboronic acid groups. Besides, the structure was disrupted by higher concentrations of glucose in the second-stage response. The two-stage glucose response model demonstrated that effective glucose control could be achieved for up to 8 h in Sprague–Dawley diabetic rats.

2.3.2.3 Reactive oxygen species–responsive microneedles

ROS have been identified as essential signals in biometabolism and intercellular communication. In various pathological processes, including diabetes, inflammation, neurodegenerative diseases, and cancer, ROS levels are markedly increased above normal levels owing to metabolic abnormalities. This has prompted researchers to design ROS-responsive MN systems to combat such diseases. ROS-responsive polymers can be categorized into two groups: ROS-induced non-cleavable hydrophobic-hydrophilic transition and ROS-induced structural cleavage. In the initial category, ROS oxidizes chalcogenide elements such as selenium, altering their valence. The formation of covalent bonds between oxygen and sulfur atoms, as well as the establishment of hydrogen bonds between polar groups and water molecules, initiates a hydrophobic–hydrophilic transition within the polymer backbone. In the second category, ROS can react with chemical structures such as polyvinyl alcohol and phenylboronic acids/esters. Various ROS-responsive MNs have been designed based on these two mechanisms.

For instance, there is evidence supporting the use of ROS-responsive polyvinyl alcohol MN patches for the treatment of acne. A comparison of the efficacy of MNs with that of commonly used antiacne creams reveals the potential of MNs to enhance the treatment of dermal lesions through enhanced dermal penetration. The level of ROS in inflamed tissues under pathological conditions can surpass 500 μM, which is significantly higher than the level in normal tissues (1–15 μM). Therefore, ROS-responsive MNs were capable of releasing drugs within the *Propionibacterium acnes*–infected follicle in a continuous manner, effectively inhibiting bacterial proliferation.[102]

2.3.2.4 Enzyme-responsive microneedles

Hyaluronic acid constitutes a significant component of the skin's extracellular matrix, functioning as a sealing element that facilitates the formation of a continuous transdermal barrier. This barrier impedes the permeation of drugs from the epidermis to the dermis, hindering systemic absorption. Accordingly, hyaluronidase (HAase)-powered MNs were developed to enhance drug delivery efficiency by hijacking sequential transdermal barriers.[103] The results demonstrated that HAase MNs effectively improved the transdermal penetration of drugs in a dose-dependent manner, with a depth and area of penetration comparable to subcutaneous injections.

Aberrant expression of enzymes has been identified at all stages of cancer, and it has been posited that this could serve as a target for oncology therapies. Therefore, drug release can be activated by the enzymatic digestion of hyaluronic acid at the tumor site, with the assistance of HAase, an enzyme that is overexpressed in the tumor microenvironment. Ye et al.[88] developed immunotherapeutic NPs formed by the self-assembly of 1-MT-modified hyaluronic acid, which was embedded in the assembly process to encapsulate aPD-1. The enzyme-responsive MN system has been shown to trigger the sustained release and retention of the checkpoint inhibitor in the B16-F10 mouse melanoma tumor microenvironment. In the presence of HAase, the release of aPD-1 was significantly elevated, whereas only a minimal amount of aPD-1 was released from HA-NPs in HAase-free phosphate-buffered saline. In addition, 1-MT exhibited a release profile from HA-NPs with only ~8% in 8 h without HAase and ~18% released in 48 h. In contrast, the cumulative release of 1-MT was elevated up to fivefold with the assistance of HAase.

Beyond their potential in antitumor therapy, enzyme-responsive MNs demonstrate remarkable applications in the domain of antibacterial therapy. Research findings indicate that the overexpression of gelatinase in *Staphylococcus aureus* wound infections signifies conducive conditions for the use of gelatinase-responsive MN-based antimicrobial therapy and wound healing strategies. Lei et al.[104] reported a degradable MN patch made of a physically inert polymer, polyvinylpyrrolidone K-30, and a recombinant type III collagen protein. This patch was designed to deliver antibacterial gelatin NPs to the dermis of infected chronic wounds. The NPs were conjugated with an antibacterial photothermal peptide, AMP-Cypate, which was released in the presence of gelatinase overexpressed by *S. aureus*. In a rat model of staphylococcal infection-induced chronic wounds, the MN patches eradicated the bacterial infection and resulted in complete healing of the wounds. These results demonstrate the outstanding therapeutic efficacy of MN patches for chronic wound healing in diabetic mice.

2.3.3 Dual stimuli-responsive microneedles

Notwithstanding the potential for on-demand drug delivery to be achieved with a single stimulus-responsive MN, the precise on-demand delivery of drugs is imperative for the treatment of certain diseases.[105] Consequently, the integration of two stimulus-responsive drug delivery functions within a singular MN system to enhance efficacy has garnered the attention of numerous researchers. Li et al.[106] developed a fully integrated wearable closed-loop system (IWCS) based on the MN platform for real-time and in situ diabetes monitoring and treatment. The IWCS consisted of three interconnected modules: (1) a glucose sensor MN, (2) a flexible printed circuit board (FPCB) serving as an integrated recording and control section, and (3) a therapeutics-loaded MN. The glucose sensor MN performs the reverse-iontophoretic extraction of interstitial fluid, followed by glucose level sensing through GOx-enzyme catalyzed environmental changes. The electric signals are then transmitted through the FPCB to the therapeutics-loaded MN, resulting in the release of insulin. The IWCS demonstrated the capacity to monitor glucose fluctuations accurately and release insulin in a controlled manner, facilitating the effective regulation of blood glucose levels in a diabetic rat model.

Another study by Yang and colleagues reported dual stimuli-responsive MNs that can deliver the drug upon application of both NIR and electric current.[107] They prepared a biocompatible polyvinylpyrrolidone-based MN patch. The back side of the MN patch was coated with gold or silver to serve as bioelectrodes, to maximize and control the efficiency of drug release. The obtained MN patch had sufficient mechanical properties to penetrate the skin successfully. The drug-release efficiency of the MN patch increased by more than 7.9 times in comparison to a flat MN patch (without applying electricity). In addition, the drug-release efficiency of the MN patches was improved by approximately 5.3 times when compared to a flat patch without NIR irradiation.

Patients with superficial tumors often experience pain of varying intensity, which can manifest from the cancer itself or be a consequence of the therapeutic interventions employed. In such conditions, dual stimuli-responsive MNs have been demonstrated to be capable of treating superficial tumors under one stimulus and managing the encountered pain under another stimulus. In support of this, Gaware et al. developed a MNs patch containing a porous carbon nanocomposite that exhibited pH and electric-current responsiveness for on-demand transdermal drug delivery.[63] The pH-dependent release function allows the initiation of drug release at superficial tumor sites, while current stimulation facilitates spatiotemporal control of the drug release for pain management.

3. Conclusions and perspectives

Remarkable progress has been accomplished in the field of MN-based drug delivery systems. MN technology can facilitate painless transdermal drug delivery by creating microchannels in the stratum corneum of the skin. Such breakthroughs represent a significant advancement beyond the limitations of traditional transdermal drug delivery methods, which are constrained by the water solubility and molecular weight of the drugs. The combination of MNs and nanotechnology has the potential to improve therapeutic efficacy by increasing the amount of drug loading, preventing biodegradation, and enhancing stability. MNs also have the potential to mitigate the necessity for cold chain storage.

Furthermore, the multifunctionality of drug delivery can be achieved through the design of MNs with intelligent response functions. Stimuli-responsive MN patches represent an emerging technology for on-demand drug delivery, with the potential to improve treatment efficacy and alleviate systemic side effects. As described here, such systems can be triggered by a variety of physiological or external stimuli. MNs triggered by physiological signals are capable of self-regulated/closed-loop drug delivery in response to physiological signals, maximizing therapeutic efficacy and enhancing patient compliance. Although physiological signal stimulus-responsive MNs do not necessitate an additional device to control drug release, certain health conditions, including irregular pain, cardiovascular disease, and neurological disease, are absent with a suitable internal stimulus. In such cases, exogenous stimulus-responsive MNs, which release the appropriate dose of medication via a manual or external trigger, play an essential role. However, more studies are required to understand the drug pharmacokinetics following MN-based delivery. Rzhevskiy et al.[108,109] and Wei et al.[110] have initiated the incorporation of mathematical modeling to address this gap in knowledge, facilitating a more profound understanding of the diffusion of the deposited drug within tissues and the impact of the MNs, vehicle, and drug properties. MNs will have a far more significant clinical impact and will further improve human health through a variety of delivery modalities.

References

1. Kim, Y.-C.; Park, J.-H.; Prausnitz, M. R. Microneedles for Drug and Vaccine Delivery. *Adv. Drug Delivery Rev.* **2012,** *64* (14), 1547–1568.
2. Naguib, Y. W.; Kumar, A.; Cui, Z. The Effect of Microneedles on the Skin Permeability and Antitumor Activity of Topical 5-Fluorouracil. *Acta Pharm. Sin. B* **2014,** *4* (1), 94–99.
3. Zhao, X.; Li, X.; Zhang, P.; Du, J.; Wang, Y. Tip-Loaded Fast-Dissolving Microneedle Patches for Photodynamic Therapy of Subcutaneous Tumor. *J. Control. Release* **2018,** *286,* 201–209.

4. Bhatnagar, S.; Bankar, N. G.; Kulkarni, M. V.; Venuganti, V. V. K. Dissolvable Microneedle Patch Containing Doxorubicin and Docetaxel Is Effective in 4T1 Xenografted Breast Cancer Mouse Model. *Int. J. Pharm.* **2019,** *556,* 263–275.
5. Lin, S.; Quan, G.; Hou, A.; Yang, P.; Peng, T.; Gu, Y.; Qin, W.; Liu, R.; Ma, X.; Pan, X. Strategy for Hypertrophic Scar Therapy: Improved Delivery of Triamcinolone Acetonide Using Mechanically Robust Tip-Concentrated Dissolving Microneedle Array. *J. Control. Release* **2019,** *306,* 69–82.
6. Yang, P.; Lu, C.; Qin, W.; Chen, M.; Quan, G.; Liu, H.; Wang, L.; Bai, X.; Pan, X.; Wu, C. Construction of a Core-Shell Microneedle System to Achieve Targeted Co-delivery of Checkpoint Inhibitors for Melanoma Immunotherapy. *Acta Biomater.* **2020,** *104,* 147–157.
7. Zheng, A.-P.; Wang, J.-C.; Lu, W.-L.; Zhang, X.; Zhang, H.; Wang, X.-Q.; Zhang, Q. Thymopentin-loaded pH-Sensitive Chitosan Nanoparticles for Oral Administration: Preparation, Characterization, and Pharmacodynamics. *J. Nanosci. Nanotechnol.* **2006,** *6* (9–10), 2936–2944.
8. Lin, S.; Cai, B.; Quan, G.; Peng, T.; Yao, G.; Zhu, C.; Wu, Q.; Ran, H.; Pan, X.; Wu, C. Novel Strategy for Immunomodulation: Dissolving Microneedle Array Encapsulating Thymopentin Fabricated by Modified Two-Step Molding Technology. *Eur. J. Pharm. Biopharm.* **2018,** *122,* 104–112.
9. You, J.; Juhng, S.; Song, J.; Park, J.; Jang, M.; Kang, G.; Yang, H.; Min, H. S.; Shin, J.; Lee, S. Egg Microneedle for Transdermal Delivery of Active Liraglutide. *Adv. Healthcare Mater.* **2023,** *12* (9), 2202473.
10. Chen, W.; Tian, R.; Xu, C.; Yung, B. C.; Wang, G.; Liu, Y.; Ni, Q.; Zhang, F.; Zhou, Z.; Wang, J. Microneedle-Array Patches Loaded with Dual Mineralized Protein/Peptide Particles for Type 2 Diabetes Therapy. *Nat. Commun.* **2017,** *8* (1), 1777.
11. Liu, J. F.; GhavamiNejad, A.; Lu, B.; Mirzaie, S.; Samarikhalaj, M.; Giacca, A.; Wu, X. Y. "Smart" Matrix Microneedle Patch Made of Self-Crosslinkable and Multifunctional Polymers for Delivering Insulin On-Demand. *Adv. Sci.* **2023,** *10* (30), 2303665.
12. You, Y.; Tian, Y.; Yang, Z.; Shi, J.; Kwak, K. J.; Tong, Y.; Estania, A. P.; Cao, J.; Hsu, W.-H.; Liu, Y. Intradermally Delivered mRNA-Encapsulating Extracellular Vesicles for Collagen-Replacement Therapy. *Nat. Biomed. Eng.* **2023,** *7* (7), 887–900.
13. Le, Z.; Ramos, M. C.; Shou, Y.; Li, R. R.; Cheng, H. S.; Jang, C. J.; Liu, L.; Xue, C.; Li, X.; Liu, H. Bioactive Sucralfate-Based Microneedles Promote Wound Healing through Reprogramming Macrophages and Protecting Endogenous Growth Factors. *Biomaterials* **2024,** *311,* 122700.
14. Ko, M. J.; Min, S.; Hong, H.; Yoo, W.; Joo, J.; Zhang, Y. S.; Kang, H.; Kim, D.-H. Magnetic Nanoparticles for Ferroptosis Cancer Therapy with Diagnostic Imaging. *Bioact. Mater.* **2024,** *32,* 66–97.
15. Jin, X.; Zhu, D. D.; Chen, B. Z.; Ashfaq, M.; Guo, X. D. Insulin Delivery Systems Combined with Microneedle Technology. *Adv. Drug Delivery Rev.* **2018,** *127,* 119–137.
16. Noh, I.; Lee, K.; Rhee, Y.-S. Microneedle Systems for Delivering Nucleic Acid Drugs. *J. Pharm. Invest.* **2022,** 1–20.
17. Mansoor, I.; Eassa, H. A.; Mohammed, K. H.; Abd El-Fattah, M. A.; Abdo, M. H.; Rashad, E.; Eassa, H. A.; Saleh, A.; Amin, O. M.; Nounou, M. I. Microneedle-Based Vaccine Delivery: Review of an Emerging Technology. *AAPS PharmSciTech* **2022,** *23* (4), 103.
18. Zhang, L.; Dong, Z.; Liu, W.; Wu, X.; He, H.; Lu, Y.; Wu, W.; Qi, J. Novel Pharmaceutical Strategies for Enhancing Skin Penetration of Biomacromolecules. *Pharmaceuticals* **2022,** *15* (7), 877.

19. Chen, M.; Quan, G.; Sun, Y.; Yang, D.; Pan, X.; Wu, C. Nanoparticles-Encapsulated Polymeric Microneedles for Transdermal Drug Delivery. *J. Control. Release* **2020,** *325,* 163–175.
20. Jiang, X.; Zhao, H.; Li, W. Microneedle-Mediated Transdermal Delivery of Drug-Carrying Nanoparticles. *Front. Bioeng. Biotechnol.* **2022,** *10,* 840395.
21. Dong, L.; Li, Y.; Li, Z.; Xu, N.; Liu, P.; Du, H.; Zhang, Y.; Huang, Y.; Zhu, J.; Ren, G. Au Nanocage-Strengthened Dissolving Microneedles for Chemo-Photothermal Combined Therapy of Superficial Skin Tumors. *ACS Appl. Mater. Interfaces* **2018,** *10* (11), 9247–9256.
22. Peng, K.; Ammar, A. A.; Himawan, A.; Dai, X.; Duncan, R.; Gilmore, B. F.; Donnelly, R. F.; Vora, L. K. Amphotericin B PLGA Nanoparticles Loaded Dissolving Microneedle Patches in Treating Cutaneous Fungal Infections. *J. Drug Delivery Sci. Technol.* **2025,** 106697.
23. Luo, S.; Zhao, Y.; Pan, K.; Zhou, Y.; Quan, G.; Wen, X.; Pan, X.; Wu, C. Microneedle-mediated Delivery of MIL-100 (Fe) as a Tumor Microenvironment-Responsive Biodegradable Nanoplatform for O_2-Evolving Chemophototherapy. *Biomater. Sci.* **2021,** *9* (23), 8051.
24. Abdelghany, S.; Tekko, I. A.; Vora, L.; Larrañeta, E.; Permana, A. D.; Donnelly, R. F. Nanosuspension-Based Dissolving Microneedle Arrays for Intradermal Delivery of Curcumin. *Pharmaceutics* **2019,** *11* (7), 308.
25. Zhao, Z.; Li, M.; Zheng, L.; Yang, Y.; Cui, X.; Xu, T.; Zhang, W.; Wang, C. Noninvasive Transdermal Delivery of Mesoporous Silica Nanoparticles Using Deep Eutectic Solvent. *J. Control. Release* **2022,** *343,* 43–56.
26. Wang, M.; Li, X.; Du, W.; Sun, M.; Ling, G.; Zhang, P. Microneedle-Mediated Treatment for Superficial Tumors by Combining Multiple Strategies. *Drug Delivery Transl. Res.* **2023,** *13* (6), 1600–1620.
27. Zhang, Z.; Tsai, P. C.; Ramezanli, T.; Michniak-Kohn, B. B. Polymeric Nanoparticles-Based Topical Delivery Systems for the Treatment of Dermatological Diseases. *Wiley Interdiscip. Rev. Nanomed. Nanobiotechnol.* **2013,** *5* (3), 205–218.
28. Abd-El-Azim, H.; Tekko, I. A.; Ali, A.; Ramadan, A.; Nafee, N.; Khalafallah, N.; Rahman, T.; Mcdaid, W.; Aly, R. G.; Vora, L. K. Hollow Microneedle Assisted Intradermal Delivery of Hypericin Lipid Nanocapsules with Light Enabled Photodynamic Therapy Against Skin Cancer. *J. Control. Release* **2022,** *348,* 849–869.
29. Ding, M.; Cai, X.; Jiang, H. Improving MOF Stability: Approaches and Applications Chem. *Chem. Sci.* **2019,** *10* (44), 10209–10230.
30. Jing, Y.; Liu, X.; Zhu, Y.; Wu, L.; Nong, W. Metal-Organic Framework Microneedles for Precision Transdermal Drug Delivery: Design Strategy and Therapeutic Potential. *Nanoscale* **2025,** *17* (10), 5571–5604.
31. Yao, S.; Chi, J.; Wang, Y.; Zhao, Y.; Luo, Y.; Wang, Y. Zn-MOF Encapsulated Antibacterial and Degradable Microneedles Array for Promoting Wound Healing. *Adv. Healthcare Mater.* **2021,** *10* (12), 2100056.
32. Yin, M.; Wu, J.; Deng, M.; Wang, P.; Ji, G.; Wang, M.; Zhou, C.; Blum, N. T.; Zhang, W.; Shi, H. Multifunctional Magnesium Organic Framework-Based Microneedle Patch for Accelerating Diabetic Wound Healing. *ACS Nano* **2021,** *15* (11), 17842–17853.
33. Vora, L. K.; Vavia, P. R.; Larrañeta, E.; Bell, S. E.; Donnelly, R. F. Novel Nanosuspension-Based Dissolving Microneedle Arrays for Transdermal Delivery of a Hydrophobic Drug. *J. Interdiscip. Nanomed.* **2018,** *3* (2), 89–101.
34. Lee, H. S.; Ryu, H. R.; Roh, J. Y.; Park, J.-H. Bleomycin-Coated Microneedles for Treatment of Warts. *Pharm. Res.* **2017,** *34,* 101–112.

35. Pamornpathomkul, B.; Ngawhirunpat, T.; Tekko, I. A.; Vora, L.; McCarthy, H. O.; Donnelly, R. F. Dissolving Polymeric Microneedle Arrays for Enhanced Site-specific Acyclovir Delivery. *Eur. J. Pharm. Sci.* **2018,** *121,* 200–209.
36. Qiu, Y.; Li, C.; Zhang, S.; Yang, G.; He, M.; Gao, Y. Systemic Delivery of Artemether by Dissolving Microneedles. *Int. J. Pharm.* **2016,** *508* (1–2), 1–9.
37. Liu, S.; Wu, D.; Quan, Y-s; Kamiyama, F.; Kusamori, K.; Katsumi, H.; Sakane, T.; Yamamoto, A. Improvement of Transdermal Delivery of Exendin-4 Using Novel Tip-Loaded Microneedle Arrays Fabricated from Hyaluronic Acid. *Mol. Pharmaceutics* **2016,** *13* (1), 272–279.
38. Tas, C.; Joyce, J. C.; Nguyen, H. X.; Eangoor, P.; Knaack, J. S.; Banga, A. K.; Prausnitz, M. R. Dihydroergotamine Mesylate-Loaded Dissolving Microneedle Patch Made of Polyvinylpyrrolidone for Management of Acute Migraine Therapy. *J. Controlled Release* **2017,** *268,* 159–165.
39. Chen, J.; Huang, W.; Huang, Z.; Liu, S.; Ye, Y.; Li, Q.; Huang, M. Fabrication of Tip-Dissolving Microneedles for Transdermal Drug Delivery of Meloxicam. *AAPS PharmSciTech* **2018,** *19,* 1141–1151.
40. Ito, Y.; Kashiwara, S.; Fukushima, K.; Takada, K. Two-Layered Dissolving Microneedles for Percutaneous Delivery of Sumatriptan in Rats. *Drug Dev. Ind. Pharm.* **2011,** *37* (12), 1387–1393.
41. Xu, Q.; Li, X.; Zhang, P.; Wang, Y. Rapidly Dissolving Microneedle Patch for Synergistic Gene and Photothermal Therapy of Subcutaneous Tumor. *J. Mater. Chem. B* **2020,** *8* (19), 4331–4339.
42. Hou, A.; Quan, G.; Yang, B.; Lu, C.; Chen, M.; Yang, D.; Wang, L.; Liu, H.; Pan, X.; Wu, C. Rational Design of Rapidly Separating Dissolving Microneedles for Precise Drug Delivery by Balancing the Mechanical Performance and Disintegration Rate. *Adv. Healthcare Mater.* **2019,** *8* (21), 1900898.
43. Peng, T.; Huang, Y.; Feng, X.; Zhu, C.; Ma, X.; Wang, X.; Bai, X.; Pan, X.; Wu, C. Dissolving Microneedles Loading TPGS Biphasic Functionalized PLGA Nanoparticles for Efficient Chemo-photothermal Combined Therapy of Melanoma. *Adv. Ther.* **2020,** *3* (6), 1900190.
44. Singh, P.; Carrier, A.; Chen, Y.; Lin, S.; Wang, J.; Cui, S.; Zhang, X. Polymeric Microneedles for Controlled Transdermal Drug Delivery. *J. Control. Release* **2019,** *315,* 97–113.
45. Darge, H. F.; Lee, C.-Y.; Lai, J.-Y.; Lin, S.-Z.; Harn, H.-J.; Chen, Y.-S.; Tsai, H.-C. Separable Double-Layered Microneedle-Based Transdermal Codelivery of DOX and LPS for Synergistic Immunochemotherapy of a Subcutaneous Glioma Tumor. *Chem. Eng. J.* **2022,** *433,* 134062.
46. Li, W.; Terry, R. N.; Tang, J.; Feng, M. R.; Schwendeman, S. P.; Prausnitz, M. R. Rapidly Separable Microneedle Patch for the Sustained Release of a Contraceptive. *Nat. Biomed. Eng.* **2019,** *3* (3), 220–229.
47. He, M.; Yang, G.; Zhang, S.; Zhao, X.; Gao, Y. Dissolving Microneedles Loaded with Etonogestrel Microcrystal Particles for Intradermal Sustained Delivery. *J. Pharm. Sci.* **2018,** *107* (4), 1037–1045.
48. Chen, M.-C.; Ling, M.-H.; Lai, K.-Y.; Pramudityo, E. Chitosan Microneedle Patches for Sustained Transdermal Delivery of Macromolecules. *Biomacromolecules* **2012,** *13* (12), 4022–4031.
49. Kim, M.; Jung, B.; Park, J.-H. Hydrogel Swelling as a Trigger to Release Biodegradable Polymer Microneedles in Skin. *Biomaterials* **2012,** *33* (2), 668–678.

50. Chen, M.-C.; Lai, K.-Y.; Ling, M.-H.; Lin, C.-W. Enhancing Immunogenicity of Antigens through Sustained Intradermal Delivery Using Chitosan Microneedles with a Patch-Dissolvable Design. *Acta Biomater.* **2018,** *65,* 66–75.
51. Courtenay, A. J.; McCrudden, M. T.; McAvoy, K. J.; McCarthy, H. O.; Donnelly, R. F. Microneedle-Mediated Transdermal Delivery of Bevacizumab. *Mol. Pharmaceutics* **2018,** *15* (8), 3545–3556.
52. Wang, S.; Zhu, M.; Zhao, L.; Kuang, D.; Kundu, S. C.; Lu, S. Insulin-Loaded Silk Fibroin Microneedles as Sustained Release System. *ACS Biomater. Sci. Eng.* **2019,** *5* (4), 1887–1894.
53. Chiu, Y.-H.; Chen, M.-C.; Wan, S.-W. Sodium Hyaluronate/Chitosan Composite Microneedles as a Single-Dose Intradermal Immunization System. *Biomacromolecules* **2018,** *19* (6), 2278–2285.
54. Tsioris, K.; Raja, W. K.; Pritchard, E. M.; Panilaitis, B.; Kaplan, D. L.; Omenetto, F. G. Fabrication of Silk Microneedles for Controlled-Release Drug Delivery. *Adv. Funct. Mater.* **2012,** *22* (2), 330–335.
55. Pillai, M. M.; Ajesh, S.; Tayalia, P. Two-Photon Polymerization Based Reusable Master Template to Fabricate Polymer Microneedles for Drug Delivery. *MethodsX* **2023,** *10,* 102025.
56. Makvandi, P.; Jamaledin, R.; Chen, G.; Baghbantaraghdari, Z.; Zare, E. N.; Di Natale, C.; Onesto, V.; Vecchione, R.; Lee, J.; Tay, F. R. Stimuli-Responsive Transdermal Microneedle Patches. *Mater. Today* **2021,** *47,* 206–222.
57. Al Sulaiman, D.; Chang, J. Y.; Bennett, N. R.; Topouzi, H.; Higgins, C. A.; Irvine, D. J.; Ladame, S. Hydrogel-Coated Microneedle Arrays for Minimally Invasive Sampling and Sensing of Specific Circulating Nucleic Acids from Skin Interstitial Fluid. *ACS Nano* **2019,** *13* (8), 9620–9628.
58. Hardy, J. G.; Larrañeta, E.; Donnelly, R. F.; McGoldrick, N.; Migalska, K.; McCrudden, M. T.; Irwin, N. J.; Donnelly, L.; McCoy, C. P. Hydrogel-Forming Microneedle Arrays Made from Light-Responsive Materials for On-Demand Transdermal Drug Delivery. *Mol. Pharmaceutics* **2016,** *13* (3), 907–914.
59. Hao, Y.; Chen, Y.; Lei, M.; Zhang, T.; Cao, Y.; Peng, J.; Chen, L.; Qian, Z. Near-Infrared Responsive PEGylated Gold Nanorod and Doxorubicin Loaded Dissolvable Hyaluronic Acid Microneedles for Human Epidermoid Cancer Therapy. *Adv. Ther.* **2018,** *1* (2), 1870004.
60. Chen, M.-C.; Ling, M.-H.; Wang, K.-W.; Lin, Z.-W.; Lai, B.-H.; Chen, D.-H. Near-Infrared Light-Responsive Composite Microneedles for On-Demand Transdermal Drug Delivery. *Biomacromolecules* **2015,** *16* (5), 1598–1607.
61. Hao, Y.; Dong, M.; Zhang, T.; Peng, J.; Jia, Y.; Cao, Y.; Qian, Z. Novel Approach of Using Near-Infrared Responsive PEGylated Gold Nanorod Coated Poly (L-lactide) Microneedles to Enhance the Antitumor Efficiency of Docetaxel-Loaded MPEG-PDLLA Micelles for Treating an A431 Tumor. *ACS Appl. Mater. Interfaces* **2017,** *9* (18), 15317–15327.
62. Zhang, Y.; Jiang, G.; Hong, W.; Gao, M.; Xu, B.; Zhu, J.; Song, G.; Liu, T. Polymeric Microneedles Integrated with Metformin-Loaded and PDA/LA-Coated Hollow Mesoporous SiO_2 for NIR-Triggered Transdermal Delivery on Diabetic Rats. *ACS Appl. Bio Mater.* **2018,** *1* (6), 1906–1917.
63. Gaware, S. A.; Rokade, K. A.; Bala, P.; Kale, S. N. Microneedles of Chitosan-porous Carbon Nanocomposites: Stimuli (pH and Electric Field)-Initiated Drug Delivery and Toxicological Studies. *J. Biomed. Mater. Res., Part A* **2019,** *107* (8), 1582–1596.

64. Sivaraman, A.; Banga, A. K. Novel In Situ Forming Hydrogel Microneedles for Transdermal Drug Delivery. *Drug Delivery Transl. Res.* **2017,** *7,* 16–26.
65. Lee, H.; Choi, T. K.; Lee, Y. B.; Cho, H. R.; Ghaffari, R.; Wang, L.; Choi, H. J.; Chung, T. D.; Lu, N.; Hyeon, T. A Graphene-Based Electrochemical Device with Thermoresponsive Microneedles for Diabetes Monitoring and Therapy. *Nat. Nanotechnol.* **2016,** *11* (6), 566–572.
66. Zhang, Y.; Wang, D.; Gao, M.; Xu, B.; Zhu, J.; Yu, W.; Liu, D.; Jiang, G. Separable Microneedles for Near-Infrared Light-Triggered Transdermal Delivery of Metformin in Diabetic Rats. *ACS Biomater. Sci. Eng.* **2018,** *4* (8), 2879–2888.
67. Choi, I.-J.; Kang, A.; Ahn, M.-H.; Jun, H.; Baek, S.-K.; Park, J.-H.; Na, W.; Choi, S.-O. Insertion-responsive Microneedles for Rapid Intradermal Delivery of Canine Influenza Vaccine. *J. Control. Release* **2018,** *286,* 460–466.
68. Zhang, Y.; Chai, D.; Gao, M.; Xu, B.; Jiang, G. Thermal Ablation of Separable Microneedles for Transdermal Delivery of Metformin on Diabetic Rats. *Int. J. Polym. Mater. Polym. Biomater.* **2019,** *68* (14), 850–858.
69. Su, Y.; Andrabi, S. M.; Shahriar, S. S.; Wong, S. L.; Wang, G.; Xie, J. Triggered Release of Antimicrobial Peptide from Microneedle Patches for Treatment of Wound Biofilms. *J. Control. Release* **2023,** *356,* 131–141.
70. Chen, M.-C.; Wang, K.-W.; Chen, D.-H.; Ling, M.-H.; Liu, C.-Y. Remotely Triggered Release of Small Molecules from LaB6@ SiO_2-Loaded Polycaprolactone Microneedles. *Acta Biomater.* **2015,** *13,* 344–353.
71. Liu, F.; Cheng, Z.; Yi, H. NIR Light-Activatable Dissolving Microneedle System for Melanoma Ablation Enabled by a Combination of ROS-Responsive Chemotherapy and Phototherapy. *J. Nanobiotechnol.* **2023,** *21* (1), 61.
72. Pan, Y.; Zhang, Y.; Shi, X.; Li, D.; Xu, X.; Xiao, B.; Piao, Y.; Xiang, J.; Shao, S.; Ho, F. C.-Y. Electrical Stimulation Induces Anti-tumor Immunomodulation via a Flexible Microneedle-Array-Integrated Interdigital Electrode. *Sci. Bull.* **2023,** *68* (22), 2779–2792.
73. Wang, Y.; Chen, Z.; Davis, B.; Lipman, W.; Xing, S.; Zhang, L.; Wang, T.; Hafiz, P.; Xie, W.; Yan, Z. Digital Automation of Transdermal Drug Delivery with High Spatiotemporal Resolution. *Nat. Commun.* **2024,** *15* (1), 511.
74. Yang, Y.; Chen, B. Z.; Zhang, X. P.; Zheng, H.; Li, Z.; Zhang, C. Y.; Guo, X. D. Conductive Microneedle Patch with Electricity-Triggered Drug Release Performance for Atopic Dermatitis Treatment. *ACS Appl. Mater. Interfaces* **2022,** *14* (28), 31645–31654.
75. Justin, R.; Chen, B. Multifunctional Chitosan–Magnetic Graphene Quantum Dot Nanocomposites for the Release of Therapeutics from Detachable and Non-detachable Biodegradable Microneedle Arrays. *Interface Focus* **2018,** *8* (3), 20170055.
76. Fang, J.-H.; Liu, C.-H.; Hsu, R.-S.; Chen, Y.-Y.; Chiang, W.-H.; Wang, H.-M. D.; Hu, S.-H. Transdermal Composite Microneedle Composed of Mesoporous Iron Oxide Nanoraspberry and PVA for Androgenetic Alopecia Treatment. *Polymers* **2020,** *12* (6), 1392.
77. Yu, J.; Zhang, Y.; Ye, Y.; DiSanto, R.; Sun, W.; Ranson, D.; Ligler, F.; Buse, J.; Gu, Z. Microneedle-array Patches Loaded with Hypoxia-Sensitive Vesicles Provide Fast Glucose-Responsive Insulin Delivery. *Diabetes Technol. Ther* **2016,** *18,* S54–S55.
78. Yu, J.; Wang, J.; Zhang, Y.; Chen, G.; Mao, W.; Ye, Y.; Kahkoska, A. R.; Buse, J. B.; Langer, R.; Gu, Z. Glucose-responsive Insulin Patch for the Regulation of Blood Glucose in Mice and Minipigs. *Nat. Biomed. Eng.* **2020,** *4* (5), 499–506.
79. Xu, B.; Jiang, G.; Yu, W.; Liu, D.; Zhang, Y.; Zhou, J.; Sun, S.; Liu, Y. H_2O_2-responsive Mesoporous Silica Nanoparticles Integrated with Microneedle Patches for the Glucose-Monitored Transdermal Delivery of Insulin. *J. Mater. Chem. B* **2017,** *5* (41), 8200–8208.

80. Hu, X.; Yu, J.; Qian, C.; Lu, Y.; Kahkoska, A. R.; Xie, Z.; Jing, X.; Buse, J. B.; Gu, Z. H_2O_2-Responsive Vesicles Integrated with Transcutaneous Patches for Glucose-Mediated Insulin Delivery. *ACS Nano* **2017,** *11* (1), 613–620.
81. Yu, J.; Qian, C.; Zhang, Y.; Cui, Z.; Zhu, Y.; Shen, Q.; Ligler, F. S.; Buse, J. B.; Gu, Z. Hypoxia and H_2O_2 Dual-Sensitive Vesicles for Enhanced Glucose-Responsive Insulin Delivery. *Nano Lett.* **2017,** *17* (2), 733–739.
82. Wang, J.; Yu, J.; Zhang, Y.; Zhang, X.; Kahkoska, A. R.; Chen, G.; Wang, Z.; Sun, W.; Cai, L.; Chen, Z. Charge-Switchable Polymeric Complex for Glucose-Responsive Insulin Delivery in Mice and Pigs. *Sci. Adv.* **2019,** *5* (7), eaaw4357.
83. Wang, J.; Ye, Y.; Yu, J.; Kahkoska, A. R.; Zhang, X.; Wang, C.; Sun, W.; Corder, R. D.; Chen, Z.; Khan, S. A. Core-Shell Microneedle Gel for Self-Regulated Insulin Delivery. *ACS Nano* **2018,** *12* (3), 2466–2473.
84. Chen, S.; Matsumoto, H.; Moro-oka, Y.; Tanaka, M.; Miyahara, Y.; Suganami, T.; Matsumoto, A. Microneedle-Array Patch Fabricated with Enzyme-Free Polymeric Components Capable of On-Demand Insulin Delivery. *Adv. Funct. Mater.* **2019,** *29* (7), 1807369.
85. GhavamiNejad, A.; Li, J.; Lu, B.; Zhou, L.; Lam, L.; Giacca, A.; Wu, X. Y. Glucose-Responsive Composite Microneedle Patch for Hypoglycemia-Triggered Delivery of Native Glucagon. *Adv. Mater.* **2019,** *31* (30), 1901051.
86. Zhang, Y.; Liu, Q.; Yu, J.; Yu, S.; Wang, J.; Qiang, L.; Gu, Z. Locally Induced Adipose Tissue Browning by Microneedle Patch for Obesity Treatment. *ACS Nano* **2017,** *11* (9), 9223–9230.
87. Wang, Z.; Li, H.; Wang, J.; Chen, Z.; Chen, G.; Wen, D.; Chan, A.; Gu, Z. Transdermal Colorimetric Patch for Hyperglycemia Sensing in Diabetic Mice. *Biomaterials* **2020,** *237,* 119782.
88. Ye, Y.; Wang, J.; Hu, Q.; Hochu, G.; Xin, H.; Wang, C.; Gu, Z. Synergistic Transcutaneous Immunotherapy Enhances Antitumor Immune Responses through Delivery of Checkpoint Inhibitors. *ACS Nano* **2016,** *10* (9), 8956–8963.
89. Li, X.; Xu, Q.; Zhang, P.; Zhao, X.; Wang, Y. Cutaneous Microenvironment Responsive Microneedle Patch for Rapid Gene Release to Treat Subdermal Tumor. *J. Control. Release* **2019,** *314,* 72–80.
90. Duong, H. T. T.; Yin, Y.; Thambi, T.; Nguyen, T. L.; Phan, V. G.; Lee, M. S.; Lee, J. E.; Kim, J.; Jeong, J. H.; Lee, D. S. Smart Vaccine Delivery Based on Microneedle Arrays Decorated with Ultra-pH-Responsive Copolymers for Cancer Immunotherapy. *Biomaterials* **2018,** *185,* 13–24.
91. Zhang, Y.; Yu, J.; Wang, J.; Hanne, N. J.; Cui, Z.; Qian, C.; Wang, C.; Xin, H.; Cole, J. H.; Gallippi, C. M. Thrombin-Responsive Transcutaneous Patch for Auto-Anticoagulant Regulation. *Adv. Mater.* **2017,** *29* (4).
92. Zhang, Y.; Feng, P.; Yu, J.; Yang, J.; Zhao, J.; Wang, J.; Shen, Q.; Gu, Z. ROS-Responsive Microneedle Patch for Acne Vulgaris Treatment. *Adv. Ther.* **2018,** *1* (3), 1870006.
93. Mir, M.; Permana, A. D.; Ahmed, N.; Khan, G. M.; ur Rehman, A.; Donnelly, R. F. Enhancement in Site-Specific Delivery of Carvacrol for Potential Treatment of Infected Wounds Using Infection Responsive Nanoparticles Loaded into Dissolving Microneedles: A Proof of Concept Study. *Eur. J. Pharm. Biopharm.* **2020,** *147,* 57–68.
94. Liu, G.; Zhao, X.; Zhang, Y.; Xu, J.; Xu, J.; Li, Y.; Min, H.; Shi, J.; Zhao, Y.; Wei, J. Engineering Biomimetic Platesomes for pH-Responsive Drug Delivery and Enhanced Antitumor Activity. *Adv. Mater.* **2019,** *31* (32), 1900795.
95. Ullah, A.; Jang, M.; Khan, H.; Choi, H. J.; An, S.; Kim, D.; Kim, Y.-R.; Kim, U.-K.; Kim, G. M. Microneedle Array with a pH-Responsive Polymer Coating and Its Application in Smart Drug Delivery for Wound Healing. *Sensor. Actuator. B Chem.* **2021,** *345,* 130441.

96. Duong, H. T. T.; Kim, N. W.; Thambi, T.; Phan, V. G.; Lee, M. S.; Yin, Y.; Jeong, J. H.; Lee, D. S. Microneedle Arrays Coated with Charge Reversal pH-Sensitive Copolymers Improve Antigen Presenting Cells-Homing DNA Vaccine Delivery and Immune Responses. *J. Control. Release* **2018,** *269,* 225–234.
97. Wang, C.; Ye, Y.; Hochu, G. M.; Sadeghifar, H.; Gu, Z. Enhanced Cancer Immunotherapy by Microneedle Patch-Assisted Delivery of Anti-PD1 Antibody. *Nano Lett.* **2016,** *16* (4), 2334–2340.
98. Liu, J.; Yi, X.; Zhang, J.; Yao, Y.; Panichayupakaranant, P.; Chen, H. Recent Advances in the Drugs and Glucose-Responsive Drug Delivery Systems for the Treatment of Diabetes: A Systematic Review. *Pharmaceutics* **2024,** *16* (10), 1343.
99. Yang, X.-X.; Feng, P.; Cao, J.; Liu, W.; Tang, Y. Composition-Engineered Metal–Organic Framework-Based Microneedles for Glucose-Mediated Transdermal Insulin Delivery. *ACS Appl. Mater. Interfaces* **2020,** *12* (12), 13613–13621.
100. Yu, J.; Zhang, Y.; Ye, Y.; DiSanto, R.; Sun, W.; Ranson, D.; Ligler, F. S.; Buse, J. B.; Gu, Z. Microneedle-Array Patches Loaded with Hypoxia-Sensitive Vesicles Provide Fast Glucose-Responsive Insulin Delivery. *Proc. Natl. Acad. Sci.* **2015,** *112* (27), 8260–8265.
101. Wang, Y.; Yu, H.; Wang, L.; Hu, J.; Feng, J.; Shen, D.; Hong, Y.; Liu, J.; Chen, D. Microneedles with Two-Stage Glucose-Sensitive Controlled Release for Long-Term Insulin Delivery. *ACS Biomater. Sci. Eng.* **2023,** *9* (5), 2534–2544.
102. Zhang, Y.; Feng, P.; Yu, J.; Yang, J.; Zhao, J.; Wang, J.; Shen, Q.; Gu, Z. Ros-responsive Microneedle Patch for Acne Vulgaris Treatment. *Adv. Ther.* **2018,** *1* (3), 1800035.
103. Hu, W.; Peng, T.; Huang, Y.; Ren, T.; Chen, H.; Chen, Y.; Feng, D.; Wu, C.; Pan, X. Hyaluronidase-Powered Microneedles for Significantly Enhanced Transdermal Delivery Efficiency. *J. Control. Release* **2023,** *353,* 380–390.
104. Lei, X.; Li, M.; Wang, C.; Cui, P.; Qiu, L.; Zhou, S.; Jiang, P.; Li, H.; Zhao, D.; Ni, X. Degradable Microneedle Patches Loaded with Antibacterial Gelatin Nanoparticles to Treat Staphylococcal Infection-Induced Chronic Wounds. *Int. J. Biol. Macromol.* **2022,** *217,* 55–65.
105. Gowda, B. J.; Ahmed, M. G.; Sahebkar, A.; Riadi, Y.; Shukla, R.; Kesharwani, P. Stimuli-responsive Microneedles as a Transdermal Drug Delivery System: A Demand-Supply Strategy. *Biomacromolecules* **2022,** *23* (4), 1519–1544.
106. Li, X.; Huang, X.; Mo, J.; Wang, H.; Huang, Q.; Yang, C.; Zhang, T.; Chen, H. J.; Hang, T.; Liu, F. A Fully Integrated Closed-Loop System Based on Mesoporous Microneedles-Iontophoresis for Diabetes Treatment. *Adv. Sci.* **2021,** *8* (16), 2100827.
107. Yang, S.-J.; Jeong, J.-O.; Lim, Y.-M.; Park, J.-S. Synthesis and Characterization of PVP Microneedle Patch Using Metal Bioelectrodes for Novel Drug Delivery System. *Mater. Des.* **2021,** *201,* 109485.
108. Rzhevskiy, A. S.; Guy, R. H.; Anissimov, Y. G. Modelling Drug Flux through Microporated Skin. *J. Control. Release* **2016,** *241,* 194–199.
109. Rzhevskiy, A. S.; Telaprolu, K.; Mohammed, Y. H.; Grice, J. E.; Roberts, M. S.; Anissimov, Y. G. Using a Simple Equation to Predict the Microporation-Enhanced Transdermal Drug Flux. *Eur. J. Pharm. Biopharm.* **2018,** *127,* 12–18.
110. Wei, J. C.; Edwards, G. A.; Martin, D. J.; Huang, H.; Crichton, M. L.; Kendall, M. A. Allometric Scaling of Skin Thickness, Elasticity, Viscoelasticity to Mass for Micro-medical Device Translation: From Mice, Rats, Rabbits, Pigs to Humans. *Sci. Rep.* **2017,** *7* (1), 15885.

Chapter 5

Strategies used to enhance the mechanical properties of microneedles

Xinyu Zhang[a, 1], Chunxian Zhou[a, 1], Tianxiang Chen[2] and Tingting Peng[1]
[1]*State Key Laboratory of Bioactive Molecules and Druggability Assessment/International Cooperative Laboratory of Traditional Chinese Medicine Modernization and Innovative Drug Development of Ministry of Education (MOE) of China/College of Pharmacy, Jinan University, Guangzhou, China;* [2]*School of Mechatronic Engineering, Guangdong Polytechnic Normal University, Guangzhou, China*

1. Introduction

The stratum corneum is widely recognized as the primary barrier to transdermal drug delivery. Microneedles (MNs) have emerged as a promising alternative to traditional transdermal delivery systems due to their ability to penetrate the stratum corneum and create mechanical micropores for drug transport (Fig. 5.1A). Compared to conventional transdermal preparations, MNs offer higher delivery efficiency and overcome the significant limitations imposed by the molecular weight, lipophilicity, and melting point of the loaded drugs. Since the pioneering work by Henry et al.[1] in 1998, which first demonstrated the significant enhancement of calcein's transdermal delivery efficiency. MN has gained increasing attention for its versatile applications in delivering various therapeutic agents. These include small-molecular chemicals,[2,3] biomacromolecules,[4,5] vaccines,[6,7] exosomes,[8,9] and even cells.[10–12] Compared to conventional injections, MNs offer distinct advantages by eliminating the generation of biohazardous waste, reducing patient discomfort, and removing the requirement for trained medical personnel during administration. These unique characteristics enable MNs to combine the benefits of both transdermal delivery systems and injectable formulations, thereby demonstrating tremendous potential for diverse biomedical applications.

---

a. The first two authors contributed equally to this work.

Microneedle Technology. https://doi.org/10.1016/B978-0-443-45544-5.00014-2

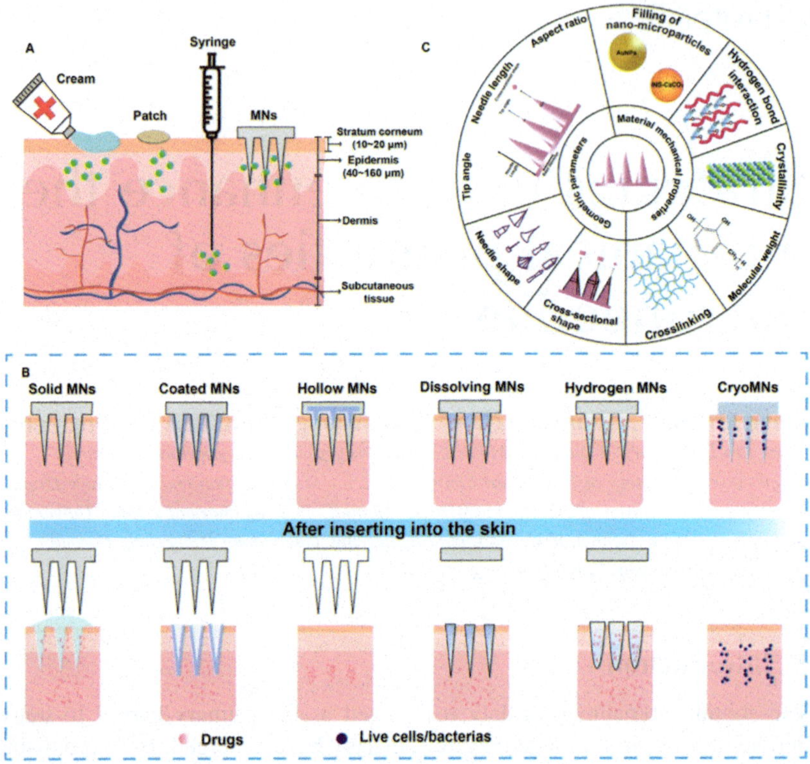

FIGURE 5.1 (A) Compared to injectable preparations, microneedles (MNs) pierce the cutaneous tissue to deliver the cargoes inside without pain, as they do not stimulate the nerves and vessels. (B) Different types of MNs. (C) An overview of the approaches used to enhance the mechanical properties of MNs.

Based on the drug delivery mode, MNs can be categorized into solid MNs,[13,14] coated MNs,[15,16] hollow MNs,[17,18] dissolving MNs,[19,20] hydrogel MNs,[21,22] and cryomicroneedles (cryoMNs)[23,24] (Fig. 5.1B). The mechanical properties of MNs are critical determinants of their ability to penetrate the skin effectively. To achieve successful skin puncture, MNs must be capable of withstanding significant mechanical stresses. The skin, which consists of multiple layers including the epidermis, dermis, and subcutaneous tissue, presents a complex barrier to penetration. The stratum corneum with dense brick-wall-like structure is recognized as the foremost barrier to MN penetration. If MNs lack adequate mechanical strength, they may buckle, bend, or fracture upon insertion into the stratum corneum.[25] Furthermore, when MNs are inserted into cutaneous tissues, the inherent elasticity of the skin can exert resistance, potentially causing needle deformation or bending. This underscores the necessity for MNs to possess sufficient mechanical strength and toughness to overcome both the rigid stratum corneum and the elastic resistance of the skin. Failure to penetrate to an effective depth can lead to inaccuracies in drug delivery dosage and inconsistent delivery efficiency. Additionally, the breakage of solid MNs made from materials such as metals, silicon, ceramics, or glass could result in needle fragments being left within the skin, potentially causing skin irritation (e.g., allergies, granulomas, infections) and posing longterm safety risks. Therefore, it is imperative to design and fabricate MNs with robust mechanical properties to ensure effective and safe transdermal drug delivery.

The mechanical properties of MNs are influenced by various factors, including the physical properties of individual polymers, formulation compositions of materials, and geometric parameters of MNs.[25] Recent studies have delved deeper into understanding how these factors impact the mechanical properties of MNs from a more theoretical standpoint.[26] Nevertheless, current strategies for improvement largely rely on extensive and often random optimization of formulation compositions and geometric parameters.

Based on the aforementioned knowledge, this chapter systematically investigates the critical factors determining the mechanical properties of MNs, which include three primary aspects: (1) the intrinsic physicochemical properties of constituent polymers, including molecular weight, crosslinking density, and crystallinity; (2) the intermolecular interactions within composite matrix materials; and (3) the geometric parameters of MNs, particularly crosssectional morphology, tip geometry, needle length, and aspect ratio (Fig. 5.1C). This comprehensive analysis establishes an evidencebased framework for the rational design of mechanically optimized MNs, providing researchers with fundamental principles for developing robust MN systems.

2. Stress analysis and fracture patterns of microneedles

When MNs are inserted into the skin, various stresses, including axial, compressive, skin resistance, buckling, bending, and shear forces, can lead to mechanical failure.[27] Among these, buckling and bending forces are the primary contributors to needle fracture. To ensure successful skin penetration, the applied force must exceed the skin resistance force, which is calculated as the product of pressure and the crosssectional area of the MNs (Eq. 5.1).[28] The stratum corneum, due to its rigid brick–wall structure, generates higher resistance compared to other skin layers.[29] Notably, the pressure required for MN insertion decreases from 3.183 to 1.6 MPa after initial puncture.[30] For MNs to remain mechanically stable and avoid breakage during insertion, the applied force must be greater than both the skin resistance force and the minimum insertion force.

The axial force is the primary load applied to the microneedles (MNs), leading to buckling at the needle tips. As defined by Eq. (5.2), the axial force is the product of the yield strength of the needle matrix material and the crosssectional area of the needle tips. Thus, reducing the radius of the needle tips can help minimize the risk of buckling. The critical buckling force represents the maximum force the MNs can endure before breaking. According to buckling analysis,[31] this force can be calculated using Euler's formula (Eq. 5.3) and depends primarily on the Young's modulus of the material, the length of the MNs, and the moment of inertia. Notably, the moment of inertia is closely related to the crosssectional area of the MNs.

In contrast, bending of MNs typically occurs due to improper needle insertion or a rough skin surface. As shown in Eq. (5.4), the maximum bending force that MNs can withstand is determined by the material's yield strength, the needle length, the moment of inertia, and the distance between the vertical axis and the outer edge of the crosssection. In summary, the mechanical performance of MNs is influenced by two key factors: the geometric parameters of MNs and the mechanical properties of the preparation materials. These aspects will be discussed in detail in the following sections.

$$F_{\text{skin}} = P_{\text{Piercing}}A \tag{5.1}$$

$$F_{\text{axial}} = \sigma_y A \tag{5.2}$$

$$F_{\text{buckling}} = \frac{\pi^2 EI}{(2L)^2} \tag{5.3}$$

$$F_{\text{bending}} = \frac{\sigma_y I}{cL} \tag{5.4}$$

where F_{skin} is the skin resistance force, F_{axial} is the axial force, F_{buckling} is the critical buckling load of MNs, F_{bending} is the maximum bending force that MNs can withstand, P_{piercing} is the pressure, A is the cross-sectional area of MNs, σ_y is the yield strength of the material, E is Young's Modulus of the material, I is the cross-sectional moment of inertia, L is the length of MNs, and c is the distance from the ordinate axis to the perimeter of MNs.

3. Influence of the physicochemical properties of single material on the mechanical properties of microneedles

The proper selection of matrix materials is crucial for ensuring that MNs have sufficient mechanical strength to successfully insert into the skin. The modulus of the matrix materials is directly correlated with the critical buckling load of the MNs. To assess the mechanical properties of materials, the stressstrain curve is widely used, as it provides key parameters such as tensile strength, elongation at break, and tensile modulus (Fig. 5.2A).[32] In the initial part of the curve, the material undergoes elastic deformation, which generally follows Hooke's law (Eq. 5.5), where the stress is proportional to the strain. The slope

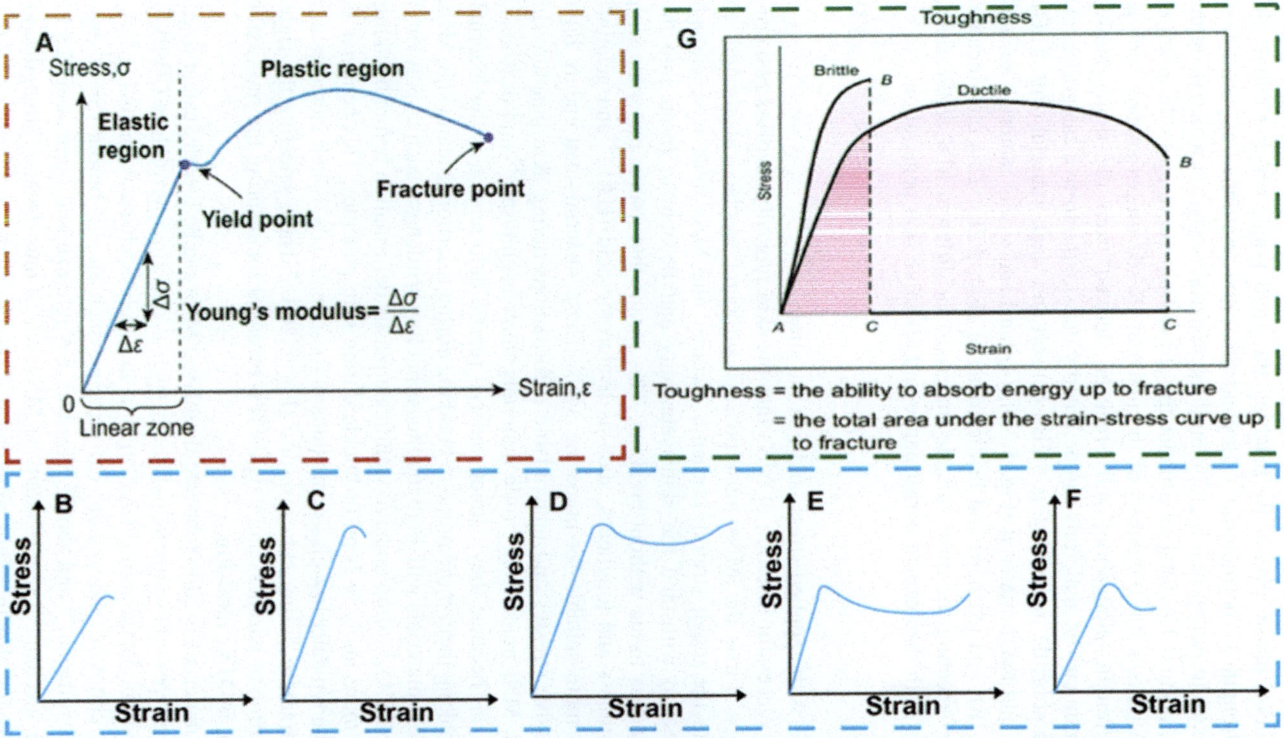

FIGURE 5.2 (A) Stressstrain curve of materials. Representative stressstrain profiles of materials with different mechanical properties: (B) Hard and crisp material; (C) hard and strong material; (D) hard and tough material; (E) soft and tough material; (F) soft and weak material; (G) Comparison of the toughness of brittle materials and ductile materials. *(A) Reprinted from Ref. 25 with permission. (G) Reprinted from Ref. 38 with permission.*

of the elastic region corresponds to the material's elastic modulus, which represents its hardness. As the material is further stressed, it enters a plastic deformation phase, eventually reaching the yield point, which signifies the yield strength of the material.[33] Beyond this point, continued application of stress leads to further deformation until the material breaks, at which point the fracture occurs. The total area under the stressstrain curve up to the point of fracture reflects the toughness of the material. Based on parameters such as yield strength, elongation, and breakage derived from the representative stressstrain curves, materials can be broadly classified into five categories: (1) strong and brittle material (Fig. 5.2B), (2) soft and weak material (Fig. 5.2C), (3) soft and tough material (Fig. 5.2D), (4) hard and strong material, such as (Fig. 5.2E), and (5) hard and brittle material (Fig. 5.2F). This classification helps in selecting the most suitable materials for different MN applications, ensuring both mechanical performance and effective skin penetration.

$$E = \frac{\sigma}{\varepsilon} \tag{5.5}$$

Where σ is the applied stress, and ε is the strain.

The successful insertion of MNs into the skin typically involves two critical steps. First, the MNs must possess sufficient mechanical strength to penetrate the tough stratum corneum. Second, the MNs must have enough toughness to withstand the skin's elasticity and be able to insert into the subcutaneous tissue without breaking. This means that matrix materials with both hardness and toughness, such as hyaluronic acid (HA),[34] and gelatin,[8] may be more suitable for preparing MNs. These materials can endure high compressive stresses and unavoidable deformations during skin puncture. In contrast, MNs made from brittle materials, such as chitosan,[35] and dextran,[36] tend to fail because they cannot undergo plastic deformation under external force, making them more prone to breaking during insertion. Meanwhile, MNs made from materials with high toughness but low brittleness, such as polyvinyl alcohol (PVA),[37] often undergo significant plastic deformation before breaking and are more likely to bend or buckle during skin puncture. This behavior is due to the typically low elastic modulus of these materials, which increases the risk of fracture during skin insertion (Fig. 5.2G).[38] Thus, selecting the right combination of hardness and toughness is essential for ensuring the successful insertion of MNs into skin.

The materials used to fabricate dissolving MNs can be broadly categorized into three types: (1) synthetic polymers, such as polyvinylpyrrolidone (PVP),[39] PVA, and polylactic acid (PLA)[40]; (2) natural polysaccharides, including HA, dextran, chitosan, and gelatin; (3) proteins, such as silk fibroin (SF),[41] hydrolyzed collagen,[42] and zein.[43] However, a single material often struggles to meet the mechanical requirements of MNs, which must balance strength and toughness for effective skin penetration. Growing evidence suggests that the mechanical properties of MNs can be significantly improved by modifying the physicochemical properties of the matrix materials, such as their degree of crosslinking, crystallinity, and molecular weight. The impact of these factors on MN performance will be discussed in detail in the following section.

3.1 Crosslinking degree

The mechanical properties of polymers, including Young's modulus and compression modulus, are generally positively correlated with their crosslinking density. Two predominant strategies for enhancing polymer crosslinking density are photocrosslinking and ionic crosslinking. Photocrosslinking is typically employed to fabricate covalently crosslinked hydrogels through UV/visible light irradiation of a solution containing photoinitiators and methacrylated polymers (Fig. 5.3A).[44] This method offers distinct advantages: (1) rapid crosslinking completion within seconds under physiologically compatible conditions, and (2) excellent compatibility with delicate bioactive agents, including proteins, vaccines, nucleic acids, and exosomes, thereby preserving their therapeutic efficacy. Ionic crosslinking, alternatively, relies on electrostatic interactions between polymer chains and multivalent metal ions. However, the resultant crosslinking density demonstrates environmental sensitivity, being modulated by ionic strength, pH, and temperature. Recent advances in MN development have strategically utilized both photocrosslinking and ionic crosslinking approaches to optimize polymer crosslinking density, consequently improving the mechanical performance of MNs (Table 5.1).

Taken GelMA MNs as an example, photocrosslinked MNs are generally prepared through the following steps (Fig. 5.3B): (1) A specific amount of GelMA and Irgacure 2959, a photoinitiator, is added to a polydimethylsiloxane (PDMS) mold. The mixture is then centrifuged to form the MNs; (2) After centrifugation, the PDMS mold is exposed to UV light, initiating the photocrosslinking of GelMA; (3) The mold is then placed in a drying oven or vacuum dryer for further drying, after which the photocrosslinked MNs are removed by peeling off the mold. Zhu et al.[45] investigated the effect of GelMA concentration (10%, 15%, 20%, and 25%) and crosslinking time (10, 50, 100, 200, and 300 s) on the mechanical properties of GelMA MNs. The authors used a lowforce mechanical testing system to obtain the compressive stressdisplacement curves (Fig. 5.3C–D) of various GelMA MNs and calculated their compressive modulus. Compared to uncrosslinked MNs, the compressive modulus of GelMA MNs increased with higher GelMA concentration and longer crosslinking time (Fig. 5.3E–F). When the GelMA concentration ranged from 15% to 25% and the crosslinking time was set to 200 s, the GelMA MNs achieved sufficient mechanical strength for effective skin insertion. After applying these MNs to rat skin for 30 s, the micropores created by the MNs were clearly visible in the trypan bluestained skin tissue.

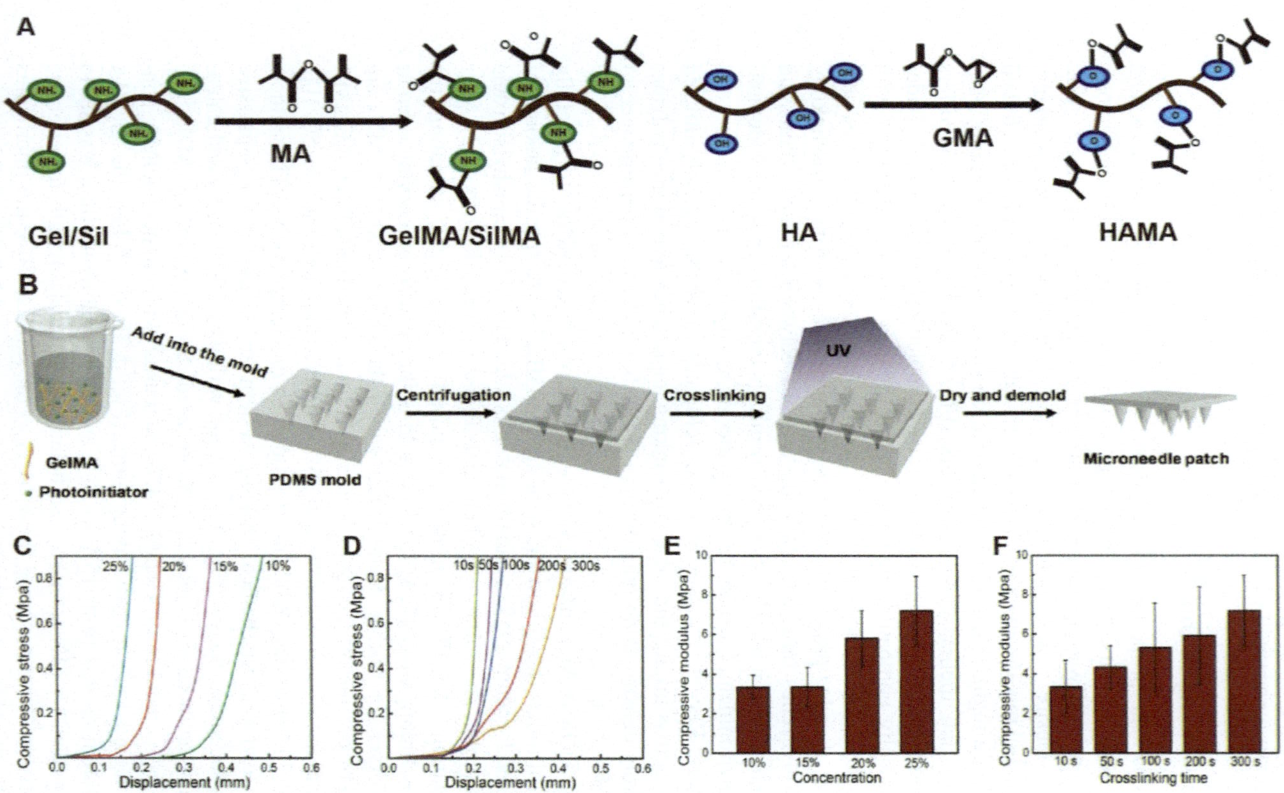

FIGURE 5.3 (A) Schematic diagram of preparing gelatin methacryloyl (GelMA), silk fibroin methacryloyl (SFMA), and hyaluronic acid methacryloyl (HAMA) via amidation or esterification reactions. (B) Schematic illustrating the preparation of photocrosslinked GelMA MNs. The compressive stressdisplacement curves of GelMA MNs prepared with (C) different GelMA concentrations and (D) different crosslinking times. (E) Influence of GelMA concentration on the compressive modulus of GelMA MNs ($n = 3$). (F) Influence of crosslinking time on the compressive modulus of GelMA MNs ($n = 3$). *Reprinted from Ref. 45 with permission.*

TABLE 5.1 Comparison of mechanical properties of crosslinked MNs and uncrosslinked MNs.

Polymer	Crosslinker	Crosslinking time	Key results	References
GelMA	Irgacure 2959	10, 50, 100, 200, and 300 seconds	With the crosslinking time increased from 10 to 300 s, the compressive modulus of GelMA MNs increased from 3.34 ± 1.33 MPa to 7.23 ± 1.66 MPa.	45
HAMA	N, N′-methyl enebisacrylamide	A short period	After crosslinking, the fracture force of HAMA MNs was increased from 0.003 N per needle to 0.06 N per needle.	46
HAMA	Irgacure 1173	0.5, 1, 2, 3, and 5 minutes	The compressive strength of HAMA MNs crosslinked for 0.5, 1, 2, 3, and 5 min was about 23.85 ± 3.0, 30.67 ± 4.2, 45.19 ± 4.0, 65.68 ± 4.4, and 74.28 ± 2.8 MPa, respectively.	50
SF-MA	Lithium phenyl-2,4,6-trimethylbenzoyl phosphonate	30 seconds	With the SFMA concentration increased from 10% to 20%, the mechanical strength of per needle increased from 0.068 ± 0.013 N to 0.090 ± 0.018 N.	51
Alginate	Calcium ions	/	The fracture force of the crosslinked alginate MNs was 1.93 times that of alginate MNs.	49

In a study conducted by Yu et al.,[46] methacrylated hyaluronic acid (HAMA) and N, N′-methylene bisacrylamide (a photoinitiator) were used to prepare photo-crosslinked HAMA MNs under UV/Vis irradiation for loading glucose-responsive vesicles. The force–displacement curve of the MNs showed that the fracture force of the crosslinked HAMA MNs was three times greater than that of the uncrosslinked HAMA MNs. Similarly, Rajput et al.[47] synthesized methacrylated silk fibroin (SFMA) by using glycidyl methacrylate (Fig. 5.3A). This was further combined with a photo-initiator (lithium phenyl (2,4,6-trimethylbenzoyl) phosphonate) to prepare photo-crosslinked SFMA hydrogels. The mechanical properties of SFMA hydrogels with different concentrations of SFMA (10%, 15%, and 25%) were analyzed using a universal testing machine. As the SFMA concentration increased from 10% to 25%, the compressive stress of the SFMA hydrogel increased by more than 100 kPa due to the enhanced crosslink density of the SF.

Ion crosslinking is another widely used method for preparing crosslinked hydrogels. A typical example is the ion crosslinking of alginate with calcium ions.[48] In a study by Chen et al.,[49] alginate MNs were prepared through ion crosslinking with dual-mineralized particles containing glucose oxidase (GOx) and exendin-4 (Ex4). First, GOx and Ex4 were integrated with copper phosphate and calcium phosphate to produce mineralized GOx (m-GOx) and mineralized Ex4 (m-Ex), respectively. Then, the dual-mineralized particles, along with alginate, were loaded into the MNs by centrifugation. The resulting crosslinked alginate MNs were formed through the ion crosslinking of alginate with the calcium ions present in the dual-mineralized particles. A tensile compression apparatus was used to record the force–displacement curves of the MNs. The fracture force of the crosslinked alginate MNs (0.29 N/needle) was 1.93 times greater than that of the uncrosslinked alginate MNs (0.15 N/needle). After applying the MNs to the dorsal skin of mice, trypan blue staining revealed the microchannels created by the MNs. The crosslinked alginate MNs demonstrated sufficient mechanical strength to penetrate the skin, as evidenced by the clear micropores observed in the tissue.

3.2 Crystallinity

The degree of crystallinity plays a key role in determining material strength and brittleness. Therefore, modulating material crystallinity could be an effective strategy to enhance the mechanical strength of MNs. In a study by Tiraton et al.,[52] MNs made from sodium alginate and gelatin were prepared for the delivery of the antibiotic clindamycin. The authors found that the addition of sodium alginate reduced the crystallinity of the matrix material, thereby improving the mechanical properties of the MNs. Specifically, the fracture force of the 25% sodium alginate/75% gelatin MNs was 1.241 ± 0.073 N, while that of the gelatin MNs was only 0.169 ± 0.013 N. This result demonstrates that tuning the crystallinity of the material can significantly enhance the strength of MNs.

The mechanical properties of SF can be enhanced by increasing the content of β-crystals. Two commonly used methods to promote the stacking of β-sheet crystallites in SF are water vapor and methanol vapor annealing treatments. For instance, Lu et al.[53] prepared triptorelin acetate-loaded SF MNs by exposing the female mold of MNs to methanol vapor for varying durations (0, 8, 24, 48, 72, and 96 h). The protein structure of SF in the MNs was analyzed using Fourier transform infrared spectroscopy (FTIR). Compared to untreated SF MNs, a shift toward the β-sheet region was observed in the infrared spectra of SF MNs treated with methanol vapor (Fig. 5.4A). The β-sheet content increased as both the SF concentration and methanol vapor treatment time

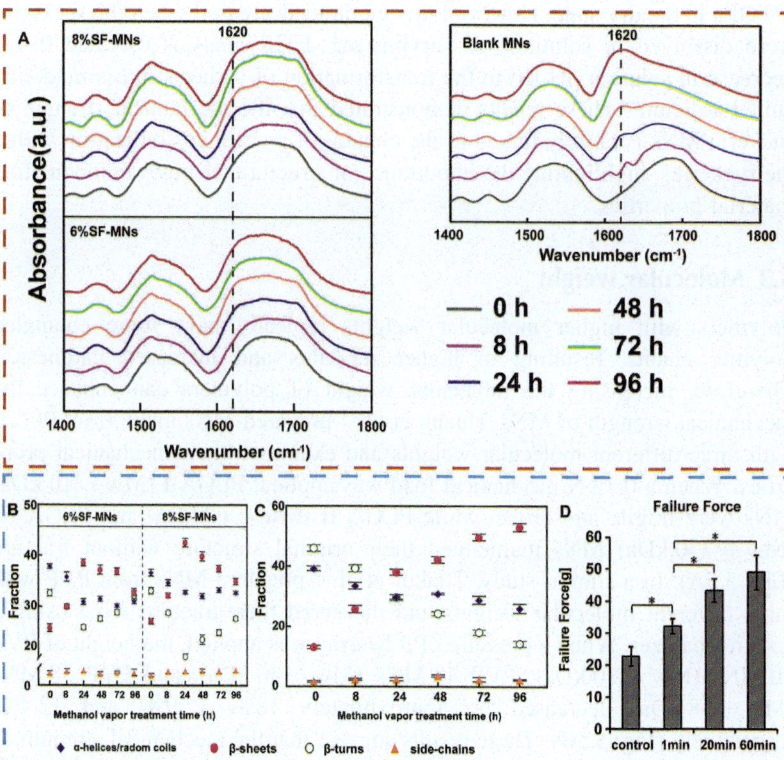

FIGURE 5.4 (A) Fourier transform infrared (FTIR) spectrum of silk fibroin microneedles (SF MNs) and blank MNs. Fraction of the secondarystructure content of (B) SF MNs and (C) blank MNs after treatment with methanol vapor for different durations. (D) Fracture force of SF MNs treated with methanol vapor for 0, 1, 20, and 60 minutes ($n = 9$, $*P < .05$). *(C) Reprinted from Ref. 53 with permission. (D) Reprinted from Ref. 54 with permission.*

were extended (Fig. 5.4B–C). A texture analyzer was used to measure the mechanical strength of the SF MNs. The results showed that SF MNs treated with methanol vapor for 24 h (with a fracture force of over 300 mN) had significantly higher fracture strength compared to untreated MNs (which had a fracture force of less than 100 mN). A similar outcome was reported by Lee et al.,[54] who used a precision balance to measure the fracture force of MNs under a pressure of 1.8 μm/s. After exposing SF to methanol vapor for 1 min, the average fracture force of SF MNs increased from 23 g/needle to 32 g/needle (Fig. 5.4D). These findings demonstrate that methanol vapor treatment significantly improves the mechanical properties of SF MNs.

Sucker ring teeth proteins (suckerins) can selfassemble into supramolecular networks with β-sheet structures, which can be used to modulate the mechanical strength of materials. In a study by Ding et al.,[55] suckerins were used to prepare MNs, and the Young's modulus of the suckerins was found to be 8.9 GPa in the dry state. However, this modulus decreased when the suckerins were dissolved in solutions of varying pH. FTIR analysis revealed that a decrease in solution pH led to the transformation of β-sheets into a randomly coiled structure. These results demonstrated that the mechanical strength of suckerin MNs is closely linked to the content of β-sheet crystallization within the suckerins, highlighting the importance of structural changes in modulating material properties.

3.3 Molecular weight

Polymers with higher molecular weights typically have more entangled polymer chains, resulting in higher modulus and increased hardness.[56] Therefore, increasing the molecular weight of polymers can enhance the mechanical strength of MNs. Huang et al.[57] prepared MNs made from PLGA with three different molecular weights and examined their mechanical properties. When a 0.12 N mechanical load was applied, PLGA I (Mw = 10 kDa) MNs were fragile and broke, while PLGA II (Mw = 60 kDa) and PLGA III (Mw = 330 kDa) MNs maintained their original structure without fracture (Fig. 5.5A). In a similar study, Thakur et al.[58] prepared MNs from PVP with three different molecular weights and measured their fracture force using a texture analyzer. When a pressure of 3 N/array was applied, the height of PVP 10 MNs (Mw = 10 kDa), PVP 40 MNs (Mw = 40 kDa), and PVP 58 MNs (Mw = 58 kDa) decreased by approximately 18%, 12.9%, and 12.4%, respectively (Fig. 5.5B). These results suggest that the mechanical strength of PVP MNs increases with the molecular weight of PVP. Lee et al.[59] used mixtures of PVP 360 (Mw = 360 kDa) and PVP 10 at weight ratios of 1:1, 1:2, 1:3, and 1:5 to prepare the first layer of MNs (needle tips), while the second layer was made with 30% PVP 360. The authors employed optical coherence tomography (OCT) to monitor the penetration depth of the bilayer MNs in realtime. MNs containing a higher percentage of PVP 360 were found to

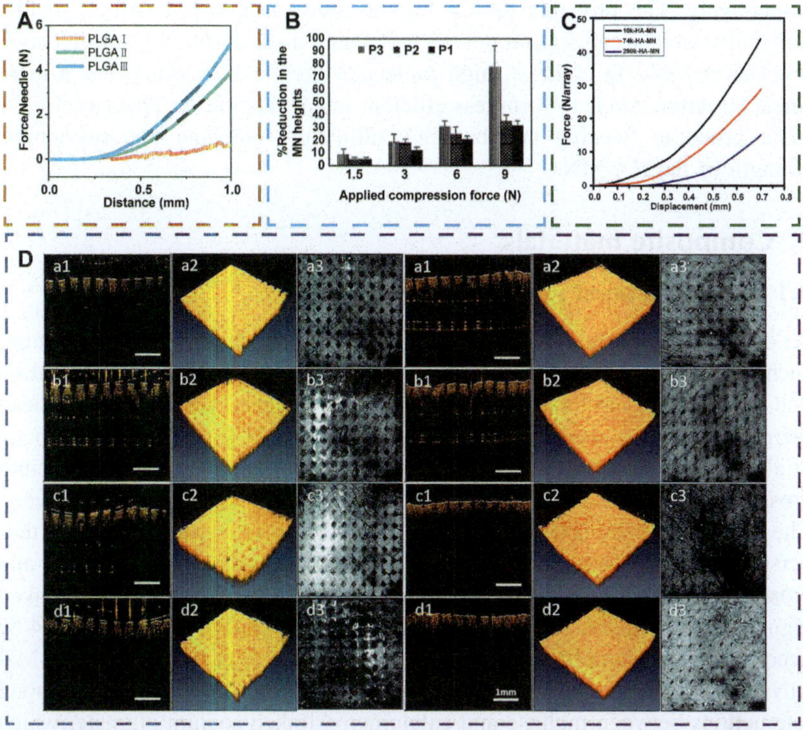

FIGURE 5.5 (A) Force–distance curves of poly(lactic-co-glycolic acid) microneedles (PLGA MNs): PLGA I (Mw = 10 kDa), PLGA II (Mw = 60 kDa), and PLGA III (Mw = 330 kDa). (B) Percentage reduction in height of different poly(vinyl pyrrolidone) (PVP) MNs under the same compression force. The MNs were made from PVP hydrogel, that is, PVP 58 (P1), PVP 40 (P2), and PVP 10 (P3). (C) Optical coherence tomography (OCT) images illustrating the puncture depth of MNs in the porcine skin: (a) PVP 360/PVP 10 = 1:1, (b) PVP 360/PVP 10 = 1:2, (c) PVP 360/ PVP 10 = 1:3, (d) PVP 360/PVP 10 = 1:5. (D) Force–displacement curves of MNs prepared from different molecular weights of hyaluronic acid (HA). *(A) Reprinted from Ref. 57 with permission. (B) Reprinted from Ref. 58 with permission. (C) Reprinted from Ref. 59 with permission. (D) Reprinted from Ref. 60 with permission.*

penetrate deeper into the skin layers (Fig. 5.5C), further confirming that increasing the molecular weight of PVP enhances the mechanical strength of MNs.

It is important to note that in some cases, the hardness of polymers may decrease with increasing molecular weight. For instance, Chi et al.[60] used three types of HA with molecular weights of 10 kDa, 74 kDa, and 290 kDa to prepare HAbased MNs and investigated how the molecular weight of HA affected the mechanical properties of the MNs. The force–displacement curves (Fig. 5.5D) showed that the compression force of 10k-HA-MNs, 74k-HA-MNs, and 290k-HA-MNs was 21.6 N, 13.9 N, and 5.9 N per 100 needles, respectively. The probable reason for this trend is that HA with a smaller molecular weight can form a denser molecular stack during the solidification process of MNs. In contrast, high molecular weight HA, which has a more linear structure, tends to form less efficient molecular stacks. This structure is more prone to twisting and bending, ultimately reducing the mechanical strength of the HA-MNs.

4. Composite materials

4.1 Hydrogen bond interactions

As displayed in Table 5.2, polysaccharides with abundant hydrophilic groups, such as hydroxyl and carboxyl groups, are commonly used in MN preparation due to their ability to form hydrogen bonds with synthetic or natural polymers, thereby enhancing the mechanical properties of MNs. For instance, Zhang et al.[61] used a template method to prepare composite MNs made of calcium crosslinked alginate and maltose (Ca^{2+}/Alg-Mal MNs) for insulin delivery. The mechanical strength of Ca^{2+}/Alg-Mal MNs was measured using a universal testing machine and compared to that of pure alginate MNs and crosslinked alginate MNs. The fracture force of alginate MNs, crosslinked alginate MNs, and Ca^{2+}/Alg-Mal MNs was about 0.02 N/needle, 0.04 N/needle, and 0.41 N/needle, respectively. This indicates that the Ca^{2+}/Alg-Mal MNs exhibit the highest mechanical strength, likely due to hydrogen bond interactions between maltose and calciumcrosslinked sodium alginate, which result in a more tightly packed composite material. Hou et al.[62] reported that the mechanical strength of HA MNs can be significantly improved by incorporating a small amount of sucrose, as HA and sucrose can form hydrogen bond interactions. Similarly, Sullivan et al.[63] synthesized a poly (vinylpyrrolidone-co-methacrylic acid) (PVP-MAA) copolymer with varying molar ratios of vinyl pyrrolidone (VP) and methacrylic acid (MAA) (VP/MAA = 100/0, 99/1, 90/10, 75/25, 50/50, and 25/75). Compared to PVP MNs, the fracture force of PVP-MAA MNs increased twofold and fourfold at 1% and 25% MAA, respectively. These results suggest that hydrogen bonding interactions between VP and MAA monomers contribute to the improved mechanical strength of PVP-MAA MNs.

The fracture of polymeric molecular chains can occur through three primary modes: disruption of chemical bonds, disruption of van der Waals forces or hydrogen bonds, and intermolecular slippage (Fig. 5.6). To effectively increase the mechanical strength of materials, it is crucial to enhance intermolecular linkages, making the molecular chains less prone to slippage. For example, Chen et al.[64] modified PVP by forming a PVP-cyclodextrin (CD) inclusion complex. The hydroxyl groups in the cavity of CD can form hydrogen bonds with the side chains of PVP, thereby restricting the movement of the PVP chains. Compared to pure PVP, the PVP-CD complex showed a

TABLE 5.2 Summary of the reinforcements used to form hydrogen-bonding interactions with polymers to improve the mechanical properties of microneedles.

Polymers used for microneedle fabrication	Reinforcements	Interaction type between polymer and reinforcement	Key results	References
Calcium-crosslinked alginate	Maltose	Hydrogen bond	With the addition of maltose, the fracture force of calcium-crosslinked sodium alginate MNs increased from 0.04 to 0.41 N.	61
Hyaluronic acid (HA)	Sucrose	Hydrogen bond	The fracture force of MNs is 338 N, much larger than the 0.1 N required for insertion	62
Poly(vinyl pyrrolidone) (PVP)	Methacrylic acid (MAA)	Hydrogen bond	The fracture force of PVP-MAA MNs containing 25 wt% MAA was almost four times higher than that of PVP MNs.	63
Hyaluronic acid (HA)	Hydroxypropyl-β-cyclodextrin (HP-β-CD)	Hydrogen bond	The addition of HP-β-CD increased the fracture force of HA MNs in a dose-dependent manner.	65

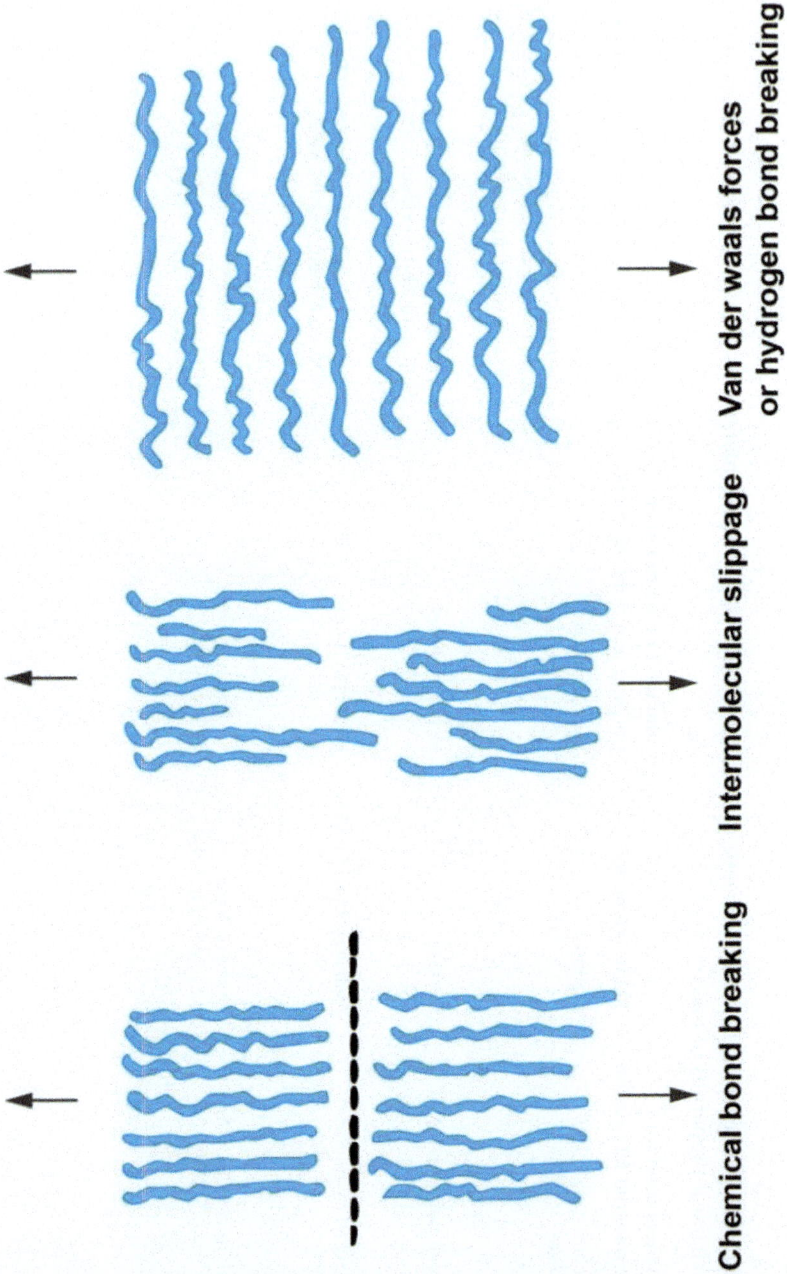

FIGURE 5.6 Schematic diagram of the microscopic mechanisms of polymer fracture. Typical fracture modes of polymers include chemical bond disruption, van der Waals force, hydrogen-bond disruption, and intermolecular slippage.

significant improvement in Young's modulus and hardness, particularly for low molecular weight PVP. A skin insertion ability test further demonstrated the effectiveness of this modification: MNs made of pure PVP were unable to puncture pig skin, while PVP-CD MNs successfully pierced the skin and created clear micropores. This result confirmed that hydrogen bond formation between the composite materials enhances the mechanical strength of MNs. Similarly, Lin et al.[65] prepared MNs by mixing HA with varying concentrations of hydroxypropyl-β-cyclodextrin (HP-β-CD) and measured their mechanical strength using a texture analyzer. They found that the fracture force of the MNs increased in a dose-dependent manner with the addition of HP-β-CD. When the concentration of HP-β-CD reached 9 mM, the mechanical strength of the MNs increased from 175 to 300 N. FTIR spectroscopy revealed the formation of hydrogen bond interactions between HA and HP-β-CD, which contributed to the enhanced mechanical strength of the MNs.

4.2 Introduction of nano-microparticles

The development of nanomicroparticle technology has opened up new possibilities for controllable and targeted drug delivery. Recently, the combination of MNs and nanomicroparticles has been explored as a strategy to overcome issues such as low drug loading and poor therapeutic efficiency often associated with conventional MNs.[66] However, the specific effects of incorporating nanomicroparticles into the MN matrix on the mechanical strength of MNs have not been thoroughly studied. In the study by Li et al. ([67] the mechanical strength of carboxymethylcellulose MNs loaded with layered double hydroxides (LDHs) nanoparticles was investigated. Using a nanoindentation test, they found that MNs loaded with LDH nanoparticles required a larger load to reach the same penetration depth, indicating that the presence of LDHs enhanced the indentation resistance of the MNs. Similarly, Zhao et al.[68] explored the incorporation of metal nanoparticles into MNs. They prepared HA MNs to load copperdoped graphene quantum dots (CuGQD) and palladium nanoparticles (PdNPs) by incorporating porous silicon (CuGQD/PdNPs@PSi). The mechanical strength of single HA MNs, initially $0.2853 \pm 0.02{,}366$ N/needle, increased to $0.9735 \pm 0.01{,}553$ N/needle when loaded with CuGQD/PdNPs@PSi. Zhang et al.[69] studied the effect of gold nanoclusters (AuNCs) on the mechanical strength of gelatin/starch MNs. Atomic force microscopy measurements showed that the Young's modulus of gelatin/starch MNs increased from 175 MPa to over 300 MPa after AuNCs were loaded. When applied to mouse skin, the AuNCsloaded MNs successfully penetrated the skin, creating visible microholes. Hematoxylin and eosin (H&E) staining confirmed that the insertion depth of the AuNCsloaded MNs was greater than that of the gelatin/starch MNs, further proving that the incorporation of AuNCs enhanced the mechanical strength of the MNs. Liu et al.[70] used a mixture of insulinloaded calcium carbonate (INS-CaCO3) microparticles and polyvinylpyrrolidone (PVP) to prepare MNs and measured their mechanical strength using a lowforce mechanical testing system. The fracture force of PVP MNs was only 0.26 N at a displacement of 600 μm, but this increased to 1.05 N when INS-CaCO3 microparticles were loaded. Similarly, Yu et al.[71] utilized gelatin and calcium sulfate hemihydrate ($CaSO_4$) nanoparticles to prepare MNs for insulin delivery. Resistance and failure strength tests showed that MNs loaded with $CaSO_4$ nanoparticles exhibited higher mechanical strength. Overall, these studies confirm that the incorporation of nanomicroparticles can significantly enhance the mechanical strength of MNs, making them more effective for skin penetration and drug delivery.

5. Optimization of geometric parameters

5.1 Cross-sectional shape

Based on Eq. (5.3), the mechanical properties of MNs are positively correlated with their crosssectional moment of inertia, which is closely linked to the crosssectional shape of the MNs. Different shapes affect the crosssectional area and the number of crosssectional vertices, which, in turn, influence the mechanical strength of the MNs. Numerous studies have explored the relationship between crosssectional shape and the mechanical strength of MNs (Fig. 5.7A). Aggarwal et al.[72] used finite element analysis (FEA) to theoretically investigate the impact of crosssectional shape on the mechanical strength of circular, rectangular, and square MNs. The mechanical strength was assessed by applying a buckling force of 11.135 N and a bending force of 2.1 N along the Y-axis. According to the FEA simulation results, the minimum stress obtained for circular MNs was 2.36 GPa, which is well below the breaking limit of silicon (7 GPa), suggesting that circular MNs can withstand forces greater than the applied bending force. In contrast, the maximum stress values for rectangular and square MNs were 6.47 and 7.11 GPa, respectively. These findings indicate that circular MNs are the most capable of withstanding external bending forces before reaching their fracture limits. Further studies have quantitatively examined how the shape of the crosssection affects the mechanical strength of MNs, considering factors such as the number of vertices, the area of the crosssection, and the geometry from the base to the needle tip. The geometry and structure of the crosssection are key determinants in enhancing the mechanical performance of MNs, and ongoing research continues to refine our understanding of how these parameters contribute to the overall strength and stability of MNs during skin insertion.

Von Mises stress is a critical parameter used to predict the mechanical properties of MNs by assessing the principal stresses in the x, y, and z dimensions. MNs with higher Von Mises stress values typically exhibit a larger maximum penetration depth.[73] Loizidou et al.[74] utilized FEA to predict the

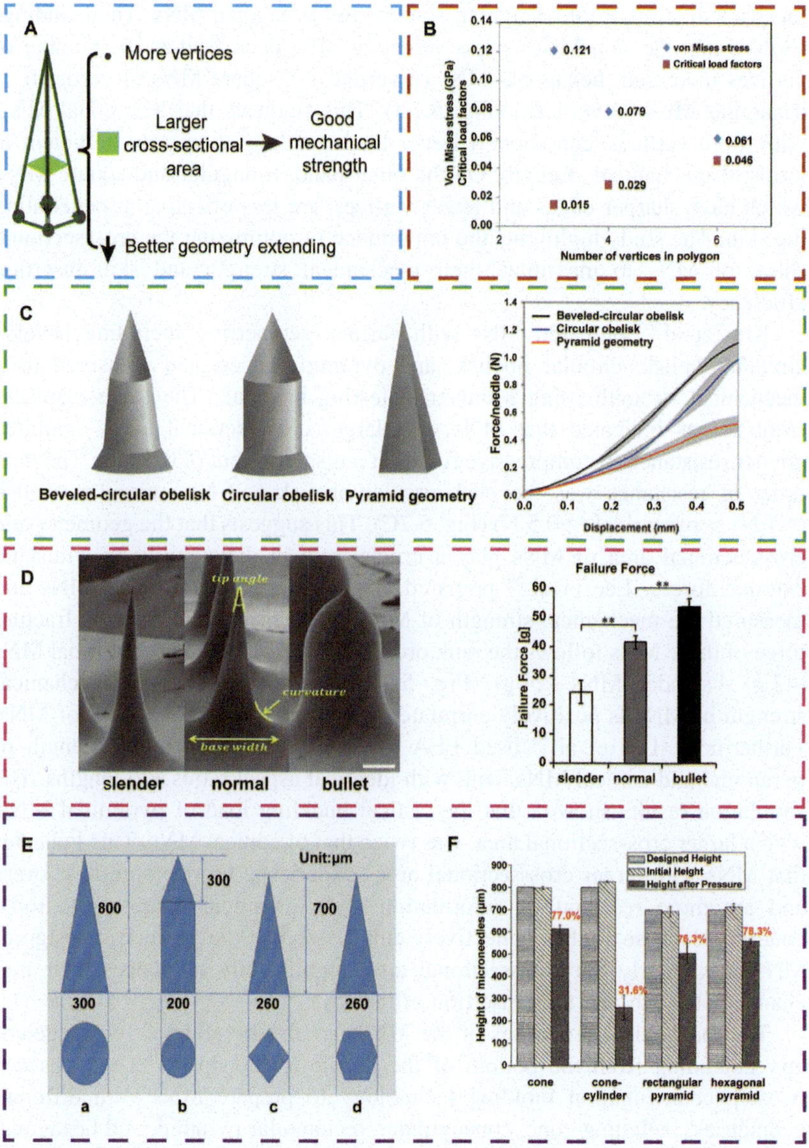

FIGURE 5.7 (A) Schematic illustrating the cross-sectional shape parameters of microneedles (MNs). (B) The von Mises stress distributions and the corresponding critical load factors were analyzed for MNs with triangular, square, and hexagonal base geometries. (C) The force–displacement curves of the beveled circular obelisk, circular obelisk, and pyramid-shaped MNs. (D) Scanning electron microscopy (SEM) images and failure force of slender, normal, and bullet MNs. (E) Front and top views of (a) conical, (b) cone–cylinder, (c) rectangular pyramid, and (d) hexagonal pyramid MNs. (F) The initial height, designed height, and remaining height after compression of different MNs. *(B) Reprinted from Ref. 74 with permission. (C) Reprinted from Ref. 75 with permission. (D) Reprinted from Ref. 54 with permission. (E) Reprinted from Ref. 77 with permission.*

mechanical strength of triangular, square, and hexagonal MNs. Their findings showed that the Von Mises stress values of MNs decreased as the number of vertices increased: hexagonal MNs (6 vertices) < square MNs (4 vertices) < triangular MNs (3 vertices) (Fig. 5.7B). This suggests that hexagonal MNs, with more vertices, can support larger loads, making them more efficient in terms of mechanical strength. On the other hand, triangular and square MNs, which have sharper edges and fewer vertices, are less effective at penetrating the skin. The study highlights the importance of optimizing the crosssectional shape of MNs to maximize their mechanical strength and skin insertion efficiency.

Kim et al.[75] prepared MNs with various geometries, including beveled circular obelisk, circular obelisk, and pyramid shapes, and measured their mechanical strength using a universal testing machine. The force–displacement curves indicated that MNs with larger crosssectional areas exhibited greater resistance to compressive forces at a displacement of 0.5 mm. The rank order of resistance was: beveled circular obelisk (>1 N) > circular obelisk (≈1 N) > pyramidal (≈0.5 N) (Fig. 5.7C). This suggests that the geometry and crosssectional area of MNs play a crucial role in their ability to withstand external forces. Lee et al.[54] prepared slender, normal, and bullet MNs and measured the mechanical strength of MNs by precision balance. The fracture force of these MNs follows the rank order of bullet MNs (54 g) > normal MNs (42 g) > slender MNs (24 g) (Fig. 5.7D), suggesting that the mechanical strength of MNs is positively correlated with the crosssectional area of MNs. Furthermore, Lee et al.[76] used FEA to predict the mechanical strength of pyramidal and conical MNs, both with identical aspect ratios and lengths. The simulation results showed that the critical buckling load of pyramidal MNs, with a larger crosssectional area, was twice that of conical MNs. This indicates that MNs with larger crosssectional areas experience less compressive stress and are more resistant to deformation under identical compressive force conditions. These studies collectively emphasize that the geometric design of MNs, particularly the crosssectional area, significantly influences their mechanical strength and skin insertion efficiency.

The mechanical properties of the MNs can also be affected by the geometry extending from the bottom of the needle to the tip. Li et al.[77] used a twostep centrifugation molding technology to prepare MNs with different geometries, including cone, conecylinder, rectangular pyramid, and hexagonal pyramid shapes (Fig. 5.7E). The mechanical strength of these MNs was evaluated by comparing the change in the needle height before and after compression. When a pressure of 100 N was applied to the MNs for 5 s, the remaining height of the MNs followed the rank order: hexagonal pyramid MNs (78.3%) > cone MNs (77.0%) > rectangular pyramid MNs (76.3%) > conecylinder MNs (31.6%) (Fig. 5.7F). The results indicate that the geometry extending from the base of the needle to the tip plays a significant role in the mechanical strength of MNs. Specifically, conecylinder MNs, which have the smallest diameter at the bottom and lack a welldefined shape from the base to the tip, were more prone to destruction under the same pressure. This highlights the importance of geometry in enhancing the mechanical robustness of MNs, as shapes that allow for a more gradual transition from the base to the tip offer better resistance to deformation and failure.

5.2 Tip angle

Tip angle is a crucial parameter influencing both the mechanical strength and skin puncture efficiency of MNs. Jung et al.[78] studied the effect of six different tip angles (Fig. 5.8A) on the insertion force of solid MNs by testing their ability to pierce chicken breast tissue. They found that MNs with a tip angle of less than 10° required a minimum force of 15 mN to puncture the tissue, whereas those with a tip angle greater than 30° required forces above 30 mN. This suggests that MNs with smaller tip angles are more effective at piercing due to their sharper tips, which concentrate the applied stress at the needle tip (Fig. 5.8B).[79] Further research by Davis et al.[80] explored the effect of tip angle on the mechanical strength of polyethylene terephthalate MNs with tip angles of 60°, 65°, and 70°. The force–displacement curves and FEA indicated that the fracture force of MNs increased with the wall angle. When the angle reached 90°, the tips became almost cylindrical, and the stress was distributed along the walls of the MNs (Fig. 5.8C). As the tip angle increases, the force required for insertion decreases, thus reducing the stress applied to the needle and the risk of fracture. However, while small tip angles can reduce the insertion force, they may increase the risk of tip fracture during skin penetration. Therefore, the tip angle of MNs should be optimized to achieve a balance between effective insertion and maintaining the mechanical strength and safety of the device.

5.3 Needle length

Researchers have investigated the impact of needle length on the mechanical strength of MNs through both experimental tests and theoretical simulations. For instance, Park et al.[81] prepared taperedcone MNs with three different needle lengths and measured their failure force using a force–displacement test station. The MNs were pressed against a hard metal surface at a rate of 1.1 mm/s to obtain stressstrain curves. The results showed that the fracture force of MNs decreased as the needle length increased: 3.79 N/needle for 700 μm, 2.8 N/needle for 1000 μm, and 1.68 N/needle for 1500 μm (Fig. 5.9A). This suggests that MNs with shorter needle lengths exhibit higher mechanical strength. In a similar study, Ando et al.[26] compared the fracture force of PVA MNs with needle lengths of 570, 851, and 1133 μm. The results revealed that MNs with a length of 570 μm showed progressive deformation without catastrophic buckling, indicating no needle fracture (Fig. 5.9B). In

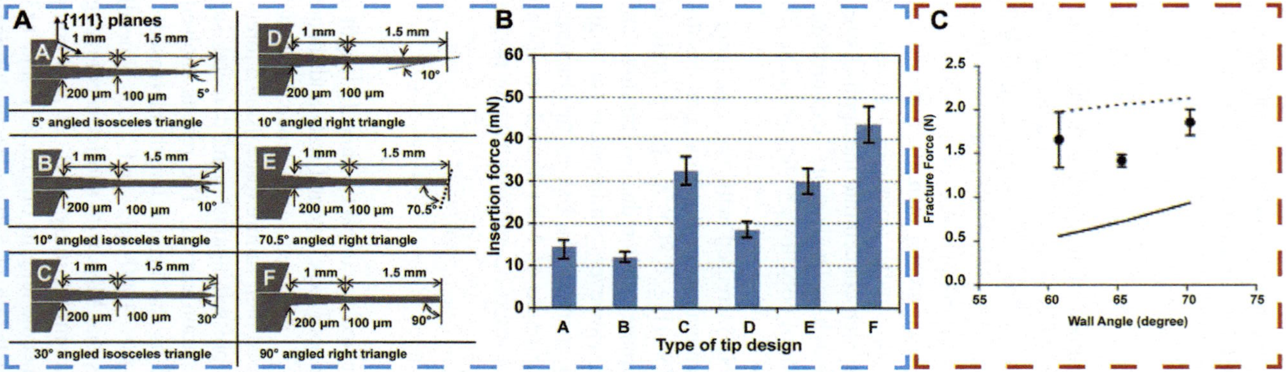

FIGURE 5.8 (A) Schematic illustration of microneedles (MNs) with different isosceles triangles (A:5°, B:10°, C:30°) and right triangles (D:10°, E:70.5°, F:90°), and (B) the insertion force required for these microneedles. (C) The influence of wall angle on the fracture force of MNs. *(B) Reprinted from Ref. 78 with permission. (C) Reprinted from Ref. 80 with permission.*

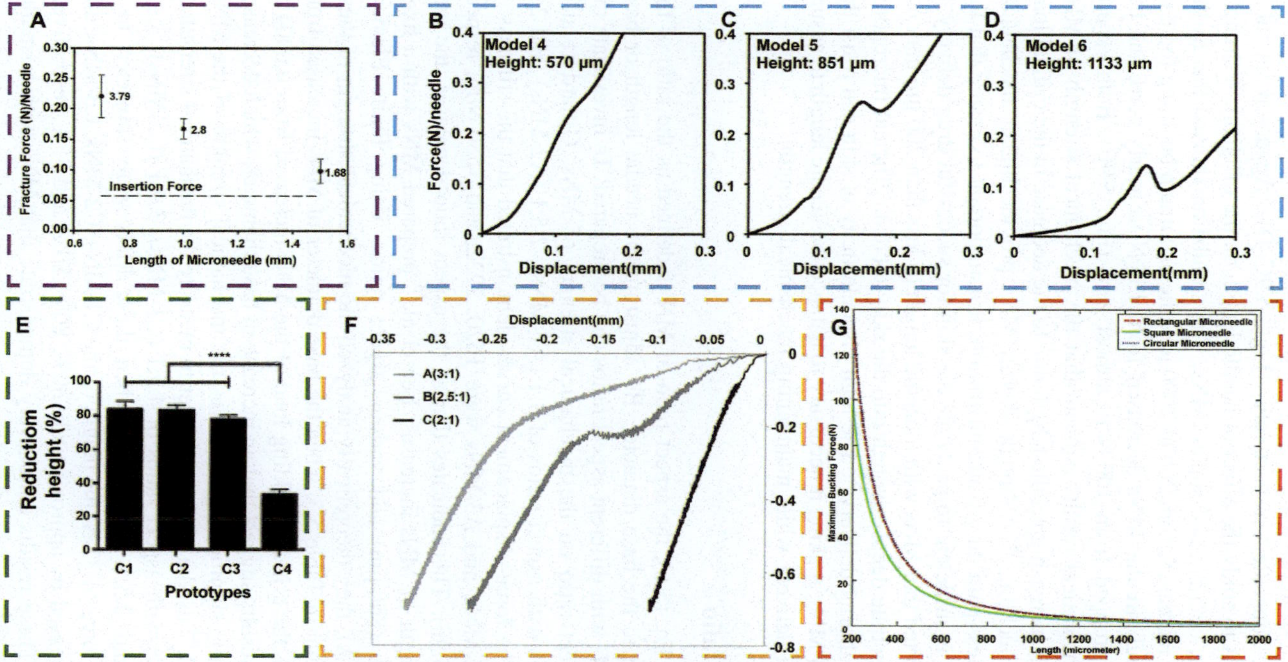

FIGURE 5.9 (A) Mechanical strength of poly(lactic-co-glycolic acid) (PLGA) microneedles (MNs) with different needle lengths. (B) Representative force–displacement curves of poly(vinyl alcohol) (PVA) MNs with lengths of (B) 570 μm, (C) 851 μm, (D) 1133 μm. (E) Variation of maximum buckling force with length of MNs. (F) Force–displacement curves and optical images of e-Shell 200 acrylatebased solid MNs with various aspect ratios after compression. (G) Mechanical properties of hydrogel MNs. *(A) Reprinted from Ref. 81 with permission. (D) Reprinted from Ref. 26 with permission. (E) Reprinted from Ref. 84 with permission. (F) Reprinted from Ref. 83 with permission. (G) Reprinted from Ref. 72 with permission.*

contrast, MNs with lengths of 851 and 1133 μm exhibited buckling deformation at compression forces of 0.253 ± 0.011 N/needle and 0.135 ± 0.015 N/needle, respectively (Fig. 5.9C–D). Aggarwal et al.[72] employed FEA to study the effect of needle length on the mechanical strength of rectangular, circular, and square MNs. The simulation showed that as the needle length increased from 200 to 2000 μm, the maximum buckling force of rectangular and circular MNs decreased from 135 N to nearly zero, and the square MNs dropped from 100 N to almost zero (Fig. 5.9E). Both experimental and theoretical results indicate that the mechanical strength of MNs decreases with increasing needle length due to the decrease in the critical load required for buckling.[28] Although shorter MNs provide higher mechanical strength and less pain during administration, it is important to ensure that the needle length is sufficient to penetrate the skin for successful drug delivery. It has been reported that MNs with a needle length of less than 200 μm may not be able to penetrate deeply enough into the skin to create efficient micropores for intradermal drug delivery.[82] The high viscoelasticity of skin can resist the penetration of MNs, making it necessary to balance the needle length to ensure effective drug delivery while maintaining mechanical strength.

5.4 Aspect ratio

As shown in Eq. (5.6), the aspect ratio of MN is defined as the ratio of the needle length to the base diameter. By adjusting the needle length or base diameter, MNs with different aspect ratios can be prepared. To investigate the effect of aspect ratio on the mechanical strength of MNs, Gittard et al.[83] prepared MNs with aspect ratios of 3:1 (A1-MNs), 2.5:1 (A2-MNs), and 2:1 (A3-MNs). The force–displacement curves demonstrated that the stiffness of A1-MNs, A2-MNs, and A3-MNs were 1620, 2222, and 7580 N/m, respectively (Fig. 5.9F). This indicates that as the aspect ratio decreases, the mechanical strength of MNs increases. Lee et al.[76] prepared MNs with a fixed needle length (800 μm) and various aspect ratios. Using Finite Element Analysis (FEA), they theoretically analyzed the mechanical strength of these MNs. As shown in Table 5.3, when the base diameter of MNs increased from 50 to 800 μm, the critical buckling load of the MNs increased from 0.0004 to 27 N. These results suggest that an increase in base diameter, which decreases the aspect ratio, significantly improves the mechanical strength of the MNs. Martinez et al.[84] used a texture analyzer to measure the mechanical strength of hydrogel MNs with aspect ratios of 4.5:1 (C1-MNs), 3.9:1 (C2-MNs), 3.1:1 (C3-MNs), and 2.5:1 (C4-MNs). When these MNs were subjected to a compression force of 32 N, the height of the C1-MNs, C2-MNs, C3-MNs, and C4-MNs decreased by approximately 80%, 80%, 80%, and 35%, respectively (Fig. 5.9G). These results collectively demonstrate that MNs with a smaller

TABLE 5.3 Summary of the effect of aspect ratio on the mechanical strength of MNs.

MNs	Needle length (μm)	Base diameter (μm)	Aspect ratio	Key results	References
A1	750	250	3:1	The average stiffness of A1, A2, and A3 MNs was: 1620, 2222, and 7580 N/m, respectively.	83
A2	750	300	2.5:1		
A3	500	250	2:1		
B1	800	50	16:1	The predicted critical buckling load of B1–B8 MNs was 0.0004, 0.0019, 0.0061, 0.10, 0.5, 1.7, 8.6, 27 N, respectively.	76
B2	800	75	32:3		
B3	800	100	8:1		
B4	800	200	4:1		
B5	800	300	8:3		
B6	800	400	2:1		
B7	800	600	4:3		
B8	800	800	1:1		
C1	2110	464	4.5:1	After MNs were compressed by the pressure probe, the reduction rate in the needle height of C1, C2, C3, and C4 MNs was approximately 80%, 80%, 80%, and 35%, respectively.	84
C2	2020	516	3.9:1		
C3	2090	681	3.1:1		
C4	2260	890	2.5:1		

aspect ratio have higher mechanical strength and are more resistant to high loads.

$$\text{Aspect ratio} = \frac{\text{Needle length}}{\text{Base diameter}} \tag{5.6}$$

In summary, both experimental and theoretical results highlight that a smaller aspect ratio enhances the mechanical strength of MNs, making them more resistant to deformation under applied forces. This is an important consideration when designing MNs for efficient drug delivery and skin penetration.

6. Conclusion

As a revolutionary transdermal delivery system, MNs have demonstrated broad applications in the biomedical field. Sufficient mechanical strength is a prerequisite for MNs to effectively penetrate the skin and achieve efficient intradermal drug delivery. This section provides an overview of the key factors influencing the mechanical strength of MNs, including their formulation compositions and geometric parameters. Generally, the mechanical strength of MNs can be enhanced through the following approaches: (1) For singlepolymer systems, mechanical strength can be improved through physical or chemical crosslinking of polymers, tuning the crystallinity of MNs, and increasing the molecular weight of polymers (with a few exceptions, such as HA). (2) Incorporating reinforcing agents or nano/microparticles into MNs is also effective, particularly for MNs made from brittle polymers, as these reinforcements can form hydrogen bonds with the polymers. (3) Optimizing the geometric parameters of MNs, such as increasing the number of vertices in the crosssection, tip angle, and crosssectional area, while reducing the length and aspect ratio, is another effective strategy to enhance mechanical strength. We believe this section offers valuable insights to guide the design of MNs with high mechanical strength.

References

1. Henry, S.; McAllister, D. V.; Allen, M. G.; Prausnitz, M. R. Microfabricated Microneedles: A Novel Approach to Transdermal Drug Delivery. *J. Pharm. Sci.* **1998,** *87* (8), 922–925.
2. Huang, Y.; Peng, T.; Hu, W.; Gao, X.; Chen, Y.; Zhang, Q.; Wu, C.; Pan, X. Fully Armed Photodynamic Therapy with Spear and Shear for Topical Deep Hypertrophic Scar Treatment. *J. Contr. Release* **2022,** *343,* 408–419.
3. Peng, T.; Huang, Y.; Feng, X.; Zhu, C.; Yin, S.; Wang, X.; Bai, X.; Pan, X.; Wu, C. TPGS/Hyaluronic Acid Dual-Functionalized PLGA Nanoparticles Delivered Through Dissolving Microneedles for Markedly Improved Chemo-photothermal Combined Therapy of Superficial Tumor. *Acta Pharm. Sin. B* **2021,** *11* (10), 3297–3309.
4. Hu, W.; Peng, T.; Huang, Y.; Ren, T.; Chen, H.; Chen, Y.; Feng, D.; Wu, C.; Pan, X. Hyaluronidase-Powered Microneedles for Significantly Enhanced Transdermal Delivery Efficiency. *J. Contr. Release* **2023,** *353,* 380–390.

5. Yang, G.; Chen, Q.; Wen, D.; Chen, Z.; Wang, J.; Chen, G.; Wang, Z.; Zhang, X.; Zhang, Y.; Hu, Q.; et al. A Therapeutic Microneedle Patch Made from Hair-Derived Keratin for Promoting Hair Regrowth. *ACS Nano* **2019,** *13* (4), 4354–4360.
6. Yang, D.; Chen, M.; Sun, Y.; Shi, C.; Wang, W.; Zhao, W.; Wen, T.; Liu, T.; Fu, J.; Lu, C.; et al. Microneedle-Assisted Vaccination Combined with Autophagy Regulation for Antitumor Immunotherapy. *J. Contr. Release* **2023,** *357,* 641–654.
7. Norman, J. J.; Arya, J. M.; McClain, M. A.; Frew, P. M.; Meltzer, M. I.; Prausnitz, M. R. Microneedle Patches: Usability and Acceptability for Self-Vaccination Against Influenza. *Vaccine* **2014,** *32* (16), 1856–1862.
8. Yuan, J.; Yang, H.; Liu, C.; Shao, L.; Zhang, H.; Lu, K.; Wang, J.; Wang, Y.; Yu, Q.; Zhang, Y.; et al. Microneedle Patch Loaded with Exosomes Containing MicroRNA-29b Prevents Cardiac Fibrosis After Myocardial Infarction. *Adv. Healthcare Mater.* **2023,** *12* (13), 2202959.
9. Meng, S.; Wei, Q.; Chen, S.; Liu, X.; Cui, S.; Huang, Q.; Chu, Z.; Ma, K.; Zhang, W.; Hu, W.; et al. MiR-141-3p-Functionalized Exosomes Loaded in Dissolvable Microneedle Arrays for Hypertrophic Scar Treatment. *Small* **2024,** *20* (8), 2305374.
10. Henry, S.; McAllister, D. V.; Allen, M. G.; Prausnitz, M. R. Microfabricated Microneedles: A Novel Approach to Transdermal Drug Delivery. *J. Pharmaceut. Sci.* **1998,** *87* (8), 922–925.
11. Zhang, X.; Wang, Y.; Chi, J.; Zhao, Y. *Smart Microneedles for Therapy and Diagnosis;* Research, 2020.
12. Zheng, M.; Hu, T.; Yang, Y.; Qie, X.; Yang, H.; Zhang, Y.; Zhang, Q.; Yong, K.-T.; Liu, W.; Xu, C. In Situ-formed Cryomicroneedles for Intradermal Cell Delivery. *NPG Asia Mater.* **2024,** *16* (1), 11.
13. Bvjc, S.; Mannayee, G. Structural Analysis and Simulation of Solid Microneedle Array for Vaccine Delivery Applications. *Materials Today. Proceedings* **2022,** *65,* 3774–3779.
14. Wei-Ze, L.; Mei-Rong, H.; Jian-Ping, Z.; Yong-Qiang, Z.; Bao-Hua, H.; Ting, L.; Yong, Z. Super-Short Solid Silicon Microneedles for Transdermal Drug Delivery Applications. *Int. J. Pharm.* **2010,** *389* (1), 122–129.
15. Shakya, A. K.; Lee, C. H.; Gill, H. S. Coated Microneedle-Based Cutaneous Immunotherapy Prevents Der p 1–Induced Airway Allergy in Mice. *J. Allergy Clin. Immunol.* **2018,** *142* (6), 2007–2011.e2003.
16. Chen, J.; Qiu, Y.; Zhang, S.; Yang, G.; Gao, Y. Controllable Coating of Microneedles for Transdermal Drug Delivery. *Drug Dev. Ind. Pharm.* **2015,** *41* (3), 415–422.
17. Chen, Y.-H.; Wang, F.-Y.; Chan, Y.-S.; Huang, C.; Huang, Y.-Y. Biofabricating Hollow Microneedle Array with Controllable Microstructure for Cell Transplantation. *J. Biomed. Mater. Res. B Appl. Biomater.* **2022,** *110* (9), 1997–2005.
18. Gade, S. S.; Pentlavalli, S.; Mishra, D.; Vora, L. K.; Waite, D.; Alvarez-Lorenzo, C. I.; Herrero Vanrell, M. R.; Laverty, G.; Larraneta, E.; Donnelly, R. F.; et al. Injectable Depot Forming Thermoresponsive Hydrogel for Sustained Intrascleral Delivery of Sunitinib Using Hollow Microneedles. *J. Ocul. Pharmacol. Therapeut.* **2022,** *38* (6), 433–448.
19. Yin, M.; Zeng, Y.; Liu, H.-Q.; Zhang, W.; Wang, C.; Chen, C.; Li, W. Dissolving Microneedle Patch Integrated with Microspheres for Long-Acting Hair Regrowth Therapy. *ACS Appl. Mater. Interfaces* **2023,** *15* (14), 17532–17542.
20. Kim, S.; Dangol, M.; Kang, G.; Lahiji, S. F.; Yang, H.; Jang, M.; Ma, Y.; Li, C.; Lee, S. G.; Kim, C. H.; et al. Enhanced Transdermal Delivery by Combined Application of Dissolving Microneedle Patch on Serum-Treated Skin. *Mol. Pharm.* **2017,** *14* (6), 2024–2031.

21. Chakraborty, A.; Alexander, S.; Luo, W.; Al-Salam, N.; Van Oirschot, M.; Ranganath, S. H.; Chakrabarti, S.; Paul, A. Engineering Multifunctional Adhesive Hydrogel Patches for Biomedical Applications. *Interdiscip. Med.* **2023,** *1* (4), e20230008.
22. Li, J. Y.; Feng, Y. H.; He, Y. T.; Hu, L. F.; Liang, L.; Zhao, Z. Q.; Chen, B. Z.; Guo, X. D. Thermosensitive Hydrogel Microneedles for Controlled Transdermal Drug Delivery. *Acta Biomater.* **2022,** *153,* 308–319.
23. Zhang, Y.; Xu, Y.; Kong, H.; Zhang, J.; Chan, H. F.; Wang, J.; Shao, D.; Tao, Y.; Li, M. Microneedle System for Tissue Engineering and Regenerative Medicine. *Exploration* **2023,** *3* (1), 20210170.
24. Chang, H.; Chew, S. W. T.; Zheng, M.; Lio, D. C. S.; Wiraja, C.; Mei, Y.; Ning, X.; Cui, M.; Than, A.; Shi, P.; et al. Cryomicroneedles for Transdermal Cell Delivery. *Nat. Biomed. Eng.* **2021,** *5* (9), 1008–1018.
25. Makvandi, P.; Kirkby, M.; Hutton, A. R. J.; Shabani, M.; Yiu, C. K. Y.; Baghbantaraghdari, Z.; Jamaledin, R.; Carlotti, M.; Mazzolai, B.; Mattoli, V.; et al. Engineering Microneedle Patches for Improved Penetration: Analysis, Skin Models and Factors Affecting Needle Insertion. *Nano-Micro Lett.* **2021,** *13* (1), 93.
26. Ando, D.; Miyatsuji, M.; Sakoda, H.; Yamamoto, E.; Miyazaki, T.; Koide, T.; Sato, Y.; Izutsu, K-i. Mechanical Characterization of Dissolving Microneedles: Factors Affecting Physical Strength of Needles. *Pharmaceutics* **2024,** *16.*
27. Mamun, A Al; Sueoka, B.; Allison, N.; Huang, Y.; Zhao, F. Design and Evaluation of In-Plane Silicon Microneedles Fabricated with Post-CMOS Compatible Processes. *Sensor Actuator Phys.* **2022,** *336,* 113407.
28. Loizidou, E. Z.; Williams, N. A.; Barrow, D. A.; Eaton, M. J.; McCrory, J.; Evans, S. L.; Allender, C. J. Structural Characterisation and Transdermal Delivery Studies on Sugar Microneedles: Experimental and Finite Element Modelling Analyses. *Eur. J. Pharm. Biopharm.* **2015,** *89,* 224–231.
29. Menon, G. K.; Cleary, G. W.; Lane, M. E. The Structure and Function of the Stratum Corneum. *Int. J. Pharm.* **2012,** *435* (1), 3–9.
30. Aggarwal, P.; Johnston, C. Geometrical Effects in Mechanical Characterizing of Microneedle for Biomedical Applications. *Sensor. Actuator. B Chem.* **2004,** *102* (2), 226–234.
31. Park, J.-H.; Prausnitz, M. R. Analysis of Mechanical Failure of Polymer Microneedles by Axial Force. *J. Kor. Phys. Soc.* **2010,** *56* (4), 1223.
32. Sadasivuni, K. K.; Saha, P.; Adhikari, J.; Deshmukh, K.; Ahamed, M. B.; Cabibihan, J.-J. Recent Advances in Mechanical Properties of Biopolymer Composites: A Review. *Polym. Compos.* **2020,** *41* (1), 32–59.
33. Qiang, N.; Liu, Z.; Lu, M.; Yang, Y.; Liao, F.; Feng, Y.; Liu, G.; Qiu, S. Preparation and Properties of Polyvinylpyrrolidone/Sodium Carboxymethyl Cellulose Soluble Microneedles. *Materials* **2023,** *16.*
34. Wu, M.; Xia, T.; Li, Y.; Wang, T.; Yang, S.; Yu, J.; Liang, Q.; Shen, T.; Yu, M.; Zhao, B. Design and Fabrication of R-Hirudin Loaded Dissolving Microneedle Patch for Minimally Invasive and Long-Term Treatment of Thromboembolic Disease. *Asian J. Pharm. Sci.* **2022,** *17* (2), 284–297.
35. Ning, T.; Yang, F.; Chen, D.; Jia, Z.; Yuan, R.; Du, Z.; Liu, S.; Yu, Y.; Dai, X.; Niu, X.; et al. Synergistically Detachable Microneedle Dressing for Programmed Treatment of Chronic Wounds. *Adv. Healthcare Mater.* **2022,** *11* (11), 2102180.
36. Wang, C.; Ye, Y.; Hochu, G. M.; Sadeghifar, H.; Gu, Z. Enhanced Cancer Immunotherapy by Microneedle Patch-Assisted Delivery of Anti-PD1 Antibody. *Nano Lett.* **2016,** *16* (4), 2334–2340.

37. Zhang, X. P.; Zhang, B. L.; Chen, B. Z.; Zhao, Z. Q.; Fei, W. M.; Cui, Y.; Guo, X. D. Dissolving Microneedle Rollers for Rapid Transdermal Drug Delivery. *Drug Deliv. Transl. Res.* **2022,** *12* (2), 459–471.
38. Sadiku, E. R.; Phiri, G.; Jayaramudu, T.; Sudhakar, K.; Moropeng, L.; Khoathane, M. C.; Adegbola, T. A.; Kupolati, W. K. Chapter 16: Mechanisms of Toughening in Nanostructured Polymer Blends. In *Design and Applications of Nanostructured Polymer Blends and Nanocomposite Systems;* Thomas, S., Shanks, R., Chandrasekharakurup, S., Eds.; William Andrew Publishing, 2016; pp 365–384.
39. Su, Y.; Andrabi, S. M.; Shahriar, S. M. S.; Wong, S. L.; Wang, G.; Xie, J. Triggered Release of Antimicrobial Peptide from Microneedle Patches for Treatment of Wound Biofilms. *J. Contr. Release* **2023,** *356,* 131–141.
40. Khosraviboroujeni, A.; Mirdamadian, S. Z.; Minaiyan, M.; Taheri, A. Preparation and Characterization of 3D Printed PLA Microneedle Arrays for Prolonged Transdermal Drug Delivery of Estradiol Valerate. *Drug Deliv. Transl. Res.* **2022,** *12* (5), 1195–1208.
41. Lu, X.; Sun, Y.; Han, M.; Chen, D.; Wang, A.; Sun, K. Silk Fibroin Double-Layer Microneedles for the Encapsulation and Controlled Release of Triptorelin. *Int. J. Pharm.* **2022,** *613,* 121433.
42. Lv, H.; Gao, N.; Zhou, Q.; Wang, Y.; Ling, G.; Zhang, P. Collagen-Based Dissolving Microneedles with Flexible Pedestals: A Transdermal Delivery System for Both Anti-aging and Skin Diseases. *Adv. Healthc. Mater.* **2023,** *12* (21), e2203295.
43. Dabholkar, N.; Gorantla, S.; Waghule, T.; Rapalli, V. K.; Kothuru, A.; Goel, S.; Singhvi, G. Biodegradable Microneedles Fabricated with Carbohydrates and Proteins: Revolutionary Approach for Transdermal Drug Delivery. *Int. J. Biol. Macromol.* **2021,** *170,* 602–621.
44. Wang, Y.; Zhang, S.; Wang, J. Photo-Crosslinkable Hydrogel and Its Biological Applications. *Chin. Chem. Lett.* **2021,** *32* (5), 1603–1614.
45. Zhu, J.; Zhou, X.; Kim, H.-J.; Qu, M.; Jiang, X.; Lee, K.; Ren, L.; Wu, Q.; Wang, C.; Zhu, X.; et al. Gelatin Methacryloyl Microneedle Patches for Minimally Invasive Extraction of Skin Interstitial Fluid. *Small* **2020,** *16* (16), 1905910.
46. Yu, J.; Zhang, Y.; Ye, Y.; DiSanto, R.; Sun, W.; Ranson, D.; Ligler, F. S.; Buse, J. B.; Gu, Z. Microneedle-Array Patches Loaded with Hypoxia-Sensitive Vesicles Provide Fast Glucose-Responsive Insulin Delivery. *Proc. Natl. Acad. Sci.* **2015,** *112* (27), 8260–8265.
47. Rajput, M.; Mondal, P.; Yadav, P.; Chatterjee, K. Light-Based 3D Bioprinting of Bone Tissue Scaffolds with Tunable Mechanical Properties and Architecture from Photocurable Silk Fibroin. *Int. J. Biol. Macromol.* **2022,** *202,* 644–656.
48. Kuo, C. K.; Ma, P. X. Ionically Crosslinked Alginate Hydrogels as Scaffolds for Tissue Engineering: Part 1. Structure, Gelation Rate and Mechanical Properties. *Biomaterials* **2001,** *22* (6), 511–521.
49. Chen, W.; Tian, R.; Xu, C.; Yung, B. C.; Wang, G.; Liu, Y.; Ni, Q.; Zhang, F.; Zhou, Z.; Wang, J.; et al. Microneedle-Array Patches Loaded with Dual Mineralized Protein/Peptide Particles for Type 2 Diabetes Therapy. *Nat. Commun.* **2017,** *8* (1), 1777.
50. Li, Z.; Zhao, P.; Ling, Z.; Zheng, Y.; Qu, F.; Chang, H. An Ultraswelling Microneedle Device for Facile and Efficient Drug Loading and Transdermal Delivery. *Adv. Healthc. Mater.* **2023,** e2302406.
51. Chong, S.; Wei, C.; Feng, L.; Guo, R. Silk Fibroin-Based Hydrogel Microneedles Deliver α-MSH to Promote Melanosome Delivery for Vitiligo Treatment. *ACS Biomater. Sci. Eng.* **2023,** *9* (6), 3368–3378.

52. Tiraton, T.; Suwantong, O.; Chuysinuan, P.; Ekabutr, P.; Niamlang, P.; Khampieng, T.; Supaphol, P. Biodegradable Microneedle Fabricated from Sodium Alginate-Gelatin for Transdermal Delivery of Clindamycin. *Mater. Today Commun.* **2022,** *32,* 104158.
53. Lu, X.; Sun, Y.; Han, M.; Chen, D.; Wang, A.; Sun, K. Silk Fibroin Double-Layer Microneedles for the Encapsulation and Controlled Release of Triptorelin. *Int. J. Pharm.* **2022,** *613,* 121433.
54. Lee, J.; Park, S. H.; Seo, I. H.; Lee, K. J.; Ryu, W. Rapid and Repeatable Fabrication of High A/R Silk Fibroin Microneedles Using Thermally-Drawn Micromolds. *Eur. J. Pharm. Biopharm.* **2015,** *94,* 11–19.
55. Ding, D.; Pan, J.; Lim, S. H.; Amini, S.; Kang, L.; Miserez, A. Squid Suckerin Microneedle Arrays for Tunable Drug Release. *J. Mater. Chem. B* **2017,** *5* (43), 8467–8478.
56. Kim, J.; Zhang, G.; Shi, M.; Suo, Z. Fracture, Fatigue, and Friction of Polymers in Which Entanglements Greatly Outnumber Cross-Links. *Science* **2021,** *374* (6564), 212–216.
57. Huang, Y.; Li, H.; Hu, T.; Li, J.; Yiu, C. K.; Zhou, J.; Li, J.; Huang, X.; Yao, K.; Qiu, X.; et al. Implantable Electronic Medicine Enabled by Bioresorbable Microneedles for Wireless Electrotherapy and Drug Delivery. *Nano Lett.* **2022,** *22* (14), 5944–5953.
58. Thakur, R. R. S.; Tekko, I. A.; Al-Shammari, F.; Ali, A. A.; McCarthy, H.; Donnelly, R. F. Rapidly Dissolving Polymeric Microneedles for Minimally Invasive Intraocular Drug Delivery. *Drug Deliv. Transl. Res.* **2016,** *6* (6), 800–815.
59. Lee, I. C.; Wu, Y.-C.; Tsai, S.-W.; Chen, C.-H.; Wu, M.-H. Fabrication of Two-Layer Dissolving Polyvinylpyrrolidone Microneedles with Different Molecular Weights for In Vivo Insulin Transdermal Delivery. *RSC Adv.* **2017,** *7* (9), 5067–5075.
60. Chi, Y.; Huang, Y.; Kang, Y.; Dai, G.; Liu, Z.; Xu, K.; Zhong, W. The Effects of Molecular Weight of Hyaluronic Acid on Transdermal Delivery Efficiencies of Dissolving Microneedles. *Eur. J. Pharmaceut. Sci.* **2022,** *168,* 106075.
61. Zhang, Y.; Jiang, G.; Yu, W.; Liu, D.; Xu, B. Microneedles Fabricated from Alginate and Maltose for Transdermal Delivery of Insulin on Diabetic Rats. *Mater. Sci. Eng. C* **2018,** *85,* 18–26.
62. Hou, A.; Quan, G.; Yang, B.; Lu, C.; Chen, M.; Yang, D.; Wang, L.; Liu, H.; Pan, X.; Wu, C. Rational Design of Rapidly Separating Dissolving Microneedles for Precise Drug Delivery by Balancing the Mechanical Performance and Disintegration Rate. *Adv. Healthcare Mater.* **2019,** *8* (21), 1900898.
63. Sullivan, S. P.; Murthy, N.; Prausnitz, M. R. Minimally Invasive Protein Delivery with Rapidly Dissolving Polymer Microneedles. *Adv. Mater. (Deerfield Beach, Fla.)* **2008,** *20* (5), 933–938.
64. Chen, W.; Wang, C.; Yan, L.; Huang, L.; Zhu, X.; Chen, B.; Sant, H. J.; Niu, X.; Zhu, G.; Yu, K. N.; et al. Improved Polyvinylpyrrolidone Microneedle Arrays with Non-stoichiometric Cyclodextrin. *J. Mater. Chem. B* **2014,** *2* (12), 1699–1705.
65. Lin, S.; Quan, G.; Hou, A.; Yang, P.; Peng, T.; Gu, Y.; Qin, W.; Liu, R.; Ma, X.; Pan, X.; et al. Strategy for Hypertrophic Scar Therapy: Improved Delivery of Triamcinolone Acetonide Using Mechanically Robust Tip-Concentrated Dissolving Microneedle Array. *J. Contr. Release* **2019,** *306,* 69–82.
66. Tekko, I. A.; Permana, A. D.; Vora, L.; Hatahet, T.; McCarthy, H. O.; Donnelly, R. F. Localised and Sustained Intradermal Delivery of Methotrexate Using Nanocrystal-Loaded Microneedle Arrays: Potential for Enhanced Treatment of Psoriasis. *Eur. J. Pharmaceut. Sci.* **2020,** *152,* 105469.
67. Yan, L.; Raphael, A. P.; Zhu, X.; Wang, B.; Chen, W.; Tang, T.; Deng, Y.; Sant, H. J.; Zhu, G.; Choy, K. W.; et al. Nanocomposite-Strengthened Dissolving Microneedles for Improved Transdermal Delivery to Human Skin. *Adv. Healthcare Mater.* **2014,** *3* (4), 555–564.

68. Zhao, J.; Duan, W.; Liu, X.; Xi, F.; Wu, J. Microneedle Patch Integrated with Porous Silicon Confined Dual Nanozymes for Synergistic and Hyperthermia-Enhanced Nanocatalytic Ferroptosis Treatment of Melanoma. *Adv. Funct. Mater.* **2023,** *33* (47), 2308183.
69. Zhang, Y.; Wu, M.; Tan, D.; Liu, Q.; Xia, R.; Chen, M.; Liu, Y.; Xue, L.; Lei, Y. A Dissolving and Glucose-Responsive Insulin-Releasing Microneedle Patch for Type 1 Diabetes Therapy. *J. Mater. Chem. B* **2021,** *9* (3), 648–657.
70. Liu, D.; Yu, B.; Jiang, G.; Yu, W.; Zhang, Y.; Xu, B. Fabrication of Composite Microneedles Integrated with Insulin-Loaded CaCO3 Microparticles and PVP for Transdermal Delivery in Diabetic Rats. *Mater. Sci. Eng. C* **2018,** *90,* 180–188.
71. Yu, W.; Jiang, G.; Liu, D.; Li, L.; Chen, H.; Liu, Y.; Huang, Q.; Tong, Z.; Yao, J.; Kong, X. Fabrication of Biodegradable Composite Microneedles Based on Calcium Sulfate and Gelatin for Transdermal Delivery of Insulin. *Mater. Sci. Eng. C* **2017,** *71,* 725–734.
72. Aggarwal, P.; Johnston, C. R. Geometrical Effects in Mechanical Characterizing of Microneedle for Biomedical Applications. *Sensor. Actuator. B Chem.* **2004,** *102* (2), 226–234.
73. Yan, Q.; Weng, J.; Shen, S.; Wang, Y.; Fang, M.; Zheng, G.; Yang, Q.; Yang, G. Finite Element Analysis for Biodegradable Dissolving Microneedle Materials on Skin Puncture and Mechanical Performance Evaluation. *Polymers* **2021,** *13.*
74. Loizidou, E. Z.; Inoue, N. T.; Ashton-Barnett, J.; Barrow, D. A.; Allender, C. J. Evaluation of Geometrical Effects of Microneedles on Skin Penetration by CT Scan and Finite Element Analysis. *Eur. J. Pharm. Biopharm.* **2016,** *107,* 1–6.
75. Kim, M. J.; Park, S. C.; Rizal, B.; Guanes, G.; Baek, S.-K.; Park, J.-H.; Betz, A. R.; Choi, S.-O. Fabrication of Circular Obelisk-Type Multilayer Microneedles Using Micro-Milling and Spray Deposition. *Front. Bioeng. Biotechnol.* **2018,** *6.*
76. Lee, J. W.; Park, J.-H.; Prausnitz, M. R. Dissolving Microneedles for Transdermal Drug Delivery. *Biomaterials* **2008,** *29* (13), 2113–2124.
77. Li, Y.; Hu, X.; Dong, Z.; Chen, Y.; Zhao, W.; Wang, Y.; Zhang, L.; Chen, M.; Wu, C.; Wang, Q. Dissolving Microneedle Arrays with Optimized Needle Geometry for Transcutaneous Immunization. *Eur. J. Pharmaceut. Sci.* **2020,** *151,* 105361.
78. Jung, M.; Jeong, D.; Yun, S.-S.; Lee, J.-H. Fabrication of a 2-D In-Plane Micro Needle Array Integrated with Microfluidic Components Using Crystalline Wet Etching of (110) Silicon. *Microsyst. Technol.* **2016,** *22* (9), 2287–2294.
79. Aoyagi, S.; Izumi, H.; Fukuda, M. Biodegradable Polymer Needle with Various Tip Angles and Consideration on Insertion Mechanism of Mosquito's Proboscis. *Sensor Actuator Phys.* **2008,** *143* (1), 20–28.
80. Davis, S. P.; Landis, B. J.; Adams, Z. H.; Allen, M. G.; Prausnitz, M. R. Insertion of Microneedles into Skin: Measurement and Prediction of Insertion Force and Needle Fracture Force. *J. Biomech.* **2004,** *37* (8), 1155–1163.
81. Park, J.-H.; Allen, M. G.; Prausnitz, M. R. Biodegradable Polymer Microneedles: Fabrication, Mechanics and Transdermal Drug Delivery. *J. Contr. Release* **2005,** *104* (1), 51–66.
82. Li, G.; Badkar, A.; Nema, S.; Kolli, C. S.; Banga, A. K. In Vitro Transdermal Delivery of Therapeutic Antibodies Using Maltose Microneedles. *Int. J. Pharm.* **2009,** *368* (1), 109–115.
83. Gittard, S. D.; Chen, B.; Xu, H.; Ovsianikov, A.; Chichkov, B. N.; Monteiro-Riviere, N. A.; Narayan, R. J. The Effects of Geometry on Skin Penetration and Failure of Polymer Microneedles. *J. Adhes. Sci. Technol.* **2013,** *27* (3), 227–243.
84. Cárcamo-Martínez, Á.; Anjani, Q. K.; Permana, A. D.; Cordeiro, A. S.; Larrañeta, E.; Donnelly, R. F. Coated Polymeric Needles for Rapid and Deep Intradermal Delivery. *Int. J. Pharm. X* **2020,** *2,* 100048.

Chapter 6

Strategies used to enhance the drug permeation of microneedles

Zeshi Jiang[1], Wentao Wu[1], Xin Pan[1] and Tingting Peng[2]
[1]*School of Pharmaceutical Science, Sun Yat-Sen University, Guangzhou, China;* [2]*State Key Laboratory of Bioactive Molecules and Druggability Assessment/International Cooperative Laboratory of Traditional Chinese Medicine Modernization and Innovative Drug Development of Ministry of Education (MOE) of China/College of Pharmacy, Jinan University, Guangzhou, China*

1. Introduction

Microneedles (MNs) represent an innovative transdermal drug delivery modality that offers significant advantages over conventional systems. Despite their potential, the actual drug delivery efficiency of MNs in clinical or experimental settings often falls short of theoretical expectations. This inefficiency is largely attributable to the presence of multiple structural and physiological skin barriers that restrict drug transport. Stratum corneum with a compact brick–wall structure has long been considered a foremost physical barrier for transdermal drug delivery.[1–4] Skin resistance is another physical barrier that is usually neglected, making the skin deformed to counteract the insertion of MNs into a desired depth, resulting in large amounts of drug residue on the skin surface. Accordingly, MNs should have sufficient mechanical strength to pierce skin and counteract rapid skin rebound, which is the first step for precise intradermal drug delivery.

After MNs pierce the skin, the loaded drug is released from the MNs and permeates into the cutaneous tissue for further absorption. To produce a therapeutic effect, the drug must overcome dermal physiological barriers before reaching the targeted cells or entering the systemic circulation.[1,5–8] Extracellular matrix (ECM), consisting of dense collagen fibers and glycosaminoglycans, forms a highly viscous hydrophilic gel and acts as a major physiological barrier to impede drug permeation and absorption.[9–11] For example, excessive ECM deposition occurs in skin disorders such as hypertrophic scars, keloids, and superficial tumors, forming a very rigid barrier to

Microneedle Technology. https://doi.org/10.1016/B978-0-443-45544-5.00008-7

hinder drug permeation into deeper skin layers.[12–14] In addition, biofilm present in infectious skin also forms a rigid barrier to impede drug permeation in the interior of biofilms. Biofilm is a collection of mono-microbial or polymicrobial communities that are encaged by highly dense extracellular polymer substance (EPS). EPS are mainly composed of polysaccharides, proteins, nucleotides, and phospholipids, accounting for more than 90% of the biofilm mass and severely hindering the exposure of drugs to microbial cells.[15–17] To achieve efficient drug delivery, MNs should be ingeniously designed to overcome sequential physical (e.g., stratum corneum, skin deformation) and physiological barriers (e.g., ECM, biofilms).

In the course of disease development, the changes in biological signals create a pathological microenvironment different from that of healthy tissues. The pathological microenvironment is usually characterized by high glucose, high reactive oxygen species (ROS), low pH, specifically expressed enzymes, and so on.[18–20] However, conventional MNs usually fail to respond to the internal stimuli and provide on-demand drug release, resulting in sufficient release and accumulation of therapeutics within target tissues.[21–23] In response to this challenge, stimuli-responsive MNs have been developed using modified natural and synthetic polymers. These systems can undergo dissolution, disintegration, shrinkage, or swelling in response to pathological stimuli, thereby facilitating site-specific drug release within lesional tissues. The development of such responsive MN platforms holds promise for significantly improving drug delivery efficiency, minimizing off-target effects, and achieving superior therapeutic outcomes.

To overcome these diverse and complex skin barriers, various engineering strategies have been proposed. This chapter focuses on design principles and technological advancements that enable MNs to penetrate physical and physiological skin barriers and harness endogenous pathological cues for targeted drug delivery. First, strategies to enhance the mechanical properties and insertion capabilities of MNs are discussed, including geometric optimization, administration techniques, and needle-body separation mechanisms. Second, both passive (e.g., hyaluronidase, α-amylase) and active *(e.g.*, light, ultrasound, electric field, magnetic field, and gas) permeation enhancement methods are reviewed, with attention to their mechanisms in improving drug diffusion through physiological barriers. Furthermore, the integration of MNs with responsive biomaterials capable of reacting to pathological stimuli such as hyperglycemia, oxidative stress, and acidic pH is introduced as a means to achieve controlled, on-demand drug release. Finally, key challenges and future perspectives are addressed, including issues of large-scale manufacturing, minimizing interindividual pharmacokinetic variability, and ensuring dosing precision. Collectively, these advances offer a comprehensive framework for the development of next-generation MNs capable of overcoming the multifaceted challenges associated with transdermal drug delivery.

2. Optimizing the geometries of microneedles

MNs' geometry is widely recognized as a key determinant influencing their mechanical strength, insertion depth into the skin, and overall drug delivery effectiveness.[24,25] However, the general law among these attributes has been scarcely reported. The maximum force for the MNs to withstand can be described using the following equation:

$$F = \frac{\pi^2 EI}{(2L)^2} \tag{6.1}$$

Where, F denotes the critical buckling load of MNs, L represents needle length, I refers to the cross-sectional moment of inertia, *and* E is the elastic modulus of the material.

As indicated in Eq. (6.1), the mechanical performance of MNs is influenced by geometric parameters such as shape, interspacing, tip and base diameters, aspect ratio, and length. The following sections detail how these factors affect both mechanical strength and drug delivery efficiency Fig. 6.1.

2.1 Microneedle shape

The shape of MNs is a crucial geometric factor that affects the mechanical strength and skin penetration capability of MNs, and proper geometric design is beneficial for improving drug delivery efficiency. Cone and pyramid are the two most commonly designed shapes of MNs. Compared with conical designs, pyramidal MNs demonstrate stronger mechanical properties, mainly resulting from their reduced aspect ratio and expanded cross-section, given an equal base diameter or width.[26,27] In another study,[28] De Martino et al. fabricated MNs with different vertices to compare their puncture performance: conical MNs with no vertices, triangular pyramidal MNs with three vertices, square

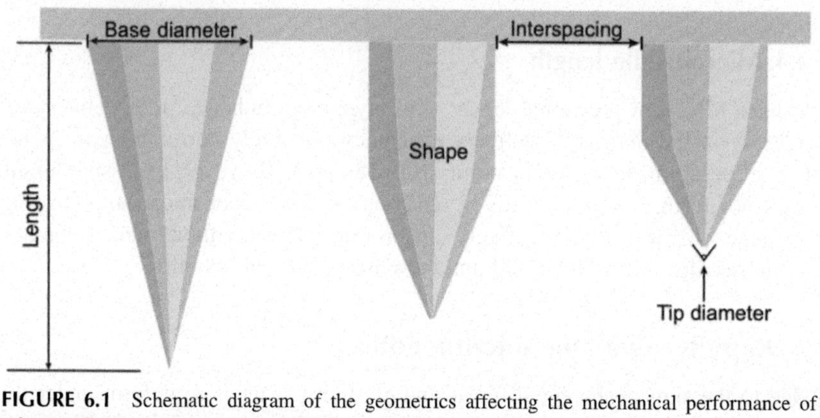

FIGURE 6.1 Schematic diagram of the geometrics affecting the mechanical performance of microneedles.

MNs with four vertices, and star-shaped MNs with 10 vertices. When the number of vertices increased, the compression force required for needle compression decreased, and the penetration efficiency increased from 30% for cMNs to about 60% for sqMNs.

2.2 Microneedle interspacing

Under the same applied force, MNs with larger tip-to-tip interspacing absorb more of the insertion force per needle, thereby enhancing skin penetration efficiency. In contrast, decreasing the spacing between MNs leads to higher needle density and intensifies the bed-of-nails effect, thereby increasing the force needed for skin insertion.[29,30]

2.3 Tip and base diameters of microneedles

An increase in tip diameter disperses the applied force over a broader contact area, thereby reducing insertion efficiency. In contrast, a smaller tip diameter concentrates the applied force at the needle apex, enhancing skin penetration performance. For microneedles with tip diameters below 80 μm, the insertion force shows a linear correlation with the tip radius (R). When the tip diameter exceeds 80 μm, the insertion force exhibits a predominantly quadratic relationship with the radius[31]

A larger base diameter enhances mechanical strength and reduces the likelihood of bending during insertion. Lee et al.[32] fabricated MNs with base diameters ranging from 50 to 800 μm to investigate the influence of base diameter on mechanical strength. MNs with base diameters below 200 μm exhibited markedly low mechanical strength. For instance, a conical MN with dimensions of 800 μm in length and 200 μm in base diameter demonstrated a critical buckling (P_{cr}) of 0.51 N. When the base diameter was halved to 100 μm, P_{cr} dropped markedly to 0.031 N.

2.4 Microneedle length

Longer MNs can penetrate deeper skin layers and enhance drug delivery efficiency in the dermis.[33] Delivery efficiency positively correlates with penetration depth up to a certain point (100–180 μm). However, excessive length does not further increase delivery and may reduce mechanical strength. For instance, 400 μm MN effectively reach the dermis without nerve contact, while needles longer than 600 μm show no additional benefit.[34]

3. Rapidly separable microneedles

Due to the skin's inherent stiffness and elasticity, it tends to deform under the applied force, which can hinder full insertion of MNs and result in drug loss and reduced delivery efficiency.[30,35] Additionally, MNs with slow dissolution rates often require hours or even days to ensure complete drug release. Prolonged wear of MN patches, however, can compromise patient compliance and cause skin irritation, such as erythema and edema.[36,37] To address these challenges, researchers have developed rapidly separable MNs over the past decade.[38–40] Based on their separation mechanism, these MNs can be broadly categorized into four types (Fig. 6.2): (1) MN patches with a rapidly dissolvable backing; (2) MN patches integrating porous or bubble structures; (3) Stimuli-responsive separable MNs; (4) Micropillar integrated separable MNs.

3.1 Introducing a rapidly dissolving backing layer

Incorporating highly water-soluble agents into the backing layer is a convenient method to fabricate rapidly separable MNs patches, as these agents

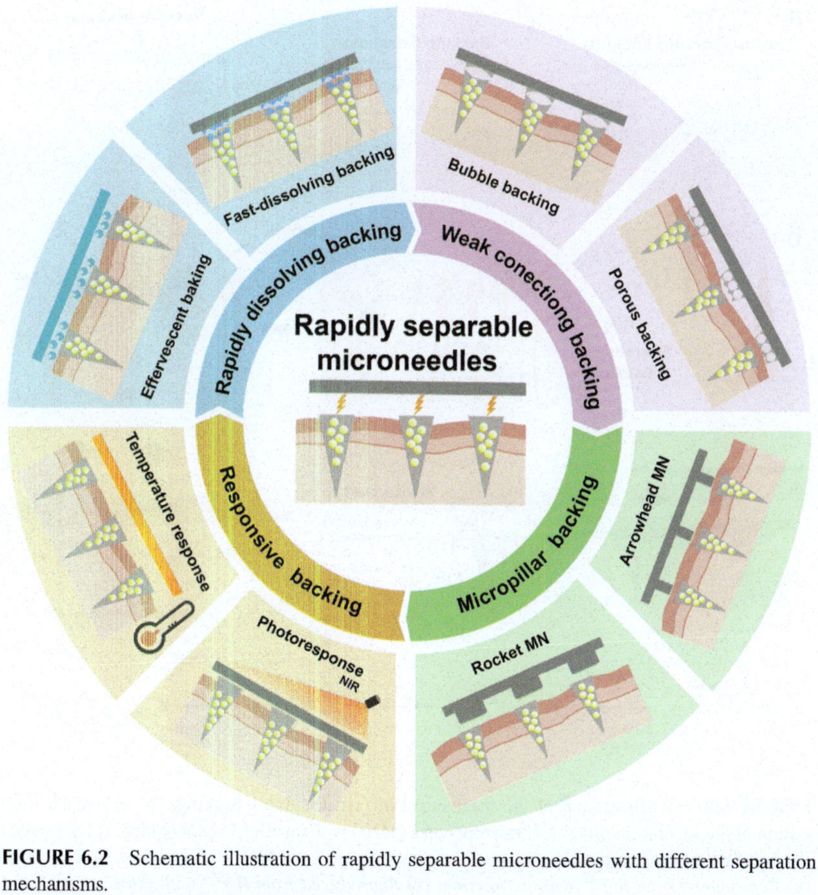

FIGURE 6.2 Schematic illustration of rapidly separable microneedles with different separation mechanisms.

readily dissolve in interstitial fluid (ISF), enabling quick detachment of the needle tips (Fig. 6.3a). Hou et al.[41] developed a rapidly separable MNs system using HA and sucrose as the backing material for delivering diclofenac sodium in rheumatoid arthritis therapy. The fast dissolution of sucrose facilitated the formation of water channels, accelerating separation. The system achieved a cumulative transdermal permeability above 75%, nearly double that of conventional MNs.

Another commonly employed strategy is the incorporation of effervescent components into the backing layer, which generate gas bubbles upon contact with ISF to drive needle separation. Liu et al.[42] fabricated MNs patches with needle tips composed of polyvinyl alcohol (PVA) and a base containing tartaric acid and sodium bicarbonate (Fig. 6.3b). Upon application, the acid–base reaction produced gas bubbles that enabled 100% needle separation within 50 s, while conventional MNs showed only 40% separation. When loaded with

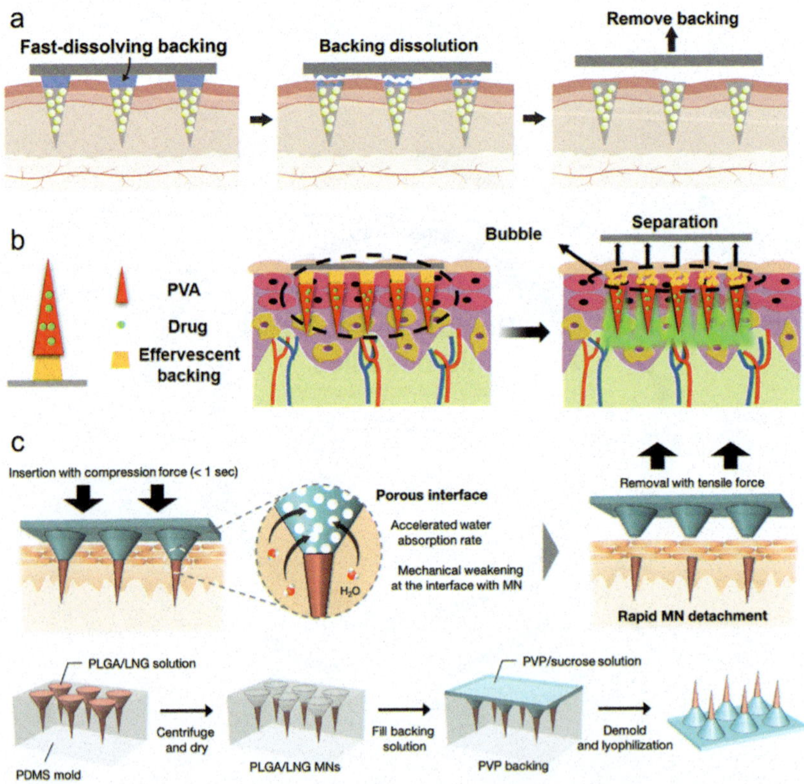

FIGURE 6.3 (a) Illustration of the mechanism of fast-dissolving backing. (b) Schematic illustration of the separation process of microneedles (MNs) with an effervescent baking. (c) Schematic illustrating the separation mechanism and fabrication process of MNs with a porous backing layer. *(b) Reproduced from Ref.*[42] *with permission. (c) Reproduced from Ref.*[43] *with permission.*

metformin, these separable MNs induced a 62.5% reduction in blood glucose levels at 4 h in diabetic rats, significantly outperforming conventional MNs (18.7% reduction).

3.2 Weakening the connection between backing and needles

While effervescent separation layers can achieve rapid needle detachment, their reliance on minimal wearing time may lead to incomplete separation in practice. To address this issue, an alternative strategy involves weakening the needle–backing connection by introducing porous or bubble structures.

Lee et al.[43] developed porous-backing MNs using a solvent casting and freeze-drying method (Fig. 6.3c), enabling rapid detachment within 1 s upon skin insertion. The PLGA tips loaded with levonorgestrel (LNG) were prepared via centrifugation, while the baseplate made of sucrose and PVA was lyophilized to form a porous structure. The MNs with a porous baseplate had a separation efficiency of 95 ± 8%, which is significantly higher than that of MNs with an entitative baseplate (1 ± 2%). Likewise, Li et al.[44] introduced a bubble structure between the PLGA/PLA needle tips and a porous baseplate composed of PVA and sucrose. Due to the limited wettability of the needle polymers, air gaps formed during casting, resulting in a bubble interface. These MNs showed over 95% tip detachment and over 90% de0000livery of Nile Red into pig skin within 5 s under mild shearing force, while MNs without bubble structures achieved only 15% detachment and 10% delivery efficiency.

3.3 Integrating a micropillar structure

Producing MNs on the micropillar can reduce the force difference of each needle to allow the entire MN array to be evenly and completely inserted into the skin, which helps to increase the penetration rate of MNs.[45,46] Depending on the size of the micropillar relative to the needle tips, micropillar detachable MNs can be classified into two types: (1) arrowhead MNs that enable rapid separation by connecting the needles to a thinner micropillar, and (2) rocket MNs with a micropillar that has a larger diameter or cross-sectional area than needles to prevent the insertion of micropillar into the skin. The micropillar can be fabricated from biocompatible polymers such as polyethylene glycol diacrylate (PEGDA)[47,48] and PLA,[39,49] or metals.[40]

The separation process of thinner micropillars from needles typically involves the following steps: Upon insertion into the skin, the needles absorb ISF and dissolve, enabling their detachment from the micropillar. This dissolution process is driven by the high solubility of the needle material in ISF. The thinner micropillar, designed to minimize frictional resistance, detaches more readily during this process. Chu et al.[40] constructed arrowhead MNs by attaching drug-loaded dissolvable tips onto a stainless steel micropillar array

(Fig. 6.4a). Upon insertion into pig skin, 70% of the tips detached within 5 s and about 90% within 60 s. Application of vibration significantly accelerated separation, achieving 90% detachment within 5 s. In contrast, rocket MNs employ a micropillar wider than the needle tip to restrict their entry into the skin. After tip insertion, a shearing force removes the backing (Fig. 6.4b).[45] Jung et al.[50] investigated the effect of different diameters and shapes of micropillars on the penetration and separation efficiency of rocket MNs. Three types of rocket MNs with a 300 μm base diameter and varying micropillar configurations (300 μm, 500 μm, and 500 μm with 45 degrees curved edges) were fabricated. The skin puncture test demonstrated that all the MNs pierced the pig skin, and the MNs with 500 μm diameter micropillar exhibited the highest separation efficiency. The MNs with 300 μm micropillars showed only about 60% separation efficiency, while the other two kinds of MNs showed 90% separation efficiency. These results suggest that wider micropillars are more easily separated by shearing forces, probably due to their larger surface area and increased resistance to insertion that prevents micropillars to penetrate penetrating into deep skin.

3.4 Triggering a responsive detachment

Although the methods mentioned above enable efficient needle-to-base detachment, there still exists the problem of uncontrollable separation. The separation process would start immediately once the MN patch is inserted into the skin. Recently, researchers have proposed the development of responsive separation MNs to attain controlled and precise detachment of the backing from the needles under the specific stimuli of temperature and light, which further ensures accurate and enhanced drug delivery.[51,52]

Fan et al.[53] proposed to design a near-infrared (NIR)-activated separable MNs patch for versatile drug delivery by introducing photothermal-responsive microspheres (MPs) as a separable layer. Briefly, the MPs made of black phosphorus (BP) and phase-transition gelatin (GT) were prepared by a simple capillary microfluidic device. Owing to the excellent photothermal performance of BP, the loaded MPs can make the local temperature elevated to trigger their solid-liquid phase transition upon NIR irradiation (Fig. 6.5a). Then, the collapsed MPs created significant cavities between the MN tips and backing, enabling controllable needle-based separation under a shear force. Benefiting from the molten MPs, the shear force required for MNs separation decreased from approximately 0.13 N/needle to around 0.06 N/needle, causing a sharp surge in the separation efficiency from 45% to about 80%. In addition, the MPs coloaded with interferon γ and dexamethasone provided NIR responsive drug release, resulting in efficient treatment of systemic lupus erythematosus.

In addition to using light as a trigger for MNs separation, Yin et al.[54] developed a thermoresponsive MNs patch for delivering DNA vaccines against

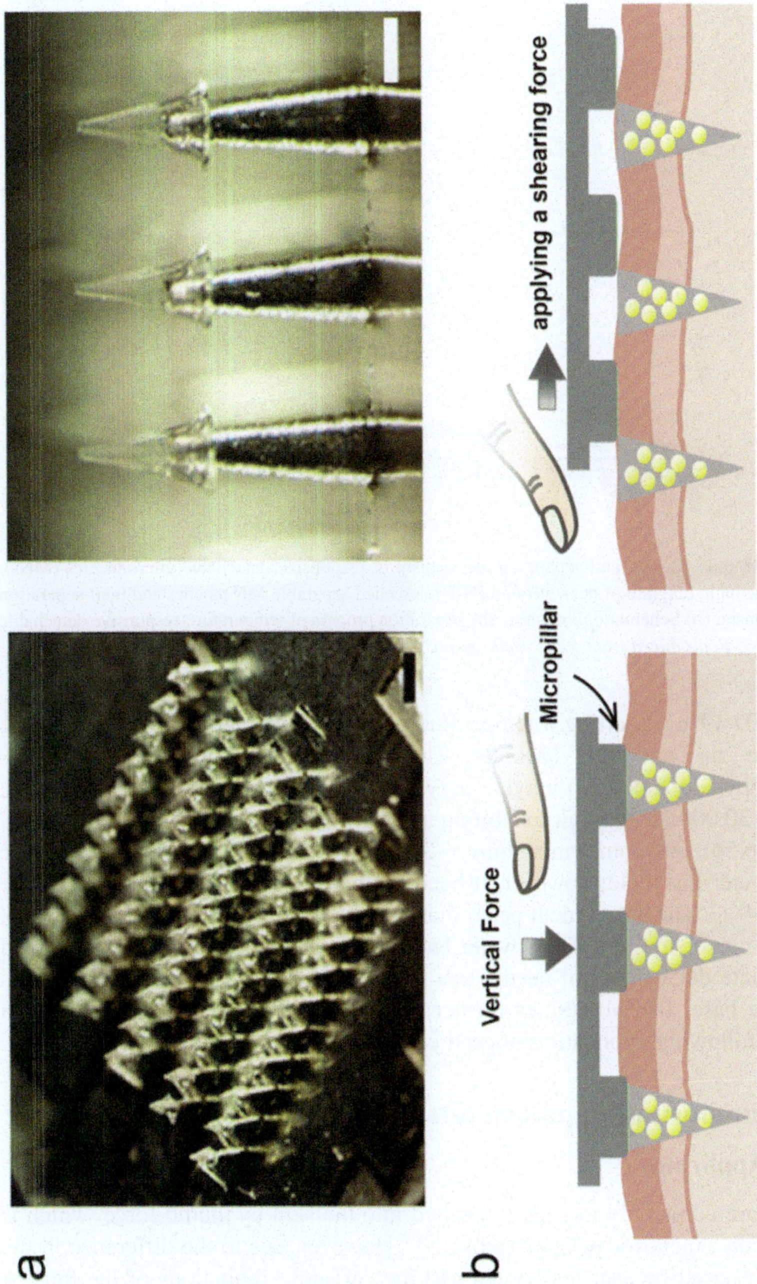

FIGURE 6.4 (a) Representative images of separable arrowhead microneedles (MNs) array. (b) Schematic diagram of separable arrowhead MNs fabrication process. *(a) Reproduced from Ref.*[40] *with permission.*

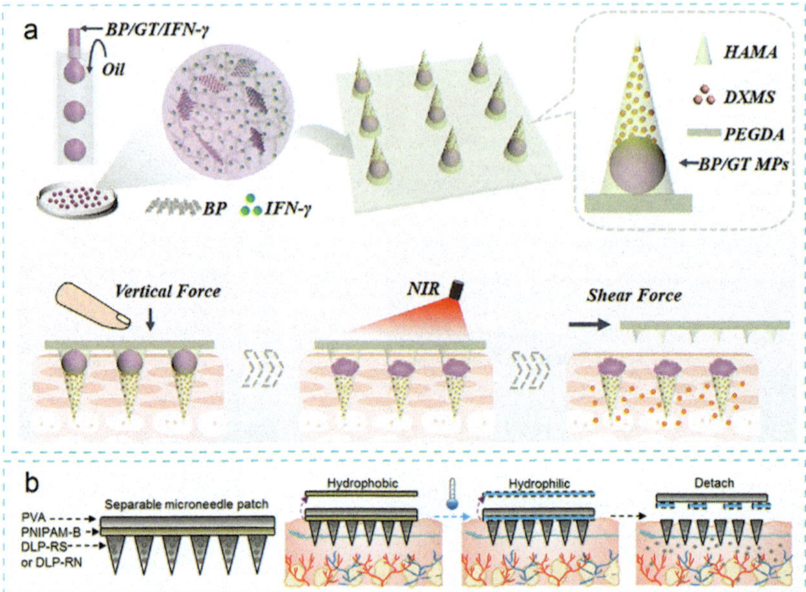

FIGURE 6.5 Design and separation mechanism of responsive detachable microneedles (MNs): (a) Schematic diagram of near-infrared (NIR)-activated separable MN patches and their separation mechanism. (b) Schematic illustrating the separation process of temperature-responsive detachable MNs. *(a) Reproduced from Ref.*[53] *with permission. (b) Reproduced from Ref.*[54] *with permission.*

COVID-19 by inserting a temperature-sensitive separation layer between the needle tip and the backing. The layer was composed of poly(N-isopropylacrylamide-co-butyl acrylate) (PNIPAM-B, NIPAM:B = 8:1, $M_n = 30{,}000$), a polymer exhibiting a solid–liquid phase transition at 14–16°C (Fig. 6.5b). At room temperature (~22 °C), PNIPAM-B remains hydrophobic and water-insoluble; however, it becomes highly hydrophilic and dissolvable at $\leq$14 °C. The detachment performance was verified both in vitro and in vivo. After immersion in a 13 °C water bath for 3 minutes, SEM images confirmed complete detachment of needle tips from the backing, leaving circular traces on the base. In vivo studies further demonstrated effective tip separation in mice following application of cold compresses (~13 °C) for 3 minutes.

4. Administration parameters

4.1 Applicator

MNs are commonly manually inserted into the skin by thumb force, which is very convenient for patients to use.[55,56] However, due to the difference in the force exerted by each individual and the curvature limitations of the fingers, direct MNs application by hand can compromise the skin insertion rate. The

MNs applicator is a device capable of precisely inserting microneedles into the skin at a predetermined angle and speed, reducing human-induced variability and enabling efficient transdermal drug delivery, Fig. 6.6. Maaden et al.[57] recruited 15 participants with no experience in MNs administration to compare the penetration success rate of manual versus applicator insertion. When participants inserted the MNs into ex vivo human skin using an impact insertion applicator, the skin penetration success rate was approaching 100%, demonstrating low inter- and intra-individual variability. In contrast, the penetration efficiency of MNs with manual administration was only 56% and showed significantly higher inter- and intra-individual variation. Many studies have reported that the usage of an applicator at a predetermined speed or impact energy significantly increased the skin puncture efficiency of MNs.[57,58] Kang et al.[59] invented a latch applicator to assist the administration of MNs, which increased the skin puncture rate of MNs from $43.0 \pm 15.5\%$ to $63.3 \pm 9.4\%$. When the insertion rate of MNs was increased from 2.4 ± 0.3 m/s to 7.1 ± 0.9 m/s by adjusting the impact distance, the skin puncture rate of MNs increased from 62.03.7% to 96.41.8%.

To achieve efficient skin puncture, some key parameters, such as insertion force and speed, are also required to be optimized. In addition, some recent studies also showed that pretreatment of the skin tissue with mechanical vibration was effective in reducing skin resistance and enhancing penetration efficiency. The detailed examples on how to enhance skin puncture and drug delivery efficiency by optimizing these key parameters and vibration process are shown below.

4.2 Insertion force

Insufficient force during MN application may result in incomplete skin insertion and significant drug loss.[60,61] Therefore, it is crucial to disclose the relationship between insertion force and drug delivery efficiency. Cheung et al.[62] used a Franz diffusion cell to determine the permeation amount of insulin into the pig skin pretreated with MNs at different insertion forces. MNs with a needle length of 1100 μm and a base width of 460 μm were applied to the pig skin at the insertion forces of 51.8 N, 60.5 N, and 69.1 N, respectively. After 4 h of administration, the amount of insulin permeated across the skin was negligible at the insertion force of 51.8 N. In contrast, the cumulative drug permeability of insulin reached 35.7% at an insertion force of 69.1 N, which was 8 times higher than that of MNs inserted at the force of 60.5 N (4.3%). These results demonstrate that the increase in insertion force helps to increase skin penetration rate and drug delivery efficiency of MNs.

4.3 Insertion speed

The thixotropic or shear-thinning effect induced by mechanical vibration can make the skin microstructure become less stiff and more viscous, thereby reducing the force required for MNs to insert skin.[58,63,64] Olatunji et al.[65]

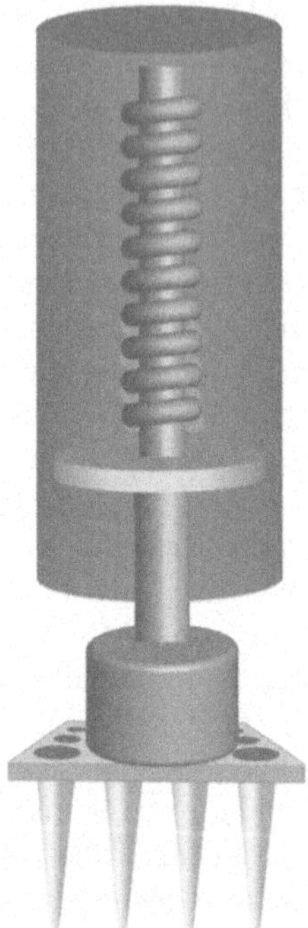

FIGURE 6.6 Schematic diagram of the microneedle applicator.

investigated the insertion resistance of MNs with different interspacing at the insertion speed of 0.5 mm/s and 1 mm/s. As the insertion speed increases, the force required for insertion decreases. The force required for MNs insertion ranges from 0.0216 to 0.0214 N/needle and from 0.03 to 0.028 N/needle at an insertion speed of 1.0 mm/s and 0.5 mm/s, respectively. However, existing studies have not reached a consensus on the effect of insertion speed on puncture force. As reported by et al., [66] no significant change in insertion force was observed at the insertion speed ranging from 10 to 250 μm/s. The negative correlation between MNs insertion resistance and insertion speed can only be established in a certain insertion rate range.[63] The detailed relationship between skin resistance and insertion speed needs to be further investigated.

4.4 Vibration

The thixotropy effect, or shear thinning, occurs when the skin is subjected to vibration. This changes the skin's microstructure, making it more viscous and less stiff.[88–90] Therefore, the force required to insert the MNs into the skin can be reduced by applying a mechanical vibration to the skin.[91,92] In a study reported by Yang et al., [93] they found that with the assistance of vibration, the insertion force applied to the MNs was reduced by approximately 70%, from about 0.26 N/needle to about 0.08 N/needle. In another study, Ebrahiminejad et al.[94] used finite element analysis to study the relationship between skin vibration frequency and MNs insertion force. The vibration frequency ranging from 0 to 150 Hz had little effect on the insertion force, whereas the insertion force decreased from 0.096 to 0.078 N when the vibration frequency was increased from 150 to 250 Hz.

5. Introduction of permeation enhancers

Conventional permeation enhancers, such as sulfoxides, alcohols, and terpenes, facilitate drug penetration primarily by modifying the structure of the stratum corneum or enhancing the solubility of the drug. After MNs pierce the skin, the cargoes liberated from the MNs are still required to traverse the physiological barrier for further absorption and exerting therapeutic purposes. For instance, ECM and biofilms are major components of the physiological barrier that restricts drug diffusion and permeation into deeper skin layers. Hence, scientists have introduced various permeation enhancers as a passive approach to breach the physiological barrier. HA is a major component of ECM that is mainly located in the dermis and forms a highly viscous gel barrier to impede drug diffusion from the epidermis to the dermis.[11,67] Therefore, degradation of HA by hyaluronidase is expected to reduce the resistance to drug permeation into deeper skin layers, thereby enhancing drug bioavailability (Fig. 6.7a). In a recent study, Hu et al.[68] developed hyaluronidase-powered MNs as a universal platform to enhance transdermal drug delivery efficiency. The ability of hyaluronidase to enhance drug permeation was verified in small molecules (RhB), peptides (melittin-Cy5), and biomacromolecules (BSA-FITC), respectively (Fig. 6.7b and c), and showed a dose-dependent manner. Compared to hyaluronidase-free MNs, hyaluronidase-powered MNs led to approximately a 30% increase in the relative bioavailability of salmon calcitonin (sCT) (Fig. 6.7d) and a significant increase in the antiinflammatory effects of melittin. Overall, these findings demonstrate the great potential of hyaluronidase-powered MNs in enhancing the transdermal delivery efficiency of peptides.

The compact ECM within tumor tissues not only diminishes vascular density but also constitutes a major component of stroma to significantly increase interstitial fluid pressure, thereby presenting a significant barrier to drug diffusion. In addition, ECM also serves as a defense barrier against the exposure of tumor cells to T cells, leading to the failure of immunotherapy. Breaching the rigid ECM barrier is a powerful approach to enhance drug delivery efficiency and infiltration of T cells in tumor tissue. Therefore, He et al.[69] developed MNs coloaded with hyaluronidase-modified semiconductor polymer nanoparticles (HSPN) and HSPNTLR3 agonist polyinosinic-polycytidylic acid (PIC) to provide combined photothermal-immunotherapy for melanoma. After the MNs pierced the local tumor tissue, hyaluronidase degraded the ECM to promote the infiltration of therapeutic agents and

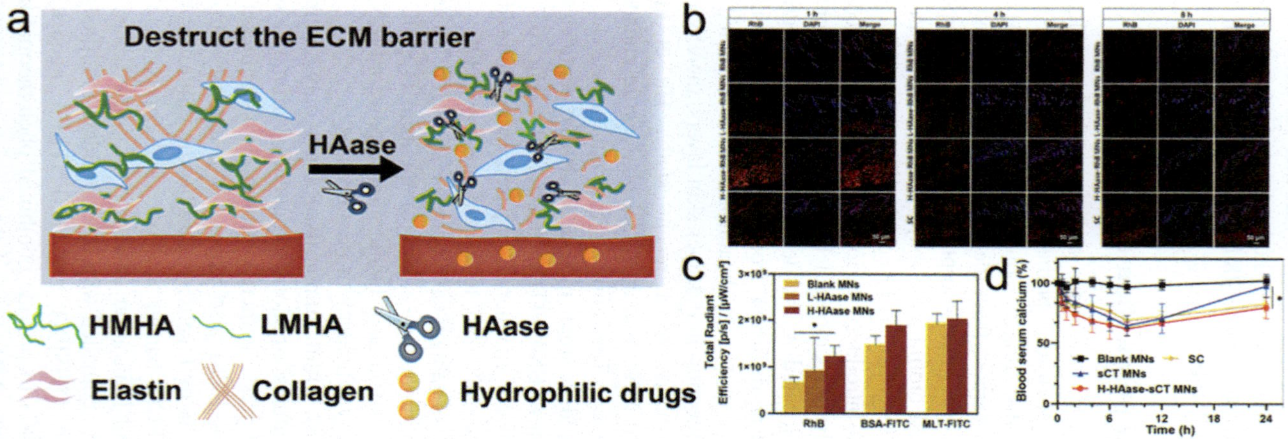

FIGURE 6.7 (a) Diagram illustrating the permeation-promoting mechanism of hyaluronidase, comparing high-molecular-weight hyaluronic acid (HMHA) and low-molecular-weight hyaluronic acid (LMHA). (b) Distribution of rhodamine B (Rhb) in the skin of the indicated groups. (c) Resultant mean fluorescence intensity of the indicated groups. (d) Pharmacokinetic profile of salmon calcitonin (sCT) from different formulations.[66] *Reproduced from Ref.[68] with permission.*

immune cells into the tumor. Upon laser irradiation, photothermal therapy (PTT) was also generated to ablate melanoma, owing to the photothermal performance of HSPN. Compared to the mice treated with hyaluronidase-free MNs, the populations of $CD3^{+}CD4^{+}$ and $CD3^{+}CD8^{+}$ T cells in the mice treated with MNs containing HSPN and PIC increased by 1.4 and 2.4 times, respectively. The findings revealed that the incorporation of hyaluronidase significantly potentiated antitumor immune response.

Biofilm is present in more than 90% of chronic wounds and forms a rigid barrier to impede drug penetration, which eventually complicates bacterial clearance and causes persistent inflammation to hinder wound healing. To disrupt the biofilm barrier, Fan et al.[70] devised an MNs patch loaded with α-amylase capable of degrading EPS to enhance bacterial susceptibility to antimicrobial treatment for accelerated wound healing. The levofloxacin-loaded polydopamine nanoparticles (PDA@Levo) were incorporated in the MNs to achieve a combined antibacterial treatment of antibiotic and mild PTT. Following incubation with PDA@Levo, the residual biomass of the *P. aeruginosa* biofilm remained $85.31 \pm 2.04\%$. In contrast, the residual biomass of the biofilm dropped to only $41.46 \pm 4.04\%$ after coincubation with α-amylase and PDA@Levo. The in vivo antibiofilm effects of the MNs were further evaluated on the rats with wounds infected by *S. aureus* and *P. aeruginosa*. With the assistance of α-amylase, PDA@Levo more efficiently eradicated the bacteria, resulting in accelerated wound healing.

6. Implementation of active permeation-enhancing strategies

6.1 Light-actuated microneedles

PTT employs photosensitizers to convert NIR energy into heat.[71–73] Renowned for its noninvasiveness, high targeting precision, and spatiotemporal controllability, PTT has found wide application in treating tumors, infections, and other diseases. Importantly, the hyperthermia generated by PTT accelerates MNs dissolution, enhances drug release, and improves transdermal permeability,[74,75] positioning light-actuated MNs as a promising strategy for topical skin disease treatment.

Take psoriasis as an example, a chronic inflammatory skin disease affecting 2%–3% of the global population, characterized by skin hyperplasia, scaling, and erythema that severely impacts patient well-being.[76–78] Using photosensitive materials like MXene, these MNs convert NIR energy into heat, rapidly elevating local temperature to trigger MNs dissolution and release therapeutic agents such as IL-17 monoclonal antibodies (Fig. 6.8a). Fluorescence imaging demonstrates that NIR-irradiated MNs achieve a drug penetration depth of 225 μm-three times deeper than nonirradiated controls (75 μm)-confirming PTT's role in enhancing transdermal drug delivery (Fig. 6.8b).[79] In psoriatic

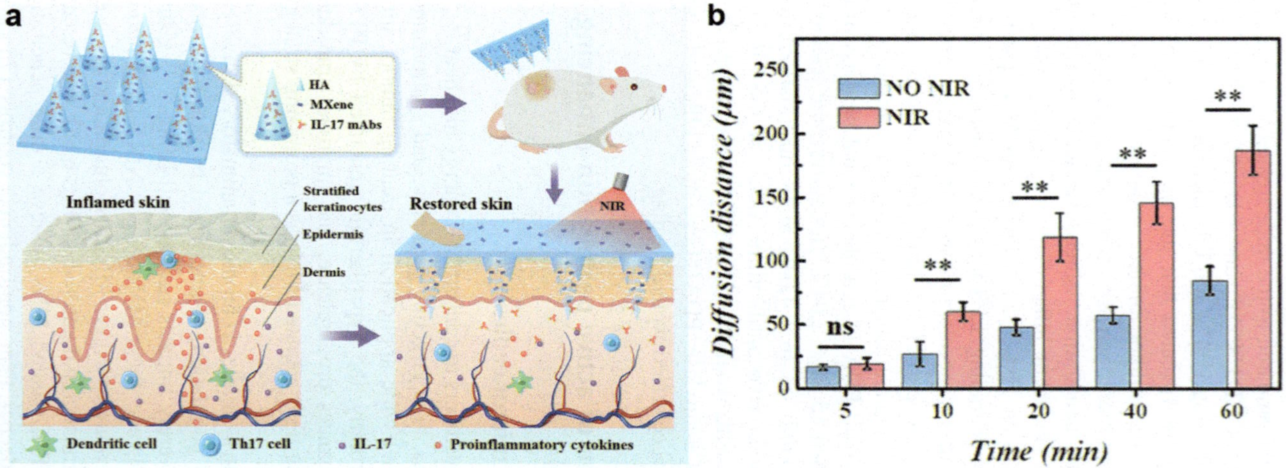

FIGURE 6.8 Development of photoresponsive microneedles (MNs) for enhanced drug delivery and therapeutic efficiency: (a) Schematic illustration of photoresponsive MN patches loaded with interleukin-17 monoclonal antibodies (IL-17 mAbs) and MXene for the treatment of psoriasis; (b) Diffusion distance of a fluorescence probe released from nonirradiated and near-infrared (NIR)-irradiated MNs in skin tissue at different time points. *Reproduced from Ref.*[79] *with permission.*

mouse models, PTT-MNs treatment outperforms intradermal drug injection in alleviating symptoms, with reduced Psoriasis Area and Severity Index (PASI) scores and epidermal thickness, indicating potent inflammation mitigation. In summary, photothermal MNs represent a robust approach to enhance drug delivery, offering significant promise for the topical treatment of inflammatory skin conditions.

The presence of biofilm in the wounds can induce excessive inflammation and form a rigid biofilm barrier against drug penetration, resulting in elevated antibiotic resistance and intractable wound healing. Although MNs can directly puncture the biofilm for enhanced drug delivery, the highly compact EPS may impede the penetration of therapeutics in the biofilm, leading to treatment failure. It has been reported that PTT can generate hyperthermia to destroy the bioactive substrates and intact structure of biofilms, thereby promoting drug penetration in the biofilm.[80] Iron is indispensable for bacterial proliferation, metabolic activities, and the survival of immune cells. In view of this, Zhu et al.[81] developed a photothermal MNs (FGO@MN) patch integrating iron-actuated Janus ion therapy to target biofilms and modulate immune cell iron metabolism. In diabetic wound models, FGO@MN with NIR achieved complete healing by day 15 and eradicated *S. aureus*, showing strong potential for treating biofilm-infected wounds.

6.2 Ultrasound-actuated microneedle delivery system

Ultrasound, a type of high-frequency mechanical wave, exhibits the capability to penetrate human skin and connective tissues to a depth exceeding 5 cm-significantly greater than that of NIR light. The penetration depth of NIR, even NIR-II, is limited to only 10 mm below the skin tissue, and the light intensity decreases progressively with increasing penetration depth.[82,83] Different from NIR-mediated PTT, the ultrasound-mediated sonodynamic therapy (SDT) has multiple therapeutic effects: (1) sonomechanical effect, which refers to the generation of liquid microjets, microstreaming shock waves, and shear forces under the action of cavitation effect occurring in the interface between liquid and sonosensitizers; (2) sonochemical effect, which refers to the production of massive ROS to kill cells or bacteria.[82,84,85] Owing to the high tissue penetration capability and multiple therapeutic effects, SDT has received extensive attention in the biomedical field. As detailed below, the integration of ultrasound to MNs offers promising potential to enhance drug delivery efficiency and improve therapeutic outcomes.[86]

Shao et al.[87] reported a nanobubble-modified MNs capable of ultrasound-responsive drug release (Fig. 6.9a). Briefly, the MNs were made of PLGA needles coated with positively charged polylysine, which enabled the adsorption of liposome-encapsulated perfluorohexane nanobubbles onto the outer layer of the MNs via electrostatic adsorption. Under ultrasound, the nano bubbles undergo a phase transition to form large bubbles, thereby generating a

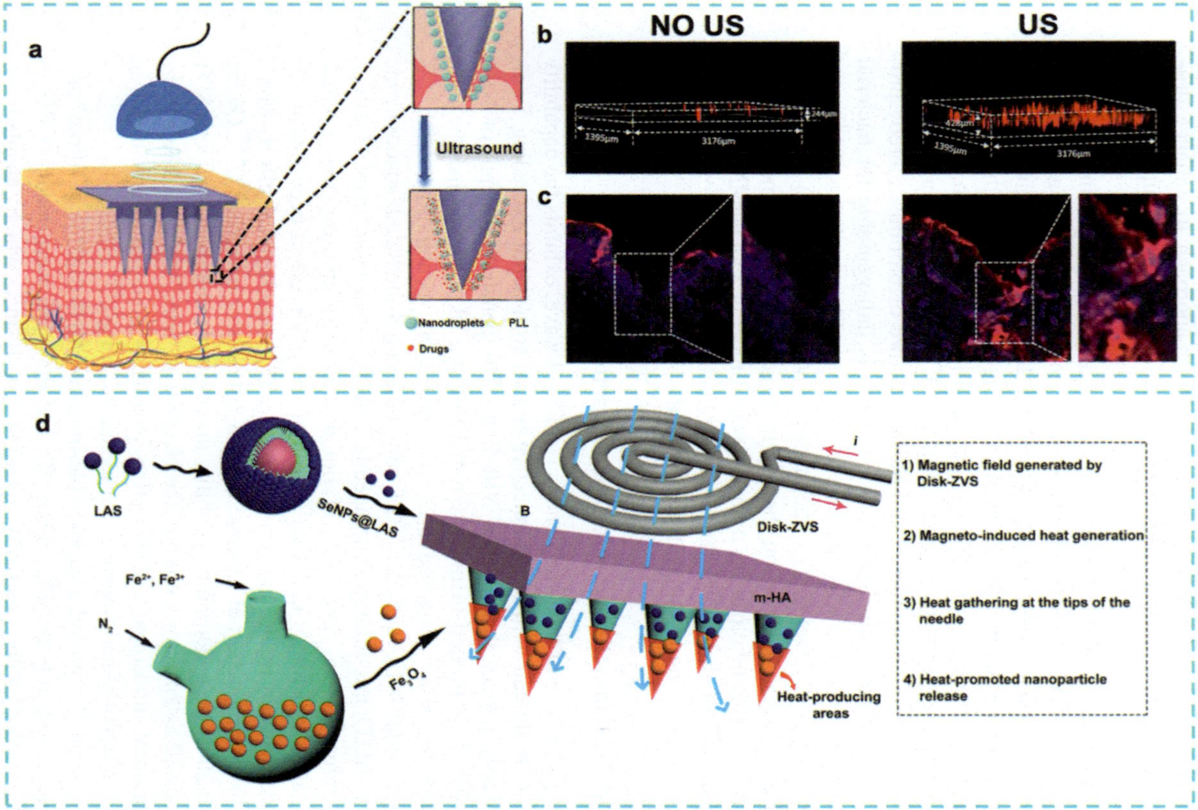

FIGURE 6.9 Ultrasonic-driven and magnetothermal responses improve drug penetration efficiency and therapeutic efficacy: (a) Schematic diagram illustrating dermal drug delivery by nanobubble ultrasound-responsive microneedles (MNs); (b) Three-dimensional (3D) projection of a vertical section of simulated skin (gelatin gel) post MN patch treatment with or without ultrasound; (c) Histological images of skin tissues treated with MNs with or without ultrasound. *(a–c) These were reproduced from Ref.*[87] *with permission. (d) Were reproduced from Ref.*[93] *with permission.*

cavitation effect to accelerate drug release. Cy5 was used as a model drug to study skin penetration. Without sonophoresis, Cy5 showed limited diffusion (244 μm) and remained mostly in MNs-created micropores. In contrast, ultrasound-responsive MNs achieved a penetration depth of 428 μm, 1.75 times greater than the control (Fig. 6.9b and c). Yang et al. [88] developed MNs loaded with the composite structure of zinc porphyrin–based MOF and zinc oxide (ZnTCPP@ZnO) to treat acne. Under ultrasound, a large amount of ROS was produced to destroy the normal structure of *P. acnes*, leading to massive protein leakage and a bacterial killing rate of 99.73%. The pharmacodynamic studies further revealed that the MNs loaded withZnTCPP@ZnO efficiently eradicated *P.acnes* under ultrasound and accelerated the recovery of acne by day 7.

6.3 Magnet-actuated microneedles

Magnetic fields, arising from synchronous microscopic particle motion, exhibit radiation-like characteristics through coordinated magnetic interactions.[89–91] Magneto-thermal therapy uses magnetic nanoparticles to generate heat under alternating magnetic fields, showing broad applications in treating tumors and infections.[91,92] Unlike NIR light or ultrasound, magnetic fields penetrate skin tissue without intensity loss, making them ideal for deep skin infection treatment.[93] However, conventional use is limited by weak field strength and bulky equipment, leading to high costs and cumbersome operation.

To address these challenges, a bilayer MNs patch integrated with a simplified electromagnetic device enables portable magneto-thermal therapy for infected diabetic wounds[93] (Fig. 6.9d). The MNs tips, loaded with selenium nanoparticles (SeNPs) for antimicrobial effects, and magnetic Fe_3O_4 nanoparticles in the tails, generate heat under a consistent magnetic field. Within 5 minutes, the MNs reach 56.1 °C, demonstrating efficient magneto-thermal conversion. In diabetic mouse models, the magnetic field-treated MNs group showed over 50% reduction in wound bacterial colonies and a significantly accelerated healing process, achieving an 88.2% wound closure rate by day 14. This highlights the dual benefits of SeNPs and magneto-thermal therapy in enhancing antimicrobial efficacy and tissue regeneration, overcoming limitations of conventional magnetic systems through simplified, efficient design.

6.4 Electricity-actuated microneedle delivery system

Electrical stimulation has gained significant attention in transdermal drug delivery and dermatological treatments, as it drives ionized/hydrophilic drug transport across the stratum corneum and activates intracellular signaling for adjuvant therapy. Electricity-actuated systems, categorized by power source into exogenous stimulation, iontophoresis, and self-powered triboelectric nanogenerators (TENG), face limitations: single electrical stimulation fails to deliver biomacromolecules >13 KDa due to the stratum corneum barrier. Integrating electrical stimulation with microneedles (MNs) bypasses this barrier, enhancing transdermal delivery efficiency and therapeutic outcomes. This synergy has made electricity-actuated MNs systems a focal point in biomedical research.[86,94,95]

6.5 Exogenous electrical stimulation

Electrical stimulation in regenerative medicine activates intracellular pathways and enhances deep tissue drug penetration, with combined electrotherapy-drug miniaturized devices outperforming monotherapies.[96,97] However, transcutaneous electrical stimulation is limited by skin barriers, reducing penetration depth and efficacy. Integrating MNs with electrical stimulation offers a promising solution for improved tissue regeneration.

Xu et al.[98] developed an implantable bioresorbable MNs (IBMN) device integrated with a drug-loaded MNs patch and a radio frequency-based wireless power transmission system for skeletal muscle regeneration (Fig. 6.10a). Ibuprofen was loaded in the needles made of PLGA for sustained drug release and long-acting antiinflammation. In vitro cell scratch assay demonstrated that the electrostimulation exerted a positive effect on the migration of myoblasts, which is beneficial for muscle regeneration (Fig. 6.10b and c). Polypyrrole (PPy) is a widely used electrically responsive material that can provide electrically-triggered drug release performance by transformation between the oxidative and reductive states.[99]

Guo et al.[100] designed a polypyrrole (PPy)-based conductive MNs patch for atopic dermatitis, where electrical stimulation triggered PPy redox changes to accelerate drug release (Fig. 6.10d). In vitro, fluorescein sodium release correlated with voltage, while in vivo imaging showed 10-fold higher fluorescent intensity in treated mice at 120 minutes. The patch outperformed controls in a dermatitis model by delivering more drug to lesions, demonstrating electrically responsive therapy.

6.6 Iontophoresis

Iontophoresis enhances transdermal delivery of charged drugs via mild electric currents, adjusting parameters like intensity/duration to tailor drug transport.[101,102] Yet single systems fail to breach the stratum corneum, limiting efficiency. Integrating iontophoresis with MNs combines their benefits: MNs bypass the skin barrier while iontophoresis drives charged molecules, achieving synergistic enhancement via the "Pork, Iontophoresis, and Penetration" strategy.[103] This approach overcomes physical and electrical transport limitations, offering a robust solution for improved transdermal delivery of charged therapeutics.

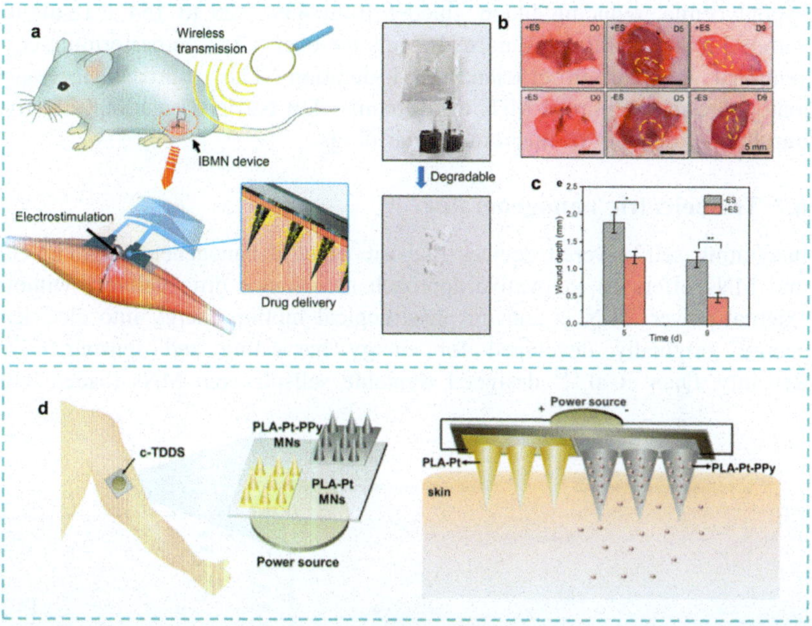

FIGURE 6.10 Electrical stimulation enhances drug delivery efficiency and therapeutic effect of microneedles (MNs): (a) Development of an implantable bioelectronic microneedle (IBMN) device for wireless electrostimulation and drug delivery; (b) Representative photographs of muscle before treatment and after treatment for 5 and 9 days; (c) Wound depth of muscle tissue following treatment with or without electrostimulation. *(a–c) Were reproduced from Ref.*[98] *with permission. (d) Were reproduced from Ref.*[100]*with permission.*

MN-mediated percutaneous immunotherapy shows promise in infectious disease prevention/treatment by enhancing vaccine delivery.[104,105] Ying et al.[103] developed a wearable iontophoresis-driven MNs patch to improve transdermal delivery of macromolecular vaccines (Fig. 6.10a). Using ovalbumen (OVA) as a model, the patch achieved 17-fold higher cumulative permeability (47.57 vs. 2.80 μg in free solution) at 30 minutes, outperforming single MNs or iontophoresis alone (Fig. 6.10b). In vivo studies demonstrated stronger immune responses than conventional intramuscular injection, highlighting the patch's ability to enhance biologic delivery. This integrated approach combines MNs' skin-penetrating capability with iontophoresis's electrical-driven transport, offering a feasible strategy for efficient transdermal vaccine delivery.

Li et al.[106] developed an iontophoresis-driven porous microneedle array patch (IDPMAP) to enhance transdermal delivery of charged macromolecules like insulin. The system combines porous MNs with electrical stimulation to drive drug release. In type 1 diabetic rats, IDPMAP increased serum insulin levels in a dose-dependent manner, achieving 89% relative bioavailability-comparable to subcutaneous injection. It induced rapid, sustained hypoglycemia, reducing blood glucose from 439 ± 26 to 154 ± 18 mg/dL within 3 h and maintaining normal levels for over 5.4 h, outperforming conventional injection in efficiency and duration. This approach leverages iontophoresis and porous MNs to overcome skin barriers, enabling effective transdermal delivery of macromolecular drugs.

6.7 Triboelectric nanogenerator

Integrating self-powered devices like triboelectric nanogenerators (TENGs) into MNs offers an innovative approach to enhance drug delivery without external power. TENGs convert physiological motion energy into electrical energy, originally developed for energy harvesting and sensing.[107–110] Recently, Qian et al.[111] designed wearable self-powered MNs loaded with

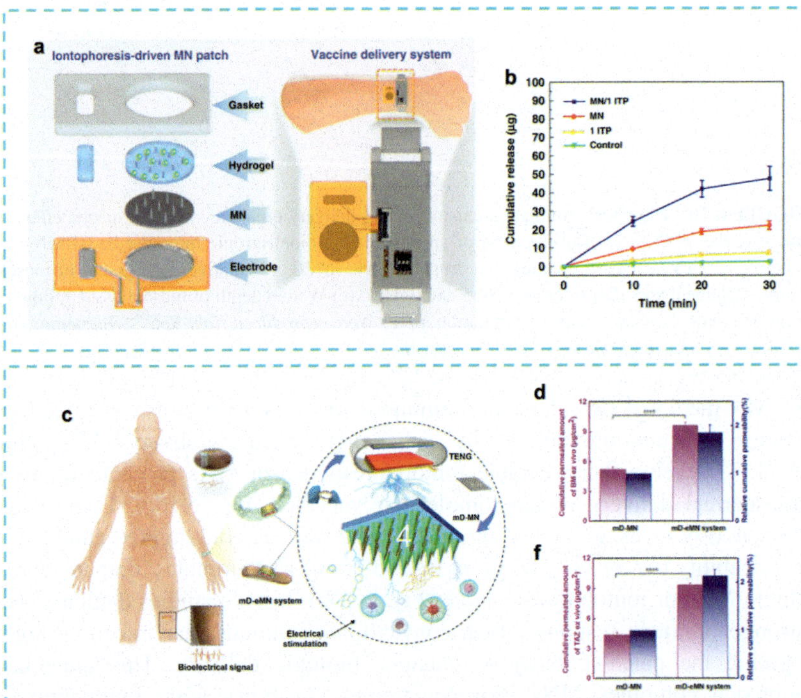

FIGURE 6.11 Implementation of iontophoresis and triboelectric nanogenerators (TENGs) to enhance drug delivery efficiency and therapeutic effects of microneedles (MNs): (a) Schematic diagram of wearable iontophoresis-driven MN patch for transdermal delivery of vaccine; (b) In vitro cumulative release amount of ovalbumin (OVA); (c) Structural diagram of the microneedle-driven electro-microneedle (mD-eMN) system for promoting drug absorption and tissue homeostasis reconstruction; (d) Cumulative permeated amount of berberine (BM) and tazarotene (TAZ) across rat skin from microneedle-driven microneedles (mD-MNs) and microneedle-driven electro-microneedles (mD-eMNs); (c–f). *(a–b) Were reproduced from Ref.[103] with permission. (c–f) Were reproduced from Ref.[111] with permission.*

tazarotene (TAZ) and betamethasone (BM) for psoriasis treatment (Fig. 6.11c). TAZ in the outer MNs layer regulates immunity and cell proliferation, while BM in the inner layer mitigates TAZ-induced irritation. TENG-generated pulsed electrons boost drug release and transdermal permeation, doubling cumulative permeability via an electroporation-like effect (Fig. 6.11d). These electrons also activate electroactive cells, promoting skin tissue repair and reducing psoriatic inflammation. This self-powered system merges mechanical energy conversion with dual-drug delivery, demonstrating enhanced therapeutic efficiency for inflammatory skin conditions through synergistic electrical and pharmacological actions.

7. Gas-propelled microneedles

7.1 Gas-propelled microneedles

Gas-propelled MNs represent an emerging class of transdermal delivery systems that harness in situ gas generation to enhance drug penetration and release. By incorporating gas-generating agents such as bicarbonates and acids into the MNs matrix, these systems produce gas (e.g., CO_2, NO, O_2, or H_2) upon contact with interstitial fluid. The resulting gas bubbles generate localized pressure, accelerating needle dissolution, drug diffusion, and tissue permeation.

Carbon dioxide (CO_2) is commonly employed in gas-propelled MNs to facilitate drug delivery.[112] For instance, Zhang et al.[113] used potassium carbonate (K_2CO_3) and citric acid (CA) as effervescent agents to prepare CO_2-propelled MNs loaded with rivastigmine (RVT) to treat Alzheimer's disease (Fig. 6.12a). Both free RVT and RVT microparticles were loaded in the MNs to provide early intervention and prolonged treatment of Alzheimer's disease, respectively. After immersing the CO_2-propelled MNs into the simulated ISF, the MNs rapidly absorbed water to trigger the chemical reaction of K_2CO_3 and CA, thereby producing massive CO_2 gas bubbles to promote the disintegration of needles and further drug release. Indocyanine green (ICG) and Rhodamine B particles (RhB particles) were used to replace free RVT and RVT microparticles to observe drug permeation in the skin. As shown in Fig. 6.12b, CO_2-propelled MNs promoted a larger amount of drug permeated into deeper skin layers than passive MNs. The pharmacokinetic studies demonstrated that compared to passive MNs, the maximum blood concentration and area under the curve of the CO_2-propelled MNs increased from approximately 152.20 to 288.00 ng/mL and 830.25 to 1558.16 h ng/mL, respectively. As a result, the CO_2-propelled MNs produced a significantly higher bioavailability than the passive MNs, better restoring the memory and learning ability of SD rats with memory deficiency.

7.2 Nitric oxide-propelled microneedles

Nitric oxide (NO) is an endogenous gas molecule that has multiple proangiogenic, antiinflammatory, and antimicrobial activities, showing a broad application in the

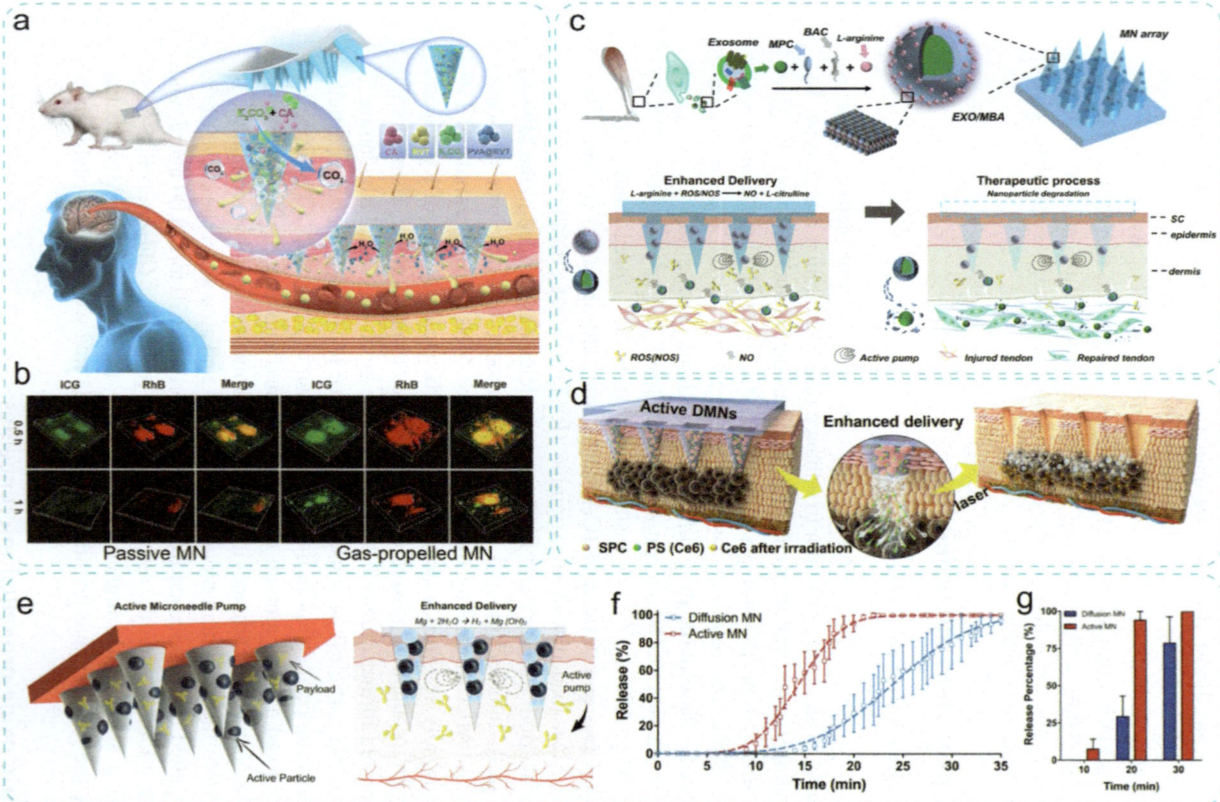

FIGURE 6.12 (a) Schematic illustration of carbon dioxide (CO_2)-propelled microneedles. (b) Comparison of the transdermal permeability of passive MNs and gas-propelled MNs in the isolated rat skin. (c) Development of nitric oxide (NO) nanomotor-propelled MNs for acute trauma (AT) healing and their therapeutic mechanism. (d) Development of oxygen (O_2)-propelled MNs for enhanced delivery of chlorin e6 (Ce6) in the melanoma and associated antitumor effects. (e) Schematic diagram of hydrogen-propelled MNs loaded with magnesium (Mg) particles. (f–g) Release profiles of immunoglobulin G (IgG) from active MNs compared with passive MNs. *(a–b) Were reproduced from Ref.*[113] *with permission. (c) Adapted from Ref.*[118] *with permission. (d) Adapted from Ref.*[122] *with permission. (e-g) Were reproduced from Ref.*[127] *with permission.*

biomedical field.[114–117] In situ generation of NO in the pathological tissue can not only produce auxiliary therapeutic effects, but also promote drug penetration into deeper skin tissues, resulting in enhanced therapeutic efficiency. For instance, Liu et al.[118] developed NO nanomotor-propelled MNs loaded with exosomes (EXO) for Achilles tendinopathy (AT) healing (Fig. 6.12c). In this study, the surface of EXOs was decorated with 2-Methacryloyloxyethyl phosphorylcholine (MPC) and N,N′-bis(acryloyl)cystamine (BAC) to form EXO/MBA particles via electrostatic interactions. Then, L-arginine was inserted on the surface of EXO/MBA particles to construct a NO nanomotor, which was further loaded in the MNs for AT healing. In the inflammatory microenvironment of AT, L-arginine can be converted into NO by endogenous ROS and NO synthase (NOS), thus driving the diffusion of EXO toward the injury site to alleviate inflammation. Compared with EXO injection and EXO/MBA-loaded MNs, the NO nanomotor-propelled MNs promoted the healing of AT most effectively, as evidenced by a higher proliferation rate of tendon cells and expression of Collagen 1a.

7.3 Oxygen-propelled microneedles

Oxygen (O_2) is an essential gas molecule involved in various physiological and pathological processes. As a critical element for cellular respiration, it plays a vital role in energy production and tissue repair. Oxygen therapy has broad applications in the biomedical field, particularly for improving oxygenation in hypoxic tissues to assist treatment of tumors and wounds.[119–121]

PDT involves the use of near-infrared laser-activated photosensitizers to generate highly cytotoxic reactive ROS for therapeutic purposes. PDT has emerged as a promising approach for tumor treatment, owing to its noninvasive nature, favorable safety profile, and high targeting efficiency. However, the therapeutic efficiency of PDT against tumors is largely limited by two major reasons: (1) the passive diffusion of photosensitizers greatly limits their penetration in tumor tissues, (2) the hypoxic tumor microenvironment further aggravates the productivity of ROS during PDT. To address the two major issues, Liu et al.[122] developed O_2-propelled MNs loaded with photosensitizer (Ce6) and sodium percarbonate (SPC) particles for the PDT of breast cancer (Fig. 6.12d). Once inserted in the skin, the MNs absorbed ISF to initiate the chemical reaction of SPC particles with water, thereby producing massive O_2. The generated O_2 bubbles can not only form a violent vortex to promote the penetration of Ce6 into deeper tumor tissues, but also increase local oxygen concentration to relieve the hypoxic tumor environment, resulting in a larger amount of ROS generation during PDT. As expected, the O_2-propelled MNs prohibited the tumor growth and prolonged the survival rate of 4T1 tumor-bearing mice more efficiently than the passive MNs.

7.4 Hydrogen-propelled microneedles

Hydrogen (H_2) is a recently identified functional gas molecule that demonstrates antioxidant, antiinflammatory, and immunoregulatory properties.[123–126]

In a recent study, Lopez-Ramirez et al.[127] developed an active MNs patch hydrogen-propelled by utilizing magnesium (Mg) particles as a built-in engine to enhance autonomous drug delivery (Fig. 6.12e). Upon insertion into the skin, the embedded Mg particles rapidly react with interstitial fluid (ISF), triggering an explosive generation of H_2 bubbles that facilitate drug release and promote penetration into deeper skin layers. COMSOL Multiphysics simulations confirmed the resulting fluid dynamics, demonstrating bubble nucleation and upward movement driven by buoyancy forces. The active MNs achieved an IgG release rate of $58.5 \pm 15.1\%$ within 15 minutes, which showed a 15-fold increase compared to passive MNs ($3.8 \pm 1.8\%$) (Fig. 6.12f and g). Furthermore, when loaded with anti-CTLA-4 checkpoint blockade as a model drug, the active MNs exhibited superior antitumor efficacy in an orthotopic melanoma model. Compared to both subcutaneous anti-CTLA4 injections and passive MNs, the active MNs elicited a more robust immune response, significantly delaying tumor progression and extending survival in melanoma-bearing mice.

References

1. Yang, D.; Chen, M.; Sun, Y.; Jin, Y.; Lu, C.; Pan, X.; Quan, G.; Wu, C. Microneedle-Mediated Transdermal Drug Delivery for Treating Diverse Skin Diseases. *Acta Biomater.* **2021,** *121,* 119–133.
2. Qu, F.; Geng, R.; Liu, Y.; Zhu, J. Advanced Nanocarrier- and Microneedle-Based Transdermal Drug Delivery Strategies for Skin Diseases Treatment. *Theranostics* **2022,** *12* (7), 3372–3406.
3. Coulman, S.; Allender, C.; Birchall, J. Microneedles and Other Physical Methods for Overcoming the Stratum Corneum Barrier for Cutaneous Gene Therapy. *Crit. Rev. Ther. Drug Carrier Syst.* **2006,** *23* (3), 205–258.
4. Leng, F.; Zheng, M.; Xu, C. 3D-Printed Microneedles with Open Groove Channels for Liquid Extraction. *Exploration* **2021,** *1* (3), 20210109.
5. Creighton, R. L.; Woodrow, K. A. Microneedle-Mediated Vaccine Delivery to the Oral Mucosa. *Adv. Healthcare Mater.* **2019,** *8* (4), e1801180.
6. Yang, Y.; Zhou, R.; Wang, Y.; Zhang, Y.; Yu, J.; Gu, Z. Recent Advances in Oral and Transdermal Protein Delivery Systems. *Angew. Chem. Int. Ed. Engl.* **2023,** *62* (10).
7. Zhang, Y.; Wang, S.; Yang, Y.; Zhao, S.; You, J.; Wang, J.; Cai, J.; Wang, H.; Wang, J.; Zhang, W.; Yu, J.; Han, C.; Zhang, Y.; Gu, Z. Scarless Wound Healing Programmed by Core-Shell Microneedles. *Nat. Commun.* **2023,** *14* (1), 3431.
8. Lee, H.; Song, C.; Baik, S.; Kim, D.; Hyeon, T.; Kim, D.-H. Device-Assisted Transdermal Drug Delivery. *Adv. Drug Deliv. Rev.* **2018,** *127,* 35–45.
9. Chaudhuri, O.; Cooper-White, J.; Janmey, P. A.; Mooney, D. J.; Shenoy, V. B. Effects of Extracellular Matrix Viscoelasticity on Cellular Behaviour. *Nature* **2020,** *584* (7822), 535–546.
10. Humphrey, J. D.; Dufresne, E. R.; Schwartz, M. A. Mechanotransduction and Extracellular Matrix Homeostasis. *Nat. Rev. Mol. Cell Biol.* **2014,** *15* (12), 802–812.
11. Sutherland, T. E.; Dyer, D. P.; Allen, J. E. The Extracellular Matrix and the Immune System: A Mutually Dependent Relationship. *Science* **2023,** *379* (6633), eabp8964.

12. Winkler, J.; Abisoye-Ogunniyan, A.; Metcalf, K. J.; Werb, Z. Concepts of Extracellular Matrix Remodelling in Tumour Progression and Metastasis. *Nat. Commun.* **2020,** *11* (1), 5120.
13. Yang, Z.-R.; Suo, H.; Fan, J.-W.; Lv, N.; Du, K.; Ma, T.; Qin, H.; Li, Y.; Yang, L.; Zhou, N.; Jiang, H.; Tao, J.; Zhu, J. Endogenous Stimuli-Responsive Separating Microneedles to Inhibit Hypertrophic Scar through Remodeling the Pathological Microenvironment. *Nat. Commun.* **2024,** *15* (1), 2038.
14. Mohanty, R. P.; Liu, X.; Ghosh, D. Electrostatic Driven Transport Enhances Penetration of Positively Charged Peptide Surfaces through Tumor Extracellular Matrix. *Acta Biomater.* **2020,** *113,* 240–251.
15. Sun, B.; Kjelleberg, S.; Sung, J. J. Y.; Zhang, L. Micro- and Nanorobots for Biofilm Eradication. *Nat. Rev. Bioeng.* **2024,** *2* (5), 367–369.
16. Yang, L.; Zhang, D.; Li, W.; Lin, H.; Ding, C.; Liu, Q.; Wang, L.; Li, Z.; Mei, L.; Chen, H.; Zhao, Y.; Zeng, X. Biofilm Microenvironment Triggered Self-Enhancing Photodynamic Immunomodulatory Microneedle for Diabetic Wound Therapy. *Nat. Commun.* **2023,** *14* (1), 7658.
17. Wang, X.; Wang, D.; Lu, H.; Wang, X.; Wang, X.; Su, J.; Xia, G. Strategies to Promote the Journey of Nanoparticles against Biofilm-Associated Infections. *Small* **2024,** *20* (10), 2305988.
18. Jing, X.; Meng, L.; Fan, S.; Yang, T.; Zhang, N.; Xu, R.; Zhao, X.; Yang, H.; Yang, Z.; Wang, D.; Liang, Y.; Zhou, G.; Ji, W.; She, J. Tumor Microenvironment Self-Regulation: Bimetallic Metal Nanozyme-Derived Multifunctional Nanodrug for Optimizable Cascade Catalytic Reaction-Synergetic Anti-tumor Theranostics. *Chem. Eng. J.* **2022,** *442,* 136096.
19. Zhang, Q.; He, M.; Zhang, X.; Yu, H.; Liu, J.; Guo, Y.; Zhang, J.; Ren, X.; Wang, H.; Zhao, Y. Tumor Microenvironment Activated Chemodynamic–Photodynamic Therapy by Multistage Self-Assembly Engineered Protein Nanomedicine. *Adv. Funct. Mater.* **2022,** *32* (17), 2112251.
20. Feng, Q.; Bennett, Z.; Grichuk, A.; Pantoja, R.; Huang, T.; Faubert, B.; Huang, G.; Chen, M.; DeBerardinis, R. J.; Sumer, B. D.; Gao, J. Severely Polarized Extracellular Acidity Around Tumour Cells. *Nat. Biomed. Eng.* **2024,** *8* (6), 787–799.
21. He, L.; Zheng, N.; Wang, Q.; Du, J.; Wang, S.; Cao, Z.; Wang, Z.; Chen, G.; Mu, J.; Liu, S.; Chen, X. Responsive Accumulation of Nanohybrids to Boost NIR-Phototheranostics for Specific Tumor Imaging and Glutathione Depletion-Enhanced Synergistic Therapy. *Adv. Sci.* **2023,** *10* (1), 2205208.
22. Zhang, Y.; Feng, P.; Yu, J.; Yang, J.; Zhao, J.; Wang, J.; Shen, Q.; Gu, Z. ROS-Responsive Microneedle Patch for Acne Vulgaris Treatment (Adv. Therap. 3/2018). *Adv. Therapeut.* **2018,** *1* (3), 1870006.
23. Wang, Z.; Fu, R.; Han, X.; Wen, D.; Wu, Y.; Li, S.; Gu, Z. Shrinking Fabrication of a Glucose-Responsive Glucagon Microneedle Patch. *Adv. Sci.* **2022,** *9* (28), 2203274.
24. Bauleth-Ramos, T.; El-Sayed, N.; Fontana, F.; Lobita, M.; Shahbazi, M.-A.; Santos, H. A. Recent Approaches for Enhancing the Performance of Dissolving Microneedles in Drug Delivery Applications. *Mater. Today* **2023,** *63,* 239–287.
25. Lu, Z.; Du, S.; Li, J.; Zhang, M.; Nie, H.; Zhou, X.; Li, F.; Wei, X.; Wang, J.; Liu, F.; He, C.; Yang, G.; Gu, Z. Langmuir–Blodgett-Mediated Formation of Antibacterial Microneedles for Long-Term Transdermal Drug Delivery. *Adv. Mater.* **2023,** *35* (38), 2303388.
26. Fentahun Darge, H.; Lee, C.-Y.; Lai, J.-Y.; Lin, S.-Z.; Harn, H.-J.; Chen, Y.-S.; Tsai, H.-C. Separable Double-Layered Microneedle-Based Transdermal Codelivery of DOX and LPS

for Synergistic Immunochemotherapy of a Subcutaneous Glioma Tumor. *Chem. Eng. J.* **2022,** *433,* 134062.

27. Gittard, S. D.; Chen, B.; Xu, H.; Ovsianikov, A.; Chichkov, B. N.; Monteiro-Riviere, N. A.; Narayan, R. J. The Effects of Geometry on Skin Penetration and Failure of Polymer Microneedles. *J. Adhes. Sci. Technol.* **2013,** *27* (3), 227–243.
28. De Martino, S.; Battisti, M.; Napolitano, F.; Palladino, A.; Serpico, L.; Amendola, E.; Martone, A.; De Girolamo, P.; Squillace, A.; Dardano, P.; De Stefano, L.; Dello Iacono, S. Effect of Microneedles Shape on Skin Penetration and Transdermal Drug Administration. *Biomater. Adv.* **2022,** *142,* 213169.
29. Kochhar, J. S.; Quek, T. C.; Soon, W. J.; Choi, J.; Zou, S.; Kang, L. Effect of Microneedle Geometry and Supporting Substrate on Microneedle Array Penetration into Skin. *J. Pharmacol. Sci.* **2013,** *102* (11), 4100–4108.
30. Zhang, Y.; A. Campbell, S.; Karthikeyan, S. Finite Element Analysis of Hollow Out-Of-Plane HfO2 Microneedles for Transdermal Drug Delivery Applications. *Biomed. Microdevices* **2018,** *20* (1), 19.
31. Davis, S. P.; Landis, B. J.; Adams, Z. H.; Allen, M. G.; Prausnitz, M. R. Insertion of Microneedles into Skin: Measurement and Prediction of Insertion Force and Needle Fracture Force. *J. Biomech.* **2004,** *37* (8), 1155–1163.
32. Lee, J. W.; Park, J.-H.; Prausnitz, M. R. Dissolving Microneedles for Transdermal Drug Delivery. *Biomaterials* **2008,** *29* (13), 2113–2124.
33. Qallaf, A.; Das, D. B.; Davidson, A. Transdermal Drug Delivery by Coated Microneedles: Geometry Effects on Drug Concentration in Blood. *Asia Pac. J. Chem. Eng.* **2009,** *4* (6), 845–857.
34. Yan, G.; Warner, K. S.; Zhang, J.; Sharma, S.; Gale, B. K. Evaluation Needle Length and Density of Microneedle Arrays in the Pretreatment of Skin for Transdermal Drug Delivery. *Int. J. Pharm.* **2010,** *391* (1–2), 7–12.
35. Shu, W.; Heimark, H.; Bertollo, N.; Tobin, D. J.; O'Cearbhaill, E. D.; Annaidh, A. N. Insights into the Mechanics of Solid Conical Microneedle Array Insertion into Skin Using the Finite Element Method. *Acta Biomater.* **2021,** *135,* 403–413.
36. Chen, B. Z.; Liu, J. L.; Li, Q. Y.; Wang, Z. N.; Zhang, X. P.; Shen, C. B.; Cui, Y.; Guo, X. D. Safety Evaluation of Solid Polymer Microneedles in Human Volunteers at Different Application Sites. *ACS Appl. Bio Mater.* **2019,** *2* (12), 5616–5625.
37. Wermeling, D. P.; Banks, S. L.; Hudson, D. A.; Gill, H. S.; Gupta, J.; Prausnitz, M. R.; Stinchcomb, A. L. Microneedles Permit Transdermal Delivery of a Skin-Impermeant Medication to Humans. *Proc. Natl. Acad. Sci.* **2008,** *105* (6), 2058–2063.
38. Wang, C.; Jiang, X.; Zeng, Y.; Terry, R. N.; Li, W. Rapidly Separable Microneedle Patches for Controlled Release of Therapeutics for Long-Acting Therapies. *Med. Drug Discov.* **2022,** *13,* 100118.
39. Zhu, D. D.; Wang, Q. L.; Liu, X. B.; Guo, X. D. Rapidly Separating Microneedles for Transdermal Drug Delivery. *Acta Biomater.* **2016,** *41,* 312–319.
40. Chu, L. Y.; Prausnitz, M. R. Separable Arrowhead Microneedles. *J. Contr. Release* **2011,** *149* (3), 242–249.
41. Hou, A.; Quan, G.; Yang, B.; Lu, C.; Chen, M.; Yang, D.; Wang, L.; Liu, H.; Pan, X.; Wu, C. Rational Design of Rapidly Separating Dissolving Microneedles for Precise Drug Delivery by Balancing the Mechanical Performance and Disintegration Rate. *Adv. Healthcare Mater.* **2019,** *8* (21), 1900898.

42. Liu, T.; Jiang, G.; Song, G.; Sun, Y.; Zhang, X.; Zeng, Z. Fabrication of Rapidly Separable Microneedles for Transdermal Delivery of Metformin on Diabetic Rats. *J. Pharmacol. Sci.* **2021,** *110* (8), 3004–3010.
43. Lee, Y.; Li, W.; Tang, J.; Schwendeman, S. P.; Prausnitz, M. R. Immediate Detachment of Microneedles by Interfacial Fracture for Sustained Delivery of a Contraceptive Hormone in the Skin. *J. Contr. Release* **2021,** *337,* 676–685.
44. Li, W.; Terry, R. N.; Tang, J.; Feng, M. R.; Schwendeman, S. P.; Prausnitz, M. R. Rapidly Separable Microneedle Patch for the Sustained Release of a Contraceptive. *Nat. Biomed. Eng.* **2019,** *3* (3), 220–229.
45. Lee, S.; Fakhraei Lahiji, S.; Jang, J.; Jang, M.; Jung, H. Micro-Pillar Integrated Dissolving Microneedles for Enhanced Transdermal Drug Delivery. *Pharmaceutics* **2019,** *11* (8), 402.
46. Chen, Y.-H.; Lai, K.-Y.; Chiu, Y.-H.; Wu, Y.-W.; Shiau, A.-L.; Chen, M.-C. Implantable Microneedles with an Immune-Boosting Function for Effective Intradermal Influenza Vaccination. *Acta Biomater.* **2019,** *97,* 230–238.
47. Xue, P.; Zhang, X.; Chuah, Y. J.; Wu, Y.; Kang, Y. Flexible PEGDA-Based Microneedle Patches with Detachable PVP–CD Arrowheads for Transdermal Drug Delivery. *RSC Adv.* **2015,** *5* (92), 75204–75209.
48. Zhou, B.; Guo, Z.; Zhao, P.; Wang, H.; Dong, S.; Cheng, B.; Yang, J.; Li, B.; Wang, X. Fabrication and Characterization of Coated Microneedle Patches Based on PEGDA for Transdermal Administration of Metformin. *Drug Deliv. Transl. Res.* **2024,** *14* (1), 131–142.
49. Zhu, D. D.; Chen, B. Z.; He, M. C.; Guo, X. D. Structural Optimization of Rapidly Separating Microneedles for Efficient Drug Delivery. *J. Ind. Eng. Chem.* **2017,** *51,* 178–184.
50. Jung, C.; Lahiji, S. F.; Kim, Y.; Kim, H.; Jung, H. Rapidly Separable Micropillar Integrated Dissolving Microneedles. *Pharmaceutics* **2020,** *12* (6), 581.
51. Chen, Y.; Yang, Y.; Zeng, X.; Carrier, A. J.; Oakes, K. D.; Zhang, X. Near-infrared Light Activated Rapidly Separable Polymeric Microneedles for Transdermal Biologics Delivery. *Next Mater.* **2024,** *4,* 100186.
52. Zhang, X.; Chen, G.; Liu, Y.; Sun, L.; Sun, L.; Zhao, Y. Black Phosphorus-Loaded Separable Microneedles as Responsive Oxygen Delivery Carriers for Wound Healing. *ACS Nano* **2020,** *14* (5), 5901–5908.
53. Fan, L.; Zhang, X.; Nie, M.; Xu, Y.; Wang, Y.; Shang, L.; Zhao, Y.; Zhao, Y. Photothermal Responsive Microspheres-Triggered Separable Microneedles for Versatile Drug Delivery. *Adv. Funct. Mater.* **2022,** *32* (13), 2110746.
54. Yin, Y.; Su, W.; Zhang, J.; Huang, W.; Li, X.; Ma, H.; Tan, M.; Song, H.; Cao, G.; Yu, S.; Yu, D.; Jeong, J. H.; Zhao, X.; Li, H.; Nie, G.; Wang, H. Separable Microneedle Patch to Protect and Deliver DNA Nanovaccines against COVID-19. *ACS Nano* **2021,** *15* (9), 14347–14359.
55. Blicharz, T. M.; Gong, P.; Bunner, B. M.; Chu, L. L.; Leonard, K. M.; Wakefield, J. A.; Williams, R. E.; Dadgar, M.; Tagliabue, C. A.; El Khaja, R.; Marlin, S. L.; Haghgooie, R.; Davis, S. P.; Chickering, D. E.; Bernstein, H. Microneedle-Based Device for the One-step Painless Collection of Capillary Blood Samples. *Nat. Biomed. Eng.* **2018,** *2* (3), 151–157.
56. Zhang, X.; Li, M.; Gao, Q.; Kang, X.; Sun, J.; Huang, Y.; Xu, H.; Xu, J.; Shu, S.; Zhuang, J.; Huang, Y. Cutting-Edge Microneedle Innovations: Transforming the Landscape of Cardiovascular and Metabolic Disease Management. *iScience* **2024,** *27* (9), 110615.
57. Van Der Maaden, K.; Sekerdag, E.; Jiskoot, W.; Bouwstra, J. Impact-Insertion Applicator Improves Reliability of Skin Penetration by Solid Microneedle Arrays. *AAPS J.* **2014,** *16* (4), 681–684.

58. Hartmann, X.; Van Der Linde, P.; Homburg, E.; Van Breemen, L.; De Jong, A.; Luttge, R. Insertion Process of Ceramic Nanoporous Microneedles by Means of a Novel Mechanical Applicator Design. *Pharmaceutics* **2015,** *7* (4), 503–522.
59. Kang, G.; Kim, M.; Yang, H.; Shin, J.; Sim, J.; Ahn, H.; Jang, M.; Kim, Y.; Min, H. S.; Jung, H. Latch Applicator for Efficient Delivery of Dissolving Microneedles Based on Rapid Release of Elastic Strain Energy by Thumb Force. *Adv. Funct. Mater.* **2023,** *33* (11), 2210805.
60. Cordeiro, A. S.; Tekko, I. A.; Jomaa, M. H.; Vora, L.; McAlister, E.; Volpe-Zanutto, F.; Nethery, M.; Baine, P. T.; Mitchell, N.; McNeill, D. W.; Donnelly, R. F. Two-Photon Polymerisation 3D Printing of Microneedle Array Templates with Versatile Designs: Application in the Development of Polymeric Drug Delivery Systems. *Pharm. Res.* **2020,** *37* (9), 174.
61. Donnelly, R. F.; Garland, M. J.; Morrow, D. I. J.; Migalska, K.; Singh, T. R. R.; Majithiya, R.; Woolfson, A. D. Optical Coherence Tomography Is a Valuable Tool in the Study of the Effects of Microneedle Geometry on Skin Penetration Characteristics and In-Skin Dissolution. *J. Contr. Release* **2010,** *147* (3), 333–341.
62. Cheung, K.; Han, T.; Das, D. B. Effect of Force of Microneedle Insertion on the Permeability of Insulin in Skin. *J. Diabetes Sci. Technol.* **2014,** *8* (3), 444–452.
63. Verbaan, F. J.; Bal, S. M.; Van Den Berg, D. J.; Dijksman, J. A.; Van Hecke, M.; Verpoorten, H.; Van Den Berg, A.; Luttge, R.; Bouwstra, J. A. Improved Piercing of Microneedle Arrays in Dermatomed Human Skin by an Impact Insertion Method. *J. Contr. Release* **2008,** *128* (1), 80–88.
64. Shek, K. T.; Lam, D. C. C. Insertion Behavior of Microneedles for Drug Delivery. *Adv. Mater. Res.* **2008,** *47–50,* 1442–1445.
65. Olatunji, O.; Das, D. B.; Garland, M. J.; Belaid, L.; Donnelly, R. F. Influence of Array Interspacing on the Force Required for Successful Microneedle Skin Penetration: Theoretical and Practical Approaches. *J. Pharmacol. Sci.* **2013,** *102* (4), 1209–1221.
66. Park, S. H.; Lee, K. J.; Lee, J.; Yoon, J. H.; Jo, D. H.; Kim, J. H.; Kang, K.; Ryu, W. Microneedle-Based Minimally-Invasive Measurement of Puncture Resistance and Fracture Toughness of Sclera. *Acta Biomater.* **2016,** *44,* 286–294.
67. Zhang, H.; Tsui, C. K.; Garcia, G.; Joe, L. K.; Wu, H.; Maruichi, A.; Fan, W.; Pandovski, S.; Yoon, P. H.; Webster, B. M.; Durieux, J.; Frankino, P. A.; Higuchi-Sanabria, R.; Dillin, A. The Extracellular Matrix Integrates Mitochondrial Homeostasis. *Cell* **2024,** *187* (16), 4289–4304.e26.
68. Hu, W.; Peng, T.; Huang, Y.; Ren, T.; Chen, H.; Chen, Y.; Feng, D.; Wu, C.; Pan, X. Hyaluronidase-Powered Microneedles for Significantly Enhanced Transdermal Delivery Efficiency | Elsevier Enhanced Reader. *J. Contr. Release* **2023,** *353,* 380–390.
69. He, T.; Luo, Y.; Zhang, Q.; Men, Z.; Su, T.; Fan, L.; Chen, H.; Shen, T. Hyalase-Mediated Cascade Degradation of a Matrix Barrier and Immune Cell Penetration by a Photothermal Microneedle for Efficient Anticancer Therapy. *ACS Appl. Mater. Interfaces* **2021,** *13* (23), 26790–26799.
70. Yu, X.; Zhao, J.; Fan, D. A Dissolving Microneedle Patch for Antibiotic/Enzymolysis/Photothermal Triple Therapy against Bacteria and Their Biofilms. *Chem. Eng. J.* **2022,** *437,* 135475.
71. Zhang, J.-Y.; Gao, W.-D.; Lin, J.-Y.; Xu, S.; Zhang, L.-J.; Lu, X.-C.; Luan, X.; Peng, J.-Q.; Chen, Y. Nanotechnology-Based Photo-Immunotherapy: A New Hope for Inhibition of Melanoma Growth and Metastasis. *J. Drug Target.* **2023,** *31* (6), 555–568.

72. Zhi, D.; Yang, T.; O'Hagan, J.; Zhang, S.; Donnelly, R. F. Photothermal Therapy. *J. Contr. Release* **2020,** *325,* 52–71.
73. Guo, Z.; Zhu, A. T.; Fang, R. H.; Zhang, L. Recent Developments in Nanoparticle-Based Photo-Immunotherapy for Cancer Treatment. *Small Methods* **2023,** *7* (5), 2300252.
74. He, L.; Di, D.; Chu, X.; Liu, X.; Wang, Z.; Lu, J.; Wang, S.; Zhao, Q. Photothermal Antibacterial Materials to Promote Wound Healing. *J. Contr. Release* **2023,** *363,* 180–200.
75. Pu, X.-Q.; Ju, X.-J.; Liu, W.-Y.; Liu, Y.-Q.; Li, X.-J.; Li, Y.; Xie, R.; Wang, W.; Liu, Z.; Chu, L.-Y. Stimulus-Responsive Nanoparticle-Integrated Dissolving Microneedles for Synergetic Chemo-Photothermal Therapy of Superficial Skin Tumors. *Ind. Eng. Chem. Res.* **2022,** *61* (23), 7982–7995.
76. Roostaeyan, O.; Kivelevitch, D.; Menter, A. A Review Article on Brodalumab in the Treatment of Moderate-To-Severe Plaque Psoriasis. *Immunotherapy* **2017,** *9* (12), 963–978.
77. Khmaladze, I.; Nandakumar, K. S.; Holmdahl, R. Reactive Oxygen Species in Psoriasis and Psoriasis Arthritis: Relevance to Human Disease. *Int. Arch. Allergy Immunol.* **2015,** *166* (2), 135–149.
78. Bellinato, F.; Chiricozzi, A.; Piaserico, S.; Targher, G.; Gisondi, P. Could Targeted Pharmacotherapies Exert a "Disease Modification Effect" in Patients with Chronic Plaque Psoriasis? *IJMS* **2022,** *23* (21), 12849.
79. Wu, D.; Shou, X.; Yu, Y.; Wang, X.; Chen, G.; Zhao, Y.; Sun, L. Biologics-Loaded Photothermally Dissolvable Hyaluronic Acid Microneedle Patch for Psoriasis Treatment. *Adv. Funct. Mater.* **2022,** *32* (47), 2205847.
80. Wang, J.; Wu, Z.; Ma, X.; Huang, Z.; Dong, H.; Zhang, J.; Liu, X.; Zhang, P.; Yao, S. NIR-II Emissive Biohybrid Nanovesicles as Mild-Temperature Photothermal Antibiofilm Agents against Acute Bacterial Skin and Skin-Structure Infections. *Interdiscip. Med.* **2024,** *1* (3), 20240053.
81. Zhu, W.; Mei, J.; Zhang, X.; Zhou, J.; Xu, D.; Su, Z.; Fang, S.; Wang, J.; Zhang, X.; Zhu, C. Photothermal Nanozyme-Based Microneedle Patch against Refractory Bacterial Biofilm Infection via Iron-Actuated Janus Ion Therapy. *Adv. Mater.* **2022,** *34* (51), 2207961.
82. Lee, S.-W.; Goo, B. High-Intensity Focused Ultrasound Enhances Drug Penetration into the Human Skin in the Franz Diffusion Cell. *CCID* **2024,** *17,* 1711–1721.
83. Wang, F.; Ren, F.; Ma, Z.; Qu, L.; Gourgues, R.; Xu, C.; Baghdasaryan, A.; Li, J.; Zadeh, I. E.; Los, J. W. N.; Fognini, A.; Qin-Dregely, J.; Dai, H. In Vivo Non-Invasive Confocal Fluorescence Imaging beyond 1,700 Nm Using Superconducting Nanowire Single-Photon Detectors. *Nat. Nanotechnol.* **2022,** *17* (6), 653–660.
84. Xu, P.-Y.; Kumar Kankala, R.; Wang, S.-B.; Chen, A.-Z. Sonodynamic Therapy-Based Nanoplatforms for Combating Bacterial Infections. *Ultrason. Sonochem.* **2023,** *100,* 106617.
85. Ouyang, J.; Tang, Z.; Farokhzad, N.; Kong, N.; Kim, N. Y.; Feng, C.; Blake, S.; Xiao, Y.; Liu, C.; Xie, T.; Tao, W. Ultrasound Mediated Therapy: Recent Progress and Challenges in Nanoscience. *Nano Today* **2020,** *35,* 100949.
86. Ouyang, Q.; Zeng, Y.; Yu, Y.; Tan, L.; Liu, X.; Zheng, Y.; Wu, S. Ultrasound-Responsive Microneedles Eradicate Deep-Layered Wound Biofilm Based on TiO_2 Crystal Phase Engineering. *Small* **2023,** *19* (3), 2205292.
87. Shao, S.; Wang, S.; Ren, L.; Wang, J.; Chen, X.; Pi, H.; Sun, Y.; Dong, C.; Weng, L.; Gao, Y.; Wang, L. Layer-by-Layer Assembly of Lipid Nanobubbles on Microneedles for Ultrasound-Assisted Transdermal Drug Delivery. *ACS Appl. Bio Mater.* **2022,** *5* (2), 562–569.

88. Xiang, Y.; Lu, J.; Mao, C.; Zhu, Y.; Wang, C.; Wu, J.; Liu, X.; Wu, S.; Kwan, K. Y. H.; Cheung, K. M. C.; Yeung, K. W. K. Ultrasound-Triggered Interfacial Engineering-Based Microneedle for Bacterial Infection Acne Treatment. *Sci. Adv.* **2023,** *9* (10), eadf0854.
89. Rumpf, K.; Granitzer, P.; Koshida, N.; Poelt, P.; Reissner, M. Magnetic Interactions between Metal Nanostructures within Porous Silicon. *Nanoscale Res. Lett.* **2014,** *9* (1), 412.
90. Flacco, A.; Vieira, J.; Lifschitz, A.; Sylla, F.; Kahaly, S.; Veltcheva, M.; Silva, L. O.; Malka, V. Persistence of Magnetic Field Driven by Relativistic Electrons in a Plasma. *Nat. Phys.* **2015,** *11* (5), 409–413.
91. Gao, M.; Feng, K.; Zhang, X.; Ruan, Y.; Zhao, G.; Liu, H.; Sun, X. Nanoassembly with Self-Regulated Magnetic Thermal Therapy and Controlled Immuno-Modulating Agent Release for Improved Immune Response. *J. Contr. Release* **2023,** *357,* 40–51.
92. Stephen, Z. R.; Zhang, M. Recent Progress in the Synergistic Combination of Nanoparticle-Mediated Hyperthermia and Immunotherapy for Treatment of Cancer. *Adv. Healthcare Mater.* **2021,** *10* (2), 2001415.
93. He, D.; Liu, X.; Jia, J.; Peng, B.; Xu, N.; Zhang, Q.; Wang, S.; Li, L.; Liu, M.; Huang, Y.; Zhang, X.; Yu, Y.; Luo, G. Magnetic Field-Directed Deep Thermal Therapy via Double-Layered Microneedle Patch for Promoting Tissue Regeneration in Infected Diabetic Skin Wounds. *Adv. Funct. Mater.* **2024,** *34* (2), 2306357.
94. Yang, J.; Li, Y.; Ye, R.; Zheng, Y.; Li, X.; Chen, Y.; Xie, X.; Jiang, L. Smartphone-Powered Iontophoresis-Microneedle Array Patch for Controlled Transdermal Delivery. *Microsyst. Nanoeng.* **2020,** *6* (1), 112.
95. Li, W.; Liu, Z.; Tan, X.; Yang, N.; Liang, Y.; Feng, D.; Li, H.; Yuan, R.; Zhang, Q.; Liu, L.; Ge, L. All-in-One Self-Powered Microneedle Device for Accelerating Infected Diabetic Wound Repair. *Adv. Healthcare Mater.* **2024,** *13* (13), 2304365.
96. Chen, C.; Bai, X.; Ding, Y.; Lee, I.-S. Electrical Stimulation as a Novel Tool for Regulating Cell Behavior in Tissue Engineering. *Biomater. Res.* **2019,** *23* (1), 25.
97. Malin, D. H.; Lake, J. R.; Hamilton, R. F.; Skolnick, M. H. Augmented Analgesic Effects of Enkephalinase Inhibitors Combined with Transcranial Electrostimulation. *Life Sci.* **1989,** *44* (19), 1371–1376.
98. Huang, Y.; Li, H.; Hu, T.; Li, J.; Yiu, C. K.; Zhou, J.; Li, J.; Huang, X.; Yao, K.; Qiu, X.; Zhou, Y.; Li, D.; Zhang, B.; Shi, R.; Liu, Y.; Wong, T. H.; Wu, M.; Jia, H.; Gao, Z.; Zhang, Z.; He, J.; Zheng, M.; Song, E.; Wang, L.; Xu, C.; Yu, X. Implantable Electronic Medicine Enabled by Bioresorbable Microneedles for Wireless Electrotherapy and Drug Delivery. *Nano Lett.* **2022,** *22* (14), 5944–5953.
99. Puiggalí-Jou, A.; Del Valle, L. J.; Alemán, C. Drug Delivery Systems Based on Intrinsically Conducting Polymers. *J. Contr. Release* **2019,** *309,* 244–264.
100. Yang, Y.; Chen, B. Z.; Zhang, X. P.; Zheng, H.; Li, Z.; Zhang, C. Y.; Guo, X. D. Conductive Microneedle Patch with Electricity-Triggered Drug Release Performance for Atopic Dermatitis Treatment. *ACS Appl. Mater. Interfaces* **2022,** *14* (28), 31645–31654.
101. Pikal, M. J. The Role of Electroosmotic Flow in Transdermal Iontophoresis. *Adv. Drug Deliv. Rev.* **2001,** *46* (1–3), 281–305.
102. Guy, R. H.; Kalia, Y. N.; Delgado-Charro, M. B.; Merino, V.; López, A.; Marro, D. Iontophoresis: Electrorepulsion and Electroosmosis. *J. Contr. Release* **2000,** *64* (1–3), 129–132.
103. Zheng, Y.; Ye, R.; Gong, X.; Yang, J.; Liu, B.; Xu, Y.; Nie, G.; Xie, X.; Jiang, L. Iontophoresis-Driven Microneedle Patch for the Active Transdermal Delivery of Vaccine Macromolecules. *Microsyst. Nanoeng.* **2023,** *9* (1), 35.

104. Gao, Z.; Sheng, T.; Zhang, W.; Feng, H.; Yu, J.; Gu, Z.; Zhang, Y. Microneedle-Mediated Cell Therapy. *Adv. Sci.* **2024,** *11* (8), 2304124.
105. Amani, H.; Shahbazi, M.-A.; D'Amico, C.; Fontana, F.; Abbaszadeh, S.; Santos, H. A. Microneedles for Painless Transdermal Immunotherapeutic Applications. *J. Contr. Release* **2021,** *330,* 185–217.
106. Li, Y.; Yang, J.; Zheng, Y.; Ye, R.; Liu, B.; Huang, Y.; Zhou, W.; Jiang, L. Iontophoresis-Driven Porous Microneedle Array Patch for Active Transdermal Drug Delivery. *Acta Biomater.* **2021,** *121,* 349–358.
107. Tian, J.; Shi, R.; Liu, Z.; Ouyang, H.; Yu, M.; Zhao, C.; Zou, Y.; Jiang, D.; Zhang, J.; Li, Z. Self-Powered Implantable Electrical Stimulator for Osteoblasts' Proliferation and Differentiation. *Nano Energy* **2019,** *59,* 705–714.
108. Zhang, H.; Wen, N.; Gong, X.; Li, X. Application of Triboelectric Nanogenerator (TENG) in Cancer Prevention and Adjuvant Therapy. *Colloids Surf. B Biointerfaces* **2024,** *242,* 114078.
109. Wang, H.; Wu, T.; Zeng, Q.; Lee, C. A Review and Perspective for the Development of Triboelectric Nanogenerator (TENG)-Based Self-Powered Neuroprosthetics. *Micromachines* **2020,** *11* (9), 865.
110. Xiang, H.; Peng, L.; Yang, Q.; Wang, Z. L.; Cao, X. Triboelectric Nanogenerator for High-Entropy Energy, Self-Powered Sensors, and Popular Education. *Sci. Adv.* **2024,** *10* (48), eads2291.
111. Qian, L.; Jin, F.; Wei, Z.; Li, T.; Sun, Z.; Lai, C.; Ma, J.; Xiong, R.; Ma, X.; Wang, F.; Sun, F.; Zheng, W.; Dong, W.; Sun, K.; Wang, T.; Feng, Z. Wearable, Self-powered, Drug-Loaded Electronic Microneedles for Accelerated Tissue Repair of Inflammatory Skin Disorders. *Adv. Funct. Mater.* **2023,** *33* (27), 2209407.
112. Zhang, M.; Yang, B.; Luan, X.; Jiang, L.; Lu, C.; Wu, C.; Pan, X.; Peng, T. State of the Art in Constructing Gas-Propelled Dissolving Microneedles for Significantly Enhanced Drug-Loading and Delivery Efficiency. *Pharmaceutics* **2023,** *15* (4), 1059.
113. Zhang, M.; Yang, B.; Ren, T.; Wang, X.; Chen, H.; Lu, C.; Wu, C.; Pan, X.; Peng, T. Dual Engine-Driven Bionic Microneedles for Early Intervention and Prolonged Treatment of Alzheimer's Disease. *J. Contr. Release* **2024,** *367,* 184–196.
114. Głowacka, U.; Brzozowski, T.; Magierowski, M. S. Discrepancies and Interactions between Hydrogen Sulfide and Carbon Monoxide in the Gastrointestinal and Digestive System Physiology, Pathophysiology and Pharmacology. *Biomolecules* **2020,** *10* (3), 445.
115. Erkoc, P.; Schmitt, M.; Ingelfinger, R.; Bischoff-Kont, I.; Kopp, L.; Bode, H. B.; Schiffmann, S.; Fürst, R. Xenocoumacin 2 Reduces Protein Biosynthesis and Inhibits Inflammatory and Angiogenesis-Related Processes in Endothelial Cells. *Biomed. Pharmacother.* **2021,** *140,* 111765.
116. Kumar, S.; Singh, R. K.; Bhardwaj, T. R. Therapeutic Role of Nitric Oxide as Emerging Molecule. *Biomed. Pharmacother.* **2017,** *85,* 182–201.
117. De Groote, M. A.; Fang, F. C. NO Inhibitions: Antimicrobial Properties of Nitric Oxide. *Clin. Infect. Dis.* **1995,** *21* (Suppl. ment_2), S162–S165.
118. Liu, A.; Wang, Q.; Zhao, Z.; Wu, R.; Wang, M.; Li, J.; Sun, K.; Sun, Z.; Lv, Z.; Xu, J.; Jiang, H.; Wan, M.; Shi, D.; Mao, C. Nitric Oxide Nanomotor Driving Exosomes-Loaded Microneedles for Achilles Tendinopathy Healing. *ACS Nano* **2021,** *15* (8), 13339–13350.
119. Ghaffari-Bohlouli, P.; Jafari, H.; Okoro, O. V.; Alimoradi, H.; Nie, L.; Jiang, G.; Kakkar, A.; Shavandi, A. Gas Therapy: Generating, Delivery, and Biomedical Applications. *Small Methods* **2024,** *8* (8), 2301349.

120. Fu, Y.; Wang, R.; Ma, C.; Wang, L.; Long, S.; Li, K.; Zhao, X.; Yang, W. Injectable Oxygen-Carrying Microsphere Hydrogel for Dynamic Regulation of Redox Microenvironment of Wounds. *Small* **2024,** *20* (42), 2403781.
121. Chen, S.; Luo, Y.; He, Y.; Li, M.; Liu, Y.; Zhou, X.; Hou, J.; Zhou, S. In-Situ-Sprayed Therapeutic Hydrogel for Oxygen-Actuated Janus Regulation of Postsurgical Tumor Recurrence/Metastasis and Wound Healing. *Nat. Commun.* **2024,** *15* (1), 814.
122. Liu, P.; Fu, Y.; Wei, F.; Ma, T.; Ren, J.; Xie, Z.; Wang, S.; Zhu, J.; Zhang, L.; Tao, J.; Zhu, J. Microneedle Patches with O2 Propellant for Deeply and Fast Delivering Photosensitizers: Towards Improved Photodynamic Therapy. *Adv. Sci.* **2022,** *9* (25), 2202591.
123. Tian, Y.; Zhang, Y.; Wang, Y.; Chen, Y.; Fan, W.; Zhou, J.; Qiao, J.; Wei, Y. Hydrogen, a Novel Therapeutic Molecule, Regulates Oxidative Stress, Inflammation, and Apoptosis. *Front. Physiol.* **2021,** *12,* 789507.
124. Johnsen, H. M.; Hiorth, M.; Klaveness, J. Molecular Hydrogen Therapy—A Review on Clinical Studies and Outcomes. *Molecules* **2023,** *28* (23), 7785.
125. Ohsawa, I.; Ishikawa, M.; Takahashi, K.; Watanabe, M.; Nishimaki, K.; Yamagata, K.; Katsura, K.; Katayama, Y.; Asoh, S.; Ohta, S. Hydrogen Acts as a Therapeutic Antioxidant by Selectively Reducing Cytotoxic Oxygen Radicals. *Nat. Med.* **2007,** *13* (6), 688–694.
126. Hu, R.; Dai, C.; Dong, C.; Ding, L.; Huang, H.; Chen, Y.; Zhang, B. Living Macrophage-Delivered Tetrapod PdH Nanoenzyme for Targeted Atherosclerosis Management by ROS Scavenging, Hydrogen Anti-inflammation, and Autophagy Activation. *ACS Nano* **2022,** *16* (10), 15959–15976.
127. Lopez-Ramirez, M. A.; Soto, F.; Wang, C.; Rueda, R.; Shukla, S.; Silva-Lopez, C.; Kupor, D.; McBride, D. A.; Pokorski, J. K.; Nourhani, A.; Steinmetz, N. F.; Shah, N. J.; Wang, J. Built-In Active Microneedle Patch with Enhanced Autonomous Drug Delivery. *Adv. Mater.* **2020,** *32* (1), 1905740.

Chapter 7

Microneedle technology for diabetes therapy

Wenhao Wang[1] and Chuanbin Wu[2]
[1]*School of Pharmaceutical Science, Sun Yat-Sen University, Guangzhou, China;* [2]*State Key Laboratory of Bioactive Molecules and Druggability Assessment/International Cooperative Laboratory of Traditional Chinese Medicine Modernization and Innovative Drug Development of Ministry of Education (MOE) of China/College of Pharmacy, Jinan University, Guangzhou, China*

1. Diabetes

Diabetes mellitus is a lifelong metabolic disease characterized by abnormally elevated blood glucose levels (BGLs), which seriously endangers human health.[1] According to the Global Burden of Disease Study, there were as many as 529 million people with diabetes in the world in 2021, with an overall prevalence of 6.1%. It is expected that by 2050, the number of patients with diabetes in the world will exceed 1.31 billion.[2] The number of patients with diabetes is increasing year by year, and the market demand for diabetes treatment drugs is huge. According to its pathogenesis and clinical manifestations, diabetes mellitus is mainly divided into type 1 diabetes mellitus, type 2 diabetes mellitus, gestational diabetes mellitus, and other special forms of diabetes mellitus.[3]

2. Pathogenesis of diabetes

2.1 Mechanism of insulin functions

Insulin plays a crucial role in the regulation of BGLs in the body. When we consume food, especially carbohydrates, our BGLs rise. In response to this increase, the beta cells of the pancreas release insulin into the bloodstream. Insulin then acts on various tissues in the body to lower BGLs.[4] One of insulin's main functions is to promote the uptake of glucose by cells, particularly muscle and fat cells. This helps to reduce the amount of glucose circulating in the bloodstream. Additionally, insulin inhibits gluconeogenesis in the liver, which is a process where new glucose is produced from noncarbohydrate

Microneedle Technology. https://doi.org/10.1016/B978-0-443-45544-5.00009-9

sources such as amino acids and glycerol.[5] Without sufficient insulin or if there is resistance to its effects (as seen in type 2 diabetes), BGLs can become elevated, leading to potential health complications over time. Therefore, understanding how insulin works and its importance in regulating blood sugar is essential for maintaining overall health and well-being.

2.2 Pathogenesis of type 1 diabetes mellitus

Type 1 diabetes, also known as juvenile diabetes, is a chronic condition that affects the body's ability to regulate blood sugar. It is characterized by the immune system mistakenly attacking and destroying insulin-producing beta cells in the pancreas. This results in an absolute lack of insulin, a hormone necessary for regulating blood sugar levels. The onset of type 1 diabetes typically occurs during childhood or adolescence, but it can develop at any age. The exact cause of this autoimmune response is not fully understood, but genetic factors, viral infections, and other environmental triggers are believed to play a role in increasing the risk of developing the condition.[6]

Managing type 1 diabetes requires lifelong dependence on exogenous insulin therapy. This involves regular monitoring of blood sugar levels and administering insulin injections to maintain stable glucose levels in the body. Additionally, individuals with type 1 diabetes must carefully manage their diet and exercise regimen to help control their blood sugar. Living with type 1 diabetes can present various challenges and require ongoing medical care and support. However, advancements in technology have led to the development of continuous glucose monitors and insulin pumps that can help individuals better manage their condition and improve their quality of life.[7]

Overall, while living with type 1 diabetes may require significant lifestyle adjustments, proper management can allow individuals to lead full and active lives despite their diagnosis.

2.3 Pathogenesis of type 2 diabetes

Type 2 diabetes is a chronic condition that affects the way our body metabolizes sugar (glucose), an important source of fuel for our body. With type 2 diabetes, our body either resists the effects of insulin or doesn't produce enough insulin to maintain normal glucose levels. The risk factors for developing type 2 diabetes include being overweight or obese, leading a sedentary lifestyle, having a family history of diabetes, and being over the age of 45. Additionally, certain ethnic groups are at higher risk for developing type 2 diabetes.[8]

In addition to insulin resistance and pancreatic β-cell dysfunction, other contributing factors to the development of type 2 diabetes include genetics and environmental factors such as exposure to certain viruses or toxins. Managing type 2 diabetes involves maintaining healthy blood sugar levels through diet, exercise, and medication. It's important for individuals with this condition to monitor their blood sugar regularly and work closely with healthcare professionals to develop an effective treatment plan. In some cases, insulin therapy may be necessary as the disease progresses to effectively manage BGLs.[9]

2.4 Gestational diabetes

Gestational diabetes is a condition that affects pregnant women, causing high blood sugar levels. It occurs when the body is unable to produce enough insulin to meet the extra needs of pregnancy. This can lead to complications for both the mother and the baby if not properly managed.[10,11]

The hormones produced by the placenta during pregnancy can make it harder for insulin to do its job, leading to insulin resistance. This means that glucose cannot be effectively transported from the bloodstream into cells, resulting in high blood sugar levels. Fortunately, most women with gestational diabetes can control their BGLs through diet and exercise during pregnancy. After giving birth, these women usually see their blood sugar levels return to normal. However, they are at an increased risk of developing type 2 diabetes later in life.[12]

It's important for women who have had gestational diabetes to continue monitoring their blood sugar levels after delivery and maintain a healthy lifestyle to reduce their risk of developing type 2 diabetes. With proper management and care, many women can prevent or delay the onset of type 2 diabetes despite having had gestational diabetes during pregnancy.

2.5 Other special forms of diabetes

It includes diabetes caused by genetic defects, endocrine diseases, drug effects, etc., such as cystic fibrosis-related diabetes, hereditary insulin deficiency diabetes, etc. These types of diabetes are rare, but their pathogenesis and clinical manifestations vary, requiring individualized treatment.[12–14]

Overall, diabetes mellitus is a complex metabolic disease, and its pathogenesis involves insufficient insulin secretion and insulin resistance. Understanding the pathogenesis and classification of diabetes mellitus is helpful to develop individualized treatment plans, effectively control blood glucose, and prevent complications.

3. Drugs for the treatment of diabetes

With the deep understanding of the pathological mechanisms of diabetes, researchers and pharmaceutical companies have been able to develop a wide range of drugs for the treatment of this condition. These drugs target various aspects of diabetes, including insulin resistance, blood sugar regulation, and pancreatic function.

3.1 Insulin and insulin analogs

Insulin plays a crucial role in managing diabetes, and the development of long-acting and ultra-long-acting insulin analogs has significantly improved treatment options for patients.[15–17] These advancements have led to more convenient dosing regimens, such as insulin glargine, which only requires one injection per day and provides 24-h blood glucose control. This not only simplifies the daily routine for patients but also helps to improve adherence to their treatment plan. In addition to long-acting insulins, premix insulin formulations have also seen improvements. Insulin degludec, for example, offers both basal and mealtime insulin coverage, providing a more physiologically similar pattern of insulin secretion. This can lead to better glycemic control throughout the day and reduce the risk of hypoglycemia.[18–20]

Overall, these developments in insulin therapy have greatly enhanced the quality of life for individuals with diabetes by offering more effective and convenient treatment options. As research continues in this field, even more innovative therapies will likely emerge, further improving outcomes for those living with diabetes.

3.2 Drugs that enhance insulin sensitivity

Insulin resistance, a key factor in the development of type 2 diabetes, is characterized by the body's inability to respond effectively to insulin. This leads to high blood sugar levels and can eventually result in diabetes.[15] Biguanides, such as metformin, work by reducing the amount of glucose produced by the liver and increasing the sensitivity of muscle cells to insulin. They are considered first-line treatment for type 2 diabetes due to their effectiveness in lowering blood sugar levels.[21,22]

Thiazolidinediones, including rosiglitazone and pioglitazone, are another class of drugs that target insulin resistance. These medications improve insulin sensitivity in fat, muscle, and liver tissues while also decreasing glucose production in the liver. However, patients and healthcare providers need to be aware of potential side effects such as edema (fluid retention) and weight gain when using these medications.[22–24]

In addition to medication therapy, lifestyle modifications such as regular exercise and a healthy diet play a crucial role in managing insulin resistance and type 2 diabetes. By incorporating these changes into daily life, individuals can further improve their body's response to insulin and better control their blood sugar levels. It is essential for individuals with type 2 diabetes or those at risk for developing it to work closely with healthcare professionals to develop a comprehensive treatment plan that addresses both medication management and lifestyle adjustments.

3.3 Drugs that promote insulin secretion

Sulfonylureas have been widely used as the first line of oral antidiabetic drugs for many years due to their ability to stimulate insulin secretion from the pancreas. However, in recent years, nonsulfonylurea secretagogues such as nateglinide and repaglinide have gained attention in clinical practice. These newer medications work by targeting the same pathway as sulfonylureas but with a more rapid onset and shorter duration of action, which closely mimics the physiological pattern of insulin secretion.[25,26]

The advantage of nonsulfonylurea secretagogues lies in their ability to lower BGLs without causing prolonged periods of low blood sugar (hypoglycemia). This makes them particularly beneficial for patients who experience frequent episodes of hypoglycemia or those with irregular meal schedules. Additionally, these medications may offer a more flexible dosing regimen compared to traditional sulfonylureas.[27]

As research continues to explore different mechanisms for managing diabetes, healthcare providers need to consider all available options when tailoring treatment plans for individual patients. The emergence of nonsulfonylurea secretagogues provides an additional tool in the management of diabetes and offers new possibilities for achieving optimal glycemic control while minimizing side effects.

3.4 α-glucosidase inhibitors

α-glucosidase inhibitors are a class of medications that play a crucial role in managing BGLs in individuals with diabetes. By inhibiting the enzyme α-glucosidase, these drugs slow down the digestion and absorption of carbohydrates, ultimately leading to a more gradual release of glucose into the bloodstream after meals. This helps prevent sharp spikes in blood sugar levels and contributes to overall glycemic control.[28,29]

Voglibose, acarbose, and miglitol are three commonly prescribed α-glucosidase inhibitors that have been proven effective in clinical practice. They work by specifically targeting the enzymes responsible for breaking down complex carbohydrates into simple sugars like glucose. As a result, these medications can help regulate postprandial BGLs and contribute to maintaining stable glycemia throughout the day.

In addition to their ability to control BGLs after meals, α-glucosidase inhibitors may also offer other potential benefits for individuals with diabetes, such as reducing the risk of cardiovascular complications associated with hyperglycemia. However, it's important for patients using these medications to be aware of potential side effects such as gastrointestinal discomfort or flatulence.[30]

Overall, α-glucosidase inhibitors represent an important therapeutic option for individuals with diabetes who struggle with postprandial hyperglycemia. When used as part of a comprehensive treatment plan that includes diet and lifestyle modifications, these medications can contribute significantly to achieving optimal glycemic control and improving long-term health outcomes for patients with diabetes.

3.5 New diabetes-treating agents

With the deepening of research on the pathogenesis of diabetes, there has been a significant development in the field of hypoglycemic drugs in recent years. In addition to glucagon-like peptide-1 (GLP-1) analogs and dipeptidyl peptidase (DPP-IV) inhibitors, other new classes of hypoglycemic drugs have also emerged. For example, sodium-glucose cotransporter 2 (SGLT2) inhibitors such as empagliflozin and dapagliflozin have shown promising results in reducing BGLs by promoting urinary glucose excretion.[31,32]

In addition to these drug developments, there is ongoing research into novel targets for diabetes treatment, including agents that target hepatic glucose production or improve β-cell function. The advancement in understanding the molecular mechanisms underlying diabetes has paved the way for innovative therapeutic approaches aimed at addressing different aspects of the disease.[33,34]

Overall, the continuous progress in hypoglycemic drug development offers hope for improving management options for individuals living with diabetes and may contribute to better outcomes in terms of glycemic control, weight management, and reduction of cardiovascular risk factors.

4. Status of diabetes drug delivery systems

Most drugs utilized in the treatment of diabetes are protein-polypeptide drugs, which have poor stability and are difficult to administer. Therefore, their delivery mode has an important impact on the treatment effect and the quality of life of patients. With the advancement of science and technology, the delivery form of diabetes drugs has undergone an evolution from traditional syringes to modern insulin pens.

4.1 Traditional syringes

The traditional insulin syringe, while being one of the first insulin delivery tools, has some drawbacks that have led to the development of alternative methods. This method requires the patient to draw the insulin solution manually and then administer it by subcutaneous injection. The syringe price is relatively low, but the operation is relatively complex, and the dose accuracy depends on the patient's operating skills. In addition, the injection process may cause pain and discomfort, leading some patients to become resistant to the injection.

4.2 Insulin pen

The insulin pen is designed to provide a convenient and efficient way for patients to administer insulin. It eliminates the need for vials, syringes, and needles, making it easier for individuals with diabetes to manage their condition. The preloaded insulin refill allows for quick and easy dosage adjustments, reducing the risk of errors in medication administration. In addition to its ease of use, the insulin pen offers precise dosage delivery, ensuring that patients receive the correct amount of insulin each time. This accuracy is crucial in maintaining blood sugar levels within a healthy range and preventing complications associated with diabetes. Furthermore, the compact size of the insulin pen makes it ideal for carrying on the go. Whether at work, school, or traveling, patients can discreetly and conveniently administer their insulin without drawing attention to themselves. Overall, the insulin pen has revolutionized the way individuals with diabetes manage their condition by providing a user-friendly and effective method of delivering insulin. With options available for short-acting, medium-acting, long-acting, and premixed insulins, patients have access to a variety of treatment options tailored to their specific needs.[35–37]

4.3 Insulin pump

An insulin pump is a small, computerized device that delivers insulin continuously throughout the day and night.[38] It closely mimics the natural pattern of insulin secretion by the pancreas, providing a steady flow of insulin to help regulate blood sugar levels. This can be especially beneficial for individuals with type 1 diabetes, as it helps to maintain more stable blood sugar levels and reduces the risk of complications associated with high or low blood sugar. One of the key advantages of using an insulin pump is its ability to provide precise dose control. This allows for more accurate management of insulin delivery, which can lead to better overall glucose control. Additionally, using an insulin pump can also reduce the number of injections needed each day compared to traditional methods of insulin administration. Another benefit is that using an insulin pump can help minimize fluctuations in blood sugar levels. By delivering a continuous supply of insulin, it can help prevent sudden spikes or drops in blood sugar that may occur with other forms of treatment.[39,40]

However, it's important to note that while there are many benefits to using an insulin pump, there are also some drawbacks. For example, these devices can be quite expensive and may not be covered by all insurance plans. In addition, they require regular maintenance and monitoring to ensure proper functioning. Patients must also have a good understanding of how to operate and manage their pump effectively for it to be successful in controlling their diabetes. In conclusion, while there are both pros and cons associated with using an insulin pump as a method for managing diabetes, it remains an important option for many individuals seeking better control over their condition.

4.4 Novel delivery systems

Although the above delivery methods can achieve blood sugar control, there are common problems, such as injection pain and poor patient compliance. Therefore, there is an urgent need to develop noninjectable diabetes drug delivery systems. With advances in biotechnology and materials science, new insulin delivery systems are emerging. For example, inhaled insulin is absorbed through the lungs without injection, improving patient comfort, but its clinical application is still limited. In addition, new delivery systems such as oral insulin and transdermal patches are being developed, which are expected to provide more options for diabetics.

5. Microneedles in the treatment of diabetes

The traditional diabetes treatment methods mainly include insulin injection, oral hypoglycemic drugs, strict diet control, and exercise therapy. Among them, insulin injection is a key means to control blood sugar in patients with type 1 diabetes and in some patients with type 2 diabetes. However, the traditional subcutaneous injection method has many limitations; for example, the injection process is often accompanied by obvious pain, which not only brings physical pain to patients but also may lead to patients' fear and resistance to treatment, thereby reducing treatment compliance. In addition, frequent subcutaneous injection may also cause local tissue damage, fat atrophy or hyperplasia, and other adverse reactions, affecting insulin absorption and its effect.

MN technology, as a new technology of transdermal drug delivery and minimally invasive detection, has been widely studied in the field of diabetes treatment in recent years. MNs have the characteristics of small size (usually in the micron to millimeter level) and can form tiny wounds or channels on the skin surface, thereby overcoming the natural barrier effect of the skin cuticle on the transmission of drugs and substances, and achieve painless or painless drug delivery. This feature makes MNs have great application potential in insulin administration in diabetes treatment, which is expected to provide a more convenient, efficient, and comfortable treatment plan for patients with diabetes.

MNs can effectively improve the efficiency of insulin absorption. The cuticle of the skin is a major obstacle to the transdermal absorption of drugs, and it has a highly ordered lipid structure that can block the penetration of most drug molecules. By forming microporous channels in the stratum corneum, MNs greatly increase the contact area between insulin and the deep tissue of the skin and promote the diffusion and absorption of insulin. In terms of insulin pharmacokinetics, MN administration also shows unique advantages. Compared with traditional subcutaneous injection, MN administration can reduce the amount of insulin and improve its bioavailability. After the traditional subcutaneous injection of insulin, the drug will quickly form a high concentration area at the injection site, and then quickly enter the blood circulation, resulting in a large peak and trough value of blood drug concentration, which not only increases the risk of hypoglycemia, but also may cause other adverse reactions due to high blood drug concentration. Due to the characteristics of slow release and continuous absorption, MN administration can precisely control the release of insulin from the tips, so that the blood concentration of insulin in the body is more stable and the fluctuation of blood sugar is reduced.

5.1 Traditional microneedles for the treatment of diabetes

The use of solid MNs, fabricated from an array of materials, including silicon, metal, and polymer, has demonstrated encouraging outcomes in the reduction of BGLs through the enhancement of insulin permeability via skin pretreatment.[41] For example, Zhou et al. conducted a study in which stainless steel MNs with varying lengths of needles were used to assess the efficacy of insulin delivery to diabetic rats.[42] The findings indicated that the permeability of the skin to insulin increased, resulting in a rapid decrease in BGLs within 1 h. Moreover, the integration of solid MNs with other techniques, such as iontophoresis, can further enhance the transdermal delivery efficiency of insulin. Additionally, hollow MNs have been demonstrated to facilitate intradermal insulin delivery, thereby accelerating the onset of insulin action. This can be achieved through passive diffusion, pressure application, or even electricity. Studies by McAllister and colleagues have shown that hollow MNs permit the penetration of small volumes of solutions into the skin, and that applying greater pressure results in a more rapid decrease in BGLs.[43] Furthermore, Roxhed et al. devised a patch system comprising an electronically controlled liquid dispenser integrated with MNs, which resulted in a markedly elevated plasma insulin concentration compared to passive diffusion at the 3-hour post-dosing interval.[44]

Insulin delivery using drug-free MNs, whether solid or hollow, is inconvenient for patients due to the multiple steps involved. However, drug-loaded MNs, such as coated, dissolving, and hydrogel-forming MNs, have shown promise in overcoming these issues. In a study conducted by Ross and colleagues, insulin polymeric layers were successfully coated onto metal MNs.[45] The thin and homogeneous layers were found to be effective in retaining insulin in an intact state and allowing for rapid release within 20 min, thereby demonstrating the feasibility of solid-state insulin delivery using coated MNs. Despite these promising results, further research on insulin-coated MNs is limited. This may be attributed to challenges related to achieving sufficient doses of coated insulin. As such, there is a need for additional studies to explore ways of optimizing the dose and effectiveness of insulin-coated MN technology. By addressing these limitations, it may be possible to enhance the practicality and efficacy of this approach for delivering insulin to patients more conveniently and efficiently.

The use of dissolving MNs to encapsulate insulin within the MN matrix represents a promising avenue of research due to the favorable biocompatibility of the technology, the relatively simple manufacturing process, and the low cost of production. Given that insulin is heat-sensitive, it is crucial to incorporate insulin at a relatively mild temperature. A variety of water-soluble polymers, including hyaluronic acid (HA), chondroitin sulfate, poly-gammaglutamic acid, and a combination of starch and gelatin, have been utilized to prepare insulin-loaded dissolving MNs at room temperature through the micromold casting method. In recent years, researchers have been investigating various methods for incorporating insulin into dissolving MNs, with the aim of improving its stability and bioavailability. One approach entails the utilization of lyophilization techniques to ensure the preservation of the insulin's structural integrity throughout the fabrication process. An alternative approach is to modify the composition of the polymer matrix in order to enhance its ability to protect and release insulin effectively. The capacity of these MNs to deliver therapeutic levels of insulin into diabetic rats has generated interest in their prospective clinical applications for the management of diabetes.

Overall, advancements in the development of dissolving MN technology for delivering insulin hold great promise for improving patient compliance and treatment outcomes in individuals with diabetes. Further research efforts are underway to optimize these innovative drug delivery systems and bring them closer to practical implementation in clinical settings.

5.2 The glucose-responsive microneedles

The abovementioned traditional MNs could effectively deliver insulin into the blood. However, given that BGLs must be precisely controlled within a specific range, unregulated insulin release from MNs may lead to hypoglycemia. Consequently, there is an urgent need to develop a "smart" insulin release system to ensure both therapeutic efficacy and safety. Currently, MN-based drug delivery systems can be improved by incorporating responsive materials that enable automatic adjustment of insulin release in accordance with real-time BGLs. This means that when blood sugar levels rise, the rate of insulin release from the MNs is accelerated in response. Conversely, when blood sugar drops back into the normal range, the rate of insulin release slows down or stops altogether. This responsive system not only improves the accuracy and safety of diabetes treatment but also reduces the risk of complications such as hypoglycemia. Furthermore, this innovative approach to insulin delivery offers greater convenience for patients by eliminating the need for frequent injections and providing a more discreet and comfortable method of administering medication. With ongoing advancements in technology and research in this field, novel MN drug delivery systems will likely continue to play a significant role in improving diabetes management and enhancing overall patient outcomes.

Yu et al. designed an MNs patch loaded with insulin by using a nondegradable glucose-responsive polymer (Fig. 7.1).[46] Under the hyperglycemic condition, the polymeric matrix swelled and weakened the electrostatic interaction between the negatively charged polymers and insulin, thereby promoting the release of insulin. When exposed to euglycemic conditions, the inhibited volume change and the restoration of electrostatic interaction slowed down the insulin release rate. After the afternoon meal, levels in the treated pig showed a small increase and quickly returned to the normal glycemic state after the mealtime. The glucose levels could be maintained at a reduced state overnight until the next meal on the second day. Similarly, Liu et al. designed the hydrogel MN patch with blood glucose-responsive swelling.[47] The MN-based matrix is composed of an HA polymer functionalized with dopamine and 4-amino-3-fluorobenylboric acid. The principle of controlled insulin release relies on the reversible crosslinking between the catechol–amine groups in dopamine and boric acid, which is glucose concentration dependent; under hyperglycemic conditions, the hydrogel exhibits a high degree of crosslinking, an increased swelling rate, and accelerated drug release. When the blood glucose concentration decreases, the degree of crosslinking decreases, and the drug release slows down, which can reduce the risk of hypoglycemia and improve the safety of insulin use. A glucose tolerance test was performed by intraperitoneal injection of 2 g/kg glucose 2 h after the application of the hydrogel MNs or insulin injection. The results showed that the blood glucose concentration of diabetic rats treated with the MNs returned to a normal blood glucose level similar to that of healthy rats after the blood glucose level reached the peak. On the contrary, sustained hyperglycemia was observed in diabetic rats 2 h after subcutaneous insulin injection, confirming the glucose-dependent drug release ability of the hydrogel MNs.

However, for such "open-loop" insulin-MNs, the continuous, accurate glucose monitoring and the reliable insulin release remain the problems that need to be solved. To address these issues, the "smart" MNs have been designed to provide a sustained and controlled release of insulin to maintain normal BGLs. By coencapsulating both insulin and glucose-responsive

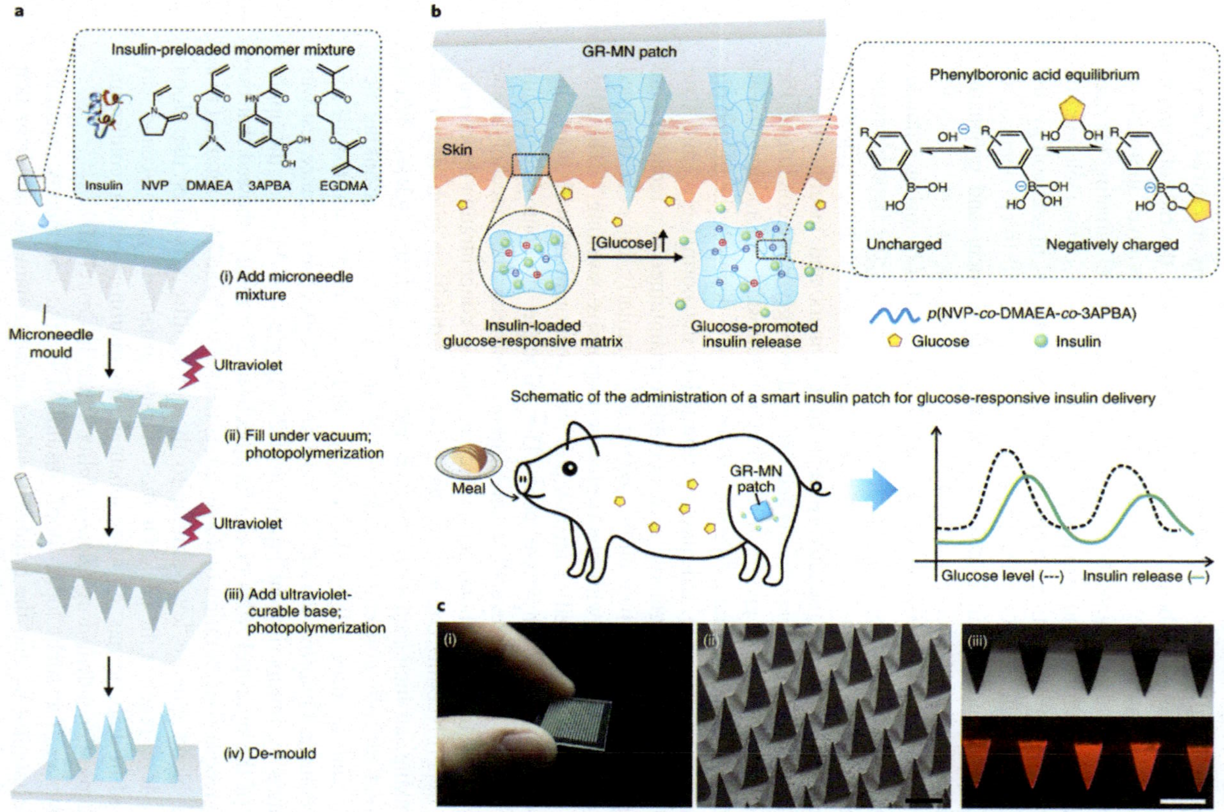

FIGURE 7.1 (a), Schematic of the fabrication process of a glucose-responsive insulin patch from a silicone mold using an in situ photopolymerization strategy. (b), Mechanism of glucose-triggered insulin release from glucose-responsive microneedles (GR-MNs). Glucose levels of diabetic pigs can be effectively regulated by the administration of a glucose-responsive insulin patch. (c), Characterization of the GR-MN. (i) Photograph of the GR-MN patch; (ii) scanning electron microscopy image of the microneedle array (scale bar, 500 μm); (iii) optical microscopy (*top*) and fluorescence microscopy (*bottom*) images of the rhodamine B-labeled insulin (*red*)-loaded MN patch (scale bar: 500 μm). *Reproduced from reference[46] with permission from Nature.*

moieties into nanoparticles and then incorporating them into MNs, the "smart" MNs can be obtained. Among them, Glucose oxidase (GOx) is typically used in the detection of glucose for the purpose of smart insulin release. GOx is an oxidoreductase that consumes oxygen (O_2) to catalyze the oxidation of glucose to gluconic acid and byproduct hydrogen peroxide (H_2O_2), inducing a hypoxic, acidic, and H_2O_2-concentrated local milieu, which thereafter can be employed as a trigger to release insulin. According to the glucose-responsive mechanism, "smart" MNs can be divided into hypoxia-responsive, pH-responsive, H_2O_2-responsive, and dual-responsive polymeric MNs.[48]

5.3 The hypoxia-responsive microneedles

The process of catalytic oxidation of glucose by GOx is a crucial step in the development of intelligent insulin delivery systems. This biochemical reaction leads to the consumption of oxygen, creating a local hypoxic environment that can be harnessed for targeted drug release. Gu and coworkers utilized this feature to design a hypoxia-responsive polymeric MN system based on glucose-responsive vesicles (GRVs) containing insulin and GOX (Fig. 7.2).[49] The GRVs were constructed using amphiphilic HA conjugated with a hypoxia-sensitive group 2-nitroimidazole, allowing them to respond to changes in oxygen levels. In hyperglycemic conditions, GOx catalyzed the conversion of glucose to gluconic acid while consuming oxygen, leading to the creation of a hypoxic microenvironment. This triggered the decomposition of GRVs and the subsequent release of preloaded insulin. When tested in mice, the administration of GRV-loaded MNs resulted in a rapid reduction in BGLs within 1 hour, reaching around 200 mg/dL. Importantly, there was no further decline into hypoglycemia (<200 mg/dL), and normoglycemic concentrations were maintained for 3 hours, even with double-dose administration. This innovative approach demonstrates the potential for utilizing bio-responsive materials and enzymatic reactions for precise control over drug delivery in response to physiological changes such as hyperglycemia. The development of such intelligent insulin delivery systems holds promise for improving diabetes management and enhancing patient outcomes.

Apart from direct insulin administration, transplantation of pancreatic β-cells has been explored for diabetes treatment to provide continuous insulin secretion. This strategy could be achieved in a prolonged delivery system using a hypoxia-responsive polymeric MNs, which incorporated "glucose-signal amplifiers" (GSAs) and insulin secretion devices, pancreatic β-cell capsules into the system tip and base of MN patches, respectively. The GSAs were also prepared from self-assembled hypoxia-sensitive material, and coloaded with three enzymes: GOx, α-amylase (AM), and glucoamylase (GA).[50] Once subjected to the elevated BGLs, the hypoxia condition resulting from glucose oxidation led to the dissociation of GSAs and the quick release of entrapped enzymes. Subsequently, the released AM catalyzed the hydrolysis

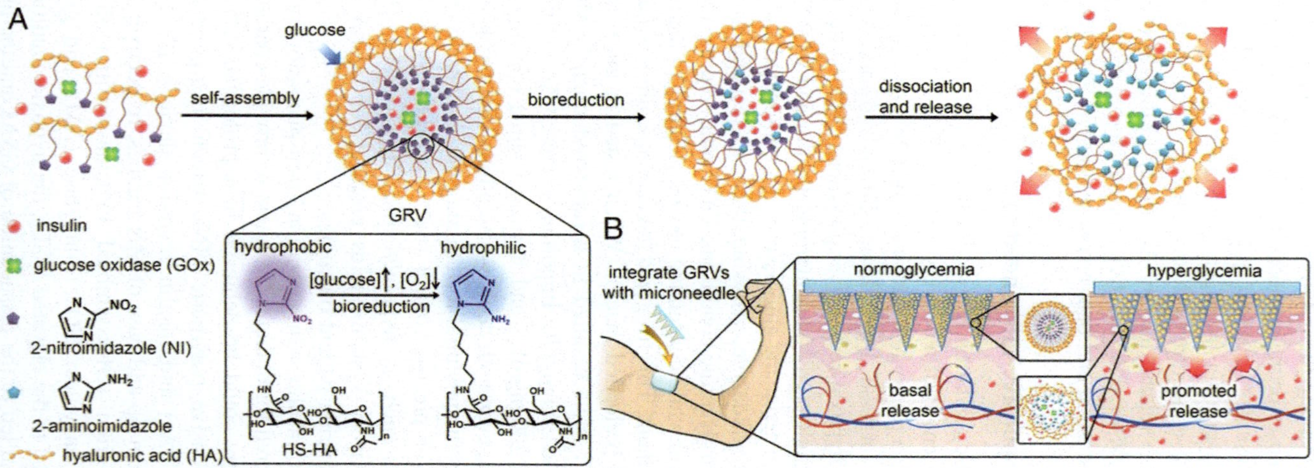

FIGURE 7.2 Schematic illustration of the glucose-responsive insulin delivery system using hypoxia-sensitive vesicle–loaded microneedle (MN) patches. (A) Formation and operating mechanism of glucose-responsive vesicles (GRVs) composed of hypoxia-sensitive hyaluronic acid (HS-HA). (B) Schematic of the GRV-containing MN patch (smart insulin patch) for in vivo insulin delivery, in which a hyperglycemic state triggers enhanced insulin release. *Reproduced from reference*[49] *with permission from PNAS.*

of α-amylose encapsulated in the MNs matrix into disaccharides and trisaccharides, which were further hydrolyzed into glucose by GA, thereby generating a localized glucose-rich site. Then, the "amplified" glucose diffused into positioned β-cell capsules and stimulated insulin secretion. The regulation of these BGLs by these innovative MNs was explored in the animal model of diabetes. Results showed that the BGLs in mice treated with this device quickly declined to a normoglycemic state within 2 h and remained at a significantly reduced level for up to 6 h without causing the potential risk of hypoglycemia.

5.4 The pH-responsive microneedles

As previously mentioned, GOx has the capability of converting glucose into gluconic acid, resulting in a reduction of local pH levels. This pH change can act as a trigger for the release of preloaded antidiabetic drugs. In a study carried out by Chen and colleagues, they developed a glucose-responsive device for regulating BGLs by using polymer MNs integrated with dual mineralized particles containing Exendin-4 (Ex4) and GOx (Fig. 7.3).[51] Ex4 is a glucagon-like peptide-1 receptor agonist that has been approved for the treatment of Type 2 diabetes. Nevertheless, its therapeutic efficacy is limited by its short circulation time in the bloodstream (60–90 min) and potential adverse reactions such as nausea and vomiting when taken in excessive amounts. By utilizing the GOx-induced changes in local pH levels as a trigger for drug release, this technology has the potential to enhance both the efficacy and safety profile of Ex4 therapy for individuals with Type 2 diabetes. Ex4 was encapsulated into pH-sensitive calcium phosphate mineralized particles (m-Ex4), which could release Ex4 on demand upon the stimulation of H^+ signals. Furthermore, GOx was innovatively trapped in copper phosphate mineralized particles (m-GOx), which was employed to prevent rapid leakage from the insulin delivery device and enhance the enzyme activity of GOx. In this design, m-GOx served as an immobilized enzyme to catalyze glucose along with a steady decrease in pH, leading to the dissolution of pH-sensitive m-Ex4 and the release of Ex4. After protein/peptide mineralization and MNs fabrication, the patch can be directly applied onto the skin. In the normoglycemic state, the relatively lower glucose concentration only induces a mild pH decrease under m-GOx catalysis, which is insufficient to induce the dissociation of m-Ex4 to release the payloads. In contrast, when the blood glucose concentration rises to the hyperglycemic level, the significant pH decrease triggers the rapid dissociation of m-Ex4 to release Ex4 for blood glucose regulation, suggesting a pathology-specific drug release. Moreover, m-GOx is stable and robust throughout the entire process, providing long-term responsiveness to glucose changes. In this way, a smart, long-acting, and on-demand Ex4 release is accomplished. This MNs-based delivery system demonstrated promising long-term glucose regulation ability and a desirable

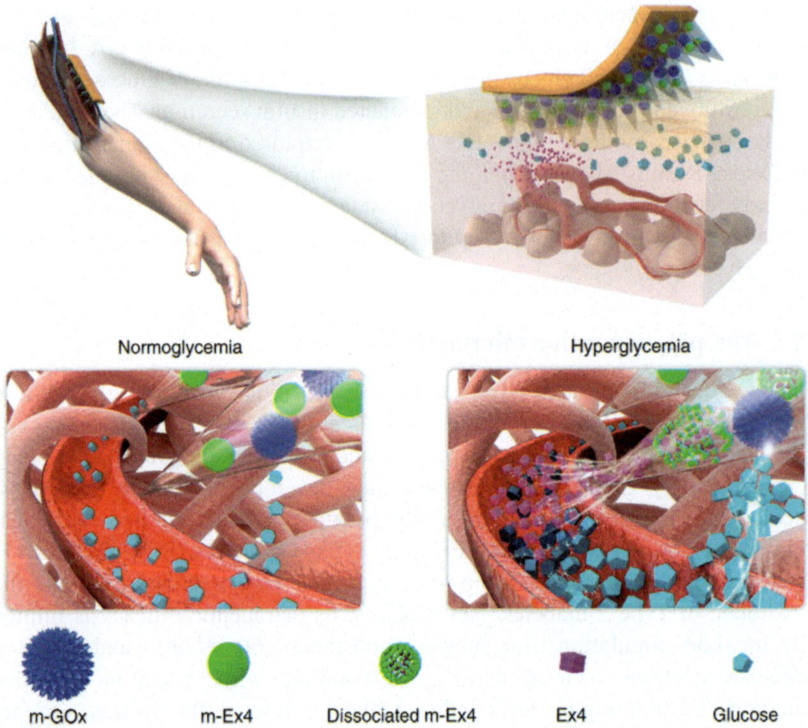

FIGURE 7.3 Schematic of glucose-responsive exendin-4 (Ex4) delivery microneedle patches. *Reproduced from reference[51] with permission from Nature.*

amount of Ex4 release in diabetic mice, which might be regarded as a potential strategy for glucose management in Type 2 diabetes.

It is important to consider the potential consequences of the byproduct H_2O_2 that is produced during the GOx-catalyzed oxidation of glucose. While this process is essential for measuring glucose levels, the presence of excessive H_2O_2 can have detrimental effects on enzymatic activity and response rates. Furthermore, prolonged exposure to high levels of H_2O_2 may lead to free radical-induced damage in skin tissue, posing a risk for individuals using glucose monitoring devices over an extended period of time. Therefore, it is crucial to carefully monitor and manage the production of H_2O_2 in order to minimize its negative impact on enzymatic activity and potential harm to skin tissue.

To solve this problem, Xu et al. depicted a smart pH-sensitive insulin delivery system by integrating the ZnO quantum dots coated mesoporous bioactive glasses (MBGs) with polymeric MNs.[52] Two enzymes, GOx and catalase (CAT), were loaded into the internal pores of MBGs together with insulin. The glucose-responsive GOx catalyzed glucose to create a local acidic environment, which could trigger the peculiar acid degradation of ZnO quantum dots to allow the preloaded insulin release. Besides, coencapsulated enzyme CAT could catalyze H_2O_2 into water (H_2O) and O_2, ultimately scavenging excess H_2O_2 to protect the enzymatic activity of GOx as well as reduce the risk of skin inflammation. After transdermal administration of the MN patch in a diabetic animal model, an excellent hypoglycemic effect with lower hypoglycemia risk was obtained. In addition, Yang and colleagues also designed multi-enzyme metal-organic framework (MOF)-based pH-responsive polymer MNs for glucose-mediated insulin delivery.[53] In this work, the local acidic environment generated by the catalytic reaction of GOx led to the degradation of MOF and the simultaneous release of insulin. Meanwhile, the byproduct H_2O_2 was secreted by the catalysis of Co^{2+} ions to avoid the associated side effects.

5.5 The H_2O_2-responsive microneedles

In addition to being consumed by enzymes, some compounds that can react with H_2O_2, such as phenylboronic ester (PBE), can also be employed to remove excess H_2O_2 generated during the GOx-mediated enzymatic reaction process. This is a crucial aspect in the development of efficient and safe insulin delivery systems. By employing nanoparticles modified with these compounds, researchers have devised a system that is responsive to the presence of H_2O_2, thereby enabling precise control over insulin release in response to glucose levels. This technology has the potential to markedly enhance the management of diabetes and mitigate the risk of hypoglycemia in patients. Moreover, this discovery paves the way for the development of targeted drug delivery systems that can respond directly to specific biomarkers or environmental cues within the body.

For example, in a study conducted by Hu et al., crosslinked HA was used to fabricate a polymeric MN patch integrated with H_2O_2-responsive polymeric vesicles (PVs) to construct a rapid hyperglycemia-response insulin delivery system.[54] The block copolymer, composed of polyethylene glycol (PEG) and PBE-conjugated polyserine, was employed to form spherical PVs, with insulin and GOx coloaded in the interior. The PVs were capable of undergoing H_2O_2-mediated depolymerization, which subsequently enabled the release of the encapsulated insulin in a rapid response to hyperglycemic conditions. In vivo testing demonstrated that this MN patch was capable of effectively regulating BGLs in a diabetic mouse model while reducing the risk of hypoglycemia. Building on this research, Xu and coworkers designed a similar MNs device incorporated with H_2O_2-responsive mesoporous silica nanoparticles (MSNs) for insulin delivery.[55] The MSNs were modified by 4-(imidazoyl carbamate) phenylboronic acid pinacol ester (ICBE), followed by a host–guest complexation between ICBE and α-cyclodextrin with insulin and GOx encapsulated inside. Under the hyperglycemic conditions, the MNs rapidly disassembled and simultaneously released the preloaded insulin to downregulate elevated glucose levels, which were induced by disruption of host–guest complexation caused by H_2O_2-mediated oxidation of the ICBE on the surface of MSNs. These advances in drug delivery systems show promise for more effective management of diabetes without increasing the risk of hypoglycemia.

5.6 The dual-responsive microneedles

To further enhance the precision and efficiency of insulin release, researchers have developed dual-responsive nanoparticles/MNs systems. For example, Tong and colleagues successfully created a dual-responsive MNs-based insulin delivery system that responds to both glucose and H_2O_2 levels.[54] This innovative system utilizes glucose- and H_2O_2-responsive polymeric vesicles (PVs) prepared from a triblock copolymer containing PEG, phenylboronic acid (PBA), and poly(phenylboronic). Both GOx and insulin are loaded into the PVs. The presence of PBA in the polymer allows for glucose sensitivity, as it can exist in an equilibrium between an uncharged form and a charged form when exposed to glucose. This transition from hydrophobic to hydrophilic under glucose stimulus leads to the disassembly of the PVs, resulting in rapid insulin release. Additionally, GOx-mediated catalytic oxidation generates H_2O_2, which triggers the hydrolysis of ester linkages within poly(phenylboronic acid pinacol ester), leading to H_2O_2-responsive insulin release.

To further enhance the glucose-responsive capacity and reduce the negative effect of H_2O_2, Yu et al. engineered an insulin-loaded polymeric MNs based on a di-block copolymer consisting of PEG and polyserine modified with 2-nitroimidazole via a thioether moiety to realize hypoxia and H_2O_2 dual-sensitivity.[56] This amphiphilic polymer was able to self-assemble into a nano-scale bilayer vesicle with insulin and GOx coencapsulated in the aqueous core. During the glucose oxidation process, the local hypoxic environment accelerated insulin release due to the bio-reduction of 2-nitroimidazole. Meanwhile, the H_2O_2-sensitive thioether moiety effectively eliminates the undesirable H_2O_2 to maintain the catalytic activity of GOx as well as reduce side effects. Another strategy was adopted by Zhang and partners; they innovated a H_2O_2 and pH dual-responsive polymeric MNs, in which insulin-entrapped degradable micelles (Ins-NCs) and GOx encapsulated nondegradable micelles (GOx-NCs) were both loaded into the crosslinked gel core matrix of MNs, and CAT nanogel (CAT-NGs) was embedded into the PVA sheath structure. In a high glucose condition, GOx catalyzed glucose to produce gluconic acid and H_2O_2, which induced the disruption of Ins-NCs as well as the charge reduction of polymer and insulin, subsequently promoting insulin release. Meanwhile, CAT-NGs could catalyze the decomposition of byproduct H_2O_2 to mitigate the injury.

5.7 Microneedles for the delivery of GLP-1 receptor agonists

GLP-1 receptor agonists are new hypoglycemic drugs that have attracted much attention in recent years. They play a role in lowering blood glucose through multiple mechanisms, such as stimulating islet β-cells to secrete insulin, inhibiting islet α-cells to secrete glucagon, inhibiting appetite, and slowing down gastric emptying. Unlike other antidiabetic drugs, the hypoglycemic effect of GLP-1 receptor agonists is dependent on glucose concentration. When the blood glucose concentration is lower than 4–5 mmol/L, it no longer exerts its effect. Therefore, GLP-1 receptor agonists alone do not cause hypoglycemia and are known as "smart" hypoglycemic drugs. In view of the long administration time of traditional MNs, You et al. developed an ultra-rapid drug release pneumatic MN for the delivery of liraglutide (Fig. 7.4).[57] In this MN, the effervescent agents and drugs are distributed at the tip of the needle through the addition of effervescent agents to the MN matrix and by centrifugation. Upon contact with the skin tissue fluid, the effervescent agents absorb water and produce a substantial quantity of gas, thereby accelerating the release and transdermal diffusion of liraglutide. This facilitates the completion of liraglutide administration within minutes and enhances patient medication compliance. The in vitro drug release results demonstrated that, in comparison with other nonpneumatic MNs, the pneumatic MNs markedly accelerated the release of liraglutide, with a drug release rate exceeding 60% within 10 min. Furthermore, the biocompatibility of the MN was validated in a mouse model of type 2 diabetes, and the therapeutic efficacy of the MN was found to be comparable to that of a subcutaneous injection. Following a 6-week treatment period, the weight gain of the mice was inhibited by 33%, and BGLs were reduced by approximately 58% in comparison to the control group. These findings demonstrate that pneumatic MNs possess the attributes of rapid and efficient drug delivery, which has significant potential for the transcutaneous delivery of biomacromolecule drugs.

6. Microneedles for blood glucose detection

Blood glucose monitoring is crucial for diabetes management, enabling timely treatment adjustments through glycemic tracking. Traditional methods relying on finger-prick blood samples with test strips and meters often cause pain and inconvenience. MN-based technology addresses these limitations by offering continuous, real-time, and painless glucose monitoring through interstitial fluid analysis. This microinvasive approach enhances measurement accuracy while eliminating discomfort, empowering patients to optimize diet, exercise, and medication regimens proactively. By providing dynamic insights into daily glucose fluctuations and their triggers, MN systems not only improve therapeutic precision but also reduce diabetes-related complications. Their user-friendly design promotes consistent monitoring compliance, thereby

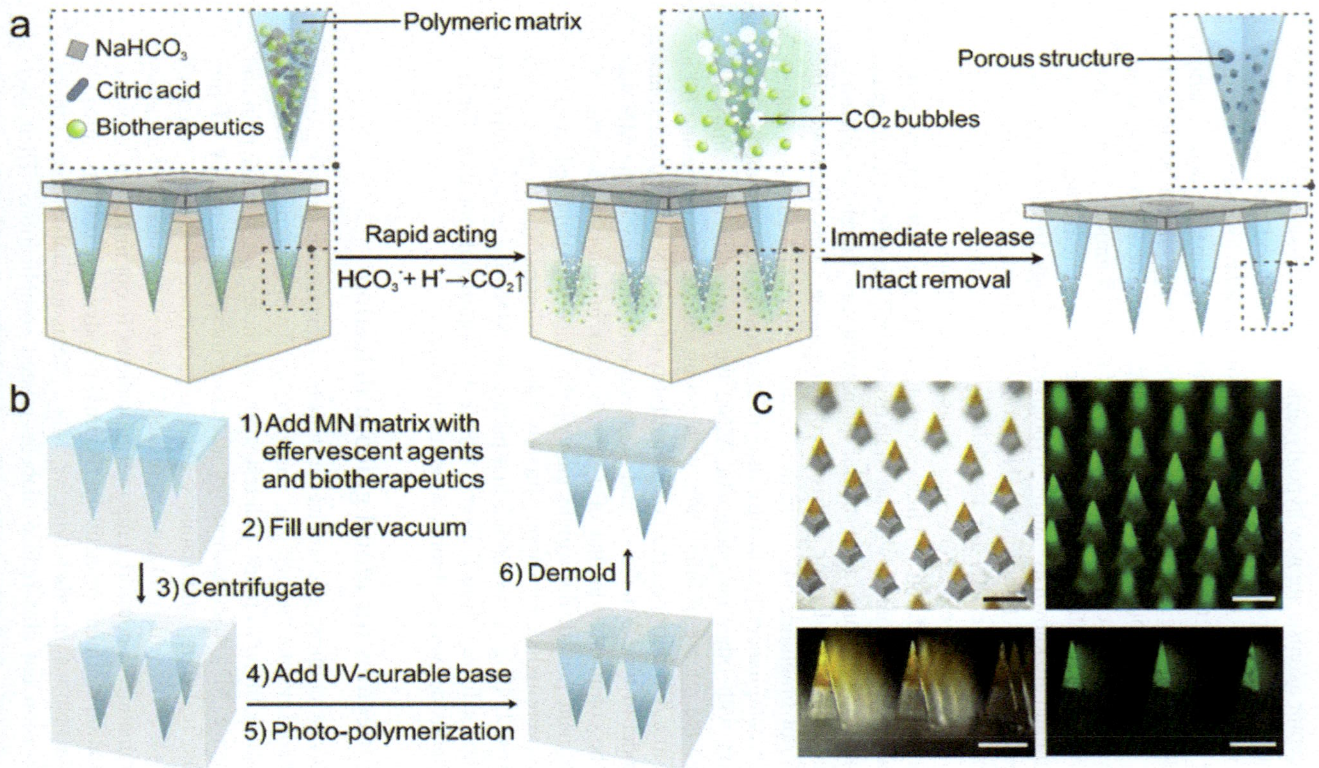

FIGURE 7.4 Schematic illustration of immediate delivery of biotherapeutics via the ultrarapid-acting microneedle (URA-MN) patch. (a) Mechanism of ultrarapid drug release from the URA-MN patch. (b) Scheme presenting the fabrication process of the URA-MN patch. (c) Fluorescence microscopy images of the URA-MN array embedded with effervescent agents and fluorescently labeled liraglutide. Scale bars: 1 mm (*top*), 500 μm (*bottom*). *Reproduced from reference*[57] *with permission from John Wiley & Sons, Inc.*

elevating long-term disease management efficacy and quality of life for diabetic individuals. Furthermore, the development of electrochemical sensor systems based on MN has opened up new possibilities for improving the accuracy and reliability of blood glucose measurements. The modification of MN surfaces with enzymes such as GOx enables precise detection and conversion of interstitial fluid glucose into corresponding blood glucose values.

Currently, there are numerous types of MN sensor devices for blood glucose monitoring, primarily comprising colorimetric MN glucose sensors and electrochemical MN glucose sensors. Colorimetric MN glucose sensors are devices that determine blood glucose concentration based on colorimetric methods. The earliest known report of such a device was in 2004 by Mukerjee et al., who designed a silicon MN array for extracting interstitial fluid (ISF) from the skin.[58] Each MN in the array has a hole at the top. Following insertion into the skin, a 20- to 30-minute waiting period is required for the ISF to fill the MN tip. The approximate blood glucose concentration is then indicated by a color change. However, due to technical limitations, this strategy is not highly accurate for determining blood glucose concentration. Currently, there are studies that integrate extraction analysis. For example, Zeng et al. developed a glucose-responsive colloidal crystal microneedle (GCC-MN) patch for minimally invasive, painless, and visually identifiable glucose colorimetric monitoring (Fig. 7.5).[59] The GCC-MN consists of a polymer core and a glucose-responsive colloidal crystal shell, which can convert the glucose concentration into a visually recognizable color change within 5 minutes (as the glucose concentration increases, the GCC-MN can exhibit purple, blue, and green colors), thereby greatly reducing the analysis time. In the electrochemical MN glucose sensor, the concentration of glucose is converted into an electrical signal by the MN electrode. Yin et al. developed a unique skin patch for blood glucose monitoring based on MN.[60] This patch employed a hollow syringe that had been transformed with GOx and carbon nanotubes, which functioned as an electrochemical sensor. Additionally, integrated circuits were used for signal processing and transmission. The patch was also equipped with Bluetooth connectivity, enabling the real-time display of BGLs on a smartphone. The MN design improves sensitivity, facilitating continuous real-time measurement of blood glucose in live animals at millimolar levels. Glucose monitoring ranges from 0 to 20 mmol/L with a sensitivity of 1 μA/(mmol/L), and exhibits a quick response time of 1 s and a monitoring lifetime of 14 days.

Despite the considerable promise of MNs in blood glucose monitoring, several obstacles remain. Firstly, the biocompatibility and stability of MN materials must be further optimized in order to guarantee the safety and accuracy of the device for long-term use. Secondly, the high cost of the MN manufacturing process represents a significant barrier to large-scale commercial production and widespread application. Furthermore, the precise conversion relationship between interstitial fluid glucose and blood glucose

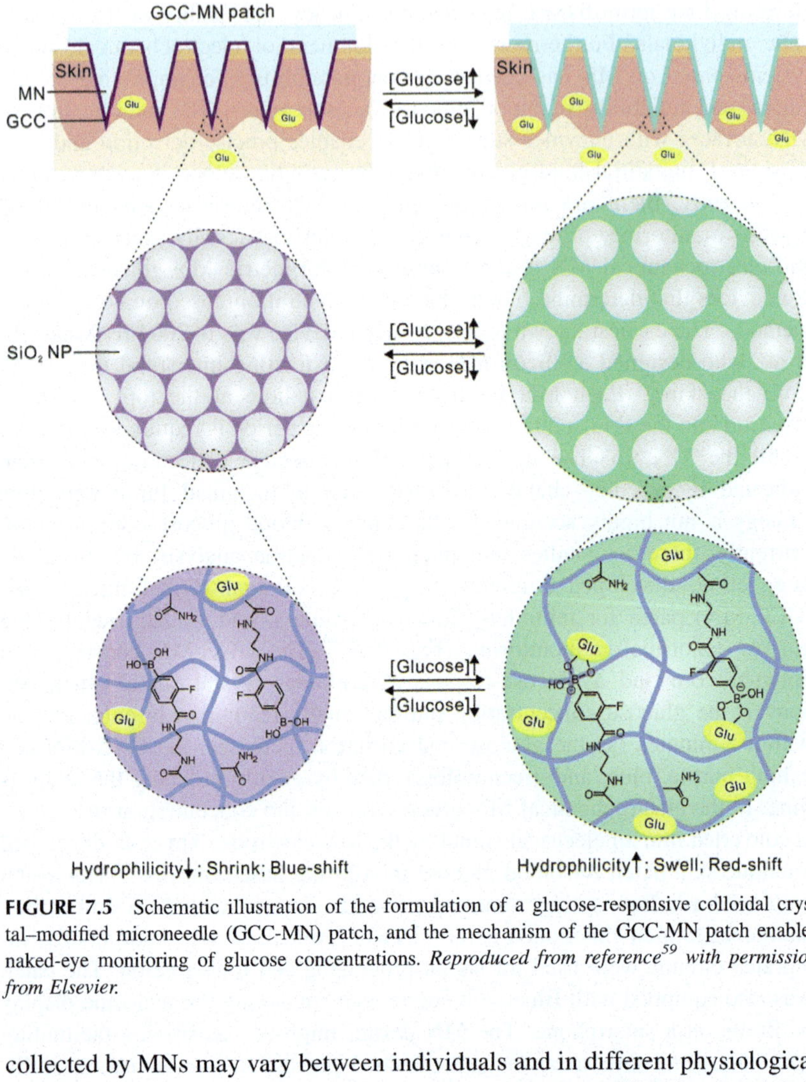

FIGURE 7.5 Schematic illustration of the formulation of a glucose-responsive colloidal crystal–modified microneedle (GCC-MN) patch, and the mechanism of the GCC-MN patch enables naked-eye monitoring of glucose concentrations. *Reproduced from reference*[59] *with permission from Elsevier.*

collected by MNs may vary between individuals and in different physiological and pathological states. Therefore, a more accurate calibration model is required.

It is anticipated that, with ongoing advances in materials science, micro- and nanomanufacturing technologies, and biosensing technologies, MN-based blood glucose monitoring technology will continue to evolve and improve. Future development directions include reducing costs, improving accuracy, enabling multi-indicator joint monitoring, and achieving deep integration with intelligent medical systems. It is anticipated that MN-based technology will completely transform current models of diabetes glucose monitoring, thereby providing convenient, efficient, and accurate health management solutions for patients with diabetes worldwide.

7. Conclusions and remarks

Although the application of MN in the treatment of diabetes shows great potential, there are still many challenges to achieving its widespread clinical application.

(1) Material selection and optimization

The choice of MN material is one of the key factors affecting its performance and application effect. In addition to ensuring biocompatibility and excellent mechanical strength, the aforementioned MN for diabetes treatment often involves complex responsive release materials, which brings new difficulties to the batch production and performance evaluation of the materials. It is urgent to conduct a systematic evaluation of the existing responsive materials to assess their possibility and potential problems for clinical trials. Thus, further optimization of material properties and formulations is needed to meet the specific needs of diabetes treatment.

(2) Long-term safety and effectiveness verification

Although current laboratory studies and partial preclinical trials show good potential for MN in the treatment of diabetes, long-term, large-scale clinical trials are still needed to verify its long-term safety and efficacy. The effects of microtargeting on skin tissue during long-term use are not fully understood. Although short-term studies have shown that wound healing after MN puncture is faster and there are no obvious adverse reactions, whether long-term repeated use of MN will lead to skin fibrosis, chronic inflammation, infection, and other problems still needs to be further studied. For example, some studies have found that long-term use of certain MN may cause thickening of the skin stratum corneum, which affects the puncture effectiveness and drug delivery efficiency of MN. In addition, more clinical data are needed to evaluate the long-term effects of MN in the overall control of diabetes, including the prevention and delay of diabetes complications. Therefore, to carry out multi-center, large sample, and long-term follow-up clinical trials is the only way for MN technology to be widely applied from laboratory to clinical. Only through sufficient clinical validation can we ensure the safety and effectiveness of MN technology in the treatment of diabetes and provide a solid basis for its wide application.

References

1. Sun, Z.; Zheng, Y. Metabolic Diseases in the East Asian Populations. *Nat. Rev. Gastroenterol. Hepatol. 2025* **2025,** 1–17.

2. Wei, J.; Fan, L.; He, Z.; Zhang, S.; Zhang, Y.; Zhu, X.; Xia, F.; Song, X.; Chen, L.; Zou, Z.; et al. The Global, Regional, and National Burden of Type 2 Diabetes Mellitus Attributable to Low Physical Activity from 1990 to 2021: A Systematic Analysis of the Global Burden of Disease Study 2021. *Int. J. Behav. Nutr. Phys. Activ.* **2025,** *22* (1), 8.
3. Wilson, L. M.; Herzig, S. J.; Marcantonio, E. R.; Steinman, M. A.; Schonberg, M. A.; Wang, B. X.; Hileman-Kaplan, E.; Anderson, T. S. Management of Diabetes and Hyperglycemia in the Hospital: A Systematic Review of Clinical Practice Guidelines. *Diabetes Care* **2025,** *48* (4), 655–664.
4. Hughes, M. S.; Addala, A.; Buckingham, B. Digital Technology for Diabetes. *N. Engl. J. Med.* **2023,** *389* (22), 2076–2086.
5. Abel, E. D.; Gloyn, A. L.; Evans-Molina, C.; Joseph, J. J.; Misra, S.; Pajvani, U. B.; Simcox, J.; Susztak, K.; Drucker, D. J. Diabetes Mellitus—Progress and Opportunities in the Evolving Epidemic. *Cell* **2024,** *25,* 3789–3820.
6. Redondo, M. J.; Morgan, N. G. Heterogeneity and Endotypes in Type 1 Diabetes Mellitus. *Nat. Rev. Endocrinol.* **2023,** *19* (9), 542–554.
7. Forbes, J. M.; Cooper, M. E. Mechanisms of Diabetic Complications. *Physiol. Rev.* **2013,** *93* (1), 137–188.
8. Cole, J. B.; Florez, J. C. Genetics of Diabetes Mellitus and Diabetes Complications. *Nat. Rev. Nephrol.* **2020,** *16* (7), 377–390.
9. Zheng, Y.; Ley, S. H.; Hu, F. B. Global Aetiology and Epidemiology of Type 2 Diabetes Mellitus and its Complications. *Nat. Rev. Endocrinol.* **2017,** *14* (2), 88–98.
10. McIntyre, H. D.; Catalano, P.; Zhang, C.; Desoye, G.; Mathiesen, E. R.; Damm, P. *Gestational Diabetes Mellitus. Nature Reviews Disease Primers;* Nature Publishing Group, July 11, 2019; pp 1–19.
11. Buchanan, T. A.; Xiang, A. H.; Page, K. A.; Watanabe, R. M. What Is Gestational Diabetes – Really? *Diabetes* **2025,** *74* (7), 1037–1046.
12. Petry, C. J. Future Prospects for Gestational Diabetes. In *Gestational Diabetes: Origins, Complications, and Treatment,* **377**; British Medical Journal Publishing Group, 2014; pp 195–222.
13. Christ-Crain, M.; Winzeler, B.; Refardt, J. Diagnosis and Management of Diabetes Insipidus for the Internist: An Update. *J. Intern. Med.* **2021,** *290* (1), 73–87.
14. Lega, I. C.; Lipscombe, L. L. Review: Diabetes, Obesity, and Cancer—Pathophysiology and Clinical Implications. *Endocr. Rev.* **2020,** *41* (1), 33–52.
15. Sims, E. K.; Carr, A. L. J.; Oram, R. A.; DiMeglio, L. A.; Evans-Molina, C. 100 Years of Insulin: Celebrating the Past, Present and Future of Diabetes Therapy. *Nat. Med.* **2021,** *27* (7), 1154–1164.
16. Petersen, M. C. GI Shulman. Mechanisms of Insulin Action and Insulin Resistance. *Physiol. Rev.* **2018,** *98* (4), 2133–2223.
17. Rosenstock, J.; Juneja, R.; Beals, J. M.; Moyers, J. S.; Ilag, L.; McCrimmon, R. J. The Basis for Weekly Insulin Therapy: Evolving Evidence with Insulin Icodec and Insulin Efsitora Alfa. *Endocr. Rev.* **2024,** *45* (3), 379–413.
18. Haeusler, R. A.; McGraw, T. E.; Accili, D. Biochemical and Cellular Properties of Insulin Receptor Signalling. *Nat. Rev. Mol. Cell Biol.* **2017,** *19* (1), 31–44.
19. Riddell, M. C.; Peters, A. L. Exercise in Adults with Type 1 Diabetes Mellitus. *Nat. Rev. Endocrinol.* **2022,** *19* (2), 98–111.
20. Hirsch, I. B.; Juneja, R.; Beals, J. M.; Antalis, C. J.; Wright, E. E. The Evolution of Insulin and How it Informs Therapy and Treatment Choices. *Endocr. Rev.* **2020,** *41* (5), 733–755.

21. McIntyre, H. D.; Catalano, P.; Zhang, C.; Desoye, G.; Mathiesen, E. R.; Damm, P. Gestational Diabetes Mellitus. *Nat. Rev. Dis. Primers* **2019,** *5* (1), 1–19.
22. Tang, C.; Hao, J.; Tao, F.; Feng, Q.; Song, Y.; Zeng, B. Association of Metformin Use with Risk of Dementia in Patients with Type 2 Diabetes: A Systematic Review and Meta-Analysis. *Diabetes Obes. Metabol.* **2025,** *27* (4), 1992–2001.
23. López-Otín, C.; Galluzzi, L.; Freije, J. M. P.; Madeo, F.; Kroemer, G. Metabolic Control of Longevity. *Cell* **2016,** *166* (4), 802–821.
24. Chang, Y. C.; Hsieh, M. L.; Lee, H. L.; Hee, S. W.; Chang, C. F.; Yen, H. Y.; Chen, Y. A.; Chen, Y. R.; Chou, Y. W.; Li, F. A.; et al. Identification of PTGR2 Inhibitors as a New Therapeutic Strategy for Diabetes and Obesity. *EMBO Mol. Med.* **2025,** *17,* 938–966.
25. Wong, Y. W.; Thomas, L.; Sun, J. L.; McMurray, J. J. V.; Krum, H.; Hernandez, A. F.; Rutten, G. E. H. M.; Leiter, L. A.; Standl, E.; Haffner, S. M.; et al. Predictors of Incident Heart Failure Hospitalizations Among Patients with Impaired Glucose Tolerance Insight from the Nateglinide and Valsartan in Impaired Glucose Tolerance Outcomes Research Study. *Circ. Heart Fail.* **2013,** *6* (2), 203–210.
26. Li, L.; Yang, M.; Li, Z.; Yan, X.; Guo, H.; Pan, H.; Liu, H.; Liao, Y.; Yang, G. Efficacy and Safety of Mitiglinide versus Nateglinide in Newly Diagnose Patients with Type 2 Diabetes Mellitus: A Randomized Double Blind Trial. *Diabetes Obes. Metabol.* **2012,** *14* (2), 187–189.
27. TNS Group. Effect of Nateglinide on the Incidence of Diabetes and Cardiovascular Events. *N. Engl. J. Med.* **2010,** *362,* 1463–1476 (16).
28. Wardrop, D. J.; Waidyarachchi, S. L. Synthesis and Biological Activity of Naturally Occurring α-Glucosidase Inhibitors. *Nat. Prod. Rep.* **2010,** *27* (10), 1431–1468.
29. Deng, S.; Xia, L.; Xiao, H. Screening of α-Glucosidase Inhibitors from Green Tea Extracts Using Immobilized Enzymes Affinity Capture Combined with UHPLC-QTOF MS Analysis. *Chem. Commun.* **2014,** *50* (20), 2582–2584.
30. Doungwichitrkul, T.; Damsud, T.; Phuwapraisirisan, P. α-Glucosidase Inhibitors from Cold-Pressed Black Sesame (Sesamum Indicum) Meal: Characterization of New Furofuran Lignans, Kinetic Study, and In Vitro Gastrointestinal Digestion. *J. Agric. Food Chem.* **2024,** *72* (2), 1044–1054.
31. Emanuel, E. J.; Dellgren, J. L.; McCoy, M. S.; Persad, G. Fair Allocation of GLP-1 and Dual GLP-1–GIP Receptor Agonists. *N. Engl. J. Med.* **2024,** *390* (20), 1839–1842.
32. Suissa, K.; Cromer, S. J.; Patorno, E. GLP-1 Receptor Agonists and Gastrointestinal Adverse Events. *JAMA* **2024,** *331* (10), 884–885.
33. Samms, R. J.; Coghlan, M. P.; Sloop, K. W. How May GIP Enhance the Therapeutic Efficacy of GLP-1? *Trends Endocrinol. Metabol.* **2020,** *31* (6), 410–421.
34. Baggio, L. L.; Drucker, D. J. Biology of Incretins: GLP-1 and GIP. *Gastroenterology* **2007,** *132* (6), 2131–2157.
35. Mueller-Schoop, J. Accidental Intravenous Self-Injection with Insulin Pen. *Lancet* **1993,** *341* (8849), 894.
36. Wolpert, H.; Rodbard, D.; Xue, J.; Johnson, J.; Dassau, E. Characterizing Insulin Dosing Behaviour and Glycaemic Excursions: Development of Metrics Using Connected Insulin Pen and Continuous Glucose Monitoring. *Diabetes Obes. Metabol.* **2025,** *27* (5), 2507–2514.
37. Gomez-Peralta, F.; Abreu, C.; Fernandez-Rubio, E.; Cotovad, L.; Pujante, P.; Gaztambide, S.; Bellido, D.; Torre, E. M.; Ruiz-Valdepeñas, S.; Bello, H.; et al. Efficacy of a Connected Insulin Pen Cap in People with Noncontrolled Type 1 Diabetes: A Multicenter Randomized Clinical Trial. *Diabetes Care* **2023,** *46* (1), 206–208.

38. Ranjan, A. G.; Schmidt, S.; Nørgaard, K. Faster-Acting Insulin Aspart versus Insulin Aspart for Adults with Type 1 Diabetes Treated with Non-automated Insulin Pump and Continuous Glucose Monitoring—A Double-Blind Randomized Controlled Crossover Trial. *Diabetes Obes. Metabol.* **2025,** *27* (6), 3145–3153.
39. Karakus, K. E.; Mason, E.; Akturk, H. K. Long-Term Glycaemic Improvement with the Initiation of an Automated Insulin Delivery System in Insulin Pump-Naïve Older Adults with Type 1 Diabetes. *Diabetes Obes. Metabol.* **2025,** *27* (1), 410–413.
40. Borel, A.-L.; Lablanche, S.; Waterlot, C.; Joffray, E.; Barra, C.; Arnol, N.; Amougay, H.; Benhamou, P.-Y. Closed-Loop Insulin Therapy for People with Type 2 Diabetes Treated with an Insulin Pump: A 12-Week Multicenter, Open-Label Randomized, Controlled, Crossover Trial. *Diabetes Care* **2024,** *47* (10), 1778–1786.
41. Liu, T.; Chen, M.; Fu, J.; Sun, Y.; Lu, C.; Quan, G.; Pan, X.; Wu, C. Recent Advances in Microneedles-Mediated Transdermal Delivery of Protein and Peptide Drugs. *Acta Pharm. Sin. B* **2021,** *11* (8), 2326–2343.
42. Zhou, C. P.; Liu, Y. L.; Wang, H. L.; Zhang, P. X.; Zhang, J. L. Transdermal Delivery of Insulin Using Microneedle Rollers In Vivo. *Int. J. Pharm.* **2010,** *392* (1–2), 127–133.
43. McAllister, D. V.; Wang, P. M.; Davis, S. P.; Park, J. H.; Canatella, P. J.; Allen, M. G.; Prausnitz, M. R. Microfabricated Needles for Transdermal Delivery of Macromolecules and Nanoparticles: Fabrication Methods and Transport Studies. *Proc. Natl. Acad. Sci. U. S. A.* **2003,** *100* (Suppl. 2), 13755–13760.
44. Roxhed, N.; Samel, B.; Nordquist, L.; Griss, P.; Stemme, G. Painless Drug Delivery through Microneedle-Based Transdermal Patches Featuring Active Infusion. *IEEE Trans. Biomed. Eng.* **2008,** *55* (3), 1063–1071.
45. Ross, S.; Scoutaris, N.; Lamprou, D.; Mallinson, D.; Douroumis, D. Inkjet Printing of Insulin Microneedles for Transdermal Delivery. *Drug Deliv. Transl. Res.* **2015,** *5* (4), 451–461.
46. Yu, J.; Wang, J.; Zhang, Y.; Chen, G.; Mao, W.; Ye, Y.; Kahkoska, A. R.; Buse, J. B.; Langer, R.; Gu, Z. Glucose-Responsive Insulin Patch for the Regulation of Blood Glucose in Mice and Minipigs. *Nat. Biomed. Eng.* **2020,** *4* (5), 499–506.
47. Liu, J. F.; GhavamiNejad, A.; Lu, B.; Mirzaie, S.; Samarikhalaj, M.; Giacca, A.; Wu, X. Y. "Smart" Matrix Microneedle Patch Made of Self-Crosslinkable and Multifunctional Polymers for Delivering Insulin On-Demand. *Adv. Sci.* **2023,** *10* (30), 2303665.
48. Chen, M.; Quan, G.; Sun, Y.; Yang, D.; Pan, X.; Wu, C. Nanoparticles-Encapsulated Polymeric Microneedles for Transdermal Drug Delivery. *J. Contr. Release* **2020,** *325* (June), 163–175.
49. Yu, J.; Zhang, Y.; Ye, Y.; DiSanto, R.; Sun, W.; Ranson, D.; Ligler, F. S.; Buse, J. B.; Gu, Z.; Ho, D. Microneedle-Array Patches Loaded with Hypoxia-Sensitive Vesicles Provide Fast Glucose-Responsive Insulin Delivery. *Proc. Natl. Acad. Sci. U. S. A.* **2015,** *112* (27), 8260–8265.
50. Ye, Y.; Yu, J.; Wang, C.; Nguyen, N.-Y.; Walker, G. M.; Buse, J. B.; Gu, Z.; Ye, Y.; Yu, J.; Wang, C.; et al. Microneedles Integrated with Pancreatic Cells and Synthetic Glucose-Signal Amplifiers for Smart Insulin Delivery. *Adv. Mater.* **2016,** *28* (16), 3115–3121.
51. Chen, W.; Tian, R.; Xu, C.; Yung, B. C.; Wang, G.; Liu, Y.; Ni, Q.; Zhang, F.; Zhou, Z.; Wang, J.; et al. Microneedle-Array Patches Loaded with Dual Mineralized Protein/Peptide Particles for Type 2 Diabetes Therapy. *Nat. Commun.* **2017,** *8* (1), 1–11.
52. Xu, B.; Cao, Q.; Zhang, Y.; Yu, W.; Zhu, J.; Liu, D.; Jiang, G. Microneedles Integrated with ZnO Quantum-Dot-Capped Mesoporous Bioactive Glasses for Glucose-Mediated Insulin Delivery. *ACS Biomater. Sci. Eng.* **2018,** *4* (7), 2473–2483.

53. Yang, X. X.; Feng, P.; Cao, J.; Liu, W.; Tang, Y. Composition-Engineered Metal-Organic Framework-Based Microneedles for Glucose-Mediated Transdermal Insulin Delivery. *ACS Appl. Mater. Interfaces* **2020,** *12* (12), 13613–13621.
54. Tong, Z.; Zhou, J.; Zhong, J.; Tang, Q.; Lei, Z.; Luo, H.; Ma, P.; Liu, X. Glucose- and H2O2-Responsive Polymeric Vesicles Integrated with Microneedle Patches for Glucose-Sensitive Transcutaneous Delivery of Insulin in Diabetic Rats. *ACS Appl. Mater. Interfaces* **2018,** *10* (23), 20014–20024.
55. Xu, B.; Jiang, G.; Yu, W.; Liu, D.; Zhang, Y.; Zhou, J.; Sun, S.; Liu, Y. H2O2-Responsive Mesoporous Silica Nanoparticles Integrated with Microneedle Patches for the Glucose-Monitored Transdermal Delivery of Insulin. *J. Mater. Chem. B* **2017,** *5* (41), 8200–8208.
56. Yu, J.; Qian, C.; Zhang, Y.; Cui, Z.; Zhu, Y.; Shen, Q.; Ligler, F. S.; Buse, J. B.; Gu, Z. Hypoxia and H2O2 Dual-Sensitive Vesicles for Enhanced Glucose-Responsive Insulin Delivery. *Nano Lett.* **2017,** *17* (2), 733–739.
57. You, J.; Yang, C.; Han, J.; Wang, H.; Zhang, W.; Zhang, Y.; Lu, Z.; Wang, S.; Cai, R.; Li, H.; et al. Ultrarapid-Acting Microneedles for Immediate Delivery of Biotherapeutics. *Adv. Mater.* **2023,** *35* (45), 2304582.
58. Mukerjee, E. V.; Collins, S. D.; Isseroff, R. R.; Smith, R. L. Microneedle Array for Transdermal Biological Fluid Extraction and In Situ Analysis. *Sens. Actuators A Phys.* **2004,** *114* (2–3), 267–275.
59. Zeng, Y.; Wang, J.; Wang, Z.; Chen, G.; Yu, J.; Li, S.; Li, Q.; Li, H.; Wen, D.; Gu, Z.; et al. Colloidal Crystal Microneedle Patch for Glucose Monitoring. *Nano Today* **2020,** *35,* 100984.
60. Yin, S.; Yu, Z.; Song, N. N.; Guo, Z.; Li, W.; Ma, J.; Wang, X.; Liu, J.; Liang, M. A Long Lifetime and Highly Sensitive Wearable Microneedle Sensor for the Continuous Real-Time Monitoring of Glucose in Interstitial Fluid. *Biosens. Bioelectron.* **2024,** *244,* 115822.

Chapter 8

Microneedle technology for antitumor therapy

Siyuan Peng[1], Guilan Quan[2] and Xin Pan[1]
[1]*School of Pharmaceutical Science, Sun Yat-Sen University, Guangzhou, China;* [2]*State Key Laboratory of Bioactive Molecules and Druggability Assessment/International Cooperative Laboratory of Traditional Chinese Medicine Modernization and Innovative Drug Development of Ministry of Education (MOE) of China/College of Pharmacy, Jinan University, Guangzhou, China*

1. Superficial cancer

Tumor disease is always a major problem that threatens human health seriously. Among them, superficial cancer, as a more common type of tumor, refers to tumors occurring in superficial parts of the body, mainly including breast cancer, malignant melanoma, squamous cell carcinoma, and basal cell carcinoma. The latest global cancer statistics indicate that the incidence of breast cancer, skin melanoma, and nonmelanoma skin cancers is 11.6%, 1.7%, and 5.8%, respectively. Correspondingly, the mortality is 6.9%, 0.6%, and 0.7%, respectively.[1] The incidence and mortality of these superficial cancers are on the rise globally, seriously jeopardizing human health and life, and bringing a heavy medical burden and psychological pressure to the patient's family as well as the whole society.[1]

2. Tumor treatment strategies

2.1 Traditional treatment strategies

Surgical resection, as the traditional sharp edge of tumor treatment, has long occupied an important position in anticancer therapy.[2] For superficial skin cancer, surgery aims to achieve the treatment goal by directly removing tumor tissues in superficial areas. However, in practice, it is difficult for surgery to remove tumor tissues completely because the boundaries of tumor tissues are often not precisely defined, especially in cases where some tumors have already undergone tiny metastasis or infiltrative growth.[3,4] Such residual tumors after surgery are like "hidden time bombs," which pose hidden dangers

Microneedle Technology. https://doi.org/10.1016/B978-0-443-45544-5.00002-6

for tumor recurrence and lead to a high recurrence rate after surgery.[5] In addition, surgical resection itself is an invasive treatment. It will cause greater trauma to the patient's body, and the postoperative recovery cycle is longer.[3] And it may be accompanied by a series of complications such as infection, scar formation, and so on, bringing additional pain to patients.

Radiotherapy, which uses high-energy rays to kill tumor cells, is another major weapon in antitumor therapy.[6] In superficial skin cancer treatment, radiotherapy destroys the DNA structure of tumor cells by focusing high-energy rays on the tumor site, thus inhibiting the growth and proliferation of the tumor.[7] However, the indistinguishable characteristic of radiotherapy will inevitably cause damage to the normal tissues around the tumor in the treatment process. For example, in radiotherapy for head and neck skin cancer, patients often suffer from radiation dermatitis, which is manifested as redness, swelling, itching, peeling, or even ulceration of the skin.[8–10] The hair in the irradiated area will also fall off as a result of radiotherapy.[11–13] In addition, it may also trigger inflammation of the oral mucous membranes, resulting in patients' difficulty in eating and painful mouth ulcers.[12,14–16] Moreover, some tumor cells will gradually develop tolerance during long-term radiotherapy, making it difficult to eliminate tumor cells.[17–19]

Chemotherapy, as a systemic treatment, works by using chemical drugs to interfere with various physiological processes of the tumor.[20] After entering the human body, drugs will spread to various tissues and organs throughout the body along with blood circulation to achieve the purpose of treating tumors.[21,22] However, while killing tumor cells, chemotherapy will also cause great damage to normal cells in the human body.[23,24] For example, the inhibition of bone marrow hematopoietic stem cells will lead to leukopenia, anemia, platelet lowering, and other blood system problems, which will cause a sharp decline in immunity and easily lead to a variety of infections and bleeding symptoms.[25–29] The damage to gastrointestinal mucosal cells will cause nausea, vomiting, loss of appetite, diarrhea, and other digestive adverse reactions, which will seriously affect the patient's nutrient intake and physical function.[30–33] These side effects not only bring great physical pain to patients, but also seriously affect the patient's psychological state and treatment compliance, and many patients are forced to interrupt treatment because they cannot tolerate the side effects of chemotherapy, which makes it difficult to ensure the therapeutic effect.[20,23]

2.2 Novel treatment strategies

Faced with the dilemmas of conventional treatments, the field of antitumor therapy is in dire need of new progress and innovation. Against this background, a series of new therapies such as phototherapy, chemodynamic therapy (CDT), gene therapy, and immunotherapy have gradually emerged as emerging forces in antitumor therapy.[34,35] Each of these methods has its own unique therapeutic principles and advantages. For example, phototherapy uses specific wavelengths of light and photosensitizers to generate reactive oxygen species (ROS) to kill tumor cells.[36,37] Or it achieves an antitumor effect through photothermal conversion performance.[38–40] It has the characteristics of strong targeting and small local damage.[41] CDT generates cytotoxic free radicals in tumor tissues through chemical reaction to achieve specific killing of tumor cells.[42–44] Gene therapy corrects the abnormal biological behavior of tumor cells by repairing or modifying their genes.[45–47] Immunotherapy focuses on activating the body's own immune system to enhance the ability of immune cells to recognize and attack tumor cells.[48–50] These therapies have brought new ideas and hopes for antitumor. However, these new therapies also face their own challenges in practical application. For example, the therapeutic effect of phototherapy is often limited by factors such as the distribution of photosensitizers and the depth of light penetration.[39,41,51] The response efficiency and selectivity of CDT still need to be further improved.[43,52,53] Gene therapy suffers from the low efficiency of gene transfection and safety issues.[46,47] And immunotherapy may trigger immune-related adverse reactions, such as an immune storm.[48,49] Therefore, the search for a more effective, safe, and convenient treatment or therapeutic strategy is still an important research direction in the field of tumor therapy.

3. Advantages of microneedles for antitumor therapy

In the research of superficial tumor therapy, MNs-mediated local drug delivery is gradually coming into the limelight as a highly promising new technological tool.[34,54] MNs have many unique advantages over traditional intravenous, intratumoral, and oral drug delivery modalities.[55] It can effectively penetrate the skin's stratum corneum, which is the main barrier for drug absorption, without significantly destroying the skin's overall structure and deeper tissues, and opens up several fine delivery channels for drugs to directly enter the superficial tumor tissues or their surrounding microenvironment.[56–58] Through the MNs, the drug can be precisely delivered to the tumor lesion site, which greatly improves the accumulation of the drug in the tumor.[57,59] This significantly improves drug delivery efficiency and enhances drug efficacy against tumors. Meanwhile, the MN delivery system decreases the drug content in the systemic blood circulation, which effectively reduces the various side effects caused by systemic drug exposure.[57,60,61] In addition, MNs cause only a minimal incision when inserted into the skin, and the patient can hardly feel any obvious pain, which has good patient compliance.[62,63] These advantages make MNs in the field of antitumor therapy, especially in superficial tumor treatment, showing a broad application prospect and great potential for development. Therefore, it has attracted the attention and research enthusiasm of many researchers and medical experts, and has become one of the hotspots in the current research field of antitumor therapy.

As research on MNs continues to develop, the scope of their application in antitumor therapy is gradually expanding. From the initial simple drug delivery to the current combined application with various therapeutic methods, MNs are continuously exhibiting their versatility and adaptability.[54,55,57] For example, the combination of MNs and immunotherapy, through which immunomodulatory drugs or tumor vaccines are delivered to the tumor locally, can activate a strong immune response.[64–67] The therapy not only can directly attack the tumor, but also may trigger systemic immune effects, and can play an inhibitory role in distant metastatic tumors. The combined application of MNs and phototherapy can use MNs to deliver the photosensitizer or photothermal materials to the tumor tissues with high efficiency, firstly, and then carry out light irradiation.[34,40,68–70] Thus, the therapeutic effect of phototherapy can be improved, and the side effects of phototherapy can be reduced. The synergistic effect of MNs and CDT can enable chemodynamic agents to react more accurately at the tumor site, enhance the generation of free radicals and the ability to kill tumor cells, and improve the efficacy and selectivity of CDT.[63,71–73] In conclusion, today's extensive research on MNs, especially MNs loaded with nanoparticles (NP@MNs), can be combined with different therapies to amplify their effects against tumors.

4. Application of microneedles in antitumor therapy

4.1 Microneedle-mediated chemotherapy

Currently, chemotherapeutic drugs are predominantly administered orally or by intravenous injection. However, these methods often fail to achieve sufficient drug accumulation at superficial tumor sites, leading to suboptimal therapeutic efficacy and significant systemic side effects due to the systemic circulation. In contrast, MNs offer a promising alternative by directly delivering chemotherapeutic drugs to the target regions of superficial tumors, thereby markedly enhancing drug enrichment and significantly improving treatment safety.[74,75]

Bhatnagar et al.[76] designed dissolving MNs for codelivery of adriamycin hydrochloride and docetaxel. The in vivo pharmacodynamics study demonstrated that the combination of MNs with adriamycin and docetaxel, either as standalone agents or in combination, resulted in a significantly higher survival rate compared to the intratumor injection approach (Fig. 8.1). Recent studies have also demonstrated that MNs loaded with nanoparticles can significantly reduce the systemic toxicity of chemotherapeutic drugs.[54,77,78] Cisplatin has proven to be effective when administered intravenously for the treatment of solid malignant tumors. However, its use is associated with the occurrence of severe adverse effects, including nephrotoxicity and neurotoxicity.[79] Lan et al.[80] developed MNs-mediated transdermal delivery of lipid-encapsulated cisplatin nanoparticles for squamous cell carcinoma treatment. The use of

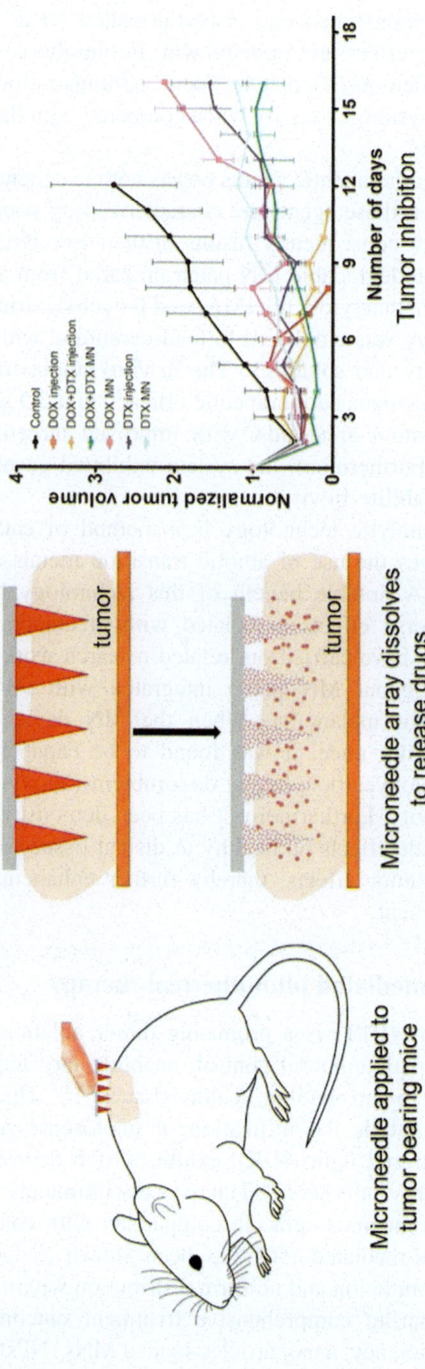

FIGURE 8.1 Dissolving microneedle patch containing doxorubicin and docetaxel to combat 4T1 xenografted breast cancer.[76]

MNs ensures the targeted delivery of the drug to the tumor location. In vivo antitumor studies demonstrated that cisplatin-loaded NPs@MNs exhibited a substantial inhibitory effect on tumor growth. Furthermore, the results of the safety assessment demonstrated that local administration of NPs@MNs resulted in minimal systemic toxicity when compared with the intravenous and local injection groups.

A subset of chemotherapeutic agents boasts both excellent efficacy and low side effects. However, these agents are characterized by poor water solubility and instability, which consequently results in their low bioavailability. Zhou et al.[81] designed a biodegradable MN patch prepared from a natural polymer coupling of gelatin methacryloyl (GelMA) and β-cyclodextrin (GelMA-β-CD). The inclusion strategy was employed to load curcumin, with the objective of enhancing its stability and solubility. The in vitro studies demonstrated that this system achieved enhanced therapeutic efficacy in a 3D skin cancer model based on B16F10 tumor spheroids, with improved targeting accuracy and deeper penetration. Furthermore, the system exhibited excellent biocompatibility and biodegradability in vivo.

Bioorthogonal catalytic technology is a method of catalyzing chemical reactions that involves the use of abiotic transition metals as substitutes for natural enzymes.[82] A notable benefit of this technology is its capacity to circumvent the adverse effects associated with precursor drugs in normal tissues. Chen et al.[83] have carried out related research work and successfully developed a biorthogonal MN patch integrated with Pd NP-doped TiO_2 nanosheets to combat melanoma. When the MN device was applied to B16–F10 tumor-bearing mice, it was found to be capable of inducing the conversion of n-allyloxycarbonyl cage doxorubicin (alloxy-DOX) into DOX with anticancer activity. Furthermore, it has been demonstrated to effectively mitigate off-target side effects on healthy or distant tissues due to drug action while exerting antitumor effects, thereby further enhancing the safety and efficacy of the treatment.

4.2 Microneedle-mediated photothermal therapy

Photothermal therapy (PTT) is a promising tumor ablation strategy characterized by precise spatiotemporal control, enabling targeted treatment while minimizing damage to surrounding healthy tissues.[39,68] The core mechanism of this technology entails the utilization of photosensitizers, which, upon exposure to near-infrared light (NIR), exhibit a high degree of efficiency in converting light energy into heat.[84] This process ultimately leads to the irreversible demise of cancerous cells. In comparison with conventional administration routes, MN-mediated PTT has been shown to facilitate enhanced photosensitizer accumulation and uniform distribution within superficial tumor tissues, thereby ensuring comprehensive treatment outcomes.[75] To further enhance delivery efficiency, nanoparticles-loaded MNs (NPs@MNs) are being extensively investigated as an innovative transdermal PTT platform, demonstrating considerable potential for advancing tumor therapy.[85–87]

For instance, Fu and his colleagues[88] fabricated an MN patch with gold nanorods delicately incorporated at the needle tips. When the patch is exposed to NIR, surface plasmon resonance occurs within the gold nanorods, thereby efficiently converting light energy into heat energy. The generated hyperthermia exerted a direct effect on tumor tissues, thereby facilitating the complete ablation of tumors. Wei et al.[89] developed a dissolving MN system that integrated pH-responsive micelles loaded with NIR950 AIEgen. This innovative approach has been demonstrated to elicit a remarkable therapeutic effect, effectively eradicating malignant melanoma with a single low-dose administration and a singular NIR irradiation.

The limitations of single PTT in the treatment of tumors can readily result in tumor recurrence. In view of this, the combination of PTT and chemotherapy is proposed as a superior cancer treatment strategy due to its powerful synergy. In this combination therapy regimen, both organic photosensitizers, such as indocyanine green,[90] IR780,[74] and inorganic photosensitizers, such as gold nanocages[91] and CuS,[92] can be meticulously loaded in MNs with chemotherapy drugs for synergistic treatment. For instance, Peng et al.[75] Specifically, functionalized poly(lactic-co-glycolic acid) (PLGA) NPs were prepared for codelivery of paclitaxel and indocyanine green, followed by encapsulation into MNs. PTT has been shown to promote apoptosis in tumor cells, thereby activating caspase-3 enzyme, significantly reducing the expression of survivin and MMP-9 protein, and enhancing the sensitivity of tumor cells to chemotherapy. The in vivo model demonstrated that the combination therapy led to a notable intensification of the antitumor effect, resulting in complete tumor ablation (Fig. 8.2).

In another study, Qin et al.[74] proposed a chemo-photothermal synergistic system for the treatment of melanoma. The researchers encapsulated paclitaxel and photothermal agents IR-780 within solid lipid nanoparticles to fabricate NPs@MNs patches exhibiting high photothermal conversion and antitumor efficiency. In addition, Ying et al.[93] designed a NIR responsive polyethylene glycol-coated gold nanorod encapsulated poly(L-lactide) microneedle system (GNR-PEG@MN) to enhance the antitumor efficiency of intravenously injected docetaxel micelles. In evaluations of antitumor efficacy, the combination of low-dose docetaxel with gold nanorods demonstrated complete eradication of A431 epidermoid tumors without recurrence. This strategy offers a promising and effective therapeutic approach for the treatment of epidermoid carcinoma.

The aggressive nature and high rate of recurrence of malignant tumors often necessitate frequent treatments. Nevertheless, this demand can result in suboptimal patient compliance.[7,57,94] To address this challenge, Chen et al.[95] developed an NIR-responsive MN device for precise and repeated drug delivery. The device consists of an embeddable polycaprolactone needle tip that is supported by a polyvinyl alcohol/polyvinyl pyrrolidone (PVA/PVP) backing layer.

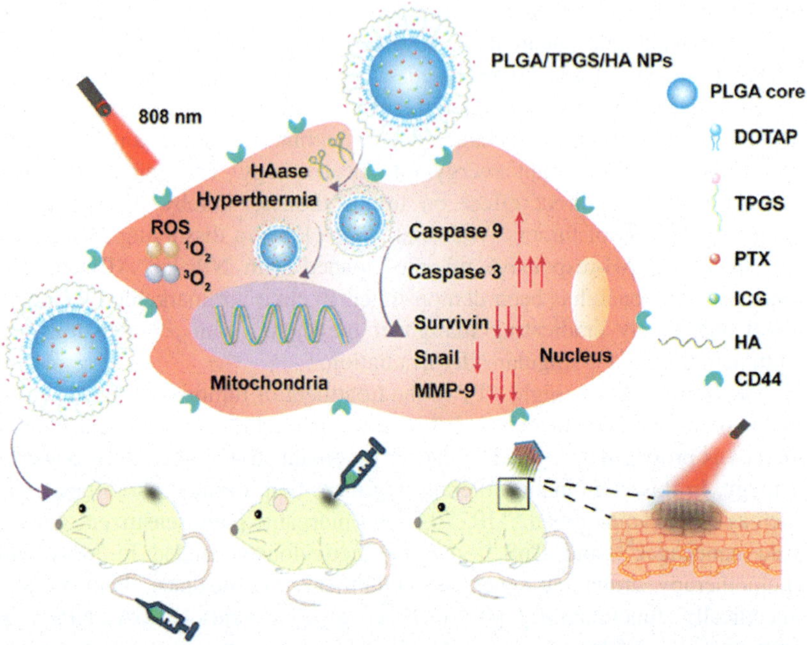

FIGURE 8.2 Functionalized poly(lactic-co-glycolic acid) (PLGA) nanoparticles delivered through dissolving microneedles for markedly improved chemo-photothermal combined therapy of superficial tumors.[75]

Lanthanum hexaboride nanoparticles and docetaxel were coencapsulated within the needles. Upon NIR exposure, the nanoparticles raised the local temperature to approximately 50 °C, inducing the melting of the needles and triggering a pulsed release of docetaxel. When the NIR light was turned off, the temperature quickly returned to baseline, halting drug release. In a preclinical study, the MN patch, when administered in a single application in conjunction with three cycles of NIR irradiation, resulted in the complete eradication of 4T1 breast tumors within a week. This outcome demonstrates the potential for precise, light-controlled dosing tailored to individual treatment needs.

4.3 Microneedle-mediated photodynamic therapy

In photodynamic therapy (PDT), photosensitizers activated by laser irradiation play a pivotal role by converting light energy into highly cytotoxic ROS, which effectively destroy cancer cells.[96] However, the clinical application of systemically administered photosensitizers is limited by their low tissue selectivity and poor stability.[97] To address these limitations, MNs have emerged as a promising targeted delivery strategy for the PDT treatment of superficial tumors.[55]

In 2008, Donnelly et al.[98] provided a comprehensive account of silicon MNs-mediated intradermal delivery of 5-aminolevulinic acid (ALA) for PDT of skin tumors. This approach proved highly effective, with silicon MN pretreatment significantly enhancing both in vitro and in vivo transdermal delivery of ALA from a bioadhesive patch. The results demonstrated that MN-assisted delivery resulted in a substantial increase in protoporphyrin IX production and enabled comparable PDT efficacy at lower ALA doses and shorter application times. In another study, Zhu et al.[99] developed ALA-loaded MNs for the treatment of breast cancer. This innovative system was meticulously engineered to deliver ALA directly and precisely to the tumor site. The targeted delivery system has been demonstrated to enhance treatment efficacy, promote long-term therapeutic stability, and minimize potential side effects. Zhao and colleagues[100] also found that ALA-loaded dissolving MNs exhibited a superior antimelanoma effect in comparison to the ALA injection.

Hypericin, a naturally derived photosensitizer, exhibits high photodynamic efficiency and a strong tendency to accumulate in tumor tissues. However, its clinical application in PDT is constrained by its hydrophobicity and poor skin permeability. To address these challenges, Abd-El-Azim et al.[101] developed an innovative approach using AdminPen hollow MNs to facilitate the intradermal delivery of hypericin-loaded lipid capsules. This technique demonstrated promising antitumor efficacy in CT-26 tumor-bearing mice, achieving a tumor volume reduction of approximately 85.84% following MN-assisted PDT (Fig. 8.3).

In the field of tumor therapy, researchers have continuously sought to improve the antitumor efficacy of PDT and have proposed several promising strategies. These include: combination chemotherapy, which enhances therapeutic outcomes through the synergistic action of multiple drugs via different mechanisms[68,102]; increasing tumor oxygen saturation to boost photosensitizer activity and amplify cytotoxic effects; glutathione (GSH) depletion to disrupt the tumor's redox balance and weaken its defense mechanisms[70]; and image-guided PDT, which enables more accurate visualization and monitoring of the treatment process.[103]

Liu et al.[68] investigated the combination of PDT with chemotherapy, with the objective of achieving enhanced precision in the regulation of drug release. They developed an NIR-activated system consisting of a dissolving polyvinylpyrrolidone MN array loaded with liposomes coencapsulating the photosensitizer indocyanine green and a ROS-activated doxorubicin prodrug (pB-DOX). Upon NIR irradiation, indocyanine green generates ROS, thereby initiating PDT. Concurrently, it activates pB-DOX by cleaving its prodrug fragments, thus converting it into active DOX for the purpose of chemotherapy. In vivo pharmacodynamic studies demonstrated that this system effectively eliminated melanoma tumors with a single dose while minimizing chemotherapy-related side effects, highlighting its strong potential for melanoma treatment.

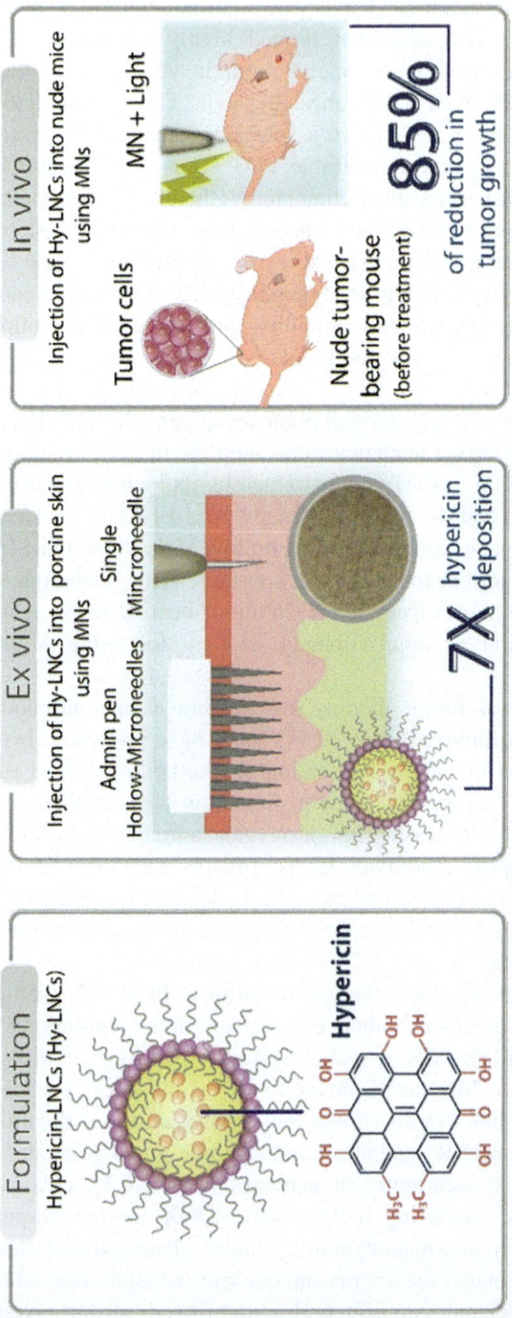

FIGURE 8.3 Hollow microneedles (MNs)-assisted intradermal delivery of hypericin lipid nanocapsules for photodynamic therapy (PDT) against skin cancer.[101]

In another study, the researchers have achieved notable progress in enhancing PDT efficacy.[104] They developed oxygen-propelled MNs by incorporating sodium percarbonate, enabling the precise and simultaneous delivery of chlorin e6 into deep tumor tissues. This strategy effectively alleviated tumor hypoxia by locally releasing oxygen, thereby enhancing the photosensitizing activity of chlorin e6 and increasing its cytotoxic effect against breast cancer cells. The findings offer substantial evidence in support of the therapeutic efficacy of this approach, underscoring notable enhancements in PDT outcomes and delineating a promising new trajectory for the treatment of breast cancer.

As indicated by the works of He et al., substantial progress has been made in the technological innovation of PDT. These advancements include the development of a synthetic biology-guided MN patch designed for traceable and repeatable treatment.[103] This system incorporated Cu^{2+}-doped calcium phosphate nanoparticles coloaded with ALA and catalase. The catalase functioned as a catalyst, facilitating the decomposition of hydrogen peroxide into oxygen. This process effectively reversed tumor hypoxia and enhanced the conversion of ALA into protoporphyrin IX. Furthermore, the combination of Cu^{2+} and Ca^{2+} ions resulted in a synergistic depletion of intracellular GSH, leading to an increase in oxidative stress. This, in turn, enhanced the therapeutic efficacy of ALA-mediated PDT. Notably, the system facilitated real-time observation of tumor oxygenation and protoporphyrin IX metabolism through fluorescence/photoacoustic dual-mode imaging, enabling precise image-guided therapy. This innovative MN patch offers a more precise and effective strategy for tumor treatment and contributes to the advancement of PDT technology.

4.4 Microneedle-mediated chemodynamic therapy

Chemodynamic therapy (CDT), a concept proposed by Zhang et al. in 2016,[105] capitalizes on the elevated levels of H_2O_2 and the moderately acidic conditions that predominate within the tumor microenvironment (TME).[53] The primary pharmaceutical agents employed in CDT are transition metal ions, which exhibit remarkable catalytic capabilities in the context of H_2O_2, leading to the generation of highly cytotoxic hydroxyl radicals (•OH).[52] These radicals trigger cell apoptosis through Fenton or Fenton-like reactions, resulting in phospholipid peroxidation, protein inactivation, and DNA damage, thereby effectively eliminating the tumor cells.[52,58] CDT is clearly distinguishable from PDT in that it does not require external laser irradiation or an additional supply of oxygen to generate ROS. Consequently, it possesses considerable potential as a form of tumor treatment. However, concerns regarding systemic toxicity and biosafety have been raised for CDT in intravenous administration.[43] To address this challenge, researchers have ingeniously combined CDT with MNs-based local therapy, a combination that has rapidly emerged as a prominent research focus.

Chen *et al.* developed an MN patch loaded with ultrathin two-dimensional bimetallic nanosheets with the aim of enhancing CDT against melanoma through the cascade biocatalyst.[106] Upon penetration into the skin, endogenous glucose was catalytically transformed to H_2O_2 by the loaded glucose oxidase (GOx). Subsequently, the Fe/Cu metal center in the nanosheet effectively converted H_2O_2 to highly toxic •OH. The consumption of glucose nutrients also resulted in the death of tumor cells. The results showed that the MN path induced a cascade catalytic reaction at the tumor site, thereby promoting the generation of •OH while conducting tumor starvation therapy. This approach effectively suppressed melanoma growth, achieving an inhibition rate of up to 97.6%.

Reduced GSH within tumors poses a significant challenge to the efficacy of CDT.[107] In addressing this challenge, Yu et al.[108] proposed a strategy through the MNs mediated deep delivery into the tumor regions. The MNs were composed of dissolving polyvinylpyrrolidone combined with a stimulus-responsive prodrug (P–NO-CA@Fe). In the acidic tumor microenvironment, this system triggered H_2O_2 generation through cinnamaldehyde release, which subsequently oxidized arginine to produce nitric oxide, consuming GSH in the process. This synergistic cascade amplified the Fenton-like reaction, thereby maximizing the cytotoxic effects of iron-mediated CDT. In the A375 melanoma mouse model, P–NO-CA@Fe MNs significantly inhibited the tumor growth, and the inhibition rate reached up to 94.5%.

CDT can also serve as an effective adjunct to other treatments such as chemotherapy and PTT, overcoming the limited efficacy often observed with monotherapies and significantly enhancing antitumor effects.[44] Ruan et al.[109] developed multifunctional CaO_2@Mn-polydopamine nanoparticles embedded in MNs to synergistically treat melanoma by integrating PTT, CDT, and hypoxia alleviation. Mn-polydopamine generated heat under NIR irradiation for PTT and catalyzes H_2O_2 to produce hydroxyl radicals (•OH). Concurrently, CaO_2 released oxygen to relieve tumor hypoxia, thereby significantly inhibiting tumor growth. In another study, Ju et al.[110] fabricated an MN patch loaded with MnO_2/Cu_2O nanosheets and combretastatin A4 to enhance skin squamous cell carcinoma treatment. Upon application, the released nanosheets mimic GOx activity, catalyzing glucose to generate H_2O_2. The released Cu ions then trigger a Fenton-like reaction with H_2O_2 to efficiently produce •OH. Combretastatin A4 inhibits tumor growth and cancer cell migration by disrupting the tumor vasculature. Additionally, MnO_2/Cu_2O exhibits photothermal conversion under NIR irradiation, killing cancer cells without impairing GOx-like activity, thereby sustaining high H_2O_2 levels for enhanced CDT. Liao et al.[111] introduced a "ROS accumulation" strategy to improve CDT efficacy in melanoma. The MN patch was utilized to deliver a nanomedicine that amplifies ROS in TME via gallic acid–mediated reduction of Fe^{3+} to Fe^{2+}, catalytic conversion of H_2O_2 to •OH, polydopamine-induced hyperthermia, and GSH depletion. In vivo studies demonstrated that this system exerted synergistic antitumor effects and markedly inhibited tumor growth.

4.5 Microneedle-mediated immunotherapy

Immunotherapy is an effective cancer treatment modality that plays a critical role in combating both primary and metastatic tumors. It functions by activating the body's innate immune system to eliminate cancer cells. To elicit a potent antitumor immune response, a series of systematic and sequential processes, collectively known as the "cancer immunity cycle," must be initiated and sustained under favorable conditions.[112] This cycle involves: (1) the presentation of tumor neoantigens, which triggers the immune response; (2) antigen capture and presentation, enabling the immune system to recognize tumor-specific markers; (3) activation and proliferation of effector T cells to ensure an adequate and effective response; (4) migration and infiltration of these T cells into tumor tissues; and (5) recognition and destruction of cancer cells, culminating in tumor eradication. The destruction of tumor cells subsequently releases additional damage-associated molecular patterns and neoantigens, which further amplify the immune response. In essence, cancer immunotherapy aims to initiate or enhance this self-sustaining immune cycle, creating a positive feedback loop that persistently combats cancer. However, most immunotherapeutic approaches face significant challenges in clinical application due to the immunosuppressive tumor microenvironment and the inherently low immunogenicity of tumor tissues.[48,65] Over the past decade, extensive research has focused on leveraging MN-mediated immunotherapy as a promising strategy to overcome these limitations in superficial cancer treatment.

4.5.1 Cancer vaccine

The exposure of cancer neoantigens can initiate an effective antitumor immune response, thereby activating the "cancer immunity cycle" and providing systemic protection against tumor recurrence and metastasis.[113,114] In recent years, tumor antigens have garnered significant attention due to their relative ease of deployment and promising therapeutic potential. Diverse cancer antigens, including peptides, proteins, DNA, and RNA derived from tumor cells, have been developed with varying degrees of success.[115,116]

Zaric et al.[117] provided a notable example by demonstrating the use of polymeric MNs to deliver poly(lactic-co-glycolic acid) NPs loaded with the model antigen ovalbumin, targeting resident dendritic cells (DCs) in the skin. Their experiments confirmed that antigen-coated NPs promote maturation of myeloid-derived DCs. In vivo results showed that MNs effectively bypassed the stratum corneum barrier to facilitate transdermal delivery of ovalbumen NPs. The resident skin DCs captured these NPs, migrated to lymph nodes, and induced robust proliferation of antigen-specific T cells, thereby initiating a potent cellular immune response against ovalbumin-expressing tumors.

Similarly, Kim et al.[118] employed Pluronic F127, an amphiphilic polymer, to fabricate MNs capable of encapsulating both antigens and the hydrophobic adjuvant R848. Upon insertion, these MNs dissolved in the intradermal fluid and self-assembled into nanomicelles (30–40 nm), which facilitated migration and retention in lymph nodes. When coformulated with ovalbumin and R848, the MNs elicited a strong antigen-specific immune response and effectively suppressed tumor growth in mice bearing ovalbumin-transfected tumors.

While cancer vaccination holds promise for enhancing antitumor immunity, several challenges remain. First, accurately identifying tumor-specific antigens is difficult. Second, developing a broadly effective vaccine that elicits robust immune responses in most patients is highly challenging. Third, single-antigen vaccines tend to induce greater immune resistance compared to multivalent formulations.[114,119] To address these issues, Ye et al.[120] recently developed a therapeutic strategy using polymeric MNs loaded with whole tumor lysates from B16F10 cells combined with granulocyte-macrophage colony-stimulating factor (GM-CSF) for cancer immunotherapy. GM-CSF can effectively recruit and activate DCs, while the slow release of tumor antigens can facilitate their uptake and presentation by DCs. Under NIR irradiation, melanin within the tumor lysate raises local temperature, further promoting DC migration. This approach enhanced T-cell infiltration and cytokine secretion, ultimately triggering a robust B16F10-specific immune response.

Given the central role of DCs as antigen-presenting cells (APCs), DC vaccines have been widely employed in cancer immunotherapy.[121] MNs can enable direct and efficient delivery of DC vaccines to skin tissues rich in APCs. Chang et al.[122] developed CryoMNs to deliver ovalbumen-pulsed DCs (OVA-DCs) for melanoma vaccination and immunotherapy. The CryoMNs were fabricated by encapsulating OVA-DCs in a low-temperature medium using a stepwise cryo-molding technique. These MNs demonstrated excellent skin penetration and maintained DC viability after storage in liquid nitrogen for 1 month. Optimizing administration frequency (twice weekly) and dose (4×10^5 OVA-DCs), CryoMNs elicited stronger antigen-specific immune responses and superior tumor suppression compared to intravenous or subcutaneous delivery.

DNA vaccines offer another promising approach to activate both humoral and cellular antitumor immunity. Duong et al.[64] developed polycarbonate MNs coated with a nanoengineered DNA vaccine designed for sustained release in skin tissues to combat metastatic melanoma. Using a layer-by-layer coating technique, MNs were alternately dipped in a super-pH-responsive copolymer and the adjuvant poly(I:C) to load the DNA vaccine. Upon pH-triggered copolymer dissociation, the adjuvant and vaccine were released into skin rich in immune cells, eliciting a stronger immune response and superior rejection of lung metastatic melanoma compared to subcutaneous injection of the DNA vaccine.

4.5.2 Immunogenic cell death induced immunotherapy

Recent studies have clearly demonstrated that immunogenic cell death (ICD) is a distinct form of cell death capable of robustly activating both innate and adaptive immune responses against cell death-associated antigens, thereby enhancing tumor immunogenicity.[123,124]

To induce potent in situ ICD effect, Yu et al.[125] developed a pH-sensitive and melanoma-targeted liposomal system loaded with DOX and siPD-L1, which was further integrated into a dissolving MN system for enhanced chemo-immunotherapy against melanoma. This system induced the ICD effect through the incorporation of DOX, thereby activating antitumor immunity. Additionally, the physical adjuvant properties of MNs contributed to further enhancing the efficacy of chemo-immunotherapy. Notably, the inclusion of siPD-L1 effectively downregulated PD-L1 expression, significantly improving the overall therapeutic outcome. In a melanoma mouse model, this system demonstrated enhanced antitumor effect with a tumor inhibition rate of 90.8%, markedly surpassing the therapeutic efficacy of intravenous injection at an equivalent dose. Bian et al.[126] developed a carrier-free chlorine e6 nanoparticle integrated rapid-dissolving MN patch (L-Ce6 MNs) for targeted anticancer therapy. This system enabled precise and efficient delivery of L-Ce6 nanoparticles directly to tumor tissue. In xenograft melanoma mouse models, L-Ce6 MNs effectively ablated tumors in situ through ROS generated by PDT. Concurrently, the treatment induced ICD and released damage-associated molecular patterns, promoting dendritic cell maturation and antigen presentation. This cascade stimulated a robust T cell-mediated immune response, significantly enhancing PDT efficacy and achieving complete tumor eradication.

Chen et al.[127] developed core-shell MNs for the localized codelivery of indocyanine green-loaded nanoparticles (shell) and 1-methyltryptophan (core), achieving a strong synergistic immune response. Upon NIR irradiation, indocyanine green efficiently converted light energy into heat, ablating the primary tumor, releasing cell death-associated antigens, and promoting DC maturation. Concurrently, 1-methyltryptophan, an indoleamine 2,3-dioxygenase (IDO) inhibitor, alleviated immunosuppression caused by IDO overexpression, ultimately eliciting a robust systemic immune response. The in vivo experiments demonstrated that these core-shell MNs eliminated 80% of the primary tumor, effectively inhibited the growth of distant tumors, and reduced the risk of tumor metastasis in a melanoma model.

Shi et al.[128] developed a dissolving MN patch loaded with a CD39 inhibitor (sodium polyoxotungstate, POM-1) and an ICD inducer (IR780) encapsulated in solid lipid nanoparticles for targeted tumor therapy. Subsequent to the application of the MN patch at the tumor site, the nanoparticles penetrated deeply into the tumor tissue and gradually disintegrated under PTT, releasing POM-1. This inhibitor has been shown to enhance immune stimulation by promoting adenosine triphosphate (ATP) release, reducing immunosuppressive extracellular adenosine, and remodeling the tumor immunosuppressive microenvironment. In vivo studies demonstrated that this system significantly amplified the antitumor efficacy of photothermal immunotherapy, effectively eradicating melanoma and colorectal adenocarcinoma in mice, thereby offering a promising cancer treatment strategy (Fig. 8.4).

4.5.3 Antibody immunotherapy

The complex tumor microenvironment often facilitates immune escape, primarily through tumor exploitation of immune checkpoint pathways to suppress tumor-specific immune responses.[66,129] Among the key inhibitory molecules are cytotoxic T lymphocyte-associated antigen 4 (CTLA-4) and programmed cell death protein 1 (PD-1), which deliver inhibitory signals that significantly impair T cell-mediated immunity.[130,131] Studies have conclusively shown that blockade of PD-1 or CTLA-4 can reactivate antigen-specific cellular immune responses. This breakthrough in understanding immune checkpoints was recognized with the 2018 Nobel Prize, marking a major advancement in immune checkpoint blockade (ICB) cancer therapy.[132,133] Notably, ICB therapies targeting CTLA-4 (e.g., ipilimumab) and PD-1/PD-L1 (e.g., nivolumab, pembrolizumab, and atezolizumab) have successfully elicited durable protective immune responses across numerous clinical settings.

Wang et al.[134] developed a self-degrading MN patch combined with pH-sensitive dextran NPs to improve immunotherapy efficacy in melanoma. GOx, catalase, and anti-PD-1 antibody (aPD1) were coencapsulated within dextran NPs, which were then loaded into photocrosslinked hyaluronic acid MNs. Upon application to tumor tissue, GOx catalyzes glucose oxidation to gluconic acid, generating a localized acidic environment that triggers the self-degradation of dextran NPs and the sustained release of aPD1, thereby enhancing immunotherapeutic effects. Concurrently, catalase converts the byproduct hydrogen peroxide into oxygen, further promoting GOx-mediated oxidation. Compared to free aPD1 injection or MNs lacking degradation triggers, this self-degrading MN system significantly amplified the immune response after a single administration, effectively inhibiting melanoma growth.

Although ICBs can elicit immune responses, their use as monotherapies rarely achieves durable antitumor immunity. This limitation is largely due to the tumor microenvironment's induction of immunosuppressive factors such as IDO, expressed by DCs and tumor cells. IDO suppresses T cell activation by depleting tryptophan and producing pro-apoptotic metabolites, thereby impairing T cell function.[135] Moreover, IDO overexpression can activate regulatory T cells, further inhibiting immune activation and promoting immune tolerance.[136] Consequently, there is a critical need for combination therapies targeting multiple stages of the cancer immunity cycle to elicit robust immune responses. Ye et al. developed polymeric MNs coloaded with αPD-1

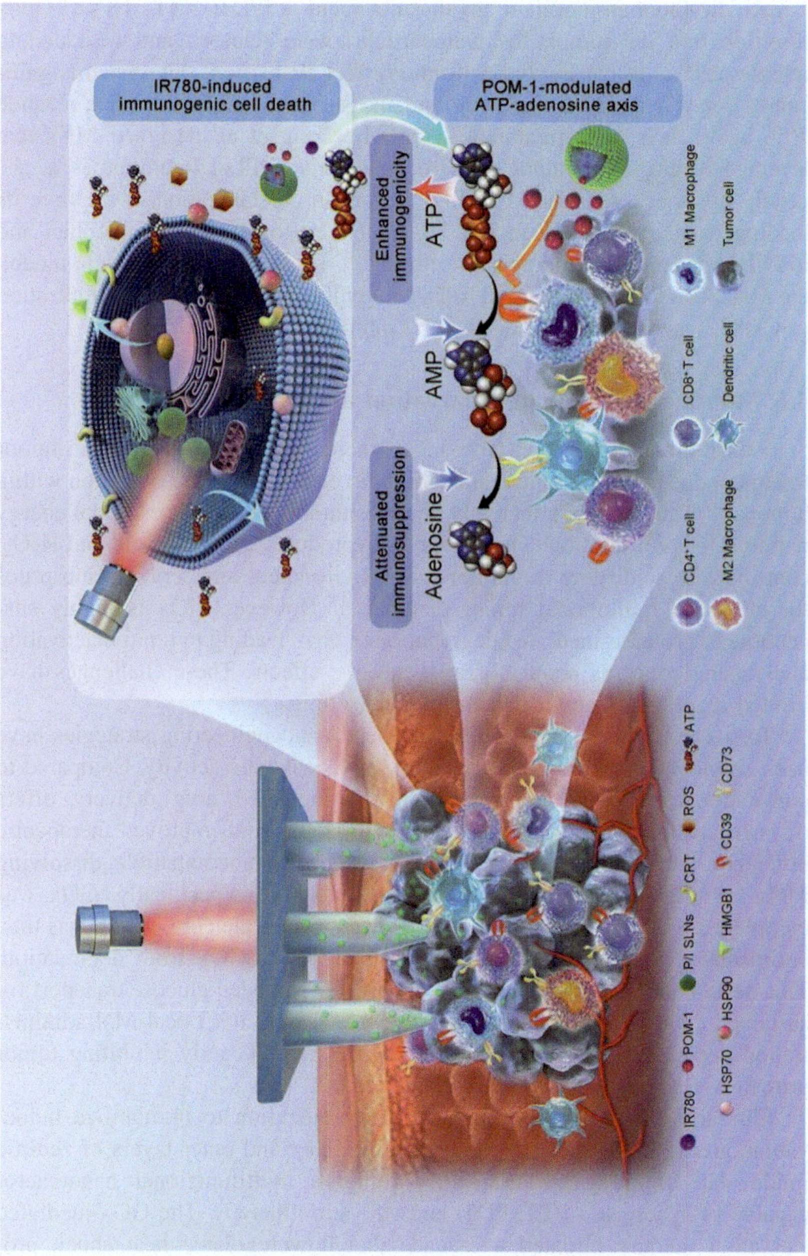

FIGURE 8.4 Adenosine triphosphate (ATP)-adenosine axis regulation combined with microneedle-assisted photoimmunotherapy to boost the immunotherapy efficiency.[128]

and 1-methyltryptophan to achieve synergistic immunotherapy.[137] In this system, αPD-1 was encapsulated within nanoparticles formed from 1-methyltryptophan-modified hyaluronic acid (αPD1/1-MT NPs). Upon insertion into the tumor, the nanoparticles were released and degraded to release αPD-1. In the in vivo antitumor effect evaluation, this system significantly delayed the tumor growth, and the survival rate of the mice reached 70% at 40 days after treatment. Similarly, Yang et al. designed MNs that leverage electrostatic interactions to concentrate αPD-L1 antibodies at the needle tips while reducing 1-methyltryptophan crystallization to improve its loading capacity, thereby facilitating codelivery of 1-methyltryptophan and αPD-L1 for enhanced therapeutic efficacy.[138] In the mouse melanoma model, the MNs loaded with 1-MT and αPD-L1 significantly increased the infiltration of $CD8^+$ T cells in the tumor, thereby inhibiting the tumor growth.

4.6 Microneedle-mediated starvation therapy

Starvation therapy based on GOx has garnered significant attention in tumor treatment due to GOx's remarkable ability to catalyze glucose oxidation within tumors.[139] This process effectively deprives tumor cells of the essential energy required for proliferation. The oxidative byproducts, gluconic acid and H_2O_2, contribute to acidifying the tumor microenvironment and serve as substrates for enhanced multimodal tumor therapies.[34] However, GOx is highly susceptible to protease-mediated degradation in vivo, leading to rapid inactivation and raising concerns about potential adverse effects. These challenges have limited its practical application in clinical settings.[63]

To overcome these challenges, advanced nanoengineering strategies have been employed to prolong and enhance GOx's catalytic activity. Compared to conventional systemic administration, MN-mediated drug delivery offers distinct advantages, including reduced side effects and improved therapeutic efficiency. For instance, Zeng et al.[140] integrated biocompatible dissolving MNs for targeted delivery in melanoma treatment. Specifically, GOx was complexed with folic acid and Zn^{2+} to form a nanoscale core, which was then encapsulated with a polydopamine shell to protect GOx from degradation. This design enabled gradual GOx release and unimpeded glucose transport for sustained catalysis, effectively prolonging GOx activity. Local MN administration depleted glucose in melanoma tissues, significantly inhibiting tumor growth while minimizing systemic side effects.

Similarly, Zhou et al.[70] utilized biomineralization to immobilize indocyanine green, GOx, and catalase within the inner and outer layers of zeolitic imidazolate framework-8 (ZIF-8), forming a multifunctional nanoreactor capable of synergistic PTT, PDT, and starvation therapy. The GOx-mediated starvation therapy initiated a cascade that downregulated heat shock proteins, which sensitized tumors to PTT, and generated H_2O_2, which was converted into oxygen by catalase to enhance PDT efficacy. The ZIF-8 nanoreactor was coated with hyaluronic acid for selective targeting of CD44-overexpressing tumor cells and loaded into MNs for intratumoral delivery. In the in vivo antitumor efficacy evaluation, this multifunctional nanoreactor-integrated MNs enhanced the therapeutic effects of starvation therapy, PDT, and PTT through cascade catalytic reactions. It significantly suppressed tumor growth with the survival rate of mice exceeding 60% at 30 days posttreatment (Fig. 8.5).

4.7 Microneedle-mediated gene therapy

Biotechnology continues to advance rapidly, deepening and refining cancer research. In this context, gene therapy has emerged as a promising approach in antitumor treatment, attracting considerable attention. Gene therapy involves the precise delivery of therapeutic genes into patient cells using specialized techniques to correct genetic defects, regulate disease-related genes, and restore normal cellular function to combat tumors at their source.[141] Various genetic materials, including small interfering RNA (siRNA) [147] and plasmid DNA (pDNA),[142] have been employed in tumor therapy. To enhance therapeutic efficacy, researchers have developed strategies that encapsulate genes within functional NPs to protect them and facilitate cellular uptake. These gene-loaded NPs are then incorporated into MNs to improve targeted gene delivery and transfection efficiency.[143]

Numerous research teams have made significant advances in gene therapy using MN technology for targeted cancer treatment. For example, Ruan et al.[144] developed nanoparticles that combine the cell-penetrating peptide octaarginine with BRAF siRNA nanocomplexes to silence the BRAF gene in melanoma. The MN system has been demonstrated to deliver BRAF siRNAs with high precision to tumors, thereby inducing tumor cell apoptosis and effectively inhibiting melanoma proliferation. Ali et al.[145] developed a

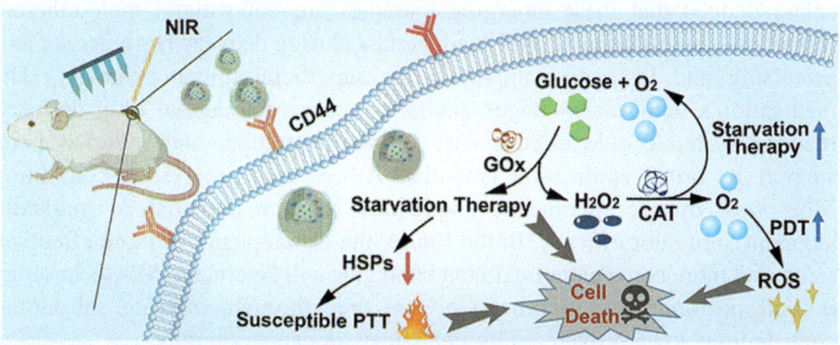

FIGURE 8.5 Multifunctional nanoreactor-integrated microneedles for cascade reaction–enhanced cancer therapy.[70]

vaccination platform by integrating NPs containing E6/E7 pDNA and RALA peptide into polyvinylpyrrolidone MNs. The RALA peptide has been shown to condense DNA into stable cationic nanoparticles, while the MNs facilitate transdermal delivery. Compared to traditional intramuscular injection, MN vaccination significantly increased E6/E7-specific IgG levels, enhanced cytotoxic T cell activity against TC-1 cells, and elevated interferon-gamma production, thereby strengthening the immune response. Importantly, MNs effectively suppressed tumor growth, thereby supporting both cervical cancer prevention and therapy.

Pan et al.[146] designed a gene delivery system by loading polyethylenimine (PEI)/STAT3 siRNA complexes into dextran/hyaluronic acid/polyvinylpyrrolidone MNs. They improved STAT3 siRNA transfection efficiency by leveraging PEI's capacity to enhance cellular uptake. The MNs were engineered to deliver siRNAs with high precision to the melanoma sites, leading to the suppression of STAT3 expression, the inhibition of tumor cell proliferation, and the control of melanoma growth. In another study, Xu et al.[147] combined PTT with gene therapy by coloading p53 DNA and IR820 into MNs. This platform facilitated the synchronous delivery of genes and photothermal agents, with the temperature-triggered gene release enhancing the therapeutic efficacy. This integrative approach has demonstrated promising antitumor effects and has opened new avenues for cancer treatment.

5. Conclusion

Superficial tumors and other neoplastic diseases pose a significant threat to public health. Conventional treatment modalities, including surgical resection, radiotherapy, and chemotherapy, encounter a multitude of challenges, including incomplete tumor removal, damage to healthy tissues, and significant side effects. The advent of novel therapeutic modalities, encompassing phototherapy, CDT, immunotherapy, and gene therapy, has emerged as a promising avenue for therapeutic intervention. However, it must be acknowledged that these emerging therapies are not without their inherent limitations. In this context, MN-mediated local drug delivery has emerged as a promising and innovative approach for superficial tumor treatment. The medication's unique advantages, namely, precise transdermal drug delivery, reduced systemic side effects, ease of administration, and minimal pain, support its broad application potential. As research progresses, combining MNs with diverse therapeutic modalities has the potential to markedly augment antitumor efficacy. In the future, the development of more effective, safer, and more convenient treatment strategies will be crucial. MN technology is well positioned to become a pivotal breakthrough, offering substantial contributions toward overcoming challenges in cancer therapy.

References

1. Bray, F.; Laversanne, M.; Sung, H.; Ferlay, J.; Siegel, R. L.; Soerjomataram, I.; Jemal, A. Global Cancer Statistics 2022: GLOBOCAN Estimates of Incidence and Mortality Worldwide for 36 Cancers in 185 Countries. *CA Cancer J. Clin.* **2024,** *74* (3), 229–263.
2. Nogueras, J. J.; Jagelman, D. G. Principles of Surgical Resection. Influence of Surgical Technique on Treatment Outcome. *Surg. Clin.* **1993,** *73* (1), 103–116.
3. Firnhaber, J. M. Basal Cell and Cutaneous Squamous Cell Carcinomas: Diagnosis and Treatment. *Am. Fam. Physician* **2020,** *102* (6), 339–346.
4. Riml, S.; Larcher, L.; Kompatscher, P. Complete Excision of Nonmelanotic Skin Cancer. *Ann. Plast. Surg.* **2013,** *70* (1), 66–69.
5. Rose, D. M. Surgical Resection for Metastatic Melanoma to the Liver. *Arch. Surg.* **2001,** *136* (8), 950.
6. Maughan, T. S. Basic Principles of Radiotherapy for Surgical Oncologists. 1–Effects of Radiotherapy on Superficial Tissues. Implications for Treating Skin Cancer. *Eur. J. Surg. Oncol.* **1994,** *20* (1), 74–79.
7. Piccinno, R.; Tavecchio, S.; Benzecry, V. Superficial Radiotherapy for Non-melanoma Skin Cancer of the Lip: A 44-Year Italian Experience. *J. Dermatol. Treat.* **2020,** *31* (4), 382–386.
8. Tungkasamit, T.; Chakrabandhu, S.; Samakgarn, V.; Kunawongkrit, N.; Jirawatwarakul, N.; Chumachote, A.; Chitapanarux, I. Reduction in Severity of Radiation-Induced Dermatitis in Head and Neck Cancer Patients Treated with Topical Aloe Vera Gel: A Randomized Multicenter Double-Blind Placebo-Controlled Trial. *Eur. J. Oncol. Nurs.* **2022,** *59,* 102164.
9. Brook, I. Late Side Effects of Radiation Treatment for Head and Neck Cancer. *Radiat. Oncol. J* **2020,** *38* (2), 84–92.
10. Russi, E. G.; Moretto, F.; Rampino, M.; Benasso, M.; Bacigalupo, A.; De Sanctis, V.; Numico, G.; Bossi, P.; Buglione, M.; Lombardo, A.; et al. Acute Skin Toxicity Management in Head and Neck Cancer Patients Treated with Radiotherapy and Chemotherapy or EGFR Inhibitors: Literature Review and Consensus. *Crit. Rev. Oncol. Hematol.* **2015,** *96* (1), 167–182.
11. Malkinson, F. D.; Keane, J. T. Radiobiology of the Skin: Review of Some Effects on Epidermis and Hair. *J. Invest. Dermatol.* **1981,** *77* (1), 133–138.
12. Brook, I. Early Side Effects of Radiation Treatment for Head and Neck Cancer. *Cancer Radiother.* **2021,** *25* (5), 507–513.
13. Severs, G. A.; Griffin, T.; Werner-Wasik, M. Cicatricial Alopecia Secondary to Radiation Therapy: Case Report and Review of the Literature. *Cutis* **2008,** *81* (2), 147–153.
14. Sciubba, J. J.; Goldenberg, D. Oral Complications of Radiotherapy. *Lancet Oncol.* **2006,** *7* (2), 175–183.
15. Shih, A.; Miaskowski, C.; Dodd, M. J.; Stotts, N. A.; MacPhail, L. Mechanisms for Radiation-Induced Oral Mucositis and the Consequences. *Cancer Nurs.* **2003,** *26* (3), 222–229.
16. Sroussi, H. Y.; Epstein, J. B.; Bensadoun, R.; Saunders, D. P.; Lalla, R. V.; Migliorati, C. A.; Heaivilin, N.; Zumsteg, Z. S. Common Oral Complications of Head and Neck Cancer Radiation Therapy: Mucositis, Infections, Saliva Change, Fibrosis, Sensory Dysfunctions, Dental Caries, Periodontal Disease, and Osteoradionecrosis. *Cancer Med.* **2017,** *6* (12), 2918–2931.
17. Barazzuol, L.; Coppes, R. P.; van Luijk, P. Prevention and Treatment of Radiotherapy-induced Side Effects. *Mol. Oncol.* **2020,** *14* (7), 1538–1554.

18. Brahme, A. Development of Radiation Therapy Optimization. *Acta Oncol. (Madr.)* **2000,** *39* (5), 579–595.
19. De Ruysscher, D.; Niedermann, G.; Burnet, N. G.; Siva, S.; Lee, A. W. M.; Hegi-Johnson, F. Radiotherapy Toxicity. *Nat. Rev. Dis. Primers* **2019,** *5* (1), 13.
20. Nygren, P. What Is Cancer Chemotherapy? *Acta Oncol. (Madr.)* **2001,** *40* (2–3), 166–174.
21. Chabner, B. A.; Roberts, T. G. Chemotherapy and the War on Cancer. *Nat. Rev. Cancer* **2005,** *5* (1), 65–72.
22. Makin, G.; Dive, C. Apoptosis and Cancer Chemotherapy. *Trends Cell Biol.* **2001,** *11* (11), S22–S26.
23. Gutteridge, W. E. Existing Chemotherapy and its Limitations. *Br. Med. Bull.* **1985,** *41* (2), 162–168.
24. Corrie, P. G. Cytotoxic Chemotherapy: Clinical Aspects. *Medicine* **2008,** *36* (1), 24–28.
25. Schirrmacher, V. From Chemotherapy to Biological Therapy: A Review of Novel Concepts to Reduce the Side Effects of Systemic Cancer Treatment (Review). *Int. J. Oncol.* **2018,** *54* (2), 407–419.
26. Groopman, J. E.; Itri, L. M. Chemotherapy-Induced Anemia in Adults: Incidence and Treatment. *JNCI J. Natl. Cancer Inst.* **1999,** *91* (19), 1616–1634.
27. Galmarini, D.; Galmarini, C. M.; Galmarini, F. C. Cancer Chemotherapy: A Critical Analysis of its 60 Years of History. *Crit. Rev. Oncol. Hematol.* **2012,** *84* (2), 181–199.
28. Zitvogel, L.; Apetoh, L.; Ghiringhelli, F.; Kroemer, G. Immunological Aspects of Cancer Chemotherapy. *Nat. Rev. Immunol.* **2008,** *8* (1), 59–73.
29. Behranvand, N.; Nasri, F.; Zolfaghari Emameh, R.; Khani, P.; Hosseini, A.; Garssen, J.; Falak, R. Chemotherapy: A Double-Edged Sword in Cancer Treatment. *Cancer Immunol. Immunother.* **2022,** *71* (3), 507–526.
30. Maloney, A. M. *Gastrointestinal Tract,* 2010; pp 353–377.
31. Santarpia, L.; Contaldo, F.; Pasanisi, F. Nutritional Screening and Early Treatment of Malnutrition in Cancer Patients. *J. Cachexia Sarcopenia Muscle* **2011,** *2* (1), 27–35.
32. Smith, P.; Lavery, A.; Turkington, R. C. An Overview of Acute Gastrointestinal Side Effects of Systemic Anti-cancer Therapy and Their Management. *Best Pract. Res. Clin. Gastroenterol.* **2020,** *48–49,* 101691.
33. He, Y.; Zheng, J.; Ye, B.; Dai, Y.; Nie, K. Chemotherapy-Induced Gastrointestinal Toxicity: Pathogenesis and Current Management. *Biochem. Pharmacol.* **2023,** *216,* 115787.
34. Peng, T.; Chen, Y.; Hu, W.; Huang, Y.; Zhang, M.; Lu, C.; Pan, X.; Wu, C. Microneedles for Enhanced Topical Treatment of Skin Disorders: Applications, Challenges, and Prospects. *Engineering* **2023,** *30,* 170–189.
35. Kaur, R.; Bhardwaj, A.; Gupta, S. Cancer Treatment Therapies: Traditional to Modern Approaches to Combat Cancers. *Mol. Biol. Rep.* **2023,** *50* (11), 9663–9676.
36. Gao, L.; Liu, R.; Gao, F.; Wang, Y.; Jiang, X.; Gao, X. Plasmon-Mediated Generation of Reactive Oxygen Species from Near-Infrared Light Excited Gold Nanocages for Photodynamic Therapy *In Vitro*. *ACS Nano* **2014,** *8* (7), 7260–7271.
37. Zhou, Z.; Song, J.; Nie, L.; Chen, X. Reactive Oxygen Species Generating Systems Meeting Challenges of Photodynamic Cancer Therapy. *Chem. Soc. Rev.* **2016,** *45* (23), 6597–6626.
38. Ge, R.; Yan, P.; Liu, Y.; Li, Z.; Shen, S.; Yu, Y. Recent Advances and Clinical Potential of Near Infrared Photothermal Conversion Materials for Photothermal Hepatocellular Carcinoma Therapy. *Adv. Funct. Mater.* **2023,** *33* (29).
39. Li, J.; Zhang, W.; Ji, W.; Wang, J.; Wang, N.; Wu, W.; Wu, Q.; Hou, X.; Hu, W.; Li, L. Near Infrared Photothermal Conversion Materials: Mechanism, Preparation, and Photothermal Cancer Therapy Applications. *J. Mater. Chem. B* **2021,** *9* (38), 7909–7926.

40. Wang, W.; Zhong, Z.; Peng, S.; Fu, J.; Chen, M.; Lang, T.; Yue, X.; Fu, Y.; He, J.; Jin, Y.; et al. "All-in-One" Metal Polyphenol Network Nanocapsules Integrated Microneedle Patches for Lipophagy Fueled Ferroptosis-Mediated Multimodal Therapy. *J. Contr. Release* **2024,** *373,* 599–616.
41. Correia, J. H.; Rodrigues, J. A.; Pimenta, S.; Dong, T.; Yang, Z. Photodynamic Therapy Review: Principles, Photosensitizers, Applications, and Future Directions. *Pharmaceutics* **2021,** *13* (9), 1332.
42. Wang, S.; Wang, Z.; Yu, G.; Zhou, Z.; Jacobson, O.; Liu, Y.; Ma, Y.; Zhang, F.; Chen, Z.; Chen, X. Tumor-Specific Drug Release and Reactive Oxygen Species Generation for Cancer Chemo/Chemodynamic Combination Therapy. *Adv. Sci.* **2019,** *6* (5).
43. Zhang, L.; Li, C.; Wan, S.; Zhang, X. Nanocatalyst-Mediated Chemodynamic Tumor Therapy. *Adv. Healthcare Mater.* **2022,** *11* (2).
44. Wang, X.; Zhong, X.; Liu, Z.; Cheng, L. Recent Progress of Chemodynamic Therapy-Induced Combination Cancer Therapy. *Nano Today* **2020,** *35,* 100946.
45. Gómez-Navarro, J.; Curiel, D. T.; Douglas, J. T. Gene Therapy for Cancer. *Eur. J. Cancer* **1999,** *35* (6), 867–885.
46. Li, H.; Yang, Y.; Hong, W.; Huang, M.; Wu, M.; Zhao, X. Applications of Genome Editing Technology in the Targeted Therapy of Human Diseases: Mechanisms, Advances and Prospects. *Signal Transduct. Targeted Ther.* **2020,** *5* (1), 1.
47. Belete, T. M. The Current Status of Gene Therapy for the Treatment of Cancer. *Biologics* **2021,** *15,* 67–77.
48. Varadé, J.; Magadán, S.; González-Fernández, Á. Human Immunology and Immunotherapy: Main Achievements and Challenges. *Cell. Mol. Immunol.* **2021,** *18* (4), 805–828.
49. Tan, S.; Li, D.; Zhu, X. Cancer Immunotherapy: Pros, Cons and beyond. *Biomed. Pharmacother.* **2020,** *124,* 109821.
50. Papaioannou, N. E.; Beniata, O. V.; Vitsos, P.; Tsitsilonis, O.; Samara, P. Harnessing the Immune System to Improve Cancer Therapy. *Ann. Transl. Med.* **2016,** *4* (14), 261.
51. Algorri, J. F.; Ochoa, M.; Roldán-Varona, P.; Rodríguez-Cobo, L.; López-Higuera, J. M. Light Technology for Efficient and Effective Photodynamic Therapy: A Critical Review. *Cancers (Basel)* **2021,** *13* (14), 3484.
52. Jia, C.; Guo, Y.; Wu, F. Chemodynamic Therapy via Fenton and Fenton-Like Nanomaterials: Strategies and Recent Advances. *Small* **2022,** *18* (6).
53. Jana, D.; Zhao, Y. Strategies for Enhancing Cancer Chemodynamic Therapy Performance. *Exploration* **2022,** *2* (2).
54. Chen, M.; Quan, G.; Sun, Y.; Yang, D.; Pan, X.; Wu, C. Nanoparticles-Encapsulated Polymeric Microneedles for Transdermal Drug Delivery. *J. Contr. Release* **2020,** *325,* 163–175.
55. Yang, D.; Chen, M.; Sun, Y.; Jin, Y.; Lu, C.; Pan, X.; Quan, G.; Wu, C. Microneedle-Mediated Transdermal Drug Delivery for Treating Diverse Skin Diseases. *Acta Biomater.* **2021,** *121,* 119–133.
56. Zhi, D.; Yang, T.; Zhang, T.; Yang, M.; Zhang, S.; Donnelly, R. F. Microneedles for Gene and Drug Delivery in Skin Cancer Therapy. *J. Contr. Release* **2021,** *335,* 158–177.
57. Wang, M.; Li, X.; Du, W.; Sun, M.; Ling, G.; Zhang, P. Microneedle-Mediated Treatment for Superficial Tumors by Combining Multiple Strategies. *Drug Deliv. Transl. Res.* **2023,** *13* (6), 1600–1620.
58. Zhao, S.; Li, Y.; Cheng, B. A Tumor Microenvironment-Responsive Microneedle Patch for Chemodynamic Therapy of Oral Squamous Cell Carcinoma. *Nanoscale Adv.* **2023,** *5* (22), 6162–6169.

59. Tang, H.; Cheng, X.; Liang, L.; Chen, B. Z.; Liu, C.; Wang, Y. A Stimulus Responsive Microneedle-Based Drug Delivery System for Cancer Therapy. *Biomater. Sci.* **2024,** *12* (24), 6274–6283.
60. Li, D.; Hu, D.; Xu, H.; Patra, H. K.; Liu, X.; Zhou, Z.; Tang, J.; Slater, N.; Shen, Y. Progress and Perspective of Microneedle System for Anti-cancer Drug Delivery. *Biomaterials* **2021,** *264,* 120410.
61. Wu, C.; Yu, Q.; Huang, C.; Li, F.; Zhang, L.; Zhu, D. Microneedles as Transdermal Drug Delivery System for Enhancing Skin Disease Treatment. *Acta Pharm. Sin. B* **2024,** *14* (12), 5161–5180.
62. Alimardani, V.; Abolmaali, S. S.; Tamaddon, A. M.; Ashfaq, M. Recent Advances on Microneedle Arrays-Mediated Technology in Cancer Diagnosis and Therapy. *Drug Deliv. Transl. Res.* **2021,** *11* (3), 788–816.
63. Li, X.; Zhao, Z.; Zhang, M.; Ling, G.; Zhang, P. Research Progress of Microneedles in the Treatment of Melanoma. *J. Contr. Release* **2022,** *348,* 631–647.
64. Duong, H. T. T.; Yin, Y.; Thambi, T.; Nguyen, T. L.; Vh, G. P.; Lee, M. S.; Lee, J. E.; Kim, J.; Jeong, J. H.; Lee, D. S. Smart Vaccine Delivery Based on Microneedle Arrays Decorated with Ultra-PH-responsive Copolymers for Cancer Immunotherapy. *Biomaterials* **2018,** *185,* 13–24.
65. Han, S.; Huang, K.; Gu, Z.; Wu, J. Tumor Immune Microenvironment Modulation-Based Drug Delivery Strategies for Cancer Immunotherapy. *Nanoscale* **2020,** *12* (2), 413–436.
66. Chen, X.; Ding, W.; Jiang, Y.; Shi, W.; Qiu, Y.; Zhao, H.; Luo, X. Emerging Strategies for Local Delivery of Immune Checkpoint Inhibitors to Potentiate Cancer Immunotherapy: Current Status and Future Prospects. *ACS Appl. Mater. Interfaces* **2024,** *16* (44), 59682–59696.
67. Wang, J.; Wen, T.; Chen, H.; Huang, S.; Guo, R.; Zheng, Y.; Xiao, Z.; Shuai, X. Microneedles-Mediated Intradermal Delivery of Paclitaxel/Anti-PD-1 for Efficient and Safe Triple-Negative Breast Cancer Therapy. *Adv. Ther.* **2024,** *7* (4).
68. Liu, F.; Cheng, Z.; Yi, H. NIR Light-Activatable Dissolving Microneedle System for Melanoma Ablation Enabled by a Combination of ROS-Responsive Chemotherapy and Phototherapy. *J. Nanobiotechnol.* **2023,** *21* (1), 61.
69. Lin, S.; Cao, Y.; Chen, J.; Tian, Z.; Zhu, Y. Recent Advances in Microneedles for Tumor Therapy and Diagnosis. *Appl. Mater. Today* **2021,** *23,* 101036.
70. Zhou, Y.; Niu, B.; Zhao, Y.; Fu, J.; Wen, T.; Liao, K.; Quan, G.; Pan, X.; Wu, C. Multifunctional Nanoreactors-Integrated Microneedles for Cascade Reaction-Enhanced Cancer Therapy. *J. Contr. Release* **2021,** *339,* 335–349.
71. Chen, J.; Cao, Y.; Lin, S.; Niu, H.; Zhang, H.; Guan, L.; Shu, C.; Wu, A.; Bian, Y.; Zhu, Y. A Responsive Microneedle System for Efficient Anti-melanoma by Combining Self-Enhanced Chemodynamic Therapy with Photothermal Therapy. *Chem. Eng. J.* **2022,** *431,* 133466.
72. Hu, Y.-Y.; Jin, Q.; Wang, J.; Wu, S.-F.; He, Y.; Jin, P.-H. Integrated Therapeutic Strategies for Various Cutaneous Malignancies: Advances and Challenges of Multifunctional Microneedle Patches toward Clinical Translation. *Chem. Eng. J.* **2024,** *494,* 153033.
73. Yu, S.; Ni, H.; Xu, X.; Cai, Y.; Feng, J.; Zhang, J. Subcutaneous Rapid Dissolution Microneedle Patch Integrated with CuO_2 and Disulfiram for Augmented Antimelanoma Efficacy through Multimodal Synergy of Photothermal Therapy, Chemodynamic Therapy, and Chemotherapy. *ACS Biomater. Sci. Eng.* **2023,** *9* (11), 6425–6437.
74. Qin, W.; Quan, G.; Sun, Y.; Chen, M.; Yang, P.; Feng, D.; Wen, T.; Hu, X.; Pan, X.; Wu, C. Dissolving Microneedles with Spatiotemporally Controlled Pulsatile Release Nanosystem

for Synergistic Chemo-Photothermal Therapy of Melanoma. *Theranostics* **2020,** *10* (18), 8179–8196.

75. Peng, T.; Huang, Y.; Feng, X.; Zhu, C.; Yin, S.; Wang, X.; Bai, X.; Pan, X.; Wu, C. TPGS/Hyaluronic Acid Dual-Functionalized PLGA Nanoparticles Delivered through Dissolving Microneedles for Markedly Improved Chemo-Photothermal Combined Therapy of Superficial Tumor. *Acta Pharm. Sin. B* **2021,** *11* (10), 3297–3309.
76. Bhatnagar, S.; Bankar, N. G.; Kulkarni, M. V.; Venuganti, V. V. K. Dissolvable Microneedle Patch Containing Doxorubicin and Docetaxel Is Effective in 4T1 Xenografted Breast Cancer Mouse Model. *Int. J. Pharm.* **2019,** *556,* 263–275.
77. Ma, G.; Wu, C. Microneedle, Bio-Microneedle and Bio-Inspired Microneedle: A Review. *J. Contr. Release* **2017,** *251,* 11–23.
78. Ingrole, R. S. J.; Gill, H. S. Microneedle Coating Methods: A Review with a Perspective. *J. Pharmacol. Exp. Therapeut.* **2019,** *370* (3), 555–569.
79. Dasari, S.; Tchounwou, P. B. Cisplatin in Cancer Therapy: Molecular Mechanisms of Action. *Eur. J. Pharmacol.* **2014,** *740,* 364–378.
80. Lan, X.; She, J.; Lin, D.; Xu, Y.; Li, X.; Yang, W.; Lui, V. W. Y.; Jin, L.; Xie, X.; Su, Y. Microneedle-Mediated Delivery of Lipid-Coated Cisplatin Nanoparticles for Efficient and Safe Cancer Therapy. *ACS Appl. Mater. Interfaces* **2018,** *10* (39), 33060–33069.
81. Zhou, X.; Luo, Z.; Baidya, A.; Kim, H.; Wang, C.; Jiang, X.; Qu, M.; Zhu, J.; Ren, L.; Vajhadin, F.; et al. Biodegradable β -Cyclodextrin Conjugated Gelatin Methacryloyl Microneedle for Delivery of Water-Insoluble Drug. *Adv. Healthcare Mater.* **2020,** *9* (11).
82. Scinto, S. L.; Bilodeau, D. A.; Hincapie, R.; Lee, W.; Nguyen, S. S.; Xu, M.; Ende, CW am; Finn, M. G.; Lang, K.; Lin, Q.; et al. Bioorthogonal Chemistry. *Nat. Rev. Methods Primers* **2021,** *1* (1), 30.
83. Chen, Z.; Li, H.; Bian, Y.; Wang, Z.; Chen, G.; Zhang, X.; Miao, Y.; Wen, D.; Wang, J.; Wan, G.; et al. Bioorthogonal Catalytic Patch. *Nat. Nanotechnol.* **2021,** *16* (8), 933–941.
84. Huang, X.; Zhang, W.; Guan, G.; Song, G.; Zou, R.; Hu, J. Design and Functionalization of the NIR-Responsive Photothermal Semiconductor Nanomaterials for Cancer Theranostics. *Acc. Chem. Res.* **2017,** *50* (10), 2529–2538.
85. Deng, R.-H.; Zou, M.-Z.; Zheng, D.; Peng, S.-Y.; Liu, W.; Bai, X.-F.; Chen, H.-S.; Sun, Y.; Zhou, P.-H.; Zhang, X.-Z. Nanoparticles from Cuttlefish Ink Inhibit Tumor Growth by Synergizing Immunotherapy and Photothermal Therapy. *ACS Nano* **2019,** *13* (8), 8618–8629.
86. Liu, J.; Liang, H.; Li, M.; Luo, Z.; Zhang, J.; Guo, X.; Cai, K. Tumor Acidity Activating Multifunctional Nanoplatform for NIR-Mediated Multiple Enhanced Photodynamic and Photothermal Tumor Therapy. *Biomaterials* **2018,** *157,* 107–124.
87. Wang, H.; Chang, J.; Shi, M.; Pan, W.; Li, N.; Tang, B. A Dual-Targeted Organic Photothermal Agent for Enhanced Photothermal Therapy. *Angew. Chem.* **2019,** *131* (4), 1069–1073.
88. T-YSC-CTY-HC.; Fu, C.-C.; Tseng, P.-L. *Photo-Thermal-Based Nonablative Dermal Remodeling Device,* 2009; pp 648–650.
89. Wei, S.; Quan, G.; Lu, C.; Pan, X.; Wu, C. Dissolving Microneedles Integrated with PH-Responsive Micelles Containing AIEgen with Ultra-photostability for Enhancing Melanoma Photothermal Therapy. *Biomater. Sci.* **2020,** *8* (20), 5739–5750.
90. Pei, P.; Yang, F.; Liu, J.; Hu, H.; Du, X.; Hanagata, N.; Zhao, S.; Zhu, Y. Composite-Dissolving Microneedle Patches for Chemotherapy and Photothermal Therapy in Superficial Tumor Treatment. *Biomater. Sci.* **2018,** *6* (6), 1414–1423.

91. Dong, L.; Li, Y.; Li, Z.; Xu, N.; Liu, P.; Du, H.; Zhang, Y.; Huang, Y.; Zhu, J.; Ren, G.; et al. Au Nanocage-Strengthened Dissolving Microneedles for Chemo-Photothermal Combined Therapy of Superficial Skin Tumors. *ACS Appl. Mater. Interfaces* **2018,** *10* (11), 9247–9256.
92. Zhao, Y.; Zhou, Y.; Yang, D.; Gao, X.; Wen, T.; Fu, J.; Wen, X.; Quan, G.; Pan, X.; Wu, C. Intelligent and Spatiotemporal Drug Release Based on Multifunctional Nanoparticle-Integrated Dissolving Microneedle System for Synergetic Chemo-Photothermal Therapy to Eradicate Melanoma. *Acta Biomater.* **2021,** *135,* 164–178.
93. Hao, Y.; Dong, M.; Zhang, T.; Peng, J.; Jia, Y.; Cao, Y.; Qian, Z. Novel Approach of Using Near-Infrared Responsive PEGylated Gold Nanorod Coated Poly(<scp>l</Scp> -Lactide) Microneedles to Enhance the Antitumor Efficiency of Docetaxel-Loaded MPEG-PDLLA Micelles for Treating an A431 Tumor. *ACS Appl. Mater. Interfaces* **2017,** *9* (18), 15317–15327.
94. Hu, J.-J.; Cheng, Y.-J.; Zhang, X.-Z. Recent Advances in Nanomaterials for Enhanced Photothermal Therapy of Tumors. *Nanoscale* **2018,** *10* (48), 22657–22672.
95. Chen, M.-C.; Lin, Z.-W.; Ling, M.-H. Near-Infrared Light-Activatable Microneedle System for Treating Superficial Tumors by Combination of Chemotherapy and Photothermal Therapy. *ACS Nano* **2016,** *10* (1), 93–101.
96. Usuda, J.; Kato, H.; Okunaka, T.; Furukawa, K.; Tsutsui, H.; Yamada, K.; Suga, Y.; Honda, H.; Nagatsuka, Y.; Ohira, T.; et al. Photodynamic Therapy (PDT) for Lung Cancers. *J. Thorac. Oncol.* **2006,** *1* (5), 489–493.
97. Kleinovink, J. W.; van Driel, P. B.; Snoeks, T. J.; Prokopi, N.; Fransen, M. F.; Cruz, L. J.; Mezzanotte, L.; Chan, A.; Löwik, C. W.; Ossendorp, F. Combination of Photodynamic Therapy and Specific Immunotherapy Efficiently Eradicates Established Tumors. *Clin. Cancer Res.* **2016,** *22* (6), 1459–1468.
98. Donnelly, R. F.; Morrow, D. I. J.; McCarron, P. A.; Woolfson, A. D.; Morrissey, A.; Juzenas, P.; Juzeniene, A.; Iani, V.; McCarthy, H. O.; Moan, J. Microneedle-Mediated Intradermal Delivery of 5-Aminolevulinic Acid: Potential for Enhanced Topical Photodynamic Therapy. *J. Contr. Release* **2008,** *129* (3), 154–162.
99. Zhu, J.; Dong, L.; Du, H.; Mao, J.; Xie, Y.; Wang, H.; Lan, J.; Lou, Y.; Fu, Y.; Wen, J.; et al. 5-Aminolevulinic Acid-Loaded Hyaluronic Acid Dissolving Microneedles for Effective Photodynamic Therapy of Superficial Tumors with Enhanced Long-Term Stability. *Adv. Healthcare Mater.* **2019,** *8* (22).
100. Zhao, X.; Li, X.; Zhang, P.; Du, J.; Wang, Y. Tip-Loaded Fast-Dissolving Microneedle Patches for Photodynamic Therapy of Subcutaneous Tumor. *J. Contr. Release* **2018,** *286,* 201–209.
101. Abd-El-Azim, H.; Tekko, I. A.; Ali, A.; Ramadan, A.; Nafee, N.; Khalafallah, N.; Rahman, T.; Mcdaid, W.; Aly, R. G.; Vora, L. K.; et al. Hollow Microneedle Assisted Intradermal Delivery of Hypericin Lipid Nanocapsules with Light Enabled Photodynamic Therapy against Skin Cancer. *J. Contr. Release* **2022,** *348,* 849–869.
102. Tham, H. P.; Xu, K.; Lim, W. Q.; Chen, H.; Zheng, M.; Thng, T. G. S.; Venkatraman, S. S.; Xu, C.; Zhao, Y. Microneedle-Assisted Topical Delivery of Photodynamically Active Mesoporous Formulation for Combination Therapy of Deep-Seated Melanoma. *ACS Nano* **2018,** *12* (12), 11936–11948.
103. He, G.; Li, Y.; Younis, M. R.; Fu, L.-H.; He, T.; Lei, S.; Lin, J.; Huang, P. Synthetic Biology-Instructed Transdermal Microneedle Patch for Traceable Photodynamic Therapy. *Nat. Commun.* **2022,** *13* (1), 6238.

104. Liu, P.; Fu, Y.; Wei, F.; Ma, T.; Ren, J.; Xie, Z.; Wang, S.; Zhu, J.; Zhang, L.; Tao, J.; et al. Microneedle Patches with O_2 Propellant for Deeply and Fast Delivering Photosensitizers: Towards Improved Photodynamic Therapy. *Adv. Sci.* **2022,** *9* (25).
105. Zhang, C.; Bu, W.; Ni, D.; Zhang, S.; Li, Q.; Yao, Z.; Zhang, J.; Yao, H.; Wang, Z.; Shi, J. Synthesis of Iron Nanometallic Glasses and Their Application in Cancer Therapy by a Localized Fenton Reaction. *Angew. Chem. Int. Ed.* **2016,** *55* (6), 2101–2106.
106. Chen, J.; Niu, H.; Guan, L.; Yang, Z.; He, Y.; Zhao, J.; Wu, C.; Wang, Y.; Lin, K.; Zhu, Y. Microneedle-Assisted Transdermal Delivery of 2D Bimetallic Metal–Organic Framework Nanosheet-Based Cascade Biocatalysts for Enhanced Catalytic Therapy of Melanoma. *Adv. Healthcare Mater.* **2023,** *12* (7).
107. Tang, Z.; Liu, Y.; He, M.; Bu, W. Chemodynamic Therapy: Tumour Microenvironment-Mediated Fenton and Fenton-like Reactions. *Angew. Chem. Int. Ed.* **2019,** *58* (4), 946–956.
108. Yu, W.; Jia, F.; Fu, J.; Chen, Y.; Huang, Y.; Jin, Q.; Wang, Y.; Ji, J. Enhanced Transcutaneous Chemodynamic Therapy for Melanoma Treatment through Cascaded Fenton-like Reactions and Nitric Oxide Delivery. *ACS Nano* **2023,** *17* (16), 15713–15723.
109. Ruan, L.; Song, G.; Zhang, X.; Liu, T.; Sun, Y.; Zhu, J.; Zeng, Z.; Jiang, G. Transdermal Delivery of Multifunctional CaO_2 @Mn-PDA Nanoformulations by Microneedles for NIR-Induced Synergistic Therapy against Skin Melanoma. *Biomater. Sci.* **2021,** *9* (20), 6830–6841.
110. Ju, E.; Peng, M.; Xu, Y.; Wang, Y.; Zhou, F.; Wang, H.; Li, M.; Zheng, Y.; Tao, Y. Nanozyme-Integrated Microneedle Patch for Enhanced Therapy of Cutaneous Squamous Cell Carcinoma by Breaking the Gap between H_2O_2 Self-Supplying Chemodynamic Therapy and Photothermal Therapy. *J. Mater. Chem. B* **2023,** *11* (28), 6595–6602.
111. Liao, K.; Niu, B.; Dong, H.; He, L.; Zhou, Y.; Sun, Y.; Yang, D.; Wu, C.; Pan, X.; Quan, G. A Spark to the Powder Keg: Microneedle-Based Antitumor Nanomedicine Targeting Reactive Oxygen Species Accumulation for Chemodynamic/Photothermal/Chemotherapy. *J. Colloid Interface Sci.* **2022,** *628,* 189–203.
112. Chen, D. S.; Mellman, I. Oncology Meets Immunology: The Cancer-Immunity Cycle. *Immunity* **2013,** *39* (1), 1–10.
113. Banchereau, J.; Palucka, K. Cancer Vaccines on the Move. *Nat. Rev. Clin. Oncol.* **2018,** *15* (1), 9–10.
114. Perez, C. R.; De Palma, M. Engineering Dendritic Cell Vaccines to Improve Cancer Immunotherapy. *Nat. Commun.* **2019,** *10* (1), 5408.
115. Butterfield, L. H. Cancer Vaccines. *BMJ* **2015,** *350* (14), h988.
116. Saxena, M.; van der Burg, S. H.; Melief, C. J. M.; Bhardwaj, N. Therapeutic Cancer Vaccines. *Nat. Rev. Cancer* **2021,** *21* (6), 360–378.
117. Zaric, M.; Lyubomska, O.; Touzelet, O.; Poux, C.; Al-Zahrani, S.; Fay, F.; Wallace, L.; Terhorst, D.; Malissen, B.; Henri, S.; et al. Skin Dendritic Cell Targeting *via* Microneedle Arrays Laden with Antigen-Encapsulated Poly-Lactide- *Co* -Glycolide Nanoparticles Induces Efficient Antitumor and Antiviral Immune Responses. *ACS Nano* **2013,** *7* (3), 2042–2055.
118. Kim, N. W.; Kim, S.-Y.; Lee, J. E.; Yin, Y.; Lee, J. H.; Lim, S. Y.; Kim, E. S.; Duong, H. T. T.; Kim, H. K.; Kim, S.; et al. Enhanced Cancer Vaccination by *In Situ* Nanomicelle-Generating Dissolving Microneedles. *ACS Nano* **2018,** *12* (10), 9702–9713.
119. van der Burg, S. H.; Arens, R.; Ossendorp, F.; van Hall, T.; Melief, C. J. M. Vaccines for Established Cancer: Overcoming the Challenges Posed by Immune Evasion. *Nat. Rev. Cancer* **2016,** *16* (4), 219–233.

120. Ye, Y.; Wang, C.; Zhang, X.; Hu, Q.; Zhang, Y.; Liu, Q.; Wen, D.; Milligan, J.; Bellotti, A.; Huang, L.; et al. A Melanin-Mediated Cancer Immunotherapy Patch. *Sci. Immunol.* **2017,** *2* (17).
121. Saxena, M.; Bhardwaj, N. Re-Emergence of Dendritic Cell Vaccines for Cancer Treatment. *Trends Cancer* **2018,** *4* (2), 119–137.
122. Chang, H.; Chew, S. W. T.; Zheng, M.; Lio, D. C. S.; Wiraja, C.; Mei, Y.; Ning, X.; Cui, M.; Than, A.; Shi, P.; et al. Cryomicroneedles for Transdermal Cell Delivery. *Nat. Biomed. Eng.* **2021,** *5* (9), 1008–1018.
123. Duan, X.; Chan, C.; Lin, W. Nanoparticle-Mediated Immunogenic Cell Death Enables and Potentiates Cancer Immunotherapy. *Angew. Chem. Int. Ed.* **2019,** *58* (3), 670–680.
124. Galluzzi, L.; Buqué, A.; Kepp, O.; Zitvogel, L.; Kroemer, G. Immunogenic Cell Death in Cancer and Infectious Disease. *Nat. Rev. Immunol.* **2017,** *17* (2), 97–111.
125. Tian, Y.; Jing, H.; Wang, Q.; Hu, S.; Wu, Z.; Duan, Y. Dissolving Microneedles-Based Programmed Delivery System for Enhanced Chemo-Immunotherapy of Melanoma. *J. Contr. Release* **2023,** *360,* 630–646.
126. Bian, Q.; Huang, L.; Xu, Y.; Wang, R.; Gu, Y.; Yuan, A.; Ma, X.; Hu, J.; Rao, Y.; Xu, D.; et al. A Facile Low-Dose Photosensitizer-Incorporated Dissolving Microneedles-Based Composite System for Eliciting Antitumor Immunity and the Abscopal Effect. *ACS Nano* **2021,** *15* (12), 19468–19479.
127. Chen, M.; Quan, G.; Wen, T.; Yang, P.; Qin, W.; Mai, H.; Sun, Y.; Lu, C.; Pan, X.; Wu, C. Cold to Hot: Binary Cooperative Microneedle Array-Amplified Photoimmunotherapy for Eliciting Antitumor Immunity and the Abscopal Effect. *ACS Appl. Mater. Interfaces* **2020,** *12* (29), 32259–32269.
128. Shi, C.; Chen, M.; Li, X.; Fu, Y.; Yang, D.; Wen, T.; Zhao, W.; Sun, Y.; Wang, W.; Lu, C.; et al. ATP-adenosine Axis Regulation Combined with Microneedle Assisted Photoimmunotherapy to Boost the Immunotherapy Efficiency. *J. Contr. Release* **2024,** *367,* 1–12.
129. Han, X.; Li, H.; Zhou, D.; Chen, Z.; Gu, Z. Local and Targeted Delivery of Immune Checkpoint Blockade Therapeutics. *Acc. Chem. Res.* **2020,** *53* (11), 2521–2533.
130. Milling, L.; Zhang, Y.; Irvine, D. J. Delivering Safer Immunotherapies for Cancer. *Adv. Drug Deliv. Rev.* **2017,** *114,* 79–101.
131. Sasidharan Nair, V.; Elkord, E. Immune Checkpoint Inhibitors in Cancer Therapy: A Focus on T-regulatory Cells. *Immunol. Cell Biol.* **2018,** *96* (1), 21–33.
132. Spranger, S.; Koblish, H. K.; Horton, B.; Scherle, P. A.; Newton, R.; Gajewski, T. F. Mechanism of Tumor Rejection with Doublets of CTLA-4, PD-1/PD-L1, or IDO Blockade Involves Restored IL-2 Production and Proliferation of CD8+ T Cells Directly within the Tumor Microenvironment. *J. Immunother. Cancer* **2014,** *2* (1), 3.
133. Swart, M.; Verbrugge, I.; Beltman, J. B. Combination Approaches with Immune-Checkpoint Blockade in Cancer Therapy. *Front. Oncol.* **2016,** *6.*
134. Wang, C.; Ye, Y.; Hochu, G. M.; Sadeghifar, H.; Gu, Z. Enhanced Cancer Immunotherapy by Microneedle Patch-Assisted Delivery of Anti-PD1 Antibody. *Nano Lett.* **2016,** *16* (4), 2334–2340.
135. Terness, P.; Chuang, J.-J.; Opelz, G. The Immunoregulatory Role of Ido-Producing Human Dendritic Cells Revisited. *Trends Immunol.* **2006,** *27* (2), 68–73.
136. Ricciuti, B.; Leonardi, G. C.; Puccetti, P.; Fallarino, F.; Bianconi, V.; Sahebkar, A.; Baglivo, S.; Chiari, R.; Pirro, M. Targeting Indoleamine-2,3-Dioxygenase in Cancer: Scientific Rationale and Clinical Evidence. *Pharmacol. Ther.* **2019,** *196,* 105–116.

137. Ye, Y.; Wang, J.; Hu, Q.; Hochu, G. M.; Xin, H.; Wang, C.; Gu, Z. Synergistic Transcutaneous Immunotherapy Enhances Antitumor Immune Responses through Delivery of Checkpoint Inhibitors. *ACS Nano* **2016,** *10* (9), 8956–8963.
138. Yang, P.; Lu, C.; Qin, W.; Chen, M.; Quan, G.; Liu, H.; Wang, L.; Bai, X.; Pan, X.; Wu, C. Construction of a Core-Shell Microneedle System to Achieve Targeted Co-delivery of Checkpoint Inhibitors for Melanoma Immunotherapy. *Acta Biomater.* **2020,** *104,* 147–157.
139. Zhang, M.; Li, C.; Wang, S.; Liu, T.; Song, X.; Yang, X.; Feng, J.; Zhang, X. Tumor Starvation Induced Spatiotemporal Control over Chemotherapy for Synergistic Therapy. *Small* **2018,** *14* (50).
140. Zeng, Y.; Zhou, H.; Ding, J.; Zhou, W. Cell Membrane Inspired Nano-Shell Enabling Long-Acting Glucose Oxidase for Melanoma Starvation Therapy via Microneedles-Based Percutaneous Delivery. *Theranostics* **2021,** *11* (17), 8270–8282.
141. McCaffrey, J.; Donnelly, R. F.; McCarthy, H. O. Microneedles: An Innovative Platform for Gene Delivery. *Drug Deliv. Transl. Res.* **2015,** *5* (4), 424–437.
142. Li, X.; Xu, Q.; Zhang, P.; Zhao, X.; Wang, Y. Cutaneous Microenvironment Responsive Microneedle Patch for Rapid Gene Release to Treat Subdermal Tumor. *J. Contr. Release* **2019,** *314,* 72–80.
143. Moreira, A. F.; Rodrigues, C. F.; Jacinto, T. A.; Miguel, S. P.; Costa, E. C.; Correia, I. J. Microneedle-Based Delivery Devices for Cancer Therapy: A Review. *Pharmacol. Res.* **2019,** *148,* 104438.
144. Ruan, W.; Zhai, Y.; Yu, K.; Wu, C.; Xu, Y. Coated Microneedles Mediated Intradermal Delivery of Octaarginine/BRAF SiRNA Nanocomplexes for Anti-melanoma Treatment. *Int. J. Pharm.* **2018,** *553* (1–2), 298–309.
145. Ali, A. A.; McCrudden, C. M.; McCaffrey, J.; McBride, J. W.; Cole, G.; Dunne, N. J.; Robson, T.; Kissenpfennig, A.; Donnelly, R. F.; McCarthy, H. O. DNA Vaccination for Cervical Cancer; a Novel Technology Platform of RALA Mediated Gene Delivery via Polymeric Microneedles. *Nanomedicine* **2017,** *13* (3), 921–932.
146. Pan, J.; Ruan, W.; Qin, M.; Long, Y.; Wan, T.; Yu, K.; Zhai, Y.; Wu, C.; Xu, Y. Intradermal Delivery of STAT3 SiRNA to Treat Melanoma via Dissolving Microneedles. *Sci. Rep.* **2018,** *8* (1), 1117.
147. Xu, Q.; Li, X.; Zhang, P.; Wang, Y. Rapidly Dissolving Microneedle Patch for Synergistic Gene and Photothermal Therapy of Subcutaneous Tumor. *J. Mater. Chem. B* **2020,** *8* (19), 4331–4339.

Chapter 9

Microneedle technology for infectious disease therapy*

Wenhao Wang[1]**, Jubo Jian**[2] **and Chuanbin Wu**[2]
[1]*School of Pharmaceutical Science, Sun Yat-Sen University, Guangzhou, China;* [2]*State Key Laboratory of Bioactive Molecules and Druggability Assessment/International Cooperative Laboratory of Traditional Chinese Medicine Modernization and Innovative Drug Development of Ministry of Education (MOE) of China/College of Pharmacy, Jinan University, Guangzhou, China*

1. Introduction

Contagious diseases remain a major threat to human health. With rising economic levels, advancements in drug research and development, and the continuous progress of next-generation sequencing technologies, some contagious diseases, such as smallpox, poliomyelitis, filariasis, leprosy, malaria, and chronic hepatitis C, have been eradicated or significantly reduced. However, new or mutated pathogens continue to emerge, leading to the spread of novel infectious diseases. The emergence of new contagious diseases, such as Coronavirus Disease 2019 (COVID-19), severe acute respiratory syndrome (SARS), and Middle East respiratory syndrome (MERS), as well as the resurgence of known pathogens (e.g., Ebola virus, Zika virus), continues to pose significant challenges to global health. Furthermore, under the influence of global climate change, rapid urbanization, increased population mobility, and modern medical practices, humanity still faces the dual threats of both traditional and emerging contagious diseases.[1]

Prevention is the cornerstone of contagious disease control, with vaccination serving as its primary tool. The success of immunization programs hinges on the delivery method, each of which carries distinct advantages and limitations that shape its suitability for different settings. Intramuscular (IM) injections are the gold standard for vaccine administration, prized for their well-documented efficacy. This method is commonly used for vaccines such as those against influenza, COVID-19, and hepatitis B, as muscle tissue's rich blood supply ensures rapid systemic

* This chapter is adapted from the Chinese paper entitled '[Advances of microneedle vaccines in transcutaneous immunization]' published in Acta Pharmaceutica Sinica, [2021], Vol.[57], No.[1], pp.[142-149], https://doi.org/10.16438/j.0513-4870.2021-1297.

Microneedle Technology. https://doi.org/10.1016/B978-0-443-45544-5.00006-3

distribution and a strong immune response. However, IM injections have significant drawbacks.[2] They require trained healthcare personnel, creating logistical challenges, particularly in low-resource settings. Additionally, their reliance on cold-chain storage complicates distribution in regions with unstable electricity. Perhaps most notably, needle-related pain and anxiety contribute to vaccine hesitancy and lower compliance, especially among children and individuals with needle phobia.[3]

Oral vaccination provides a needle-free alternative that significantly enhances patient compliance, particularly in pediatric populations. Successful examples include the oral polio vaccine and rotavirus vaccine, both of which have been instrumental in global disease eradication efforts.[4] A major advantage of oral immunization is its ability to stimulate mucosal immunity, which is critical for combating pathogens that invade through the gastrointestinal tract, such as cholera and typhoid. However, this approach faces challenges, including susceptibility to degradation by stomach acid and digestive enzymes, which may compromise vaccine efficacy before reaching intestinal immune cells. To address this, enteric coatings or repeated dosing are often required.[5] Additionally, oral vaccines typically demand strict temperature control during storage and transportation, as inadequate refrigeration can reduce potency. These logistical constraints may hinder their use in remote or resource-limited settings. Alternatively, intranasal vaccines offer a noninvasive, painless option that bypasses injections. This route is especially effective against respiratory pathogens, as it triggers mucosal immunity in the nasal passages, the primary entry point for many viruses. The simplicity of administration makes intranasal vaccines well-suited for mass vaccination campaigns.[6] Nevertheless, their efficacy may be diminished in individuals with nasal congestion or allergies, which can hinder proper absorption. There is also a rare risk of unintended systemic absorption, potentially leading to adverse effects. Despite these limitations, intranasal vaccines hold significant promise for future immunization strategies, particularly against respiratory diseases like COVID-19 and Respiratory syncytial virus (RSV).[7]

Compared with the routes mentioned above, transdermal immunization is commonly employed for live-attenuated vaccines. Its slower absorption rate ensures sustained antigen exposure, promoting long-term immune responses. Additionally, this method is less invasive than injections, as it delivers antigens to the fatty tissue layer beneath the skin rather than deep muscle.[8] Notably, microneedles (MNs) are a promising tool for transdermal immunization due to their solid-state structure and ease of administration, offering significant advantages such as painless delivery, potent immunogenicity, and high storage stability. This section will summarize and analyze the current status and prospects of MNs in transdermal immunization.[9]

2. The mechanism and advantages of microneedle vaccination

2.1 The physiological basis of percutaneous immune responses

The skin is the largest organ of the human body, serving not only as the first line of defense against physical and chemical damage as well as microbial invasion but also as a critical immune barrier. Under natural conditions, it is one of the primary pathways through which foreign antigens stimulate immune responses in the body. The skin efficiently initiates these responses while maintaining immune homeostasis to prevent pathological damage. Structurally, the skin is extensively interconnected with other parts of the body through blood vessels and draining lymph nodes. The skin-associated lymphoid tissue forms one of the most dynamic immune networks in the human body.[10] The epidermis and dermis are rich in professional antigen-presenting cells (APCs), which actively migrate to draining lymph nodes to interact with T cells and B cells. This process creates an effective interface for systemic immune regulation.[11]

The skin's immune response involves multiple cell types, including keratinocytes and Langerhans cells (LCs) in the epidermis, as well as dendritic cells (DCs) and T cells in the dermis. As the most abundant epidermal cells, keratinocytes constitute over 90% of the epidermis. Their primary roles include forming the skin's physical barrier and actively modulating the local immune microenvironment.[12] Keratinocytes express various pattern recognition receptors, such as Toll-like receptors (TLRs) and nucleotide-binding oligomerization domain (NOD)-like receptors, which detect pathogen-associated molecular patterns (PAMPs) from invading pathogens. Upon antigen exposure, keratinocytes secrete diverse cytokines and chemokines to establish a proinflammatory microenvironment, thereby recruiting immune cells, including T cells, neutrophils, and dendritic cells, to the site of infection.

Langerhans cells are professional APCs residing in the epidermis, distinguished by their expression of langerin (CD207) and the presence of Birbeck granules. Comprising 1%–3% of epidermal cells, LCs extend numerous dendrites, enabling them to cover approximately 25% of the skin surface. Like keratinocytes, they express TLRs that recognize PAMPs and secrete cytokines to sustain a pro-inflammatory microenvironment.[13] While most LCs remain localized to the epidermis, a small subset (2%–3%) is highly mobile, continuously migrating from the skin to draining lymph nodes. Upon antigen exposure, LCs transition from an immature to a mature state: they capture, process, and present antigens, differentiate from the immature state to the mature state, detach from keratinocytes, and migrate to lymph nodes to initiate immune responses. In the dermis, DCs serve as the predominant professional APCs. Similar to LCs, DCs recognize and internalize invading antigens, subsequently presenting them to immune cells in lymph nodes.[14] However, dermal DCs exhibit greater motility than LCs. Multiple DC subsets have been identified in mice, with several potentially homologous subtypes described in humans. Skin-resident T cells outnumber circulating T cells by more than twofold and are distributed throughout both the epidermis and dermis. The majority are memory T cells, with balanced proportions of $CD4^+$ and $CD8^+$ populations. Notably, $CD4^+$ T helper (Th) cell subsets (e.g., Th1, Th2, and Th17) play pivotal roles in skin immune regulation.[15]

Transdermal immunization, an emerging vaccination strategy, induces systemic immune responses by topically applying antigens and adjuvants to the skin. This approach leverages the skin's rich population of APCs and their migratory pathways to draining lymph nodes, forming the biological foundation of transdermal vaccination. The process involves four key steps (Fig. 9.1): (1) Antigen penetration: The antigen must first traverse the stratum corneum; (2) Inflammatory microenvironment establishment: A conducive milieu promotes LCs and DCs maturation; (3) Antigen uptake: LCs and DCs recognize, internalize, and process the antigen; (4) Immune activation: Matured LCs and DCs migrate to draining lymph nodes, presenting antigens to T and B cells. Upon antigen exposure, keratinocytes detect PAMPs via TLRs and NOD-like receptors, triggering cytokine secretion (e.g., IL-1, TNF-α). This establishes a pro-inflammatory microenvironment essential for DCs' maturation. Immature LCs and DCs in the epidermis or dermis capture and process the antigen while secreting cytokines to sustain inflammation and promote their own maturation.[16] Mature DCs upregulate major histocompatibility complex (MHC) and costimulatory molecules (e.g., CD40, CD80, CD86), then migrate to lymph nodes. Then, they interact with T and B cells, coordinating cytotoxic T lymphocyte development, antibody production, and macrophage/NK cell activation, ultimately eliciting robust humoral and cellular immune responses.[17]

The primary challenge hindering the application of transdermal immunization is enabling macromolecular antigens to effectively penetrate the stratum corneum barrier and reach the APCs-rich epidermis and dermis, thereby facilitating efficient antigen uptake and presentation by APCs. To address this, penetration-enhancement techniques aimed at improving the transdermal delivery efficiency of macromolecular antigens have become a key focus in percutaneous immunization research. Currently, the main transdermal penetration enhancement methods include chemical penetration enhancers, formulation-based penetration enhancement techniques (e.g., liposomes, nanoparticles), and physical penetration enhancement techniques (e.g., iontophoresis, electroporation, thermal poration, laser stratum corneum ablation).[18] Chemical penetration enhancers and formulation-based penetration enhancement methods often fail to achieve the desired transdermal efficiency. Although physical penetration enhancement techniques can increase the

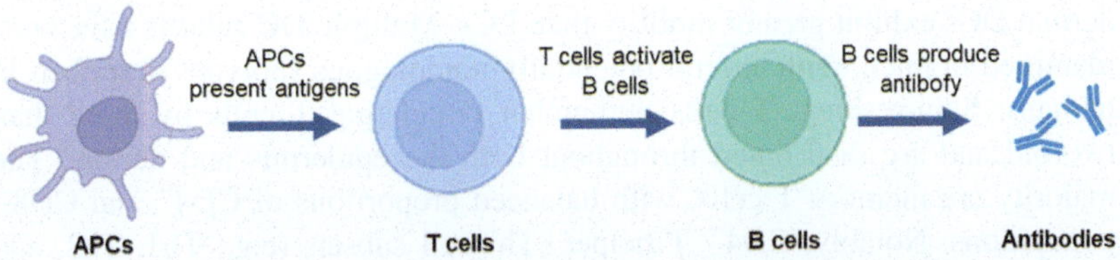

FIGURE 9.1 Schematic diagram of transcutaneous immunization.

transdermal efficiency of macromolecular antigens to a certain extent, they cannot achieve effective control of the vaccination dosage, and some techniques may cause varying degrees of damage to the skin. Given these limitations, developing novel transdermal vaccine delivery strategies that are safer, more reliable, and dose-controllable remains an urgent priority.

2.2 Advantages of microneedle-mediated percutaneous vaccination

MNs create multiple microchannels in the epidermis, significantly enhancing the transdermal delivery efficiency of vaccines. Upon reaching the epidermis and dermis, the vaccines are internalized by APCs, which then migrate to lymphoid tissues to present processed antigens to T cells, thereby initiating an adaptive immune response. Subsequently, activated T cells stimulate B cells to differentiate into antibody-secreting plasma cells, producing antigen-specific antibodies. Concurrently, memory B cells are generated to establish long-term immunological memory. Upon re-exposure to the same antigen or pathogen, these memory cells rapidly proliferate and mount a robust, targeted antibody response.[19]

Compared with traditional vaccination methods, MN-based transcutaneous immunization exhibits several significant advantages: (1) The MN administration process resembles applying a bandage, enabling self-administration by patients and thereby reducing the reliance on healthcare professionals and associated labor costs. Furthermore, MN vaccines are predominantly delivered in solid form, which minimizes the stringent low-temperature cold chain requirements during transportation and storage, thus lowering logistical expenses. (2) MNs are small solid formulations that facilitate ease of transport and purchase. Additionally, given that the length of MNs typically ranges from 25 to 1000 μm, they do not stimulate dermal nerve endings or blood vessels, resulting in minimal or no noticeable pain during use. (3) As the skin is the largest immune organ in the human body and houses various immune cells, MN transdermal immunization enhances the probability and duration of contact between the vaccine and immune cells, thereby significantly improving the immune effect.[20] (4) MN technology reduces the reliance on syringes, thereby decreasing resource consumption and environmental pollution. Moreover, due to its rapid and efficient vaccination capabilities, MN can markedly increase vaccination rates and expedite the establishment of herd immunity barriers. Given these substantial advantages of MNs in vaccine delivery, MN-based transdermal immunization has emerged as a research focus in the biomedical field, demonstrating promising clinical application prospects.

2.3 Classification and application of microneedle-based percutaneous immunization

At present, MNs have been used to deliver a variety of vaccines, such as live attenuated vaccines, inactivated vaccines, DNA vaccines, and subunit vaccines, and have shown good percutaneous immune effects. They play an important role in the treatment of multiple contagious diseases.[21]

2.3.1 Live attenuated vaccines and their applications

Live attenuated vaccines are biological preparations derived from pathogens (viruses or bacteria) that have undergone laboratory modifications to eliminate or reduce their disease-inducing potential while retaining their ability to replicate and stimulate immunity. These vaccines function by mimicking natural infections, as the attenuated pathogen colonizes host tissues and presents antigens to the immune system without causing overt illness. The preserved antigenic structure of the attenuated strain allows the immune system to recognize and mount defenses similar to those triggered by the original pathogen, often resulting in durable, long-term protection. This intrinsic immunogenicity eliminates the need for adjuvants, chemical additives required by inactivated vaccines to enhance immune responses. A defining feature of live attenuated vaccines is their capacity to induce comprehensive immune responses. They expose the immune system to a broad array of antigens over an extended period, activating both antibody-producing B cells and pathogen-specific T cells. This dual activation generates immunological memory, which underpins the long-lasting protection observed with vaccines such as those for measles and yellow fever.

However, there are still some challenges regarding the applications of live attenuated vaccines, which lie in the thermolability of live attenuated vaccines. Most viral strains degrade rapidly at temperatures above 8°C, necessitating continuous cold chain storage and transportation. This requirement imposes substantial logistical and financial burdens, particularly in low-resource regions with limited refrigeration infrastructure. Temperature excursions during distribution can render vaccines ineffective, contributing to vaccine wastage and gaps in immunization coverage.

To address these issues, the MN-mediated live attenuated vaccines have been developed. As a thermostable solid formulation, the MNs could significantly improve the stability of live attenuated vaccines, thus maintaining pathogen viability under ambient conditions. For instance, studies have demonstrated that constructing the coating of MNs with carboxymethylcellulose and trehalose could significantly improve the stability of the attenuated live dengue fever vaccine. This formulation has been shown to extend the shelf life of dengue vaccines from hours to weeks at room temperature without compromising immunogenicity. Animal studies confirm that such stabilized vaccines elicit neutralizing antibody levels comparable to traditional liquid formulations administered via subcutaneous injection.[22] Mechanistically, trehalose, a natural sugar, forms a glass-like matrix during desiccation, immobilizing viral components and preventing structural degradation. In addition, converting vaccines into freeze-dried powders is a common method to maintain vaccine stability, such as the *Bacillus Calmette-Guérin* (BCG) vaccine for tuberculosis, the varicella-zoster vaccine, and the hepatitis A vaccine. However, this formulation requires professionals to dissolve the powder and inject it within a few hours before use, which brings many inconveniences to vaccine administration. To avoid this process, the powder of the BCG vaccine is directly delivered by soluble MNs. After penetrating the skin, the needle tip dissolves, exposing the BCG vaccine powder to the interstitial fluid of the skin tissue. The powder gradually dissolves and spreads, resulting in immune responses similar to intradermal immunization. Moreover, after being stored at room temperature for more than 2 months, the survival rate of the BCG vaccine and the mechanical strength of the MNs show no significant changes in this product.

The shift toward thermostable formulations and simplified delivery systems holds transformative potential for global immunization programs. By reducing reliance on cold chain infrastructure, these MNs-based innovations could lower distribution costs, minimize vaccine wastage, and expand access to underserved populations. Despite these advancements, challenges persist in optimizing MNs' consistency and ensuring real-world applicability. Variations in humidity, temperature fluctuations, and exposure to light during storage can still impact vaccine stability, necessitating robust packaging solutions. Furthermore, regulatory frameworks must adapt to evaluate MNs, ensuring they meet safety and efficacy standards without prolonging approval timelines. Economic considerations also play a role, as the initial cost of developing and scaling advanced formulations may hinder adoption in resource-limited settings. However, the long-term benefits, including reduced logistical complexity and broader vaccine coverage, justify continued investment in these technologies.

However, the genetic instability of attenuated strains poses another critical limitation. In rare cases, genetic mutations during replication can restore virulence, particularly in immunocompromised individuals. For example, vaccine-associated paralytic polio, caused by reversion of the oral polio vaccine strain, highlights the importance of rigorous safety monitoring. Such risks necessitate meticulous manufacturing protocols and postvaccination surveillance to ensure genetic stability and minimize public health hazards. Thus, other kinds of vaccines have been developed as alternatives.

2.3.2 Inactivated vaccines and their applications

Inactivated virus vaccines are meticulously developed through a precise and complex process. First, the virus is cultivated in specific substrates like cell lines, tissues, or embryonic cells. Subsequently, it undergoes inactivation using chemical, physical, or radiation approaches. This inactivation procedure is crucial as it effectively destroys the virus's ability to replicate within the human body. Simultaneously, it ensures that all the antigens present in the viral structure are retained intact. These antigens are essential as they stimulate the immune system to recognize and respond to the virus. It's important to note that inactivated vaccines typically have lower immunogenicity compared to live viruses. This means they often require multiple administrations or need to be combined with powerful adjuvants to achieve the desired level of immunity. However, they offer a significant advantage in that they do not have the potential to restore pathogenicity. This makes inactivated virus vaccines a safer option, especially in cases where the use of live vaccines might pose certain risks.

Given that the epidermis and dermis are abundant in APCs, the application of MNs for delivering inactivated viruses shows distinct benefits.[23] For instance, chitosan microcoating and the delivery of inactivated influenza virus can maintain antigen release in the skin for up to 28 days. This leads to a significantly higher influenza virus-specific IgG response in mice when compared to intramuscular injection, thereby significantly improving the humoral immune response induced by the vaccine. Moreover, transcutaneous immunization using MNs coated with inactivated virus vaccines is capable of inducing stronger cellular immune responses. It can successfully stimulate antigen-specific memory B cell responses, thereby providing long-term and effective protection. In addition, Van Der Maaden et al.[24] employed hydrofluoric acid to etch molten silica capillary tubes for the fabrication of hollow MNs used in the administration of inactivated poliovirus vaccine (IPV). The results demonstrated that these MNs could successfully penetrate rat skin without experiencing needle body fracture or tip blockage. When compared to intramuscular injection, the vaccine delivered via hollow MNs elicited comparable immune responses. Furthermore, the same research team utilized hollow MNs for the delivery of human papillomavirus (HPV) vaccines in cancer treatment. In comparison to intramuscular injection, a reduced dosage of the vaccine administered through hollow MNs effectively induced the proliferation and differentiation of $CD4^+$ T cells and $CD8^+$ T cells in mice, thereby significantly enhancing cellular immunity and promoting the generation of cytotoxic T cells and Th cells.

Dissolving MNs have also been utilized to deliver the inactivated vaccines. For example, Sullivan et al.[25] fabricated dissolving MNs using polyvinylpyrrolidone as the material for influenza virus vaccination. These MNs dissolved approximately 85% within 5 min after penetrating the skin of mice on their backs, and they could induce effective humoral immune responses. There was no statistically significant difference in the IgG titer in the mice's bodies 28 days after immunization compared with that from intramuscular injection; T lymphocytes secreted more interleukin-4 (IL-4) and interferon-γ (IFN-γ); and the memory immune response could quickly take effect 90 days later, reducing the virus titer and increasing the rate of pulmonary virus clearance. In a similar vein, Kaplan et al.[26] created a silk fibroin MNs patch designed for delivering vaccines against influenza, *Clostridium difficile*, and *Shigella*. Each patch consisted of a 20 × 20 grid of conical MNs, each measuring 700 μm in height, spanning an area of 1.5 × 1.5 cm. When the MN surfaces were coated with a 1% silk fibroin vaccine solution, the antigen showed increased resistance to environmental variations and temperature shifts during the drying phase and subsequent storage. Silk fibroin materials possess the capacity to stabilize heat-sensitive biomacromolecules, which in turn improves the potential for designing vaccination systems that do not rely on cold-chain transportation.

2.3.3 Subunit vaccines and their applications

Subunit vaccine is a vaccine that can induce the body to produce antibodies by extracting the special protein structure of the virus and screening out the immune active fragments, which can trigger the required immune response without causing harm to the subject after being introduced into the human body. Similar to inactivated virus vaccines, subunit vaccines are safer because they are less immunogenic than live virus, but require multiple injections to achieve the desired level of immune response. Initially, Weldon et al.[27] used MNs patches to deliver an influenza subunit vaccine and demonstrated that MNs could effectively enhance the percutaneous immune response to this vaccine. Intradermally administered a fusion protein of influenza virus, composed of the repetitive extracellular region of influenza virus and the Toll-like receptor five agonist domain of flagellin, provides broad crossprotection against influenza virus. H1N1 influenza subunit vaccine coated on stainless steel MNs can trigger immune responses in young mice, producing higher antibody titers than intramuscular vaccination, and can increase the level of follicular helper T cells, thereby mediating the differentiation of B cells into antibody-secreting cells or memory B cells, and enhancing the level of the body's immune response. In response to SARS-CoV-2, a novel coronavirus that is currently circulating worldwide, researchers have tried to use MNs to deliver vaccines against SARS-CoV-2 to improve the body's immune response. Kim et al.[28] designed a 10 × 10 dissolving MNs for the delivery of SARS-CoV-2 S1 subunit vaccine, which was also constituted of carboxymethyl cellulose. The MNs were designed with a height of 750 μm, a width of 225 μm, and an apex angle of 30 degrees. The results showed that the MNs induced a significant increase in effective antigen-specific antibodies in mice 2 weeks postapplication. Kuwentrai et al.[29] prepared hyaluronic acid soluble MNs for delivery of SARS-CoV-2S1 subunit protein containing immunogenic epitopes (Fig. 9.2). In a mouse model, MNs immunization elicited significant B cell antibody responses and interferon-γ (IFN-γ) -based T cell responses compared with the subcutaneous injection control group. Therefore, delivery of subunit vaccines by MNs is an effective strategy to establish a herd immune barrier and reduce 2019-nCoV infection.

In addition to the above-mentioned dissolving MNs, the coated MNs have also been employed for the delivery of subunit vaccines. Kommareddy et al.[30] introduced an MN patch coated with an influenza subunit vaccine. In this study, the authors describe a process involving the concentration, formulation,

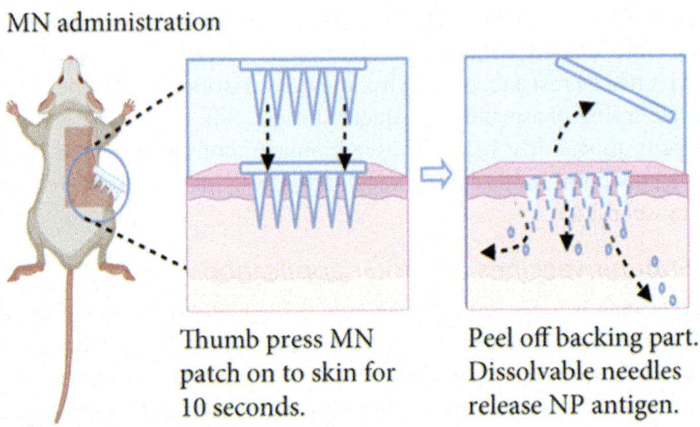

FIGURE 9.2 Schematic illustration of the vaccination with microneedles. *Reproduced from reference*[29] *with permission from John Wiley & Sons, Inc.*

and coating of monobulks derived from the 2008/2009 seasonal influenza vaccine onto 3M's solid microstructured transdermal system (sMTS) using a method scalable under Good Manufacturing Practice (GMP) conditions. The stability of the coated antigen was assessed during storage at 4 °C and room temperature in the presence of a desiccant for at least 8 weeks. Freeze-thaw stability data demonstrated the robustness of the coated antigen under stressed conditions. Furthermore, the vaccine-coated microstructures were evaluated in vivo using a guinea pig model, resulting in immune titers comparable to those achieved with the traditional intramuscularly administered trivalent vaccine. In conclusion, MNs with different types represent a strong candidate for the subunit vaccine-based vaccination.

2.3.4 DNA vaccines and their applications

A DNA vaccine is a recombinant eukaryotic plasmid expression vector that encodes a specific protein antigen. Once injected into the body, the plasmid that enters the cells proceeds with gene transcription and generates antigen proteins, thereby activating the body to produce specific humoral and cellular immunity. The fact that the antigen-encoded DNA can be directly synthesized and expressed within the host cells means that the immunity induced by the DNA vaccine bears a remarkable resemblance to the immune response brought about by natural pathogen infection. When compared with traditional vaccines, DNA vaccines have many advantages. They exhibit excellent stability, are relatively straightforward to prepare, and can successfully induce antigen-specific immune responses in small animals. However, the immunogenicity of DNA vaccines in nonhuman primates and humans is not satisfactory. The limited transfection ability of DNA vaccines is widely considered to be one of the most crucial factors contributing to the low immunity. To improve the vaccination effects, percutaneous MNs vaccination emerges as an effective means.

As early as 2002, Mikszta et al.[31] employed MNs to introduce naked plasmid DNA into the skin. Specifically, by immersing the MN arrays in the DNA solution and applying pressure on the skin of mice multiple times, the expression of plasmid DNA encoding firefly luciferase increased by an astonishing 2800 times when using MNs compared to local injection. Furthermore, when the plasmid DNA encoding the hepatitis B surface antigen was delivered via MNs, the immune response elicited by the administration of MNs was significantly stronger than that induced by subcutaneous injection. Moreover, the required serum antibody titer could be attained with a reduced number of immunization sessions. In addition, other studies using coated MNs have been reported. Zhang et al.[32] demonstrated that the method of applying DNA vaccines to the skin by coating with MNs outperforms the traditional syringe injection method in terms of antigen expression and immunogenicity. They uniformly distributed plasmid DNA on the surface of 36 MNs (with a height of 220 μm) (on a 2.6 × 2.6 mm patch). Compared with the syringe injection group, the DNA vaccine delivered through MNs increased GFP expression in the back skin by more than four times, and the duration of protein expression was further prolonged. This indicates that the use of MNs to deliver DNA vaccines can effectively enhance the expression of the encoded antigen and significantly improve immunogenicity. Four vaccinations were conducted on mice within 2 weeks, using MN-mediated DNA vaccines encoding the OVA antigen. Both humoral and cellular immune responses were significantly improved. Particularly, after injection with MNs, the level of IgG2c antibodies (an indicator reflecting cellular immune response) was 11 times higher than when injected with a syringe. This enhanced immune response may be due to the large-scale vaccination by MNs.

Besides, the combination of nanoparticles and MNs has been developed to effectively enhance the stability of DNA vaccines. For example, Yang et al.[33] coated Ebola DNA vaccine onto the surface of PLGA-poly-L-lysine (PLL)/γ-polyglycolic acid (PGA) nanoparticles to enhance the stability, which were encapsulated in dissolving MNs. They found that the Ebola DNA vaccine was inactivated in 2 weeks and completely degraded within 4–6 weeks after storage. However, when encapsulated in MNs, 94.3% + 0.9% of the DNA vaccines still exist in supercoiled form and can still promote an appropriate immune response after being stored at 37°C for 6 weeks. This might be due to the electrostatic interaction between the nanoparticles and the DNA vaccines, which reduces the degradation of the DNA vaccines during the manufacturing process and storage.

3. Strategies to enhance the vaccination effects of microneedles

Vaccine MNs have been shown to activate the immune response of the body; however, the immune response stimulated by vaccine MNs containing only antigens is often limited. Consequently, researchers have proposed several auxiliary measures from various perspectives to enhance the immune effect of vaccine MNs. These measures include (1) adjusting the physical parameters of MNs (e.g., tip length, shape, array surface volume, density); (2) codelivery of antigen and adjuvant; (3) combination of nano drug delivery system; and (4) combined use of physical infiltration technology.

3.1 Optimizing microneedle geometries

Research findings suggest that increasing the width of the needles, the length of the needle tips, and the density of the MN arrays can lead to a remarkable enhancement in transdermal drug penetration.[34] The primary reason behind this lies in the fact that these alterations result in an increase in the size, depth, and density of the channels formed after piercing the skin. When the needles are wider and the tips longer, they create larger and deeper entry points for the drugs to pass through. The increased density of the MN arrays means there are more channels available for the drugs to traverse.[35] This collective effect significantly improves the efficiency of drug penetration through the skin.

However, it's crucial to understand that there is a limit to how dense the MNs array should be. If the density is overly high, it could give rise to the "bed of nails" effect. Similarly, in the context of MN arrays, overly dense channels would compress one another, thereby reducing their diameter. This could potentially impede the smooth flow of drugs and undermine the intended enhancement in transdermal delivery. Consequently, by fine-tuning the geometry of the MNs array, the efficacy of vaccine MNs in stimulating an immune response within the body can be significantly improved. For example, Romgens et al.[36] successfully developed a sophisticated computational model. This model was specifically designed to describe the diffusion and kinetics of antigens in the skin following the delivery of the MNs array. It's like having a detailed map that shows how the antigens spread and interact within the complex terrain of the skin. Through this model, they were able to optimize the most favorable geometric parameters and arrangement of the MNs. The ultimate aim was to enhance their capacity to activate APCs and induce an immune response within the body. This is not just a matter of scientific curiosity; it's a critical step toward developing more effective vaccines and improving global health.

Currently, the COVID-19 vaccination process heavily relies on medical professionals. The need for innovative solutions that can reduce the reliance and ensure that vaccines reach every corner of the population has never been more urgent. Inspired by the structure of mushrooms, Li et al.[37] ingeniously designed MNs with a "thin neck" structure. This design was intended to facilitate the separation of the patch base layer from the MNs that had pierced the skin through dissolution and external force. As a result, it reduced the discomfort experienced by the vaccinated individuals and minimized the risk of infection. The MNs that remained within the skin would gradually and spontaneously release the vaccine through degradation. This slow and controlled release is like a carefully timed delivery system, ensuring that the body receives the vaccine in a manner that maximizes its effectiveness and minimizes potential side effects. Furthermore, through fluorescent labeling, the MN patches were crafted into specific patterns as per the requirements to achieve in-situ imprinting for data storage and convenient access. This is a revolutionary approach that combines technology and medicine in a seamless way. The patches not only deliver the vaccine but also store valuable information right at the point of administration. This innovative approach enables the storage of vaccination information directly at the patient's vaccination site, without the need for complex storage devices. This not only alleviates the workload of medical staff but also significantly reduces the likelihood of errors.

3.2 Introduction of adjuvants

Adjuvants are specific immune enhancers that, when injected into the body simultaneously with or before an antigen, can enhance the body's immune response to the antigen or alter the type of immune response. Codelivery of adjuvants and vaccines is a commonly used method to improve the immune response of vaccines at present.[38] There are many types of adjuvants, such as aluminum hydroxide adjuvant, Corynebacterium parvum, lipopolysaccharide, cytokines, alum, peptides, and nanoparticles. Currently, the adjuvants used in MNs-based vaccines are mainly some molecular immune modulators, such as CpG, poly(I: C), monophosphorylated lipid A, and imiquimod.[39] Generally, except for attenuated live vaccines that do not require adjuvants, adding adjuvants to MN-based vaccines can significantly enhance the immune response of the body. Studies have shown that adding adjuvants to MNs vaccines can effectively enhance the immune effects of inactivated virus vaccines,[40] subunit vaccines,[41] and DNA vaccines.[42]

3.3 Integration with nanotechnology

A large number of studies have shown that nanotechnology is an effective method to enhance the immune response of vaccines. Therefore, by combining MNs with nanotechnology, it is expected to improve the immune response effect of vaccines through multiple pathways: (1) Encapsulating antigens in nanoparticles can enhance the stability of vaccines in MNs, transdermal delivery efficiency and skin retention, while reducing antigen degradation in the skin and enhancing the antigen uptake ability of APCs, thereby improving the immune response of the body; (2) Nanoparticles act as antigen carriers and can also serve as adjuvants to trigger their inherent immune stimulation characteristics, and work synergistically with vaccines to enhance the immune response effect. A study by Tamayo et al.[43] demonstrated that poly(anhydride) nanoparticles can act as active Th1 adjuvants in immunotherapy, exerting the function of Toll-like receptor agonists and triggering the release of Th1-type cytokines. In addition, the physicochemical properties of nanoparticles, such as particle size, potential, shape, and surface hydrophilicity/hydrophobicity, are closely related to the immune effects they produce.[44] Ding et al.[45] successfully prepared a thiolated chitosan nanoparticle (NCM VLPs@TCS NPs) loaded with recombinant protein (NP-Cap-M2e, NCM VLPs) (Fig. 9.3). The thiol groups in this nanoparticle can form mucosal crosslinking structures, exhibiting good adhesion and stability. Compared with unmodified chitosan nanoparticles, the protein adsorption capacity of NCM VLPs@TCS NPs increased by 2.2 times, significantly prolonging the retention time of nanoparticles on the mucosal surface and enhancing the antigen delivery efficiency. This nanostructure helps induce strong humoral and cellular immune responses, thereby providing broad-spectrum protection against different influenza A viruses. Therefore, by regulating the physicochemical properties of nanoparticles in MN vaccines, the desired immune response can be induced in the body to achieve better immune effects.

3.4 Integration with physical permeation enhancement methods

Vaccines are mostly hydrophilic, biologically large-molecule substances, and their skin permeability and transdermal absorption rates are often poor. This poses a challenge in ensuring that vaccines are effectively delivered and absorbed by the body. Thus, integrating physical transdermal penetration enhancement methods into MNs has been proven to be an effective way to improve the efficiency of drug transdermal delivery, enhance drug distribution, and increase drug transdermal absorption. This approach holds great potential in the field of drug administration and has attracted significant attention from researchers.

Currently, some studies have combined MNs with methods such as ultrasound-assisted transdermal delivery,[46] electroporation,[47] and ionic electroosmosis technology[48] for drug transdermal delivery. Among them, ultrasound-assisted transdermal delivery affects the size of skin pores through cavitation, thereby increasing skin permeability. The process is similar to a gentle force that creates openings in the skin's natural barriers, allowing drugs to pass through more easily. Electroporation can generate tiny, short-lived water channels in the lipid zone of the stratum corneum barrier, enhancing drug penetration within a short period (nanosecond level). It's like creating temporary passageways for the drugs to rush through in an instant. Ionic electroosmosis technology uses low-intensity direct current electric fields to promote the passage of charged drug components through the tissue barrier into the body, achieving the effect of improving the transdermal penetration ability of drugs. This is like giving the drugs a gentle push to help them overcome the obstacles of the body's defense systems. Petchsangsai et al.[49]

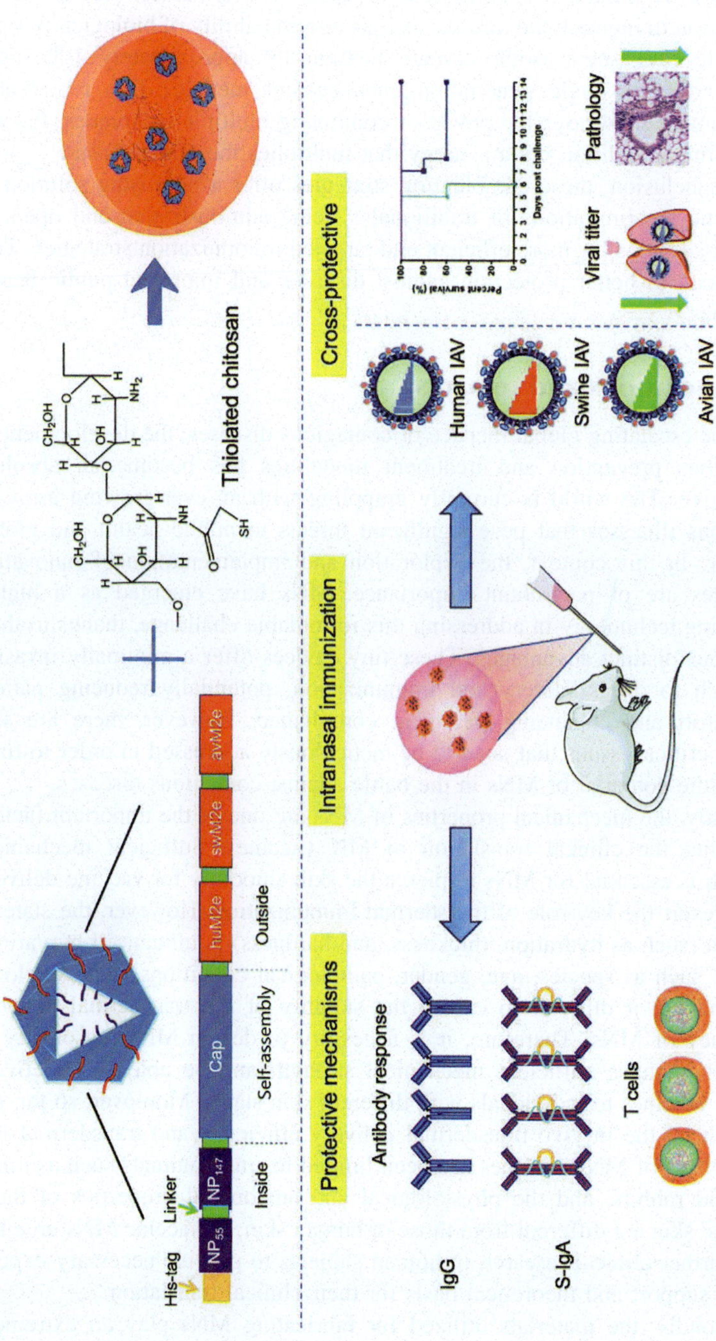

FIGURE 9.3 Schematic illustration of the preparation of nanocellulose membrane virus-like particles loaded with thermoresponsive chitosan (NCM VLPs@TCS). *Reprinted from Ref.*[45] *with permission from Elsevier.*

used MNs combined with ultrasound-assisted transdermal delivery and electroporation to increase the transdermal penetration ability of biologically large molecule fluorescein isothiocyanate dextran by approximately 100 times compared to a single penetration enhancement method. This remarkable achievement showcases the power of combining multiple techniques. It's not just a simple addition but a synergy that multiplies the effectiveness.

In conclusion, these combination strategies offer a promising solution to overcome the limitations of traditional vaccine administration and open up new possibilities for more efficient and targeted immunization strategies. This could lead to better protection against diseases and improved public health outcomes.

4. Conclusions and remarks

With the escalating global menace of contagious diseases, the development of efficacious prevention and treatment modalities has become an absolute imperative. The world is currently grappling with an ever-growing array of infectious illnesses that pose significant threats to public health and global stability. In this context, the exploration and implementation of innovative strategies are of paramount importance. MNs have emerged as a highly promising technology in addressing this formidable challenge, thanks to their numerous distinct advantages. These tiny devices offer a minimally invasive approach to drug delivery and immunization, potentially reducing patient discomfort and enhancing treatment compliance. However, there are still several critical issues that need to be meticulously addressed in order to fully unlock the potential of MNs in the battle against contagious diseases.

Firstly, the mechanical properties of MNs are one of the important factors restricting the clinical translation of MN vaccines. Sufficient mechanical strength is essential for MNs to pierce the skin smoothly for vaccine delivery and to exert the key role of transdermal immunization. However, the state of the skin (such as hydration, thickness, and hardness) is influenced by various factors, such as species, age, gender, pathological conditions, and skin location, making it difficult to ensure the stability of the transdermal delivery efficiency of MNs. Therefore, it is necessary to design MNs reasonably to make them have sufficient mechanical strength and be able to effectively deliver vaccines to individuals with different skin states. Moreover, so far, the research on the in vivo transdermal delivery efficiency and transdermal immune effect of MNs vaccines has been limited to small animals such as mice, rats, and rabbits; and the physiological and anatomical properties of these animals' skin are different from those of human skin, so vaccine MNs urgently need further clinical research in human subjects to provide necessary experimental support and theoretical basis for their clinical translation.

Secondly, the materials utilized for fabricating MNs play an extremely significant role in determining their safety and efficacy. The current range of available materials is rather limited, and there is an urgent need for the creation of novel substances with enhanced safety profiles and multifunctional attributes. Intensive research endeavors should be concentrated on identifying materials that can not only boost drug delivery efficiency but also minimize any potential adverse effects. This might involve exploring advanced biomaterials with tailored properties such as controlled biodegradability, targeted drug release capabilities, and improved biocompatibility.

Thirdly, the industrialized production of MNs is the main limiting step for the clinical translation and future commercialization of MN vaccines. Currently, the preparation methods of MN vaccines mainly include the micromolding method, photopolymerization method, stretching method, microdroplet blowing method, and 3D printing method.[50] Among them, the micromolding method is widely used due to its mild preparation conditions. It involves injecting polymer solutions into molds and solidifying them, which usually requires centrifugation or vacuum pumping to remove bubbles from the solutions, and then filling the solutions into micropores. However, this method is only suitable for small-scale laboratory preparation. To improve the industrial production efficiency of the centrifugation method, Wu et al.[51] developed a double-penetration female mold (DPFM) combined with a positive-pressure microperfusion technique (PPPT), effectively avoiding the presence of bubbles or cavities in MNs, and based on this, they designed an automated MNs production equipment. On the basis of the patch production line, they adopted a "coating" production process and developed an automatic microperfusion equipment, which can continuously produce MNs through simple and repeatable operations. This equipment can produce 126 MN arrays per batch, has a high degree of mechanization, and is easy to integrate for industrialization. This automated MNs production equipment, based on PPPT and DPFM, is expected to achieve the transformation of MNs from laboratory research to large-scale manufacturing in a controllable, efficient, and scalable manner. To expand the production scale, during the production process, the number of production steps should be minimized as much as possible, and the transfer between different equipment should be reduced to the greatest extent. The production conditions for aseptic manufacturing or terminal sterilization should be created. Besides, issues such as quality control testing, consistency, and safety of manufacturing methods, and characteristics of packaging materials also need to be considered to establish unified production norms and quality evaluation standards. These are indispensable for the final commercialization of MN's products.

In addition, the lack of a vaccine MN quality standard system is another key factor limiting its clinical translation.[52] The focus of vaccine MNs is on dose uniformity, reasonable cost, large-scale production, and good production norms (GMP).[53] Vaccines for human use require sterility and no pyrogens. Currently, most vaccines on the market use filtration for sterilization and disinfection, but due to the solid form of MNs, it is impossible to use the filtration sterilization method. Radiation sterilization may cause the vaccine to lose its antigenicity. Therefore, the preparation of vaccine MNs may need to be carried out under sterile conditions, which greatly increases the production cost of vaccine MNs. From the regulatory perspective, in the latest document of the Food and Drug Administration (FDA), vaccine MNs are classified as a combined product of biological products and mechanical equipment. Therefore, the regulation of vaccine MNs should consider the safety and effectiveness issues related to each component and the entire product. In addition, vaccine MNs should also meet the current GMP and postmarket safety requirements from laboratory to clinical use transition. In conclusion, the transformation of MN vaccines not only requires a large amount of scientific research and process design, but also needs to develop an effective quality assurance system for drugs, establish good production norms, and quality risk management.

Overall, addressing these unresolved issues is of paramount significance for advancing the clinical translation of MNs technology in the prevention and treatment of contagious diseases. Continued research efforts focused on material development, regulatory guidance, delivery optimization, and large-scale production will undoubtedly lay a solid foundation for realizing the full potential of MNs as a powerful and transformative tool in global public health initiatives.

References

1. Bartlett, J. G. Update in Infectious Diseases. *Ann. Intern. Med.* **1997,** *127* (2), 165.
2. Kaneda, Y.; Kaneda, U.; Ozaki, A.; Tanimoto, T. Intramuscular vs. Subcutaneous: Rethinking Influenza Vaccination Strategy in Japan. *JMA journal* **2024,** *7* (1), 111–113.
3. Bergeson, P. S.; Singer, S. A.; Kaplan, A. M. Intramuscular Injections in Children. *Pediatrics* **1982,** *70* (6), 944–948.
4. Kim, L.; Martinez, C. J.; Hodgson, K. A.; Trager, G. R.; Brandl, J. R.; Sandefer, E. P.; Doll, W. J.; Liebowitz, D.; Tucker, S. N. Systemic and Mucosal Immune Responses Following Oral Adenoviral Delivery of Influenza Vaccine to the Human Intestine by Radio Controlled Capsule. *Sci. Rep.* **2016,** *6* (1), 37295.
5. Zhang, Y.; Li, M.; Du, G.; Chen, X.; Sun, X. Advanced Oral Vaccine Delivery Strategies for Improving the Immunity. *Adv. Drug Deliv. Rev.* **2021,** *177,* 113928.
6. Lakhrif, Z.; Poupée-Beaugé, A.; Boursin, F.; Ducournau, C.; Lantier, L.; Moiré, N.; Carpentier, R.; Rossignol, C.; Maquart, M.; Jospin, F. Intranasal Spike and Nucleoprotein Fusion Protein-Based Vaccine Provides Cross-Protection and Reduced Transmission against SARS-CoV-2 Variants. *npj Vaccines* **2025,** *10* (1), 75.
7. Zhu, F.; Huang, S.; Liu, X.; Chen, Q.; Zhuang, C.; Zhao, H.; Han, J.; Jaen, A. M.; Do, T. H.; Peter, J. G. Safety and Efficacy of the Intranasal Spray SARS-CoV-2 Vaccine dNS1-RBD: a Multicentre, Randomised, Double-Blind, Placebo-Controlled, Phase 3 Trial. *Lancet Respir. Med.* **2023,** *11* (12).
8. Menon, I.; Bagwe, P.; Gomes, K. B.; Bajaj, L.; Gala, R.; Uddin, M. N.; D'souza, M. J.; Zughaier, S. M. Microneedles: A New Generation Vaccine Delivery System. *Micromachines* **2021,** *12* (4), 435.

9. Amani, H.; Shahbazi, M. A.; D'Amico, C.; Fontana, F.; Abbaszadeh, S.; Santos, H. A. Microneedles for Painless Transdermal Immunotherapeutic Applications. *J. Contr. Release* **2021,** *330,* 185–217.
10. Kupper, T. S.; Fuhlbrigge, R. C. Immune Surveillance in the Skin: Mechanisms and Clinical Consequences. *Nat. Rev. Immunol.* **2004,** *4* (3), 211–222.
11. Honda, T.; Egawa, G.; Kabashima, K. Antigen Presentation and Adaptive Immune Responses in Skin. *Int. Immunol.* **2019,** *31* (7), 423–429.
12. Matejuk, A. Skin Immunity. *Arch. Immunol. Ther. Exp.* **2018,** *66* (1), 45–54.
13. Peiser, M.; Koeck, J.; Kirschning, C. J.; Wittig, B.; Wanner, R. Human Langerhans Cells Selectively Activated via Toll-like Receptor 2 Agonists Acquire Migratory and CD4+ T Cell Stimulatory Capacity. *J. Leucocyte Biol.* **2008,** *83* (5), 1118–1127.
14. Kaplan, D. H. In Vivo Function of Langerhans Cells and Dermal Dendritic Cells. *Trends Immunol.* **2010,** *31* (12), 446–451.
15. Fujita, H.; Shemer, A.; Suárez-Farinas, M.; Johnson-Huang, L. M.; Tintle, S.; Cardinale, I.; Fuentes-Duculan, J.; Novitskaya, I.; Carucci, J. A.; Krueger, J. G. Lesional Dendritic Cells in Patients with Chronic Atopic Dermatitis and Psoriasis Exhibit Parallel Ability to Activate T-Cell Subsets. *J. Allergy Clin. Immunol.* **2011,** *128* (3), 574–582.
16. Fehres, C. M.; Garcia-Vallejo, J. J.; Unger, W. W.; Van Kooyk, Y. Skin-resident Antigen-Presenting Cells: Instruction Manual for Vaccine Development. *Front. Immunol.* **2013,** *4,* 157.
17. De Jong, J. M.; Schuurhuis, D. H.; Ioan-Facsinay, A.; Welling, M. M.; Camps, M. G.; Van Der Voort, E. I.; Huizinga, T. W.; Ossendorp, F.; Verbeek, J. S.; Toes, R. E. Dendritic Cells, but Not Macrophages or B Cells, Activate Major Histocompatibility Complex Class II-restricted CD4+ T Cells upon Immune-complex Uptake In Vivo. *Immunology* **2006,** *119* (4), 499–506.
18. Dragicevic, N.; Maibach, H. Combined Use of Nanocarriers and Physical Methods for Percutaneous Penetration Enhancement. *Adv. Drug Deliv. Rev.* **2018,** *127,* 58–84.
19. Kim, Y.-C.; Park, J.-H.; Prausnitz, M. R. Microneedles for Drug and Vaccine Delivery. *Adv. Drug Deliv. Rev.* **2012,** *64* (14), 1547–1568.
20. D'Amico, C.; Fontana, F.; Cheng, R.; Santos, H. A. Development of Vaccine Formulations: Past, Present, and Future. *Drug Deliv. Transl. Res.* **2021,** *11,* 353–372.
21. Nguyen, H. X. Beyond the Needle: Innovative Microneedle-Based Transdermal Vaccination. *Medicines* **2025,** *12* (1), 4.
22. Zhang, X.; Chen, G.; Yu, Y.; Sun, L.; Zhao, Y. Bioinspired Adhesive and Antibacterial Microneedles for Versatile Transdermal Drug Delivery. *Research* **2020**, *2020,* Article ID: 3672120.
23. Chen, M.-C.; Huang, S.-F.; Lai, K.-Y.; Ling, M.-H. Fully Embeddable Chitosan Microneedles as a Sustained Release Depot for Intradermal Vaccination. *Biomaterials* **2013,** *34* (12), 3077–3086.
24. van der Maaden, K.; Trietsch, S. J.; Kraan, H.; Varypataki, E. M.; Romeijn, S.; Zwier, R.; van der Linden, H. J.; Kersten, G.; Hankemeier, T.; Jiskoot, W. Novel Hollow Microneedle Technology for Depth-Controlled Microinjection-Mediated Dermal Vaccination: a Study with Polio Vaccine in Rats. *Pharm. Res.* **2014,** *31,* 1846–1854.
25. Sullivan, S. P.; Koutsonanos, D. G.; Martin, M. del P.; Lee, J. W.; Zarnitsyn, V.; Choi, S.-O.; Murthy, N.; Compans, R. W.; Skountzou, I.; Prausnitz, M. R. Dissolving Polymer Microneedle Patches for Influenza Vaccination. *Nat. Med.* **2010,** *16* (8), 915–920.
26. Stinson, J. A.; Raja, W. K.; Lee, S.; Kim, H. B.; Diwan, I.; Tutunjian, S.; Panilaitis, B.; Omenetto, F. G.; Tzipori, S.; Kaplan, D. L. Silk Fibroin Microneedles for Transdermal Vaccine Delivery. *ACS Biomater. Sci. Eng.* **2017,** *3* (3), 360–369.

27. Weldon, W. C.; Martin, M. P.; Zarnitsyn, V.; Wang, B.; Koutsonanos, D.; Skountzou, I.; Prausnitz, M. R.; Compans, R. W. Microneedle Vaccination with Stabilized Recombinant Influenza Virus Hemagglutinin Induces Improved Protective Immunity. *Clin. Vaccine Immunol.* **2011,** *18* (4), 647–654.
28. Kim, E.; Erdos, G.; Huang, S.; Kenniston, T. W.; Balmert, S. C.; Carey, C. D.; Raj, V. S.; Epperly, M. W.; Klimstra, W. B.; Haagmans, B. L. Microneedle Array Delivered Recombinant Coronavirus Vaccines: Immunogenicity and Rapid Translational Development. *EBioMedicine* **2020,** *55.*
29. Kuwentrai, C.; Yu, J.; Zhang, B-z; Hu, Y-f; Dou, Y.; Gong, H-r; Huang, J.-D.; Xu, C. Induction of Humoral and Cellular Immunity by Intradermal Delivery of SARS-CoV-2 Nucleocapsid Protein Using Dissolvable Microneedles. *J. Immunol. Res.* **2021,** *2021* (1), 5531220.
30. Kommareddy, S.; Baudner, B.; Bonificio, A.; Gallorini, S.; Palladino, G.; Determan, A.; Dohmeier, D.; Kroells, K.; Sternjohn, J.; Singh, M. Influenza Subunit Vaccine Coated Microneedle Patches Elicit Comparable Immune Responses to Intramuscular Injection in Guinea Pigs. *Vaccine* **2013,** *31* (34), 3435–3441.
31. Mikszta, J. A.; Alarcon, J. B.; Brittingham, J. M.; Sutter, D. E.; Pettis, R. J.; Harvey, N. G. *Improved Genetic Immunization via Micromechanical Disruption of Skin-Barrier Function and Targeted Epidermal Delivery;* Nature Publishing Group: US New York, 2002.
32. Zhang, S.; Zhao, S.; Jin, X.; Wang, B.; Zhao, G. Microneedles Improve the Immunogenicity of DNA Vaccines. *Hum. Gene Ther.* **2018,** *29* (9), 1004–1010.
33. Yang, H. W.; Ye, L.; Guo, X. D.; Yang, C.; Compans, R. W.; Prausnitz, M. R. Ebola Vaccination Using a DNA Vaccine Coated on PLGA-PLL/γPGA Nanoparticles Administered Using a Microneedle Patch. *Adv. Healthcare Mater.* **2017,** *6* (1), 1600750.
34. Li, Y.; Hu, X.; Dong, Z.; Chen, Y.; Zhao, W.; Wang, Y.; Zhang, L.; Chen, M.; Wu, C.; Wang, Q. Dissolving Microneedle Arrays with Optimized Needle Geometry for Transcutaneous Immunization. *Eur. J. Pharmaceut. Sci.* **2020,** *151,* 105361.
35. Makvandi, P.; Kirkby, M.; Hutton, A. R.; Shabani, M.; Yiu, C. K.; Baghbantaraghdari, Z.; Jamaledin, R.; Carlotti, M.; Mazzolai, B.; Mattoli, V. Engineering Microneedle Patches for Improved Penetration: Analysis, Skin Models and Factors Affecting Needle Insertion. *Nano-Micro Lett.* **2021,** *13,* 1–41.
36. Römgens, A. M.; Bader, D. L.; Bouwstra, J. A.; Oomens, C. W. Predicting the Optimal Geometry of Microneedles and Their Array for Dermal Vaccination Using a Computational Model. *Comput. Methods Biomech. Biomed. Eng.* **2016,** *19* (15), 1599–1609.
37. Li, Q.; Xu, R.; Fan, H.; Xu, J.; Xu, Y.; Cao, P.; Zhang, Y.; Liang, T.; Zhang, Y.; Chen, W. Smart Mushroom-Inspired Imprintable and Lightly Detachable (MILD) Microneedle Patterns for Effective COVID-19 Vaccination and Decentralized Information Storage. *ACS Nano* **2022,** *16* (5), 7512–7524.
38. Di Pasquale, A.; Preiss, S.; Tavares Da Silva, F.; Garçon, N. Vaccine Adjuvants: from 1920 to 2015 and beyond. *Vaccines* **2015,** *3* (2), 320–343.
39. Leone, M.; Mönkäre, J.; Bouwstra, J. A.; Kersten, G. Dissolving Microneedle Patches for Dermal Vaccination. *Pharm. Res.* **2017,** *34,* 2223–2240.
40. Park, S.; Lee, Y.; Kwon, Y.-M.; Lee, Y.-T.; Kim, K.-H.; Ko, E.-J.; Jung, J. H.; Song, M.; Graham, B.; Prausnitz, M. R. Vaccination by Microneedle Patch with Inactivated Respiratory Syncytial Virus and Monophosphoryl Lipid A Enhances the Protective Efficacy and Diminishes Inflammatory Disease after Challenge. *PLoS One* **2018,** *13* (10), e0205071.
41. Weldon, W. C.; Zarnitsyn, V. G.; Esser, E. S.; Taherbhai, M. T.; Koutsonanos, D. G.; Vassilieva, E. V.; Skountzou, I.; Prausnitz, M. R.; Compans, R. W. Effect of Adjuvants on

Responses to Skin Immunization by Microneedles Coated with Influenza Subunit Vaccine. *PLoS One* **2012,** *7* (7), e41501.

42. Qiu, Y.; Guo, L.; Zhang, S.; Xu, B.; Gao, Y.; Hu, Y.; Hou, J.; Bai, B.; Shen, H.; Mao, P. DNA-Based Vaccination against Hepatitis B Virus Using Dissolving Microneedle Arrays Adjuvanted by Cationic Liposomes and CpG ODN. *Drug Deliv.* **2016,** *23* (7), 2391–2398.
43. Tamayo, I.; Irache, J. M.; Mansilla, C.; Ochoa-Repáraz, J.; Lasarte, J. J.; Gamazo, C. Poly (Anhydride) Nanoparticles Act as Active Th1 Adjuvants through Toll-like Receptor Exploitation. *Clin. Vaccine Immunol.* **2010,** *17* (9), 1356–1362.
44. Jacobson, R. M.; Swan, A.; Adegbenro, A.; Ludington, S. L.; Wollan, P. C. GA Poland, VR Group. Making Vaccines More Acceptable—Methods to Prevent and Minimize Pain and Other Common Adverse Events Associated with Vaccines. *Vaccine* **2001,** *19* (17–19), 2418–2427.
45. Ding, P.; Liu, H.; Zhu, X.; Chen, Y.; Zhou, J.; Chai, S.; Wang, A.; Zhang, G. Thiolated Chitosan Encapsulation Constituted Mucoadhesive Nanovaccine Confers Broad Protection against Divergent Influenza A Viruses. *Carbohydr. Polym.* **2024,** *328,* 121689.
46. Han, T.; Das, D. B. Permeability Enhancement for Transdermal Delivery of Large Molecule Using Low-Frequency Sonophoresis Combined with Microneedles. *J. Pharmaceut. Sci.* **2013,** *102* (10), 3614–3622.
47. Yan, K.; Todo, H.; Sugibayashi, K. Transdermal Drug Delivery by In-Skin Electroporation Using a Microneedle Array. *Int. J. Pharm.* **2010,** *397* (1–2), 77–83.
48. Wu, X.-M.; Todo, H.; Sugibayashi, K. Enhancement of Skin Permeation of High Molecular Compounds by a Combination of Microneedle Pretreatment and Iontophoresis. *J. Contr. Release* **2007,** *118* (2), 189–195.
49. Petchsangsai, M.; Rojanarata, T.; Opanasopit, P.; Ngawhirunpat, T. The Combination of Microneedles with Electroporation and Sonophoresis to Enhance Hydrophilic Macromolecule Skin Penetration. *Biol. Pharm. Bull.* **2014,** *37* (8), 1373–1382.
50. Zhang, L.; Guo, R.; Wang, S.; Yang, X.; Ling, G.; Zhang, P. Fabrication, Evaluation and Applications of Dissolving Microneedles. *Int. J. Pharm.* **2021,** *604,* 120749.
51. Chen, H.; Wu, B.; Zhang, M.; Yang, P.; Yang, B.; Qin, W.; Wang, Q.; Wen, X.; Chen, M.; Quan, G. A Novel Scalable Fabrication Process for the Production of Dissolving Microneedle Arrays. *Drug Deliv. Transl. Res.* **2019,** *9,* 240–248.
52. Li, R.-R.; Wang, Y.; Liu, Z.; Xiu, X-l; Liu, Y.; Wang, Y-n. Influencing Factors and Evaluation Methods of Skin Microchannels Formation and Closure after Microneedles Application. *Acta Pharm. Sin.* **2021,** 1293–1300.
53. Nguyen, T. T.; Oh, Y.; Kim, Y.; Shin, Y.; Baek, S.-K.; Park, J.-H. Progress in Microneedle Array Patch (MAP) for Vaccine Delivery. *Hum. Vaccines Immunother.* **2021,** *17* (1), 316–327.

Chapter 10

Microneedle technology for immune-mediated inflammatory disease therapy

Chunxian Zhou[a], **Mingyu Gong**[a] **and Tingting Peng**
State Key Laboratory of Bioactive Molecules and Druggability Assessment/International Cooperative Laboratory of Traditional Chinese Medicine Modernization and Innovative Drug Development of Ministry of Education (MOE) of China/College of Pharmacy, Jinan University, Guangzhou, China

1. Introduction

Immune-mediated inflammatory diseases (IMIDs) comprise a group of systemic disorders with complex, multifactorial etiologies that commonly result in terminal organ damage.[1] These conditions are driven by shared inflammatory pathways, which may either originate from or contribute to dysregulation of both innate and adaptive immune responses.[2] Although the precise etiology of IMIDs remains elusive, genetic susceptibility and environmental triggers such as trauma, infections, and allergens are recognized as key contributing factors.[3] IMIDs affect 5%–7% of populations in developed Western countries and are increasingly observed among immigrants from developing countries.[4] These diseases can impair any organ system, leading to severe morbidity, markedly reduced quality of life, and shortened lifespan. The most common IMIDs include inflammatory skin disorders (e.g., psoriasis, atopic dermatitis), hypersensitivity reactions (e.g., allergic rhinitis, asthma), osteoarthropathies (e.g., rheumatoid arthritis, osteoarthritis), oral diseases (e.g., oral ulcers, periodontitis), and type 1 diabetes (T1D).

Inflammation, as the central link in the pathogenesis and clinical manifestations of various IMIDs, leads to the treatment strategies for different IMIDs exhibiting the characteristic of "treating different diseases with a unified approach". Extensive research has demonstrated that early and effective

[a] Authors contributed equally.

Microneedle Technology. https://doi.org/10.1016/B978-0-443-45544-5.01010-1

control of the inflammatory response, along with the restoration of immune balance, is critical for improving the long-term prognosis of patients with IMIDs.[5] Over the past 4 decades, therapeutic approaches for IMIDs have evolved iteratively, progressing from glucocorticoids and broad-spectrum immunomodulators to biologics and subsequently small-molecule targeted drugs, thereby providing the potential to realize this treatment paradigm15. Nevertheless, while current therapies can alleviate symptoms and delay disease progression to some extent, traditional drug administration methods (including oral, transdermal, and injectable routes) often encounter challenges such as difficulties in overcoming drug delivery barriers or insufficient targeting. Consequently, these methods generally face limitations, including low drug delivery efficiency, poor patient compliance, and significant systemic side effects.[6–8] Therefore, the development of a novel drug delivery system that combines efficient delivery capabilities with high patient compliance is essential for enhancing the therapeutic efficacy of IMIDs.

MN, as an innovative transdermal drug delivery system, possesses the dual advantages of both transdermal and injectable drug delivery. They can penetrate the stratum corneum barrier to precisely deliver drugs to targeted immune cells within skin tissue, thereby modulating systemic immune responses. Alternatively, they can promote the transdermal absorption of drugs into the bloodstream, exerting systemic immune regulatory effects.[9] Unlike conventional drug administration methods, MNs impose no strict limitations on the molecular weight, hydrophilicity, or hydrophobicity, and particle size of the therapeutics.[10] Based on these characteristics, MNs exhibit broad application prospects in the treatment of IMIDs.

In recent years, with a deeper understanding of the pathogenesis of IMIDs, treatment strategies have evolved from single-pathway blocking to multi-target collaborative intervention.[11] To circumvent potential side effects associated with traditional drug therapies, novel treatment modalities such as nanoenzymes,[12] probiotics,[13] exosomes,[14] and antigen inoculation[15] are garnering significant attention within the academic community. This chapter will comprehensively review the latest advancements in MN technology for the treatment of IMIDs, with a focus on the delivery of various therapeutic agents for managing inflammatory skin disorders, hypersensitivity reactions, osteoarthropathies, oral diseases, and T1D.

2. Inflammatory skin diseases

2.1 Psoriasis

Psoriasis is a chronic, recurrent inflammatory skin disease characterized by localized or widespread erythematous plaques covered with silvery-white scales.[16] Its pathogenesis involves abnormal activation of immune cells, such as dendritic cells and T cells, which release proinflammatory cytokines (e.g., TNF-α,

IL-17, IL-23, IFN-γ). These cytokines drive inflammation and induce excessive keratinocyte proliferation and abnormal differentiation, leading to the formation of psoriatic lesions.[17] Topical or systemic immunosuppressants are the mainstay treatment for psoriasis, but their low delivery efficiency often necessitates frequent administration to achieve therapeutic efficacy.[18] However, long-term use of immunosuppressants causes severe side effects such as skin atrophy, local irritation, bone marrow suppression, and nephrotoxicity. As a novel transdermal drug delivery system (TDDS), MNs may offer a promising alternative by mitigating these drawbacks and improving therapeutic outcomes.

2.1.1 Drug therapy

Calcipotriol, a vitamin D3 analog, is widely employed in the topical treatment of psoriasis. However, its therapeutic efficacy is limited by the poor transdermal penetration through the stratum corneum barrier. To enhance calcipotriol's skin permeation, Liang et al.[19] fabricated polylactic acid-based solid MNs via a hot-pressing method. In a psoriasis-like mouse model, pretreatment with these MNs significantly improved the transdermal delivery efficacy of calcipotriol. Compared with direct administration of calcipotriol cream, MN pretreatment demonstrated superior therapeutic outcomes, including more pronounced reductions in inflammatory cell infiltration, epidermal thickness, and clinical manifestations of erythema and scaling.

Tacrolimus is an immunosuppressant used to treat psoriasis, but its transdermal permeability is limited by the thickened epidermis of psoriatic lesions, and its poor water solubility further restricts its clinical application. To overcome these limitations, Men et al.[20] developed tacrolimus nanocrystals via a bottom-up approach to improve solubility and incorporated them into hyaluronate-based MNs for plaque psoriasis treatment. The tacrolimus-loaded MNs demonstrated significantly higher intradermal retention (8.40 ± 0.33 μg/cm^2) than commercial ointments (1.40 ± 0.12 μg/cm^2). In a psoriatic mouse model, the nanocrystal-loaded MNs markedly reduced the Psoriasis Area and Severity Index (PASI) score and alleviated histopathological symptoms (erythema, desquamation, and immune infiltration). Additionally, they downregulated proinflammatory cytokines (TNF-α, IL-17A, and IL-23). Notably, epidermal thickness in MN-treated mice (31.93 ± 0.44 μm) was significantly lower than that of the commercial ointment (57.57 ± 1.53 μm). These findings demonstrate that tacrolimus nanocrystal-loaded MNs enhance the transdermal delivery efficiency of tacrolimus, offering a promising strategy for improved psoriasis treatment.

Methotrexate (MTX) is a first-line medication for treating moderate to severe psoriasis. However, oral or injectable MTX often causes adverse effects, including nausea, vomiting, myelosuppression, and hepatotoxicity. Transdermal delivery of MTX could help mitigate these side effects, but its application is limited by the drug's poor water solubility and dissociation at physiological pH. To overcome these challenges, Du et al.[21] developed MTX-loaded dissolving MNs for the treatment of psoriasis. The MTX-loaded MNs demonstrated significant inhibitory effects on HaCaT cell proliferation. In a psoriatic mouse model, the MNs effectively alleviated symptoms, reducing epidermal hyperplasia and suppressing the production of proinflammatory cytokines (IL-23 and IL-17). Notably, the MNs achieved comparable therapeutic efficacy to oral MTX but at a lower dose, thereby minimizing side effects. These findings suggest that MTX-loaded MNs could represent a promising therapy for psoriasis.

2.1.2 Novel treatments

Recently, studies have explored combining MNs with emerging therapeutic strategies. Monoclonal antibodies (mAbs), as promising biologic agents, are gaining attention for psoriasis treatment. However, their topical delivery via conventional transdermal preparations to inflamed skin remains challenging due to their large molecular weight and high hydrophilicity. To address this limitation, Wu et al.[22] developed a photothermal MN system coloaded with IL-17 mAbs and MXene for targeted psoriasis therapy (Fig. 10.1A). The IL-17 mAbs selectively bind to IL-17, blocking its interaction with receptors and thereby suppressing IL-17-driven inflammation, which alleviates characteristic psoriatic symptoms such as scaling and erythema. MXene, a graphene-like two-dimensional material, exhibits exceptional photothermal conversion under near-infrared irradiation, enabling rapid MN dissolution and subsequent release of IL-17 mAbs into the dermal layer of psoriatic lesions. In an imiquimod-induced mouse model, the photothermal MN demonstrated significant antiinflammatory effects and markedly improved psoriasis-like symptoms.

Regulatory T cell (Treg) therapy represents a promising therapeutic approach for psoriasis by modulating immune homeostasis and restoring immune tolerance. However, systemic Treg delivery faces several limitations, including cell instability, poor tissue targeting, and insufficient accumulation in psoriatic lesions. To address these challenges, Zhang et al.[23] developed perforated MNs with channels on the shell to encapsulate Treg cells in the core for the amelioration of psoriasis (Fig. 10.1B). The MNs' hydrophobic shell protected the cells from nutrient depletion and external harmful factors, while surface channels facilitated cell migration. Additionally, the Treg cells-loaded gel in the core of MNs ensured cell viability and functionality. The MN matrix, composed of poly (vinyl propionate-co-methyl methacrylate), underwent enzymatic degradation in highly inflammatory psoriatic regions, releasing fatty acids. These fatty acids enhanced Treg suppression through fatty acid oxidation-mediated metabolic reprogramming. Pharmacodynamic studies demonstrated that Treg-loaded MNs significantly reduced leukocyte infiltration, suppressed inflammatory cytokine expression, and alleviated psoriasis symptoms. These findings highlight the potential of MN-mediated cell delivery for psoriasis therapy.

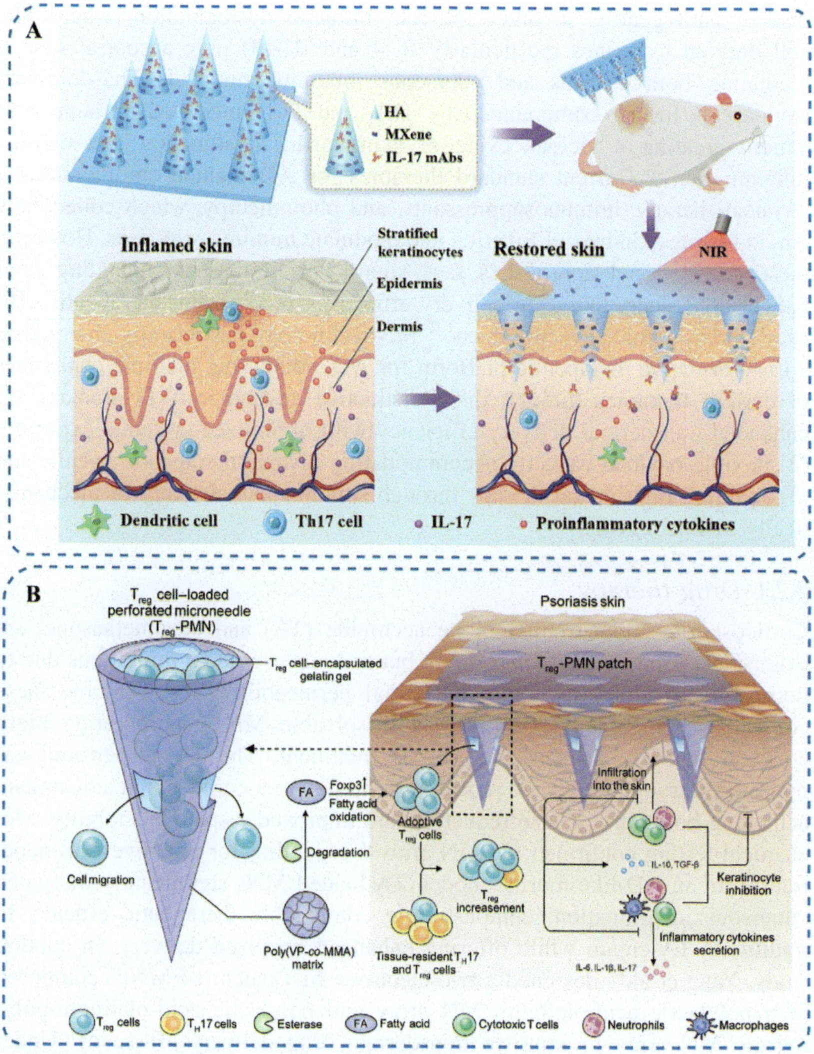

FIGURE 10.1 (A) Schematic illustration of a photothermal-responsive microneedle (MN) patch loaded with mAbs for psoriasis treatment. (B) Design and therapeutic mechanism of perforated microneedle (MN)-mediated adoptive Treg therapy for psoriasis. *(A) Reproduced from Ref. 22 with permission of **Wiley**, Copyright (**2022**). (B) Reproduced from Ref. 23 with permission of **AAAS**, Copyright (**2023**).*

2.2 Atopic dermatitis

Atopic dermatitis (AD) is a chronic, relapsing inflammatory skin disease. The typical manifestations of AD include xerosis, intense pruritus, and recurrent eczematous lesions, which significantly impair patients' quality of life. The disease pathogenesis involves intricate immune dysregulation, where Th2 cell–derived cytokines (particularly IL-4 and IL-13) play a central role in mediating both pruritus and cutaneous inflammation. This Th2-dominant response is further compounded by Th1- and Th17-mediated immune activation, creating a vicious cycle of skin barrier dysfunction and chronic inflammation.[24] Current standard therapies for AD include emollient-based physical therapy, immunosuppressants, and phototherapy, which collectively aim to restore skin barrier function and modulate immune responses. However, these conventional approaches face significant limitations, including poor patient adherence, low drug delivery efficiency, inconsistent therapeutic efficacy, and systemic adverse effects.[25] MN technology has recently emerged as a transformative therapeutic platform for AD, addressing multiple challenges of current therapies through three distinctive advantages: (1) dramatically enhanced transdermal delivery efficiency with minimized systemic exposure, (2) flexible payload capacity accommodating diverse therapeutic agents, and (3) enabling multimodal therapy through simultaneous delivery of mechanistically distinct agents.

2.2.1 Drug therapy

Corticosteroids, like triamcinolone acetonide (TA) and dexamethasone, are extensively used in AD management but encounter clinical limitations due to poor aqueous solubility and transdermal permeability. To overcome these constraints, Jang et al.[26] engineered dissolvable MNs incorporating high-concentration TA suspensions for AD treatment. The TA suspension was optimized through ultrasonic processing and polymer composition adjustment, achieving both particle size reduction and improved dispersion stability. Mechanical testing confirmed the MN arrays' capability for effective skin penetration. In an AD-like murine model, TA-loaded MNs significantly mitigated cutaneous inflammation, demonstrating comparable therapeutic efficacy to traditional TA cream while offering enhanced localized delivery. In another study, Yang et al.[27] designed a two-electrode MN patch (t-EMNP) composed of a polylactic acid-platinum MN array and polylactic acid-platinum-polypyrrole MN array to improve transdermal drug delivery efficiency. Under electrical stimulation, the system enabled controlled drug release into the skin, achieving on-demand transdermal delivery. Pharmacodynamic results demonstrated that the t-EMNP outperformed dexamethasone cream, delivering a higher drug dose into the skin and more effectively alleviating AD symptoms.

Poly-γ-glutamate (γ-PGA) is a biopolymer that can inhibit the production of dendritic cell–derived Th1 cytokines, showing immunomodulatory potential for AD treatment. Hence, Chen et al.[28] developed dissolvable γ-PGA MNs targeted delivery of γ-PGA to the dendritic cell-rich dermal layer (Fig. 10.2A). In murine AD models, 8-week MN treatment significantly reduced serum

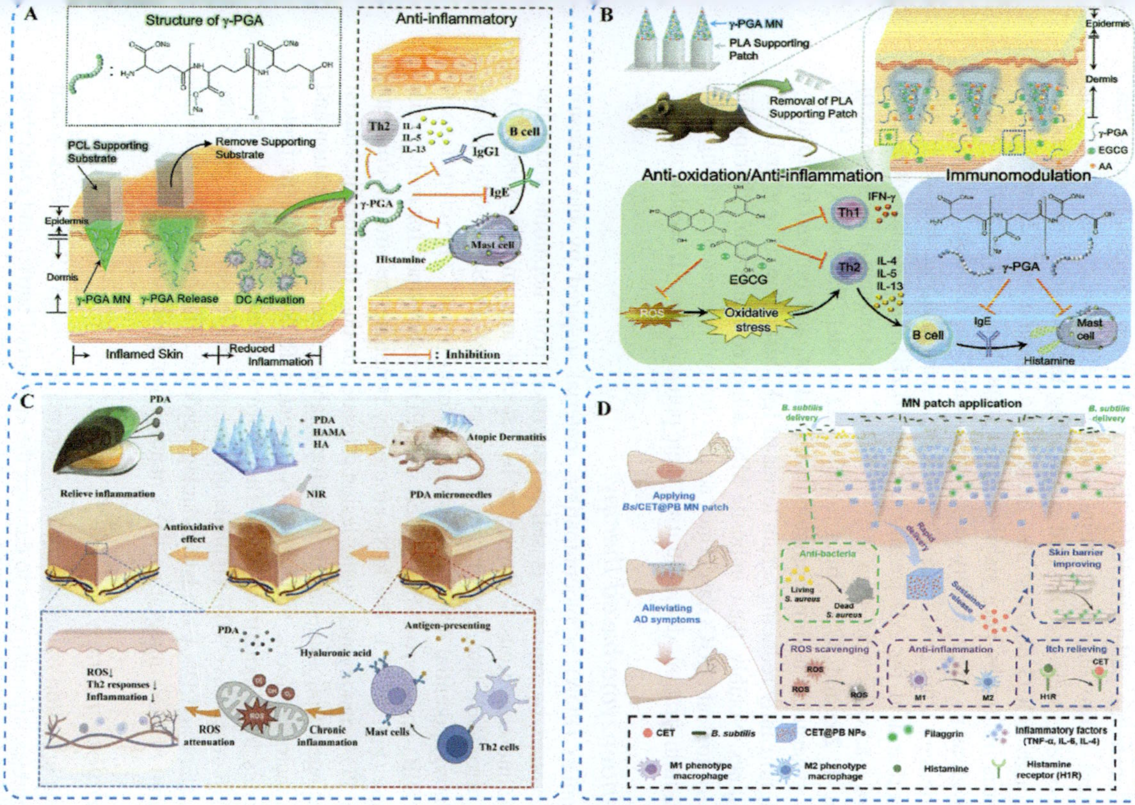

FIGURE 10.2 (A) Schematic illustration showing how poly-γ-glutamate (γ-PGA) microneedles (MNs) alleviate skin inflammation by downregulating immunoglobulin E (IgE) and immunoglobulin G1 (IgG1) production while reducing mast cell infiltration. (B) Schematic illustration of epigallocatechin gallate/L-ascorbic acid (EGCG/L-AA)–loaded γ-PGA MNs ameliorating atopic dermatitis (AD)-like symptoms in mice. (C) Schematic illustration of polydopamine (PDA) nanozymes–loaded MNs for the treatment of AD, which reduce reactive oxygen species (ROS) and oxidative stress in the affected region while suppressing inflammatory responses. (D) Schematic illustration of the bilayer MNs for AD treatment, which exhibit multiple therapeutic functions including antibacterial activity, ROS scavenging, antiinflammatory effects, itch relief, and skin barrier improvement. *(A) Reproduced from Ref. 28 with permission of* ***Elsevier****, Copyright (****2020****). (B) Reproduced from Ref. 30 with permission of* ***Elsevier****, Copyright (****2021****). (C) Reproduced from Ref. 32 with permission of* ***Elsevier****, Copyright (2024). (D) Reproduced from Ref. 35 with permission of* ***Wiley****, Copyright (2025).*

immunoglobulin (IgE and IgG1) levels while improving clinical symptoms, including epidermal hyperplasia, mast cell infiltration, and skin barrier dysfunction. Notably, high-molecular-weight γ-PGA (1100 kDa) MNs showed better therapeutic efficacy than their low-molecular-weight (200–400 kDa) counterparts.

Epigallocatechin gallate (EGCG), the most abundant catechin in green tea, exerts potent antioxidant and antiinflammatory effects through scavenging reactive oxygen species (ROS) and suppressing inflammatory factor production. Although EGCG demonstrates therapeutic potential for AD, its clinical translation is limited by poor stability under light, heat, and oxidative conditions, as well as its susceptibility to gastrointestinal degradation.[29] To address these challenges, Chiu et al.[30] developed γ-PGA-based MNs coloading EGCG and L-ascorbic acid for AD treatment (Fig. 10.2B). Due to the antioxidant property of L-ascorbic acid, over 90% of EGCG was retained after 4 weeks of storage at 4, 25, and 37°C. In a murine AD model, the composite MNs exhibited synergistic antiinflammatory, antioxidant, and immunomodulatory effects, significantly reducing serum levels of IgE, histamine, IFN-γ, and Th2-type cytokines. Notably, the composite MNs achieved comparable therapeutic efficacy to topical EGCG/ascorbic acid solutions despite requiring less frequent administration, demonstrating their potential for AD treatment.

2.2.2 Nanoenzyme-based therapy

Polydopamine (PDA), a biocompatible material inspired by the adhesion protein secreted by mussels, is self-polymerized from dopamine monomer under alkaline conditions.[31] Owing to its catechol and imine groups, PDA exhibits peroxidase-like activity, enabling effective ROS scavenging and demonstrating potential for AD treatment. Zhang et al.[32] developed near-infrared (NIR)-responsive MNs loaded with PDA nanozymes for multimodal AD therapy (Fig. 10.2C). The MNs efficiently penetrate the stratum corneum to deliver PDA nanozymes to the lesional tissue for dual therapeutic effects: (1) NIR-triggered photothermal conversion enabling bacterial eradication and antiinflammatory effects, and (2) PDA-mediated ROS clearance to protect keratinocytes from oxidative stress. The pharmacodynamic studies revealed that the NIR-responsive MNs significantly reduced the levels of Th2 cytokine, IgE infiltration and DNA oxidative damage product. These findings highlight the potential of MNs integrating PDA nanozyme for AD management.

2.2.3 Probiotic therapy

AD lesions are often colonized by *Staphylococcus aureus* (*S. aureus*), which further secretes superantigens to damage the epidermal barrier and exacerbate inflammation.[33] Although antibiotic eradication of *S. aureus* may offer therapeutic benefits for AD patients, frequent antibiotic use could increase the risk of bacterial resistance and disrupt the natural skin microbiome. Notably, studies suggest that the probiotic *Bacillus subtilis* (*B. subtilis*) can significantly reduce *S. aureus* colonization without significantly affecting commensal microbiota.[34] Building on this evidence, Zhang et al.[35] developed bilayer MNs coloaded with *B. subtilis* and cetirizine hydrochloride for AD treatment (Fig. 10.2D). The needle tips contained cetirizine hydrochloride-loaded Prussian blue nanoparticles, while the backing layer was embedded with *B. subtilis*. Upon skin insertion, the MNs rapidly dissolved, enabling sustained cetirizine release for over 2 weeks, while *B. subtilis* colonized the skin for more than 9 days. Pharmacodynamic studies in AD murine models demonstrated that the nanoparticles scavenged ROS, alleviating oxidative stress and inflammation while promoting skin barrier repair. Furthermore, *B. subtilis* suppressed *S. aureus* biofilm formation on AD lesions, thereby restoring microbial balance. The multifunctional MNs also demonstrated excellent biosafety, highlighting the potential of probiotic-integrated MNs as a promising long-term therapeutic approach for AD management.

3. Hypersensitivity reaction

Anaphylaxis disorder refers to a spectrum of clinical conditions induced by an aberrant immune response to typically harmless exogenous substances or allergens, such as pollen, dust, tobacco smoke, and animal dander.[36,37] The pathogenesis involves IgE-mediated degranulation of mast cells and basophils upon antigen exposure, leading to the release of inflammatory mediators including histamine, leukotrienes, and platelet-activating factor, which ultimately induces a type I hypersensitivity reaction.[38] Both allergic rhinitis and allergic asthma are common type I hypersensitivity-mediated disorders. Allergen-specific immunotherapy (AIT) represents a disease-modifying treatment that alters the immune system's response to allergens.[39] Although subcutaneous and sublingual immunotherapy are widely used AIT modalities, their clinical utility is frequently constrained by drawbacks including injection-related pain, low patient adherence, and variable therapeutic efficacy.[40] In this context, MN-mediated AIT has emerged as a promising minimally invasive alternative, garnering substantial attention in recent research.

3.1 Allergic rhinitis

Allergic rhinitis is clinically characterized by nasal obstruction, paroxysmal sneezing, watery rhinorrhea, and postnasal drip, significantly impairing patients' quality of life. For individuals who do not respond adequately to pharmacotherapy, allergen immunotherapy (AIT) presents a disease-modifying alternative.[41] To overcome the limitations of transdermal delivery in epicutaneous AIT, Choi et al.[42] developed a coated MN system loaded with house dust mite (HDM) antigen for allergic rhinitis treatment. In vitro skin permeation studies demonstrated efficient allergen deposition within the skin layer by bypassing the stratum corneum, while pharmacodynamic studies in mice showed reduced IgE levels and nasal eosinophil infiltration after administration of 0.1–0.5 μg HDM-loaded MNs. These findings suggest that MN-mediated AIT could represent a promising therapeutic strategy for allergic rhinitis.

3.2 Allergic asthma

Allergic asthma is characterized by heightened airway responsiveness to various allergens and environmental irritants, manifesting as recurrent wheezing, shortness of breath, chest tightness, and coughing. Conventional subcutaneous AIT faces two major limitations: the risk of systemic hypersensitivity and suboptimal patient compliance. To overcome these challenges, Park et al.[43] developed biodegradable MNs loaded with HDM allergens for asthma treatment. In murine studies, transdermal delivery of HDM via MNs demonstrated superior therapeutic efficacy to subcutaneous injection, significantly reducing HDM-specific IgE levels while promoting regulatory T-cell (Treg) responses and Th1 cytokine induction. Safety assessments confirmed excellent cutaneous tolerability with no systemic adverse effects, suggesting that MN-based AIT could be a promising therapeutic option for asthma. Similarly, Shakya et al.[44] fabricated MNs coated with ovalbumin and cytosine-phosphodiester-guanosine (CpG) for airway allergy immunotherapy. Pharmacodynamic evaluations in murine airway allergy models demonstrated that MN-mediated immunotherapy significantly increased the levels of ovalbumin-specific IgG and IL-10, while reducing the levels of IL-5/IL-13 cytokines and pulmonary eosinophil infiltration, producing comparative immunoregulatory effects to subcutaneous administration. Taken together, these findings highlight the dual advantages of MN-based AIT in effective immunomodulation and enhanced treatment adherence, positioning it as a clinically viable, safer alternative to conventional allergy management strategies.

4. Osteoarthropathy

4.1 Rheumatoid arthritis

Rheumatoid arthritis (RA) is a chronic autoimmune disorder characterized by symmetrical polyarticular inflammation, progressive joint destruction, and systemic complications. Clinically, it manifests as persistent joint swelling, pain, and irreversible damage to articular cartilage and subchondral bone, often leading to significant functional impairment and reduced quality of life. While the exact etiology remains elusive, current understanding implicates dysregulated adaptive immune responses, particularly synovium-targeted autoreactive T-cell and B-cell activity.[45]

The first-line pharmacological management of RA primarily involves two approaches: (1) nonsteroidal antiinflammatory drugs (NSAIDs; e.g., naproxen, ibuprofen) for symptomatic relief, and (2) glucocorticoids (e.g., prednisolone, dexamethasone) for immunosuppression. However, these therapies present significant safety concerns: NSAIDs are associated with gastrointestinal complications (e.g., peptic ulcers, hemorrhage), while prolonged glucocorticoid use may induce metabolic disorders (e.g., osteoporosis, steroid-induced myopathy) and Cushing's syndrome. These limitations are particularly problematic for elderly patients with comorbidities, highlighting the critical need for novel therapeutics with enhanced efficacy and safety profiles.[46]

As a first-line treatment for RA, methotrexate (MTX) exerts its therapeutic effects through two primary mechanisms: (1) suppression of proinflammatory cytokine production (e.g., TNF-α, IL-1β) in monocytes and macrophages, and (2) inhibition of synovial hyperplasia and osteoclastogenesis.[47,48] oral administration remains the most common route, its clinical application is limited by inconsistent intestinal absorption and dose-dependent gastrointestinal toxicity. While subcutaneous injections offer an alternative, they present challenges such as injection-site reactions and medical waste generation. To address these limitations, Tekko et al.[49] developed a hydrogel-forming MN system composed of poly(vinyl alcohol) and poly(vinyl pyrrolidone), incorporating an MTX reservoir formulated with hydroxypropyl methylcellulose and glycerol (Fig. 10.3A). Ex vivo permeation studies demonstrated that the hydrogel MNs enabled sustained MTX release, achieving over 1000-fold increase in drug flux compared to the MTX reservoir alone. Pharmacokinetic evaluations in Sprague–Dawley rats revealed superior performance relative to oral administration, with approximately 20% higher systemic exposure and 40% lower peak plasma concentration. This steady plasma concentration-time profile reduces concentration-dependent cytotoxicity while maintaining therapeutic efficacy. Collectively, this study underscores MN-mediated transdermal delivery as a promising strategy to enhance RA treatment by improving dosing precision and patient compliance.

Etanercept, a recombinant human tumor necrosis factor (TNF) receptor–antibody fusion protein, is a biologic agent widely used in rheumatoid arthritis (RA) treatment. Conventional subcutaneous administration of this biomacromolecule, however, is associated with injection-related complications such as localized pain, infection risk, and suboptimal patient adherence. To address these challenges, Cao et al.[50] developed dissolving MNs fabricated from acrylate-modified hyaluronic acid (mHA) for transdermal etanercept delivery. The mHA-based MNs exhibited significantly enhanced mechanical strength following UV cross-linking, facilitating robust skin penetration and efficient drug release. In a murine RA model, etanercept-loaded MNs achieved therapeutic outcomes comparable to subcutaneous injection, as demonstrated by attenuated paw edema and decreased serum levels of proinflammatory cytokines (TNF-α and IL-6). This MN-based platform represents a promising

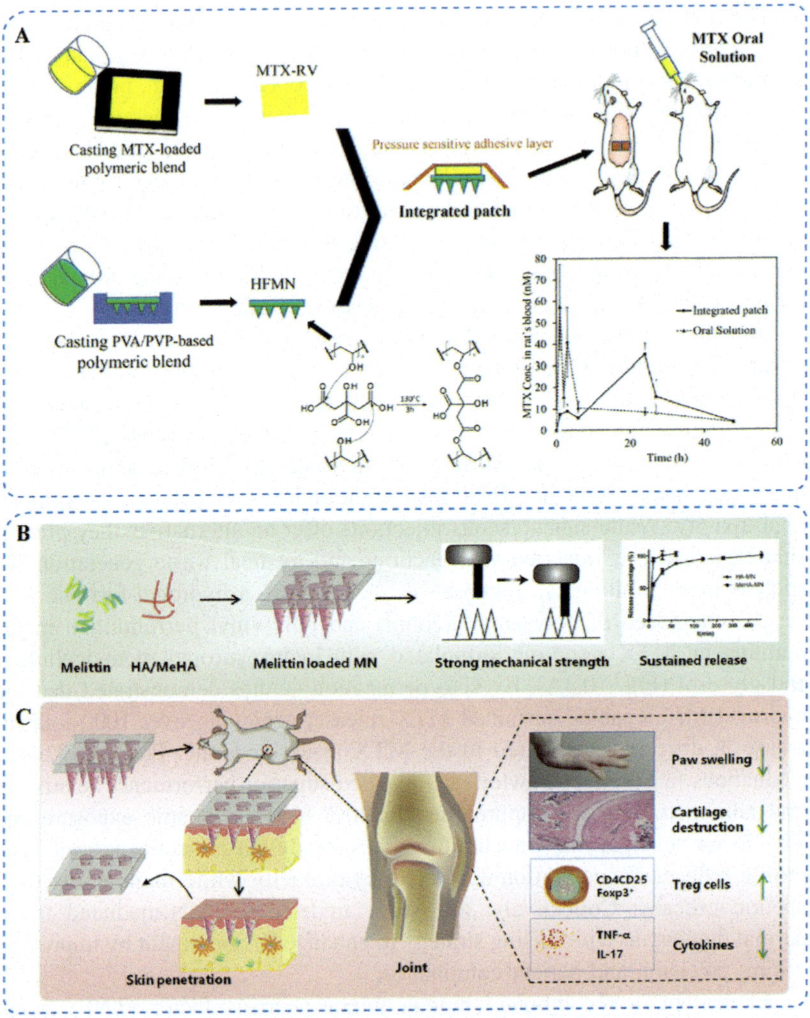

FIGURE 10.3 (A) Development of poly (vinyl alcohol)/poly (vinyl pyrrolidone)-based hydrogel-forming microneedles (MNs) for enhanced and sustained transdermal delivery of methotrexate (MTX). (B) Development of melittin-loaded methacrylate (MeHA) MNs with enhanced mechanical strength and sustained drug release behavior for rheumatoid arthritis (RA) treatment. (C) Therapeutic mechanism of melittin-loaded MeHA MNs to treat RA. *(A) Reproduced from Ref. 49 with permission of* ***Elsevier****, Copyright (****2020****). (C) Reproduced from Ref. 51 with permission of* ***Elsevier****, Copyright (****2021****).*

noninvasive alternative to conventional injections, potentially improving treatment adherence in RA therapy.

Melittin, a key bioactive component of bee venom, exhibits promising therapeutic potential for RA by modulating immune responses and suppressing proinflammatory cytokine release. However, its intravenous administration is clinically limited by hemolytic risks, necessitating alternative delivery approaches such as transdermal systems. Recently, Du et al.[51] developed photo-crosslinked hyaluronic acid methacrylate (MeHA) MNs for sustained melittin release (Fig. 10.3B and C). The photo-crosslinking process not only enhanced the mechanical strength of the MNs but also prolonged drug release, leading to significantly improved therapeutic efficacy in an arthritis mouse model. Compared to noncrosslinked controls, the photo-crosslinked MNs demonstrated superior therapeutic outcomes, including a 50% reduction in paw swelling (vs. 30%), increased splenic regulatory T (Treg) cell populations, reduced inflammatory cell infiltration, and decreased TNF-α levels in paw tissues. Importantly, the system showed no significant hematologic toxicity. These results underscore the potential of crosslinked MNs as a safe and effective transdermal delivery platform for melittin in RA treatment.

Neurotoxin, a short-chain postsynaptic α-neurotoxin and the major peptide component of *Naja naja atra* venom, has been traditionally used in medicine for analgesia and arthritis treatment. However, frequent injections are typically required to achieve therapeutic effects, which significantly reduces patient compliance. Recently, Yao et al.[52] developed dissolving MNs for the transdermal delivery of neurotoxin in RA therapy. Pharmacodynamic evaluations demonstrated that a 15-day MN-mediated neurotoxin regimen significantly reduced serum IL-1β and TNF-α levels, decreased joint swelling, and alleviated ankle joint damage, highlighting its promising therapeutic potential for RA.

4.2 Osteoarthritis

Osteoarthritis (OA), a chronic degenerative joint disorder, is characterized by articular cartilage degradation and osteophyte formation. Clinically, it manifests as pain, stiffness, and joint deformity.[53] Current pharmacological treatments for OA primarily fall into three categories: (1) oral or topical NSAIDs (e.g., diclofenac sodium, celecoxib, ibuprofen), (2) oral analgesics (e.g., tramadol), and (3) intraarticular corticosteroid injections (e.g., prednisone). However, prolonged use of topical NSAIDs may cause cutaneous irritation (e.g., rash, pruritus), while oral NSAIDs carry gastrointestinal and cardiovascular risks. Long-term corticosteroid use can also exacerbate osteoporosis and hyperglycemia.[54] MN technology addresses these limitations by bypassing the stratum corneum to deliver drugs directly into joints, thereby reducing systemic exposure, enhancing local bioavailability, and improving therapeutic efficacy. These advantages have driven significant research advances in MN-based OA therapies.

To improve the therapeutic efficiency of NSAIDs for OA treatment, Chen et al.[55] developed lubricating MNs loaded with diclofenac (Fig. 10.4A and B). The needle tips were composed of hyaluronic acid covalently conjugated with

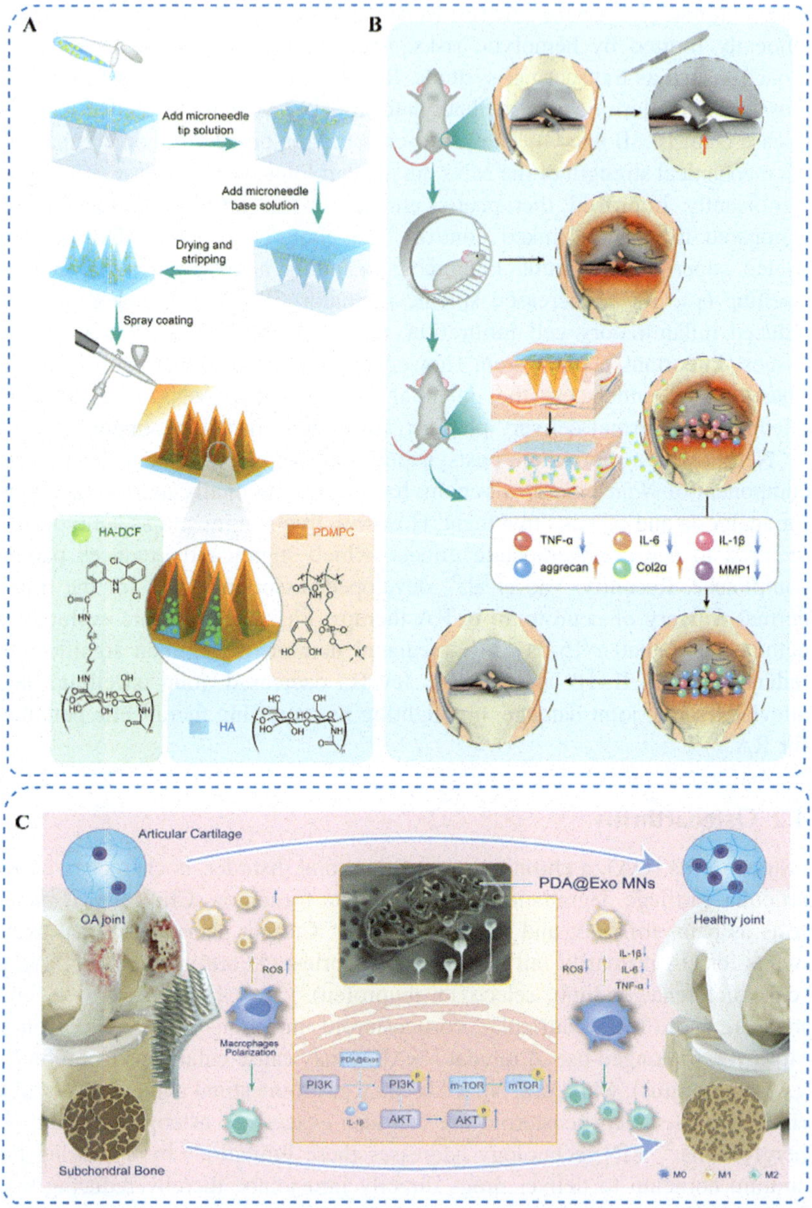

FIGURE 10.4 (A) Schematic illustration of the fabrication process for the bilayer dissolvable microneedles (MNs) system with self-adhesive lubrication coating. (B) Therapeutic mechanism of the lubricating MN system for osteoarthritis (OA) treatment, demonstrating enhanced antiinflammatory effects and reduced cartilage wear. (C) Development of polydopamine-coated exosome (PDA@Exo) MN to treat OA by inhibiting cartilage degradation and promoting osteogenesis and macrophage polarization via the PI3K–AKT–mTOR signaling pathway. *(B) Reproduced from Ref. 55 with permission of* ***Wiley****, Copyright* ***(2024)****. (C) Reproduced from Ref. 63 with permission of* ***Wiley****, Copyright* ***(2024)****.*

diclofenac, enabling sustained drug release. Additionally, the tips were coated with a self-adhesive lubricating copolymer to minimize MN-induced skin damage and further prolong diclofenac release. Mechanical and skin recovery studies confirmed that the lubricating MNs effectively penetrated the stratum corneum, delivering diclofenac while reducing skin irritation. In vitro and in vivo experiments demonstrated that this MN system significantly suppressed the release of proinflammatory cytokines (TNF-α, IL-6, and IL-1β) and alleviated articular cartilage degeneration. These findings suggest that the lubricating MN system is a promising, safe, and effective approach for OA therapy.

Oral celecoxib exhibits limited therapeutic efficacy in OA treatment due to its poor aqueous solubility. To address this challenge, Wang et al.[56] developed dissolving MNs loaded with celecoxib nanocrystals, which significantly enhanced both drug solubility and loading capacity. Pharmacodynamic evaluations revealed that these nanocrystal-based MNs achieved enhanced therapeutic outcomes at reduced doses compared to oral administration. Additionally, α-linolenic acid exhibits antiinflammatory and joint-protective properties by inhibiting proinflammatory factor release.[57] Li et al.[58] coencapsulated celecoxib and α-linolenic acid in a microemulsion to augment solubility, subsequently incorporating the formulation into dissolving MNs for efficient transdermal OA therapy. This dual-drug system synergistically promoted macrophage polarization from M1 to M2 phenotypes while attenuating chondrocyte apoptosis. In OA rat models, MN delivery at half the oral dose significantly alleviated paw edema, lowered proinflammatory cytokine levels, and attenuated cartilage degeneration, showcasing pronounced advantages over conventional oral therapy.

Triptolide, a promising therapeutic agent for OA, exerts its effects by suppressing key inflammatory cytokines (e.g., TNF-α, IL-1β, IL-6) and matrix metalloproteinases (MMPs), which are critical drivers of cartilage degradation.[59] However, its clinical translation is limited by severe systemic toxicity, poor solubility, and a short half-life. To address these challenges, Zhou et al.[60] developed a dissolving MN system incorporating triptolide-encapsulated liposomes for OA treatment. Encapsulation of triptolide within liposomes effectively prolonged drug release, resulting in a 30% reduction in cumulative drug release at 10 h. Moreover, the MN system significantly enhanced the transdermal permeability of triptolide-liposomes, achieving about 90% cumulative drug release rate within 10 h. Unlike conventional intraarticular injections, this MN system reduced injection-related pain, tissue trauma, and infection risks, thereby enhancing patient compliance. In vivo pharmacodynamic studies confirmed that the composite MNs effectively mitigated OA progression, as evidenced by decreased knee swelling, attenuated cartilage destruction, and lower inflammatory cytokine levels.

Mesenchymal stem cell–derived exosomes (MSC-Exos) are nanoscale vesicles secreted by MSCs through paracrine signaling. These exosomes contain bioactive molecules, such as proteins, lipids, mRNAs, and miRNAs, which confer antiinflammatory, immunomodulatory, and tissue-repair capabilities, positioning them as a promising therapeutic tool in regenerative medicine.[61] Studies have shown that MSC-Exos deliver specific miRNAs, including miR-136-5p and miR-320c, which upregulate type II collagen expression and inhibit MMP-13 activity, thereby maintaining chondrocyte extracellular matrix homeostasis and facilitating cartilage repair.[62] To harness this therapeutic potential, Li et al.[63] developed a core-shell MN patch loaded with MSC-Exos and PDA NPs for OA treatment (Fig. 10.4C). The PDA NPs incorporated into the MN shell efficiently scavenged reactive oxygen species (ROS), alleviating ROS-induced inflammation. Simultaneously, the MSC-Exos encapsulated in the MN core promoted chondrocyte proliferation and inhibited apoptosis. Together, the combined action of PDA NPs and MSC-Exos synergistically promoted M2-like macrophage polarization, attenuated inflammatory responses, and accelerated cartilage regeneration, ultimately mitigating OA progression.

Human placental preparations (HPPs) are regenerative products derived from placental tissue components, including growth factors, cytokines, and hormones.[64] These preparations demonstrate significant potential as a regenerative therapy for OA. Protein docking studies indicate that HPP ligands exhibit strong binding affinity to MMP13, suppressing its expression and thereby mitigating cartilage degeneration.[65] Currently, HPPs are typically administered via intraarterial injection for OA treatment, which can cause patient discomfort and carrier risks of infection. Additionally, the high molecular weight and hydrophilic nature of these biologics hinder their transdermal delivery. To address these challenges, James et al.[66] developed MNs loaded with lyophilized human placental extract (LHP). The LHP-MNs exhibited a sustained drug release profile in vitro, excellent biocompatibility, and antiinflammatory effects in both human dermal fibroblasts and IL-1β-induced synoviocytes. In OA rat models, LHP-MN treatment significantly reduced TNF-α and cartilage oligomeric matrix protein levels, alleviated knee/ankle joint damage, and preserved cartilage integrity. This study highlights the therapeutic potential of MN-based delivery systems for placental-derived biologics in OA regenerative therapy.

5. Oral disease

The oral mucosa has emerged as an attractive site for drug delivery owing to its distinct advantages. Its extensive surface area and highly vascularized structure facilitate efficient drug absorption while circumventing hepatic first-pass metabolism, thereby enhancing therapeutic effect.[67,68] Conventional dosage forms, such as mucoadhesive patches, buccal tablets, films, and hydrogels, have been widely explored. Nevertheless, challenges such as masticatory forces, enzymatic degradation, and salivary washout frequently compromise drug stability and delivery efficiency.[69,70] Furthermore, the oral mucosa's multilayered structure, comprising stratified squamous epithelium, basement membrane, lamina propria, and submucosa, poses a significant barrier to drug penetration, further limiting bioavailability.[71]

MNs address these limitations by creating transient microchannels that improve drug penetration and enhance localized drug retention. This approach has shown clinical promise in managing oral ulcers, periodontitis, and other oral mucosal disorders. The oral mucosa is also rich in immune components, with Langerhans cells (LCs) playing a key role in innate immunity and γδT cells facilitating adaptive immune responses.[72] Given this immunologically active environment, MNs can be particularly effective for localized treatments in mucosal inflammatory disease.[73]

5.1 Oral ulcers

Oral ulcers represent a prevalent oral disorder with multifactorial etiology, including trauma, infections, and immune dysfunction. These lesions manifest as localized epithelial defects or erosions in the oral mucosa, frequently causing substantial pain during mastication that significantly compromises patients' quality of life. Current therapeutic strategies predominantly rely on topical and systemic administration of antiinflammatory corticosteroids and antimicrobial agents. Nevertheless, corticosteroid therapy carries risks of adverse effects, including hyperglycemia and gastrointestinal disturbances, while excessive antibiotic use exacerbates the critical issue of antimicrobial resistance.[74,75]

MNs have emerged as a promising alternative for treating oral ulcers. For example, Meng et al.[76] developed a bilayered MN coloaded with betamethasone and lidocaine for oral ulcer therapy (Fig. 10.5). The MN design featured a photo-crosslinked HAMA shell encapsulating betamethasone and an HA core containing lidocaine. Upon application, the HA core rapidly dissolved, releasing lidocaine for immediate pain relief, while the HAMA shell enabled sustained betamethasone release for long-term antiinflammatory effects. In drug release studies, fluorescein isothiocyanate (FITC, mimicking lidocaine) was released faster than Rhodamine B (mimicking betamethasone), confirming the sequential release of the two drugs. Pharmacodynamic evaluations demonstrated that the core-shell MNs achieved superior therapeutic efficacy compared to monotherapy and commercial betamethasone, promoting complete healing of oral ulcers within 7 days. Additionally, both in vitro and in vivo biosafety assessments revealed no significant cytotoxicity or organ damage, highlighting the system's biocompatibility. In summary, the core-shell MNs with biphasic drug release offer a safe and effective strategy for targeted oral ulcer therapy.

Recently, Guo et al.[77] developed hyaluronic acid-based MNs (BSP-BDP@HA MNs) coloaded with betamethasone 21-phosphate sodium (BSP) and betamethasone 17,21-dipropionate (BDP) for oral ulcer treatment. The hydrophilic BSP enables rapid absorption at the administration site, while the

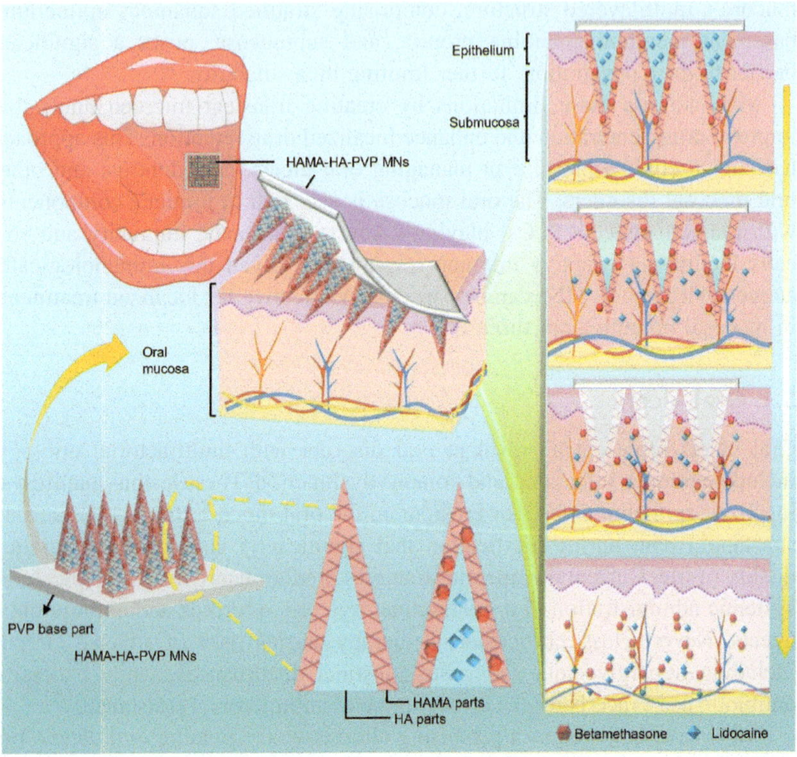

FIGURE 10.5 Schematic illustration of a core-shell microneedle (MN) patch designed for sequential release of lidocaine and betamethasone to treat oral ulcers. *Reproduced from Ref. 76 with permission of* ***ACS****, Copyright* ***(2023)****.*

lipophilic BDP provides sustained release. Both compounds metabolize into betamethasone, ensuring prolonged antiinflammatory effects at the target site.[78] Compared to the control group, BSP-BDP@HA MNs significantly promoted the proliferation of primary human gingival fibroblasts (hGFs) and suppressed TNF-α release in lipopolysaccharide (LPS)-induced hGFs, demonstrating their dual role in mitigating inflammation and enhancing wound healing. To evaluate therapeutic efficacy, a phenol-induced oral ulcer mouse model was established. The results showed that BSP-BDP@HA MNs outperformed both drug-loaded HA films and the commercial ointment Ning Zhi Zhu. The results demonstrated that BSP-BDP@HA MNs outperformed both drug-loaded HA films and the commercial ointment Ning Zhi Zhu. The MN-treated group exhibited robust collagen deposition, neovascularization, and an 87.4% ulcer healing rate. These results highlight the ability of MNs to enhance drug penetration, prolong mucosal retention, and accelerate tissue regeneration, underscoring their potential as an effective therapy for oral ulcers.

Triamcinolone acetonide (TA), a synthetic glucocorticoid, exerts its anti-inflammatory effects primarily through suppression of the nuclear factor kappa B (NF-κB) pathway and downstream inhibition of proinflammatory cytokines, including cyclooxygenase-2 (COX-2) and TNF-α.[79] Although clinically employed in oral ulcer management (e.g., as ointments or mouthwashes), TA's therapeutic potential is constrained by its poor aqueous solubility, resulting in suboptimal mucosal bioavailability.[80] To address this limitation, Qu et al.[81] developed an innovative drug delivery system by encapsulating TA within mesoporous polydopamine nanoparticles (TA@MPDA) through π–π stacking interactions, subsequently incorporating these nanoparticles into a hyaluronic acid/Bletilla striata polysaccharide (HA/BSP) dissolving MN patch (TA@MPDA-HA/BSP MN). The MPDA nanocarrier not only improves TA's local bioavailability through enhanced mucosal adhesion but also enables sustained drug release. Importantly, this system exhibits synergistic therapeutic effects: while MPDA nanoparticles possess intrinsic antiinflammatory and antioxidant properties, their combination with HA/BSP further potentiates wound healing and antiinflammatory responses in oral ulcer treatment. In vitro scratch assays demonstrated superior wound closure efficacy of TA@MPDA-HA/BSP MNs (69.40% ± 0.93%) compared to both drug-free MPDA-HA/BSP MNs (65.83% ± 1.24%) and untreated controls (54.01% ± 3.05%). The composite MN system also showed optimal performance in downregulating TNF-α expression in lipopolysaccharide (LPS)-stimulated human gingival fibroblasts (hGFs), with the integrated components exhibiting enhanced anti-inflammatory effect. Notably, in vivo studies revealed that TA@MPDA-HA/BSP MNs achieved superior ulcer healing compared to the commercial Ning Zhi Zhu ointment, despite utilizing only 10% of the conventional TA dosage. Collectively, this multifunctional MN platform represents a significant therapeutic advance in oral ulcer treatment.

5.2 Periodontitis

Periodontitis is a chronic inflammatory disease caused by dental plaque, which can destroy the supporting structures of the teeth, such as the periodontal ligament and alveolar bone.[82] The emergence of periodontitis can cause tooth loss and affect chewing functions, seriously impairing the quality of life and mental health of patients.[83] According to the global burden of disease study (GBD) statistics, approximately 1.1 billion individuals globally were afflicted with severe periodontitis in 2019, resulting in an incidence rate of 17%. To sum up, periodontitis imposes a substantial health and economic burden on society. At present, the main treatment of periodontitis is to eliminate pathogenic bacteria by oral antibiotics or periodontal flap surgery, which fails to effectively modulate the inflammatory microenvironment of periodontal tissue and prevent the recurrence of periodontitis.[84,85] Restoring the homeostasis of the periodontal immune microenvironment to enhance tissue regeneration could offer a promising therapeutic approach for periodontitis.

Interleukin-4 (IL-4) has been identified as a pivotal antiinflammatory mediator in the management of periodontitis, as it can induce the polarization of macrophages from the M1 to the M2 phenotype.[86,87] This shift helps to alleviate chronic inflammation and promote bone tissue repair. In addition, MNs can directly penetrate gingival tissues to prolong the retention of therapeutic agents in the local periodontal tissue, showing higher drug delivery efficiency than conventional oral tablets. Therefore, Zhang et al.[88] designed an immunomodulatory MN patch loaded with antibiotics and cytokines for periodontal tissue regeneration. Briefly, the MN patch consists of a rapidly dissolving gelatin baseplate and biodegradable gelatin methacryloyl needles. The needles are loaded with poly (lactic-co-glycolic acid) nanoparticles that encapsulate tetracycline and silica microparticles that encapsulate IL-4, providing continuous delivery of both antibiotics and cytokines. Lipopolysaccharide (LPS)-stimulated bone marrow-derived macrophages were used as a cellular model to evaluate the antiinflammatory effects of microneedles. Compared with drug-free microneedles, MNs loaded with IL-4 significantly downregulated the expression levels of proinflammatory genes, such as the enzyme inducible nitric oxide synthase and the cytokine interleukin-1β (IL-1β) by more than 50%, but up-regulated the expression levels of antiinflammatory genes, such as the genes mannose receptor C type 1 (MRC1) and arginase 1 (ARG-1) by more than 150%. These results indicate that the MNs loaded with IL-4 have excellent immune-modulating activity and can effectively alleviate chronic inflammation. In vivo pharmacodynamic studies showed that tetracycline released from the immunomodulatory MN patch effectively inhibited the growth of bacteria, with the bacterial burden in the periodontal tissues reduced by about 80% compared to the untreated group. Meanwhile, the immunomodulatory MN patch effectively downregulated proinflammatory cytokines such as tumor necrosis factor-alpha (TNF-α), IL-1β, and interferon-gamma (IFN-γ), while upregulating antiinflammatory factors such as interleukin-10 (IL-10) and forkhead box P3 (FOXP3). After 8 weeks of treatment, the rats receiving MN therapy showed a significant reduction in the distance between the alveolar bone crest and the cementoenamel junction, as well as a dramatic increase in the alveolar bone area. Taken together, the developed MN patch with dual antimicrobial and immunomodulatory functions holds great potential for periodontal tissue regeneration.

6. Type 1 diabetes

Type 1 diabetes (T1D) is a chronic autoimmune disease characterized by T cell-mediated destruction of pancreatic β cells, leading to absolute insulin deficiency, hyperglycemia, and ketoacidosis.[89,90] Although exogenous insulin administration remains the standard therapy, it has significant limitations, such as poor patient compliance, lifelong dependency, and imprecise blood glucose control. Moreover, poor glycemic management can result in severe complications, including cardiovascular disease, nephropathy, and retinopathy.[91,92] Thus, there is an urgent need to develop advanced drug delivery systems capable of precisely releasing insulin to enhance self-management and metabolic control in patients with T1D.

To reduce the risk of insulin-induced hypoglycemia, Wu et al.[93] developed a glucose-responsive MN system for T1D treatment by combining extrusion-based 3D printing with poststretching technology. The system incorporated phenylboronic acid (PBA)-modified insulin into the MNs. When blood glucose levels exceed the normal range, PBA preferentially binds to the hydroxyl groups of glucose, triggering the release of insulin from the MNs. Both in vitro and in vivo studies demonstrated the system's glucose-responsive release behavior, with insulin release rates directly correlated to glucose concentrations. Pharmacodynamic evaluations showed that mice treated with PBA-modified insulin MNs achieved normoglycemia within 1 h and maintained stable blood glucose levels for up to 22 h. In contrast, mice receiving unmodified insulin MNs experienced an initial rapid insulin burst, causing transient hypoglycemia before returning to hyperglycemia within 24 h. These findings highlight the potential of glucose-responsive MNs as a self-regulating insulin delivery platform, offering improved glycemic control and reduced hypoglycemia risk for T1D patients.

Luo et al.[94] developed a smart MN patch for T1D treatment, coloading pH-sensitive insulin-loaded nanoparticles (SNP(I)) with glucose oxidase (GOx)– and catalase (CAT)–loaded pH-insensitive nanoparticles (iSNP(G + C)) (Fig. 10.6A and B). The insulin-loaded nanoparticles were synthesized from pH-sensitive amphiphilic block copolymers using a double emulsion method. Under hyperglycemic conditions, GOx catalyzes glucose oxidation, generating gluconic acid and hydrogen peroxide to lower the local pH and trigger insulin release. Meanwhile, CAT decomposes hydrogen peroxide, mitigating oxidative damage to the surrounding skin. In vivo pharmacodynamic studies demonstrated that this glucose- and pH-responsive MN patch effectively regulated blood glucose levels in T1D mice, maintaining them below 200 mg/dL for 8 h. These results underscore the potential of this dual-responsive MN platform as a safe and efficient strategy for closed-loop insulin delivery in T1D.

Although insulin effectively regulates blood glucose levels in patients with T1D, it does not address the underlying autoimmune destruction of pancreatic β cells. Immunotherapy, by contrast, offers a promising approach through immune system modulation, either by inducing immune tolerance or suppressing autoreactive T cells. Such interventions could potentially delay disease onset, decrease reliance on exogenous insulin, and enhance long-term glycemic control. Consequently, developing effective immunotherapies is

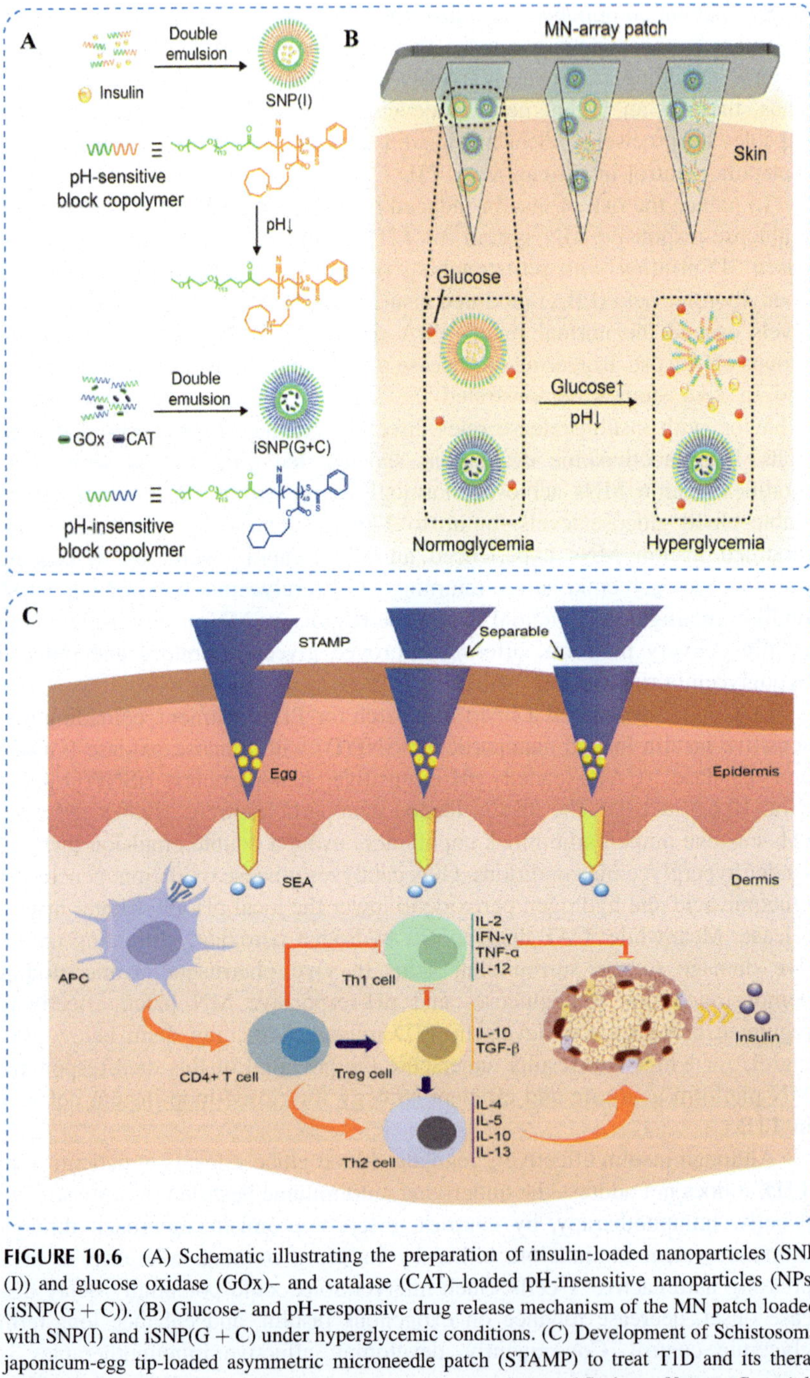

FIGURE 10.6 (A) Schematic illustrating the preparation of insulin-loaded nanoparticles (SNP (I)) and glucose oxidase (GOx)– and catalase (CAT)–loaded pH-insensitive nanoparticles (NPs) (iSNP(G + C)). (B) Glucose- and pH-responsive drug release mechanism of the MN patch loaded with SNP(I) and iSNP(G + C) under hyperglycemic conditions. (C) Development of Schistosoma japonicum-egg tip-loaded asymmetric microneedle patch (STAMP) to treat T1D and its therapeutic mechanism. *(B) Reproduced from Ref. 94 with permission of* ***Springer Nature****, Copyright* ***(2021)****. (C) Reproduced from Ref. 97 with permission of* ***Springer Nature****, Copyright* ***(2022)****.*

crucial for arresting disease progression and preserving remaining β-cell function.

Recent studies investigating the "hygiene hypothesis" propose that reduced childhood exposure to infectious agents, resulting from improved sanitation, may contribute to the increasing prevalence of autoimmune disorders such as T1D. Notably, helminths and their derivatives have emerged as potential therapeutic agents for T1D.[95] For instance, *Schistosoma* spp. infection has been found to prevent T1D progression by inducing a Th2-polarized immune response while suppressing Th1 cell activity.[96] To harness this immunomodulatory potential, Huang et al.[97] developed a *Schistosoma japonicum* egg tip-loaded asymmetric microneedle patch (STAMP) for T1D immunotherapy (Fig. 10.6C). Pharmacodynamic evaluations demonstrated that STAMP-treated T1D mice maintained stable blood glucose levels and exhibited significantly elevated serum insulin concentrations compared to untreated controls over a 2-week period. Moreover, STAMP therapy modulated cytokine profiles, characterized by reduced levels of Th1-associated cytokines (IFN-γ and IL-2) and increased Th2-associated cytokines (IL-4 and IL-5), indicating an immune-mediated therapeutic mechanism. Critically, STAMP treatment showed no adverse effects on body weight, blood biochemistry, or histopathology of major organs, confirming its biosafety. Residual *Schistosoma japonicum* egg components were completely metabolized within the epidermal layer. Together, these results position STAMP as a safe and effective immunotherapeutic strategy for T1D, presenting a viable alternative to traditional insulin-based therapies.

References

1. Kaymaz, K.; Beikler, T. Th17 Cells and the IL-23/IL-17 Axis in the Pathogenesis of Periodontitis and Immune-Mediated Inflammatory Diseases. *Int. J. Mol. Sci.* **2019,** *20* (14), 3394.
2. McInnes, I. B.; Gravallese, E. M. Immune-Mediated Inflammatory Disease Therapeutics: Past, Present, and Future. *Nat. Rev. Immunol.* **2021,** *21* (10), 680–686.
3. Monteleone, G.; Moscardelli, A.; Colella, A.; Marafini, I.; Salvatori, S. Immune-Mediated Inflammatory Diseases: Common and Different Pathogenic and Clinical Features. *Autoimmun. Rev.* **2023,** *22* (10), 103410.
4. Agrawal, M.; Shah, S.; Patel, A.; Pinotti, R.; Colombel, J.-F.; Burisch, J. Changing Epidemiology of Immune-Mediated Inflammatory Diseases in Immigrants: A Systematic Review of Population-Based Studies. *J. Autoimmun.* **2019,** *105*, 102303.
5. Kuek, A.; Hazleman, B. L.; Ostör, A. J. Immune-Mediated Inflammatory Diseases (IMIDs) and Biologic Therapy: A Medical Revolution. *Postgrad. Med. J.* **2007,** *83* (978), 251–260.
6. Pofi, R.; Caratti, G.; Ray, D. W.; Tomlinson, J. W. Treating the Side Effects of Exogenous Glucocorticoids; Can We Separate the Good From the Bad? *Endocr. Rev.* **2023,** *44* (6), 975–1011.
7. Wong, W. F.; Ang, K. P.; Sethi, G.; Looi, C. Y. Recent Advancement of Medical Patch for Transdermal Drug Delivery. *Medicina* **2023,** *59* (4).

8. Karve, T.; Dandekar, A.; Agrahari, V.; Peet, M.; Banga, A. K.; Doncel, G. F. Long-Acting Transdermal Drug Delivery Formulations: Current Developments and Innovative Pharmaceutical Approaches. *Adv. Drug Deliv. Rev.* **2024,** *210*, 115326.
9. Yu, Y.; Wang, H.; Guo, B.; Wang, B.; Wan, Z.; Zhang, Y.; Sun, L.; Yang, F. Microneedle-Based Two-Step Transdermal Delivery of Langerhans Cell-Targeting Immunoliposomes Induces a Th1-Biased Immune Response. *Eur. J. Pharm. Biopharm.* **2022,** *177,* 68–80.
10. Jiang, X.; Zhao, H.; Li, W. Microneedle-Mediated Transdermal Delivery of Drug-Carrying Nanoparticles. *Front. Bioeng. Biotechnol.* **2022,** *10*, 840395.
11. Solitano, V.; Yuan, Y.; Singh, S.; Ma, C.; Nardone, O. M.; Fiorino, G.; Felquer, M. L. A.; Barra, L.; D'Agostino, M.-A.; Pope, J. Efficacy and Safety of Advanced Combination Treatment in Immune-Mediated Inflammatory Disease: A Systematic Review and Meta-Analysis of Randomized Controlled Trials. *J. Autoimmun.* **2024,** *149*, 103331.
12. Kurian, A. G.; Singh, R. K.; Sagar, V.; Lee, J.-H.; Kim, H.-W. Nanozyme-Engineered Hydrogels for Anti-Inflammation and Skin Regeneration. *Nano-Micro Lett.* **2024,** *16* (1), 110.
13. Zeng, L.; Yang, K.; He, Q.; Zhu, X.; Long, Z.; Wu, Y.; Chen, J.; Li, Y.; Zeng, J.; Cui, G. Efficacy and Safety of Gut Microbiota-Based Therapies in Autoimmune and Rheumatic Diseases: A Systematic Review and Meta-Analysis of 80 Randomized Controlled Trials. *BMC Med.* **2024,** *22* (1), 110.
14. Saleem, M.; Shahzad, K. A.; Marryum, M.; Singh, S.; Zhou, Q.; Du, S.; Wang, S.; Shao, C.; Shaikh, I. I. Exosome-Based Therapies for Inflammatory Disorders: a Review of Recent Advances. *Stem Cell Res. Ther.* **2024,** *15* (1), 477.
15. Yepes-Nuñez, J. J.; Guyatt, G. H.; Gómez-Escobar, L. G.; Pérez-Herrera, L. C.; Chu, A. W.; Ceccaci, R.; Acosta-Madiedo, A. S.; Wen, A.; Moreno-López, S.; MacDonald, M. Allergen Immunotherapy for Atopic Dermatitis: Systematic Review and Meta-Analysis of Benefits and Harms. *J. Allergy Clin. Immunol.* **2023,** *151* (1), 147–158.
16. Griffiths, C. E.; Armstrong, A. W.; Gudjonsson, J. E.; Barker, J. N. Psoriasis. *Lancet* **2021,** *397* (10281), 1301–1315.
17. Rendon, A.; Schäkel, K. Psoriasis Pathogenesis and Treatment. *Int. J. Mol. Sci.* **2019,** *20* (6).
18. Kaufman, B. P.; Alexis, A. F. Psoriasis in Skin of Color: Insights Into the Epidemiology, Clinical Presentation, Genetics, Quality-Of-Life Impact, and Treatment of Psoriasis in Non-White Racial/Ethnic Groups. *Am. J. Clin. Dermatol.* **2018,** *19* (3), 405–423.
19. Liang, L.; Fei, W. M.; Zhao, Z. Q.; Hao, Y. Y.; Zhang, C.; Cui, Y.; Guo, X. D. Improved Imiquimod-Induced Psoriasis Like Dermatitis Using Microneedles in Mice. *Eur. J. Pharm. Biopharm.* **2021,** *164,* 20–27.
20. Men, Z.; Su, T.; Tang, Z.; Liang, J.; Shen, T. Tacrolimus Nanocrystals Microneedle Patch for Plaque Psoriasis. *Int. J. Pharm.* **2022,** *627*, 122207.
21. Du, H.; Liu, P.; Zhu, J.; Lan, J.; Li, Y.; Zhang, L.; Zhu, J.; Tao, J. Hyaluronic Acid-Based Dissolving Microneedle Patch Loaded With Methotrexate for Improved Treatment of Psoriasis. *ACS Appl. Mater. Interf.* **2019,** *11* (46), 43588–43598.
22. Wu, D.; Shou, X.; Yu, Y.; Wang, X.; Chen, G.; Zhao, Y.; Sun, L. Biologics-loaded Photothermally Dissolvable Hyaluronic Acid Microneedle Patch for Psoriasis Treatment. *Adv. Funct. Mater.* **2022,** *32* (47), 2205847.
23. Zhang, W.; Chen, Y.; Zhao, Z.; Zheng, H.; Wang, S.; Liao, Z.; Sheng, T.; Zhao, S.; Hou, W.; Yu, X.; et al. Adoptive T(reg) Therapy With Metabolic Intervention via Perforated Microneedles Ameliorates Psoriasis Syndrome. *Sci. Adv.* **2023,** *9* (20), eadg6007.
24. Sroka-Tomaszewska, J.; Trzeciak, M. Molecular Mechanisms of Atopic Dermatitis Pathogenesis. *Int. J. Mol. Sci.* **2021,** *22* (8).

25. Luger, T.; Paller, A. S.; Irvine, A. D.; Sidbury, R.; Eichenfield, L. F.; Werfel, T.; Bieber, T. Topical Therapy of Atopic Dermatitis With a Focus on Pimecrolimus. *J. Eur. Acad. Dermatol. Venereol.* **2021,** *35* (7), 1505–1518.
26. Jang, M.; Kang, B. M.; Yang, H.; Ohn, J.; Kwon, O.; Jung, H. High-Dose Steroid Dissolving Microneedle for Relieving Atopic Dermatitis. *Adv. Healthc. Mater.* **2021,** *10* (7), e2001691.
27. Yang, Y.; Chen, B. Z.; Zhang, X. P.; Zheng, H.; Li, Z.; Zhang, C. Y.; Guo, X. D. Conductive Microneedle Patch With Electricity-Triggered Drug Release Performance for Atopic Dermatitis Treatment. *ACS Appl. Mater. Interf.* **2022,** *14* (28), 31645–31654.
28. Chen, M. C.; Chen, C. S.; Wu, Y. W.; Yang, Y. Y. Poly-γ-Glutamate Microneedles as Transdermal Immunomodulators for Ameliorating Atopic Dermatitis-Like Skin Lesions in Nc/Nga Mice. *Acta Biomater.* **2020,** *114,* 183–192.
29. Mandel, S. A.; Avramovich-Tirosh, Y.; Reznichenko, L.; Zheng, H.; Weinreb, O.; Amit, T.; Youdim, M. B. Multifunctional Activities of Green Tea Catechins in Neuroprotection. *Neurosignals* **2005,** *14* (1–2), 46–60.
30. Chiu, Y. H.; Wu, Y. W.; Hung, J. I.; Chen, M. C. Epigallocatechin Gallate/L-Ascorbic Acid-Loaded Poly-γ-Glutamate Microneedles With Antioxidant, Anti-inflammatory, and Immunomodulatory Effects for the Treatment of Atopic Dermatitis. *Acta Biomater.* **2021,** *130,* 223–233.
31. Lee, H.; Dellatore, S. M.; Miller, W. M.; Messersmith, P. B. Mussel-Inspired Surface Chemistry for Multifunctional Coatings. *Science* **2007,** *318* (5849), 426–430.
32. Zhang, Y.; Zhang, X.; Wu, X.; Zhao, Y. Photo-Responsive Polydopamine Nanoenzyme Microneedles With Oxidative Stress Regulation Ability for Atopic Dermatitis Treatment. *Nano Today* **2024,** *56,* 102241.
33. Weiss, A.; Delavenne, E.; Matias, C.; Lagler, H.; Simon, D.; Li, P.; Hansen, J. U.; Dos Santos, T. P.; Jana, B.; Priemel, P. Topical Niclosamide (ATx201) Reduces *Staphylococcus aureus* Colonization and Increases Shannon Diversity of the Skin Microbiome in Atopic Dermatitis Patients in a Randomized, Double-Blind, Placebo-Controlled Phase 2 Trial. *Clin. Transl. Med.* **2022,** *12* (5), e790.
34. Piewngam, P.; Khongthong, S.; Roekngam, N.; Theapparat, Y.; Sunpaweravong, S.; Faroongsarng, D.; Otto, M. Probiotic for Pathogen-Specific *Staphylococcus aureus* Decolonisation in Thailand: A Phase 2, Double-Blind, Randomised, Placebo-Controlled Trial. *Lancet Microbe* **2023,** *4* (2), e75–e83.
35. Zhang, W.; Lei, J.; Jiang, P.; Hao, T.; Yuan, Y.; Hu, H.; Li, W. Double-Layered Microneedle Patch Integrated With Multifunctional Nanoparticles and Live Bacteria for Long-Term Treatment of Atopic Dermatitis. *Small* **2025,** *21* (3), 2409121.
36. Boonpiyathad, T.; Sözener, Z. C.; Akdis, M.; Akdis, C. A. The Role of Treg Cell Subsets in Allergic Disease. *Asian Pac. J. Allergy Immunol.* **2020,** *38* (3), 139–149.
37. Kong, Z.-L.; Sudirman, S.; Lin, H.-J.; Chen, W.-N. In Vitro Anti-Inflammatory Effects of Curcumin on Mast Cell-Mediated Allergic Responses via Inhibiting FcεRI Protein Expression and Protein Kinase C Delta Translocation. *Cytotechnology* **2020,** *72,* 81–95.
38. Nakamura, T. The Roles of Lipid Mediators in Type I Hypersensitivity. *J. Pharmacol. Sci.* **2021,** *147* (1), 126–131.
39. Kim, J. H.; Shin, J. U.; Kim, S. H.; Noh, J. Y.; Kim, H. R.; Lee, J.; Chu, H.; Jeong, K. Y.; Park, K. H.; Kim, J. D. Successful Transdermal Allergen Delivery and Allergen-Specific Immunotherapy Using Biodegradable Microneedle Patches. *Biomaterials* **2018,** *150,* 38–48.
40. Yu, Y.; Kumar, M. N. K.; Wu, M. X. Delivery of Allergen Powder for Safe and Effective Epicutaneous Immunotherapy. *J. Allergy Clin. Immunol.* **2020,** *145* (2), 597–609.

41. Klimek, L.; Pfaar, O.; Bousquet, J.; Senti, G.; Kündig, T. Allergen Immunotherapy in Allergic Rhinitis: Current Use and Future Trends. *Expet Rev. Clin. Immunol.* **2017,** *13* (9), 897–906.
42. Choi, Y.-J.; Kim, K.-A.; Jung, J.-H.; Choi, Y.-S.; Baek, S.-K.; Kim, S.-T.; Park, J.-H. Epicutaneous Allergen Administration with Microneedles as a Novel Method of Immunotherapy for House Dust Mite (HDM) Allergic Rhinitis. *Pharm. Res.* **2021,** *38* (7), 1199–1207.
43. Park, K. H.; Oh, E. Y.; Han, H.; Kim, J. D.; Kim, S. J.; Jeong, K. Y.; Kim, J. H.; Park, C. O.; Kim, S. R.; Lee, J. H. Efficacy of Transdermal Immunotherapy with Biodegradable Microneedle Patches in a Murine Asthma Model. *Clin. Exp. Allergy* **2020,** *50* (9), 1084–1092.
44. Shakya, A. K.; Lee, C. H.; Gill, H. S. Microneedle-Mediated Allergen-Specific Immunotherapy for the Treatment of Airway Allergy in Mice. *Mol. Pharm.* **2020,** *17* (8), 3033–3042.
45. Scherer, H. U.; Häupl, T.; Burmester, G. R. The Etiology of Rheumatoid Arthritis. *J. Autoimmun.* **2020,** *110*, 102400.
46. Radu, A.-F.; Bungau, S. G. Management of Rheumatoid Arthritis: An Overview. *Cells* **2021,** *10* (11), 2857.
47. Huang, J.; Fu, X.; Chen, X.; Li, Z.; Huang, Y.; Liang, C. Promising Therapeutic Targets for Treatment of Rheumatoid Arthritis. *Front. Immunol.* **2021,** *12*, 686155.
48. Zhao, Z.; Hua, Z.; Luo, X.; Li, Y.; Yu, L.; Li, M.; Lu, C.; Zhao, T.; Liu, Y. Application and Pharmacological Mechanism of Methotrexate in Rheumatoid Arthritis. *Biomed. Pharmacother.* **2022,** *150*, 113074.
49. Tekko, I. A.; Chen, G.; Domínguez-Robles, J.; Thakur, R. R. S.; Hamdan, I. M.; Vora, L.; Larrañeta, E.; McElnay, J. C.; McCarthy, H. O.; Rooney, M. Development and Characterisation of Novel Poly (Vinyl Alcohol)/Poly (Vinyl Pyrrolidone)-Based Hydrogel-Forming Microneedle Arrays for Enhanced and Sustained Transdermal Delivery of Methotrexate. *Int. J. Pharm.* **2020,** *586*, 119580.
50. Cao, J.; Zhang, N.; Wang, Z.; Su, J.; Yang, J.; Han, J.; Zhao, Y. Microneedle-Assisted Transdermal Delivery of Etanercept for Rheumatoid Arthritis Treatment. *Pharmaceutics* **2019,** *11* (5), 235.
51. Du, G.; He, P.; Zhao, J.; He, C.; Jiang, M.; Zhang, Z.; Zhang, Z.; Sun, X. Polymeric Microneedle-Mediated Transdermal Delivery of Melittin for Rheumatoid Arthritis Treatment. *J. Contr. Release* **2021,** *336,* 537–548.
52. Yao, W.; Tao, C.; Zou, J.; Zheng, H.; Zhu, J.; Zhu, Z.; Zhu, J.; Liu, L.; Li, F.; Song, X. Flexible Two-Layer Dissolving and Safing Microneedle Transdermal of Neurotoxin: A Biocomfortable Attempt to Treat Rheumatoid Arthritis. *Int. J. Pharm.* **2019,** *563,* 91–100.
53. Tang, S.; Zhang, C.; Oo, W. M.; Fu, K.; Risberg, M. A.; Bierma-Zeinstra, S. M.; Neogi, T.; Atukorala, I.; Malfait, A.-M.; Ding, C. Osteoarthritis. *Nat. Rev. Dis. Prim.* **2025,** *11* (1), 1–22.
54. Bruyère, O.; Cooper, C.; Pelletier, J. P.; Maheu, E.; Rannou, F.; Branco, J.; Brandi, M. L.; Kanis, J. A.; Altman, R. D.; Hochberg, M. C.; et al. A Consensus Statement on the European Society for Clinical and Economic Aspects of Osteoporosis and Osteoarthritis (ESCEO) Algorithm for the Management of Knee Osteoarthritis-From Evidence-Based Medicine to the Real-Life Setting. *Semin. Arthrit. Rheum.* **2016,** *45* (4 Suppl. 1), S3–11.
55. Chen, K.; Zhao, Y.; Zhao, W.; Mao, X.; Li, D.; Wang, Y.; Shang, S.; Zhang, H. Lubricating Microneedles System With Multistage Sustained Drug Delivery for the Treatment of Osteoarthritis. *Small* **2024;**, 2307281.
56. Wang, Q.; Yang, X.; Gu, X.; Wei, F.; Cao, W.; Zheng, L.; Li, Y.; Ma, T.; Wu, C.; Wang, Q. Celecoxib Nanocrystal-Loaded Dissolving Microneedles With Highly Efficient for Osteoarthritis Treatment. *Int. J. Pharm.* **2022,** *625*, 122108.

57. Deng, Y.; Li, W.; Zhang, Y.; Li, J.; He, F.; Dong, K.; Hong, Z.; Luo, R.; Pei, X. α-Linolenic Acid Inhibits RANKL-Induced Osteoclastogenesis In Vitro and Prevents Inflammation In Vivo. *Foods* **2023,** *12* (3).
58. Li, J.; Tian, X.; Wang, K.; Jia, Y.; Ma, F. Transdermal Delivery of Celecoxib and α-Linolenic Acid From Microemulsion-Incorporated Dissolving Microneedles for Enhanced Osteoarthritis Therapy. *J. Drug Target.* **2023,** *31* (2), 206–216.
59. Li, E. Q.; Zhang, J. L. Therapeutic Effects of Triptolide From Tripterygium Wilfordii Hook. F. On Interleukin-1-Beta-Induced Osteoarthritis in Rats. *Eur. J. Pharmacol.* **2020,** *883*, 173341.
60. Zhou, P.; Chen, C.; Yue, X.; Zhang, J.; Huang, C.; Zhao, S.; Wu, A.; Li, X.; Qu, Y.; Zhang, C. Strategy for Osteoarthritis Therapy: Improved the Delivery of Triptolide Using Liposome-Loaded Dissolving Microneedle Arrays. *Int. J. Pharm.* **2021,** *609*, 121211.
61. Wang, S.; Lei, B.; Zhang, E.; Gong, P.; Gu, J.; He, L.; Han, L.; Yuan, Z. Targeted Therapy for Inflammatory Diseases With Mesenchymal Stem Cells and Their Derived Exosomes: From Basic to Clinics. *Int. J. Nanomed.* **2022,** *17,* 1757–1781.
62. Zhang, Z.; Zhao, S.; Sun, Z.; Zhai, C.; Xia, J.; Wen, C.; Zhang, Y.; Zhang, Y. Enhancement of the Therapeutic Efficacy of Mesenchymal Stem Cell-Derived Exosomes in Osteoarthritis. *Cell. Mol. Biol. Lett.* **2023,** *28* (1), 75.
63. Li, Z.; Lu, H.; Fan, L.; Ma, X.; Duan, Z.; Zhang, Y.; Fu, Y.; Wang, S.; Guan, Y.; Yang, D. Microneedle-Delivered PDA@ Exo for Multifaceted Osteoarthritis Treatment via PI3K-Akt-mTOR Pathway. *Adv. Sci.* **2024,** *11* (42), 2406942.
64. Gwam, C.; Ohanele, C.; Hamby, J.; Chughtai, N.; Mufti, Z.; Ma, X. Human Placental Extract: A Potential Therapeutic in Treating Osteoarthritis. *Ann. Transl. Med.* **2023,** *11* (9), 322.
65. Jj, J.; Sandhya, K.; Sridhar, K.; Sudarson, S.; Basavaraj, B.; Bharath, S. Proteomic Characterization of Human Placenta: Insights Into Potential Therapeutic Applications for Osteoarthritis. *AAPS PharmSciTech* **2024,** *25* (6), 139.
66. James, J. J.; Pahal, S.; Jayaraman, A.; D Nayak, A.; Koteshwar Narasimhachar, S.; Sundarrajan, S.; Basappa Veerabhadraiah, B.; Srinivasan, B.; Vemula, P. K.; Nguyen, T. D. Polymeric Microneedles for Transdermal Delivery of Human Placental Tissue for the Treatment of Osteoarthritis. *Macromol. Biosci.* **2024;**, 2400485.
67. Shirvan, A. R.; Bashari, A.; Hemmatinejad, N. New Insight Into the Fabrication of Smart Mucoadhesive Buccal Patches as a Novel Controlled-Drug Delivery System. *Eur. Polym. J.* **2019,** *119,* 541–550.
68. Hearnden, V.; Sankar, V.; Hull, K.; Juras, D. V.; Greenberg, M.; Kerr, A. R.; Lockhart, P. B.; Patton, L. L.; Porter, S.; Thornhill, M. H. New Developments and Opportunities in Oral Mucosal Drug Delivery for Local and Systemic Disease. *Adv. Drug Deliv. Rev.* **2012,** *64* (1), 16–28.
69. Jacob, S.; Nair, A. B.; Boddu, S. H. S.; Gorain, B.; Sreeharsha, N.; Shah, J. An Updated Overview of the Emerging Role of Patch and Film-Based Buccal Delivery Systems. *Pharmaceutics* **2021,** *13* (8).
70. Yermak, I. M.; Davydova, V. N.; Volod'ko, A. V. Mucoadhesive Marine Polysaccharides. *Mar. Drugs* **2022,** *20* (8).
71. Johnston, T. P. *Anatomy and Physiology of the Oral Mucosa. Oral Mucosal Drug Delivery and Therapy,* 2015; pp 1–15.
72. Hovav, A. H.; Wilensky, A. The Role of the Epithelial Sentinels, Langerhans Cells and γδT Cells, in Oral Squamous Cell Carcinoma. *Periodontol 2000* **2024,** *96* (1), 221–228.

73. McNeilly, C. L.; Crichton, M. L.; Primiero, C. A.; Frazer, I. H.; Roberts, M. S.; Kendall, M. A. Microprojection Arrays to Immunise at Mucosal Surfaces. *J. Contr. Release* **2014,** *196,* 252–260.
74. Lee, J.; Shin, D.; Roh, J. L. Treatment of Intractable Oral Ulceration With an Oral Mucosa Equivalent. *J. Biomed. Mater. Res. B Appl. Biomater.* **2019,** *107* (6), 1779–1785.
75. Zeng, X.; Jin, X.; Zhong, L.; Zhou, G.; Zhong, M.; Wang, W.; Fan, Y.; Liu, Q.; Qi, X.; Guan, X.; et al. Difficult and Complicated Oral Ulceration: An Expert Consensus Guideline for Diagnosis. *Int. J. Oral Sci.* **2022,** *14* (1), 28.
76. Meng, Y.; Li, X. J.; Li, Y.; Zhang, T. Y.; Liu, D.; Wu, Y. Q.; Hou, F. F.; Ye, L.; Wu, C. J.; Feng, X. D. Novel Double-Layer Dissolving Microneedles for Transmucosal Sequential Delivery of Multiple Drugs in the Treatment of Oral Mucosa Diseases. *ACS Appl. Mater. Interf.* **2023,** *15* (11), 13892–13906.
77. Guo, X.; Zhu, T.; Yu, X.; Yi, X.; Li, L.; Qu, X.; Zhang, Z.; Hao, Y.; Wang, W. Betamethasone-Loaded Dissolvable Microneedle Patch for Oral Ulcer Treatment. *Colloids Surf. B Biointerf.* **2023,** *222*, 113100.
78. Chen, M.-Y.; Tang, Y.-J.; Wang, Y.-C.; Wang, C.-Z.; Yuan, C.-S.; Chen, Y.; Tan, Z.-R.; Huang, W.-H.; Zhou, H.-H. Quantitative Determination of Betamethasone Sodium Phosphate and Betamethasone Dipropionate in Human Plasma by UPLC-MS/MS and a Bioequivalence Study. *Anal. Methods* **2016,** *8* (17), 3550–3563.
79. Naniwa, M.; Nakatomi, C.; Hitomi, S.; Matsuda, K.; Tabuchi, T.; Sugiyama, D.; Kubo, S.; Miyamura, Y.; Yoshino, K.; Akifusa, S.; et al. Analgesic Mechanisms of Steroid Ointment Against Oral Ulcerative Mucositis in a Rat Model. *Int. J. Mol. Sci.* **2021,** *22* (22).
80. Valls, M.; Martínez Labrador, A.; Halbaut Bellowa, L.; Torres, D. B.; Granda, P. C.; Carmona, M.; Limón, D.; Campmany, A. C. C. Biopharmaceutical Study of Triamcinolone Acetonide Semisolid Formulations for Sublingual and Buccal Administration. *Pharmaceutics* **2021,** *13* (7), 1080.
81. Qu, X.; Guo, X.; Zhu, T.; Zhang, Z.; Wang, W.; Hao, Y. Microneedle Patches Containing Mesoporous Polydopamine Nanoparticles Loaded With Triamcinolone Acetonide for the Treatment of Oral Mucositis. *Front. Bioeng. Biotechnol.* **2023,** *11*, 1203709.
82. Kwon, T.; Lamster, I. B.; Levin, L. Current Concepts in the Management of Periodontitis. *Int. Dent. J.* **2021,** *71* (6), 462–476.
83. Slots, J. Periodontitis: Facts, Fallacies and the Future. *Periodontology* **2000 2017,** *75* (1), 7–23.
84. Yang, B.; Pang, X.; Li, Z.; Chen, Z.; Wang, Y. Immunomodulation in the Treatment of Periodontitis: Progress and Perspectives. *Front. Immunol.* **2021,** *12*, 781378.
85. Sczepanik, F. S. C.; Grossi, M. L.; Casati, M.; Goldberg, M.; Glogauer, M.; Fine, N.; Tenenbaum, H. C. Periodontitis Is an Inflammatory Disease of Oxidative Stress: We Should Treat it that Way. *Periodontology 2000* **2020,** *84* (1), 45–68.
86. Bastos, M.; Lima, J.; Vieira, P.; Mestnik, M.; Faveri, M.; Duarte, P. TNF-α and IL-4 Levels in Generalized Aggressive Periodontitis Subjects. *Oral Dis.* **2009,** *15* (1), 82–87.
87. Jia, X.-W.; Yuan, Y.-D.; Yao, Z.-X.; Wu, C.-J.; Chen, X.; Chen, X.-H.; Lin, Y.-M.; Meng, X.-Y.; Zeng, X.-T.; Shao, J. Association Between IL-4 and IL-4R Polymorphisms and Periodontitis: A Meta-analysis. *Dis. Mark.* **2017,** *2017* (1), 8021279.
88. Zhang, X.; Hasani-Sadrabadi, M. M.; Zarubova, J.; Dashtimighadam, E.; Haghniaz, R.; Khademhosseini, A.; Butte, M. J.; Moshaverinia, A.; Aghaloo, T.; Li, S. Immunomodulatory Microneedle Patch for Periodontal Tissue Regeneration. *Matter* **2022,** *5* (2), 666–682.
89. Anderson, M.; Bluestone, J.; Warshauer, J. New Frontiers in the Treatment of Type 1 Diabetes. *Cell Metab.* **2020,** *31* (1).

90. Norris, J. M.; Johnson, R. K.; Stene, L. C. Type 1 Diabetes—Early Life Origins and Changing Epidemiology. *Lancet Diabetes Endocrinol.* **2020,** *8* (3), 226–238.
91. Primavera, R.; Kevadiya, B. D.; Swaminathan, G.; Wilson, R. J.; De Pascale, A.; Decuzzi, P.; Thakor, A. S. Emerging Nano-and Micro-Technologies Used in the Treatment of Type-1 Diabetes. *Nanomaterials* **2020,** *10* (4), 789.
92. Bronczek, G. A.; Vettorazzi, J. F.; Soares, G. M.; Kurauti, M. A.; Santos, C.; Bonfim, M. F.; Carneiro, E. M.; Balbo, S. L.; Boschero, A. C.; Costa Júnior, J. M. The Bile Acid TUDCA Improves Beta-Cell Mass and Reduces Insulin Degradation in Mice With Early-Stage of Type-1 Diabetes. *Front. Physiol.* **2019,** *10,* 561.
93. Wu, M.; Zhang, Y.; Huang, H.; Li, J.; Liu, H.; Guo, Z.; Xue, L.; Liu, S.; Lei, Y. Assisted 3D Printing of Microneedle Patches for Minimally Invasive Glucose Control in Diabetes. *Mater. Sci. Eng. C* **2020,** *117,* 111299.
94. Luo, F.-Q.; Chen, G.; Xu, W.; Zhou, D.; Li, J.-X.; Huang, Y.-C.; Lin, R.; Gu, Z.; Du, J.-Z. Microneedle-Array Patch With pH-Sensitive Formulation for Glucose-Responsive Insulin Delivery. *Nano Res.* **2021,** *14,* 2689–2696.
95. Tang, C-l; Zou, J-n; Zhang, R-h; Liu, Z-m; Mao, C-l. Helminths Protect Against Type 1 Diabetes: Effects and Mechanisms. *Parasitol. Res.* **2019,** *118,* 1087–1094.
96. Shah, F. A.; Wang, C. S.; Butt, T. M. Nutrition Influences Growth and Virulence of the Insect-Pathogenic Fungus Metarhizium Anisopliae. *Fed. Eur. Microbiol. Soc. Microbiol. Lett.* **2005,** *251* (2), 259–266.
97. Huang, H.; Hu, D.; Chen, Z.; Xu, J.; Xu, R.; Gong, Y.; Fang, Z.; Wang, T.; Chen, W. Immunotherapy for Type 1 Diabetes Mellitus by Adjuvant-Free Schistosoma Japonicum-Egg Tip-Loaded Asymmetric Microneedle Patch (STAMP). *J. Nanobiotechnol.* **2022,** *20* (1), 377.

Chapter 11

Microneedle technology for treating skin bacterial infections

Tingting Peng[1], Weishi Ye[2] and Xin Pan[2]
[1]*State Key Laboratory of Bioactive Molecules and Druggability Assessment/International Cooperative Laboratory of Traditional Chinese Medicine Modernization and Innovative Drug Development of Ministry of Education (MOE) of China/College of Pharmacy, Jinan University, Guangzhou, China;* [2]*School of Pharmaceutical Science, Sun Yat-Sen University, Guangzhou, China*

1. Introduction

Skin is the largest organ of the human body and plays a vital role in protecting against external pathogens. Skin infections can result from external factors such as trauma, burns, and scalds, as well as internal conditions like diabetes and pressure ulcers.[1] When the skin's structural and functional integrity is compromised, bacteria can migrate from the surface into subcutaneous tissues, leading to infections ranging from mild superficial cases to severe deep soft tissue infections. According to data from the 2022 Antimicrobial Resistance Collaborators report, bacterial pathogens were responsible for 7.7 million deaths in 2019, accounting for 13.6% of global mortality that year.[2] Therefore, bacterial infection has placed a significant burden on medical resources.

Although topical treatment of skin infections is effective in minimizing adverse effects, its therapeutic efficacy is often limited by multiple drug delivery barriers. The stratum corneum acts as the primary barrier to transdermal drug permeation.[3] Furthermore, cell debris and necrotic tissue can obstruct drug diffusion to infected sites.[4] Notably, biofilms are a collection of monomicrobial or polymicrobial communities, posing an additional challenge by severely impeding drug penetration and therapeutic action. Biofilms consist of densely packed extracellular polymeric substances (EPS), including polysaccharides, proteins, extracellular DNA (eDNA), lipids, amyloids, and cellulose, which constitute over 90% of the biofilm mass.[5] These EPS matrices shield microbial cells from antibiotics and host immune defenses, leading to

Microneedle Technology. https://doi.org/10.1016/B978-0-443-45544-5.00001-4

treatment failure. Consequently, biofilm-embedded bacteria exhibit 10- to 1000-fold greater antibiotic resistance compared to their planktonic counterparts.[6]

Recently, microneedles (MNs) have emerged as an effective tool to overcome sequential drug transport barriers for improved topical treatment of skin infections. By piercing the skin in a minimally invasive manner, MNs can deliver antimicrobials directly to lesional tissue without stimulating dermal nerves or blood vessels, thereby minimizing patient discomfort. Additionally, MNs physically disrupt biofilms, enhancing antimicrobial penetration. Their efficacy can be further improved by incorporating agents that target the EPS, such as specific enzymes (e.g., α-amylase[7]), external stimuli (e.g., light,[8] ultrasound[9]), or gas nanomotors.[10] Thus, MN technology facilitates the penetration of antimicrobials across multiple biological barriers, thereby enhancing therapeutic efficiency.

To date, MNs have been increasingly utilized for the delivery of various therapeutics to treat bacterial skin infections. In this chapter, we summarize recent advances in MN-based treatments for such infections, as well as the integration of MN technology with emerging antibacterial strategies to address intractable skin infections such as skin soft tissue infections (SSTIs) and chronic wound infections.

2. The applications of microneedles in treating skin bacterial infection

Bacteria represent the primary cause of skin infections. Over the past decade, extensive research has explored the application of MNs in managing bacterial skin infections, including impetigo, infected wounds, and acne vulgaris. MNs have been employed to deliver diverse therapeutic agents, enabling targeted treatment of both acute and chronic bacterial infections through multiple modalities. These include pharmacotherapy (e.g., antibiotics, antimicrobial peptides), photodynamic therapy (PDT), photothermal therapy (PTT), metal-based antibacterial therapy (MABT), chemodynamic therapy (CDT), sonodynamic therapy (SDT), starvation therapy (ST), nanoenzyme antibacterial therapy (NABT), poly(ionic liquid) (PIL), and combination therapy. This section summarizes recent advances in these applications.

2.1 Pharmacotherapy

2.1.1 Antibiotics

Antibiotics have played a pivotal role in modern medicine, revolutionizing the treatment of bacterial infections and saving countless lives since their discovery. The advent of penicillin by Alexander Fleming in 1928 ushered in a new medical era, spurring the development of diverse antibiotics that are now indispensable in clinical practice. In recent years, researchers have increasingly focused on exploring alternative antibiotic sources, including marine organisms and soil microbiomes, while simultaneously investigating innovative strategies to combat bacterial resistance. Antibiotics are typically classified into five types by their mechanisms of action:

(1) Cell wall synthesis inhibitors (e.g., vancomycin, telavancin) disrupt peptidoglycan assembly, causing bacterial lysis in low-osmotic environments.[11] These agents primarily target actively replicating bacteria rather than dormant cells.
(2) Membrane disruptors (e.g., carvacrol) compromise bacterial membrane integrity and permeability, leading to cell death.[12]
(3) Protein synthesis inhibitors (e.g., chloramphenicol, clindamycin, erythromycin, azithromycin) bind to ribosomes, halting bacterial protein production.[13]
(4) Nucleic acid synthesis blockers (e.g., gentamicin sulfate) interfere with topoisomerase II or RNA polymerase to prevent DNA/RNA replication.[14]
(5) Metabolic pathway inhibitors (e.g., sulfonamides, isoniazid) selectively target[15] bacterial folate synthesis, a process absent in humans, making them nontoxic to human cells.

As the frontline therapy for bacterial infections, antibiotics remain clinically irreplaceable, though their prudent use is critical to addressing emerging resistance challenges.

Compared to conventional transdermal or injectable administration routes, MNs directly puncture infected skin tissue, offering targeted drug delivery, enhanced antibacterial effects, and fewer side effects. Gittard et al.[16] developed gentamicin sulfate–doped antibacterial microneedles using a photopolymerization–micromolding technique for the first time, demonstrating their effectiveness against *S. aureus* in an agar plate assay. Since then, MNs have been widely employed to deliver various antibiotics, including vancomycin,[17] carvacrol,[18] and chloramphenicol,[19] exhibiting strong antibacterial and antibiofilm activity both in vitro and in vivo. By utilizing microneedles, a higher concentration of antibiotics can be delivered precisely to the infection site, significantly improving therapeutic outcomes. This approach is particularly valuable for reducing antibiotic dosage, shortening treatment duration, and minimizing the risk of bacterial resistance. For instance, Ziesmer et al.[20] reported that vancomycin delivered via MNs achieved a local drug concentration more than 500 times higher than that achieved through intravenous injection.

In addition to delivering higher antibiotic concentrations, MNs have been reported to enhance the antibiofilm effect of antibiotics by physically disrupting biofilms and improving drug penetration.[19,21] Moreover, infected tissues often exhibit a pathological microenvironment characterized by excessive reactive oxygen species (ROS), low pH, and overexpression of specific enzymes. To address this challenge, integrating MNs with drug-loaded responsive nanoparticles (NPs) could enable targeted NP degradation and on-demand drug release at the infection site.[22] This innovative strategy has shown promising potential to boost the antibacterial efficacy of antibiotics.

2.1.2 Antimicrobial peptides

Antimicrobial peptides, also known as host defense peptides, are a diverse group of molecules that play a crucial role in the innate immune response of organisms.[23] They are produced by various cells and tissues, including epithelial cells, leukocytes, and mucosal surfaces. Antimicrobial peptides are typically composed of 10–50 amino acids and represent a promising class of antimicrobial agents. A defining feature of antimicrobial peptides is their unique antibacterial mechanism. Most natural and synthetic antimicrobial peptides are positively charged and exhibit an amphiphilic structure, enabling them to bind to negatively charged bacterial membranes through electrostatic interactions. This binding often leads to membrane disruption and subsequent bacterial lysis, which is considered a primary bactericidal mode of action.[23,24] Beyond membrane targeting, certain antimicrobial peptides can penetrate the lipid bilayer to interfere with intracellular functions, such as inactivating essential substrates (e.g., nucleic acids, proteins, and enzymes). Notably, antimicrobial peptides also exhibit immunomodulatory properties. Some peptides enhance host defense by recruiting or activating immune cells to phagocytose pathogens, while others modulate Toll-like receptor (TLR) recognition of microbial metabolites or suppress excessive inflammatory responses.[24] Due to their multifaceted mechanisms, antimicrobial peptides are regarded as potential alternatives or adjuncts to conventional antibiotics. Their versatility offers significant therapeutic potential in addressing antibiotic resistance and developing novel treatments for infectious diseases.

The antibacterial effect of antimicrobial peptides is influenced by multiple structural characteristics, including their net charge, molecular weight, hydrophobicity, amino acid composition and sequence, as well as secondary structures (e.g., α-helix, β-sheet, mixed α/β motifs, or random coils).[25] Notably, antimicrobial peptides exhibit distinct advantages over conventional antibiotics due to their multi-target mechanism of action. These include rapid microbial killing, broad-spectrum activity, lower propensity for resistance development, and the ability to modulate host immune responses.[26]

Antimicrobial peptides, despite their therapeutic potential, face limited skin permeability due to their relatively large molecular weight, which hinders penetration through the stratum corneum via conventional transdermal delivery systems. MN technology has emerged as an effective strategy to overcome this barrier, enabling targeted delivery of antimicrobial peptides to infected skin tissues. A study by Su et al.[27] showed this approach using antimicrobial peptides-loaded MNs combined with a Janus-type antibacterial dressing to treat chronic wound biofilms (Fig. 11.1). The system delivered the W379 peptide, which exhibited potent activity against both gram-positive and

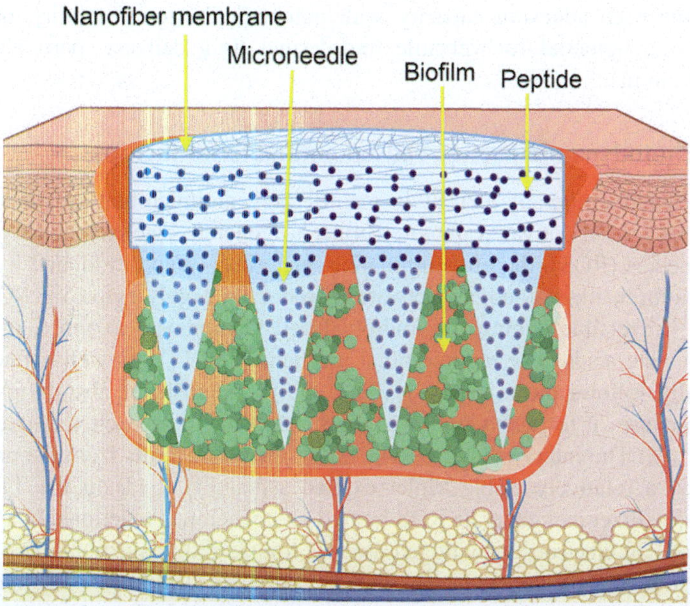

FIGURE 11.1 Schematic illustrating Janus-type antimicrobial dressings consisting of molecularly engineered peptide-loaded electrospun nanofiber membranes and microneedle arrays for the treatment of biofilms in chronic wounds. *Reproduced from Ref. 27 with permission from the American Chemical Society.*

gram-negative bacteria by disrupting their membranes. Additionally, the electrospun nanofibrous membranes in the dressing effectively eradicated MRSA biofilms in diabetic (type II) mouse wounds and *P. aeruginosa*/MRSA mixed biofilms in ex vivo human skin models. These results underscore the promise of antimicrobial peptide-functionalized MNs as a novel therapeutic platform for chronic infected wounds. By integrating MNs with antimicrobial peptides, this technology not only bypasses the stratum corneum barrier but also addresses polymicrobial infections resistant to conventional antibiotics. Such advancements could revolutionize wound care management, offering a solution to antibiotic resistance while improving therapeutic outcomes.

Polymyxin, a natural antimicrobial peptide produced by *P. polymyxa*, exhibits broad-spectrum antibacterial activity and can mitigate the development of bacterial resistance. Recently, Zhang et al.[28] developed bioinspired MNs loaded with polymyxin to prevent bacterial infection during administration. These microneedles feature a suction-cup-like structure inspired by octopus tentacles, enabling strong adhesion in diverse environments. Additionally, the team incorporated polydopamine (PDA) into the MN base to enhance skin adhesion and confer selfhealing properties. Polymyxin is distributed across both the tips and base of the microneedles, providing effective protection against common bacteria such as *E. coli*. Collectively, these bioinspired MNs combine high adhesion capacity with potent antibacterial activity, offering promising potential for versatile transdermal drug delivery, particularly in mobile joints.

2.2 Photodynamic therapy–based antibacterial therapy

Photodynamic therapy (PDT) is an effective antimicrobial treatment that employs a laser-activated photosensitizer to generate cytotoxic reactive oxygen species (ROS), thereby eliminating bacteria.[29] When irradiated by light, the photosensitizer becomes activated, producing highly cytotoxic ROS that inflict irreversible damage on microbial cell membranes, enzymes, proteins, and nucleic acids. The antimicrobial mechanism of PDT is well-defined and typically follows these steps: (1) First, the photosensitizer absorbs photons, which causes it to transit from a ground state (S0) to an excited singlet state (Sn)[30]; (2) The intersystem crossing (ISC), then transforms from the unstable Sn into a relatively stable triplet excited state (T1)[31]; (3) In the T1 state, photosensitizer can generate ROS by two photochemical reactions. In a Type I reaction, an electron or hydrogen atom is transferred to surrounding oxygen, generating oxygenated radicals, such as superoxide anion ($\bullet O^{2-}$), hydrogen peroxide (H_2O_2), and free hydroxyl radicals ($\bullet OH$).[32] In Type II reaction, the direct reaction between the photosensitizer in the T1 state and the triplet ground state oxygen ($^3O_2$) leads to the production of highly reactive singlet oxygen ($^1O_2$).[32] While both reactions can occur simultaneously during PDT, Type II-derived $^1O_2$ is often considered the dominant antibacterial agent due to its simpler reaction mechanism compared to Type I. However, the efficacy of Type II-based PDT is highly oxygen-dependent, which may limit its practical use. Unlike conventional antibiotics, PDT exhibits rapid bactericidal activity and holds promise for treating refractory infections caused by multidrug-resistant bacteria and biofilms. Its minimally invasive nature, high efficiency, and low risk of adverse effects have made PDT an increasingly attractive option in antimicrobial therapy. Below, we detail the practical applications of both single PDT and PDT-based multimodal antibacterial strategies.

2.2.1 Single photodynamic therapy

Caffarel-Salvador et al.[33] developed dissolving MNs loaded with methylene blue, which exhibited sufficient mechanical strength to effectively treat infectious wounds. The methylene blue concentration was adjusted to match the dose range delivered by the MNs (0.1–2.5 mg/mL). To evaluate its antibacterial efficacy, biofilms treated with methylene blue were irradiated at 635 nm using a Paterson lamp. The results demonstrated that methylene blue-mediated PDT achieved >96% killing of *S. aureus* and >99% of *E. coli* and *C. albicans*, showing potent antibacterial effects.

As most of the photosensitizers are water-insoluble and have poor penetration in dense biofilms, the antibiofilm effects of PDT are usually very limited. To address the challenges of photosensitizer solubility and biofilm penetration, Wang et al.[34] designed supramolecular photosensitizers using sulfobutylether-β-cyclodextrin (SCD) and tetra(4-pyridyl)-porphine (TPyP), which were then loaded into MNs to enhance PDT's antibiofilm efficacy. The TPyP/SCD complex significantly reduced photosensitizer aggregation and improved biofilm penetration, leading to a nearly tenfold increase in ROS production. The TPyP/SCD-based MNs possessed adequate mechanical strength to pierce the biofilm's EPS, enabling localized photosensitizer delivery and subsequent PDT-mediated biofilm ablation. Furthermore, in vivo studies confirmed the antibiofilm activity of TPyP/SCD MNs in *S. aureus*-infected mice with visible abscesses.

Compared to organic photosensitizers, inorganic photosensitizers demonstrate superior photostability and unique photocatalytic properties, making them ideal for repeated PDT applications and photocatalytic-enhanced antibacterial therapy. Notably, some inorganic photosensitizers can overcome oxygen dependency, enabling repeatable PDT and continuous ROS generation. For instance, Gong et al.[35] developed Cu^{2+}-doped Zn_2GeO_4 (ZGC) luminescent nanorods with prolonged photocatalytic activity for sustained ROS production. These ZGC nanorods were integrated into MN patches (ZGC@MNs) for targeted biofilm penetration and treatment of MRSA-infected wounds. After preillumination, the released ZGC exhibited persistent photocatalytic effects, maintaining continuous ROS generation for over 48 h with potent antibacterial outcomes. In vivo studies showed that preilluminated ZGC@MNs effectively eradicated MRSA biofilms in BALB/c mice, simultaneously reducing inflammation and accelerating wound healing.

2.2.2 *Photodynamic therapy–based multimodal therapy*

As the therapeutic efficiency of a single PDT is often limited by the penetration depth of the laser and insufficient oxygen supply, the combination of PDT and other antibacterial agents (e.g., metal ions, antimicrobial peptides) has emerged as a preferred approach to improve antibacterial effect by initiating complementary antibacterial activities. Acne vulgaris, a prevalent chronic inflammatory skin disease, affects a substantial proportion of the global population, especially adolescents. In a study by Wen et al.[36] researchers developed microneedles integrated with zeolitic imidazolate framework-8 (ZIF-8) NPs for chemo-photodynamic acne treatment. The ZIF-8 NPs were engineered to encapsulate indocyanine green (ICG), enhancing its photostability, and then combined with hyaluronic acid (HA) gel to form hybrid MNs. These MNs effectively penetrated the EPS of bacterial biofilms, delivering ICG-loaded ZIF-8 NPs directly into infected tissues. Upon irradiation with an 808 nm laser, the system generated substantial ROS, significantly suppressing *P. acnes* proliferation. Additionally, the sustained release of zinc ions (Zn^{2+}) from ZIF-8 further enhanced antibacterial effects by disrupting bacterial membranes through electrostatic interactions and electron transfer-induced ROS production. The combined action of PDT and Zn^{2+} release achieved a remarkable bactericidal rate of 99.76% against *P. acnes*. In an acne mouse model, this approach reduced skin thickening, erythema, and inflammation. Overall, this innovative strategy highlights the potential of combining PDT with complementary antibacterial mechanisms to improve therapeutic outcomes for acne vulgaris and similar conditions.

Lei et al.[37] reported the chemical conjugation of chlorin e6 (Ce6) with antimicrobial peptide (AmP-Ce6, AC) to exert synergistic antibacterial effects. AC was formulated into zwitterionic micelles (DPP@AC) by self-assembling with a pH-responsive triblock polymer (DMA–PEI–PLGA) to improve the solubility of Ce6. The DPP@AC micelle was further mixed with hyaluronic acid methacryloyl (HAMA) gel to fabricate MN patches for the responsive release of antibacterial agents in the skin. Under acidic conditions, the DAP group was protonated to induce the charge conversion of the micelle from negative to positive charge, which enabled the exposure of positively charged AmP for targeted bacterial killing. DPP@AC demonstrated a higher bactericidal activity against *S. aureus* biofilms at a lower pH in vitro, resulting in an MIC_{90} value of <5 μM and a biofilm eradication rate of 90% under laser irradiation. The synergistic antibacterial effect of AmP and Ce6-mediated PDT was also confirmed in diabetic mice bearing *S. aureus* biofilm-infected wounds, which completely healed within 15 days.

2.3 Photothermal therapy–based antibacterial therapy

Photothermal therapy (PTT) is a promising approach for combating bacterial infections. By using laser-irradiated photothermal agents, PTT converts light energy into heat, effectively killing bacteria through two primary mechanisms. First, the generated hyperthermia disrupts bacterial integrity by damaging cell membranes and critical bioactive components such as DNA, proteins, and enzymes, leading to bacterial destruction.[38] Second, PTT enhances the host immune response by recruiting immune cells (e.g., macrophages and neutrophils) to the infection site.[39] Compared to conventional antibiotics, PTT offers distinct advantages by simultaneously targeting multiple cellular components and modulating the immune microenvironment. These advantages include: (1) broad-spectrum antimicrobial activity against diverse pathogens; (2) rapid treatment process (typically completed within minutes); and (3) minimal risk of inducing bacterial resistance.[40] Furthermore, PTT can effectively disrupt biofilms through localized hyperthermia that enhances biofilm permeability.

Notably, light irradiation in the second near-infrared window (NIR-II, 1000–1700 nm) demonstrates superior tissue penetration depth (reaching 2–3 cm) compared to the first near-infrared window (NIR-I, 650–900 nm;

penetration depth: 0.5–1 cm)[41,42]. This characteristic suggests that optimizing the light source wavelength, particularly through NIR–II–based PTT, could significantly enhance antibacterial efficacy, making it particularly suitable for treating deep-seated skin infections. The integration of MNs with photothermal antibacterial therapy has recently gained considerable attention in skin infection management, as it combines the respective advantages of both technologies. Representative applications are discussed below.

2.3.1 Single photothermal therapy

Research has shown that over 90% of chronic wounds contain biofilms, which trigger excessive and prolonged release of proinflammatory cytokines to impede the healing process. *S. aureus*, a common pathogen in chronic wounds, secretes gelatinase, an enzyme that contributes to biofilm persistence. To address this challenge, Lei et al.[43] developed an antibacterial MN patch capable of gelatinase-responsive release of a photothermal peptide (AMP-Cypate) (Fig. 11.2). The researchers first synthesized gelatin NPs (GNPs) and conjugated them with AMP-Cypate to form AMP-Cypate@GNPs. These NPs are specifically degraded by gelatinase, enabling targeted AMP-Cypate release at infected wound sites. The AMP-Cypate@GNPs were then loaded into microneedle tips fabricated from poly(vinylpyrrolidone) (PVP K30) and recombinant Type III collagen. Upon application, the microneedles penetrated the skin, delivering Cypate@GNPs directly into the biofilm. The overexpressed gelatinase in the infected tissue subsequently degraded the nanoparticles, releasing AMP-Cypate for photothermal antibacterial therapy. The therapeutic potential of this approach was demonstrated in a murine model of *S. aureus*-infected diabetic foot ulcers. However, conventional PTT often requires high temperatures (>60 °C) and prolonged irradiation to ensure bacterial eradication, which can cause collateral damage to healthy tissues. To

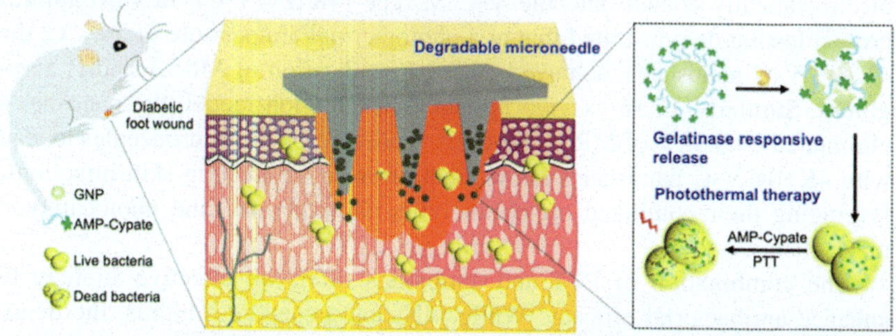

FIGURE 11.2 The degradable microneedles made of a physically inert polymer, polyvinylpyrrolidone (PVP) K-30, and a recombinant type III collagen protein to deliver antibacterial gold nanoparticles (GNPs) to the dermis of infected chronic wounds. *Reproduced from Ref. 44 with permission from Elsevier.*

mitigate this issue, mild PTT (maintained below 50 °C) has been explored as a safer alternative. Yet, its efficacy is limited by laser penetration depth and specificity. Consequently, recent studies propose combining mild PTT with complementary antimicrobial agents (e.g., antimicrobial peptides, antibiotics, or nanomotors) to enhance synergistic effects while minimizing adverse reactions.

2.3.2 Photothermal therapy–based multimodal therapy

Benefiting from the unique antibacterial activity of antimicrobial peptides, the integration of photothermal MNs with antimicrobial peptides has demonstrated great potential in affording synergistic antibacterial effects. In a recent study by Su et al., [45] a photothermal MN patch loaded with antimicrobial peptides (W379 peptide) was developed for the treatment of biofilm-infected wounds. The researchers coloaded IR780 and W379 peptides into polyvinylpyrrolidone (PVP) MNs, followed by coating them with 1-tetradecanol (TD), a phase-transition material. When exposed to 808 nm NIR irradiation, IR780 generated localized hyperthermia, melting the TD coating and accelerating the dissolution of the MNs, thereby promoting the release of the W379 peptide into the infected wound site. Compared to conventional MNs, these NIR-responsive MNs exhibited enhanced antibiofilm activity both in vitro and in vivo, demonstrating significant potential for managing biofilm-infected wounds.

In another study by Feng et al., [46] researchers developed a photothermal and photocatalytic MN patch augmented with antimicrobial peptidomimetics (4K10) for treating skin infections (Fig. 11.3). The study employed a vanadium carbide (V_2C) MXene-based two-dimensional nanosheet as a novel photothermal agent with dual photothermal and photocatalytic capabilities. Notably, the degradation products of V_2C exhibited persistent photocatalytic activity, preventing potential bacterial rebound. To further enhance antibacterial efficacy, 4K10, a peptidomimetic with membranolytic properties, was electrostatically coated onto the V_2C MXene (4K10@V_2C). In vitro antimicrobial testing demonstrated that at a photothermal temperature of 54.1°C, the 4K10@V_2C nanosheets achieved >99.9% eradication of MRSA and P. aeruginosa. Similarly, in an ex vivo human skin infection model, the nanosheets eliminated >99.98% of MRSA biofilms. Collectively, the microneedles loaded with 4K10@V_2C nanosheets show great promise for treating skin infections, leveraging their combined membranolytic, photothermal, and photocatalytic mechanisms.

The combination of PTT and antibiotics offers an effective strategy to enhance antibacterial efficacy. While MNs can penetrate biofilms, the dense and viscous EPS within biofilms hinders the diffusion and permeation of active agents, limiting their access to bacterial cells and reducing therapeutic effectiveness. To address this challenge, Yu et al.[47] developed dissolving MNs

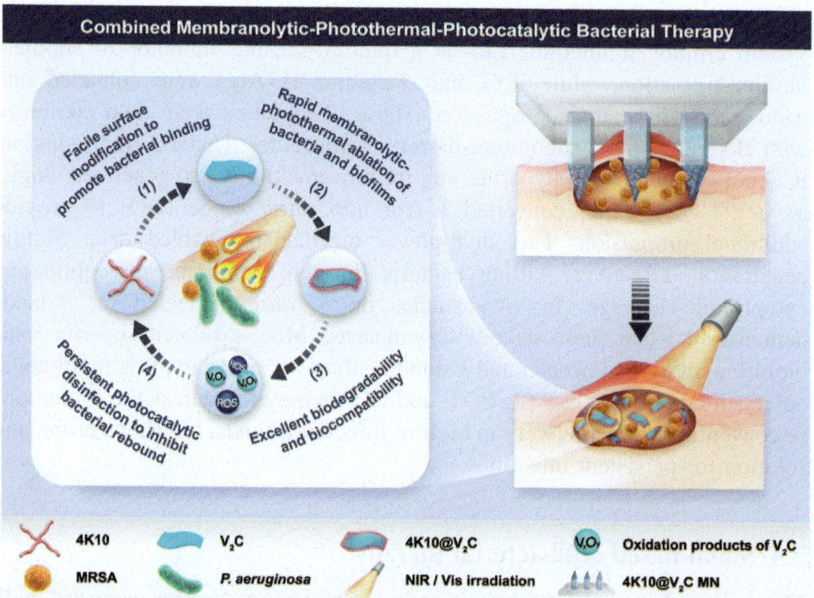

FIGURE 11.3 Microneedle-mediated transdermal delivery of 4K10@V_2C to an infected site and treatment with light and heat for focal bacterial infections. *Reproduced from Ref. 46 with permission from Elsevier.*

coloaded with α-amylase and levofloxacin-laden polydopamine NPs (PDA@Levo) for synergistic enzymolytic, antibiotic, and photothermal therapy against wound biofilms. In vitro antibiofilm assays demonstrated that this triple therapy reduced the biomass of *S. aureus* and *P. aeruginosa* biofilms to 12.6% and 31.3%, respectively. Upon application, the multifunctional MNs dissolved within 10 min under 808-nm NIR irradiation, releasing α-amylase to degrade the EPS matrix. This disruption exposed the embedded bacteria to both levofloxacin and PTT, enhancing biofilm eradication and reducing inflammation to promote wound healing. After 11 days of MN-mediated multimodal therapy, the infected wounds exhibited almost complete healing.

Although PTT and α-amylase can disrupt EPS and enhance antimicrobial penetration into biofilms, effectively reaching all bacteria within the biofilm remains challenging. The physical gap between MNs and bacterial cells limits the ability of antimicrobials to fully access and eradicate the entire bacterial population. Consequently, any surviving bacteria can regrow, leading to treatment failure. To overcome this limitation, strategies that maximize the contact area between antimicrobials and bacteria are critical. Enhancing this interaction could significantly improve the efficacy of antibiofilm therapies and yield better outcomes in treating biofilm-associated infections. To address this challenge, Chen et al.[10] developed nanomotor-integrated MN patches to actively propel antimicrobials toward bacterial cells within biofilms. The system employed luteolin (Le) as a quorum-sensing inhibitor to suppress biofilm formation, while ICG and L-arginine (L-Arg) were coloaded into nanomotors to generate propulsion. These nanomotors were then combined with HA gel to form nanomotor-driven microneedles. Under NIR irradiation, ICG not only produced hyperthermia for movement but also generated singlet oxygen ($^1O_2$), which converted L-Arg into nitric oxide (NO) to provide additional propulsion. This dual-power mechanism enabled deep biofilm penetration, effectively killing bacteria through membrane disruption and cytoplasmic leakage. In vivo studies on *S. aureus*-infected rat wounds demonstrated that these nanomotor-enhanced MNs exhibited superior anti-biofilm activity and accelerated wound healing. In summary, this multimodal approach combines Le, PTT, PDT, and NO therapy to address the limitations of conventional drug delivery in biofilm-infected wounds, offering a promising solution for persistent infections.

2.4 Metal-based antibacterial therapy

Metals have been valued for their antimicrobial properties for centuries, with silver standing as the most prominent example. The antimicrobial use of silver (Ag) dates back to ancient civilizations, where it was incorporated into coins and cutlery to inhibit bacterial and microbial growth.[48] Beyond silver, other transition metals, including copper (Cu), zinc (Zn), ferrous (Fe), and gold (Au), have also demonstrated significant antimicrobial activity. As d-block elements, these metals have found diverse applications, from medical implants and devices to water treatment systems. In the mid-20th century, the discovery of organic antibiotics marked a turning point in medicine, diminishing the reliance on metal-based antimicrobial agents.[49] However, the rise of antibiotic resistance has spurred renewed interest in metallic alternatives. Their long-standing role in infection control underscores their enduring relevance in public health. Continued research into their efficacy, mechanisms, and safety profiles will further define their place in contemporary healthcare strategies. In many cases, metals exert antibacterial effects in the form of ions, nanoparticles (NPs), or oxides through the following mechanisms: (1) Electric field adsorption-induced sterilization: Metal cations or NPs (e.g., Ag NPs) electrostatically adsorb onto the negatively charged bacterial cell wall and membrane, causing structural damage, cytoplasmic shrinkage, and leakage of cellular contents.[50] They may also penetrate cells and disrupt intracellular bioactive substances, impairing cellular function. (2) Oxidative stress induction: Metal ions or NPs catalyze oxygen to generate free radicals (e.g., $\bullet O^{2-}$, $HO\bullet$, H_2O_2) and impair reactive oxygen species (ROS) scavenging by binding to thiol groups in enzymes and glutathione disulfide (GSSG), leading to oxidative damage.[51] (3) Photocatalytic sterilization: Under UV irradiation, metal oxide NPs (e.g., ZnO NPs, TiO2 NPs) generate abundant free radicals at their surface, thereby exerting antimicrobial effects.[52]

2.4.1 Single metal-based antibacterial therapy

Both Zn^{2+} and zinc oxide (ZnO) are widely used in antibacterial therapy due to their ability to generate reactive oxygen species (ROS), which can damage bacterial proteins and cell walls. For instance, Gittard et al.[53] fabricated solid MNs from acrylate and coated them with silver or ZnO films. Agar diffusion tests demonstrated that these coated MNs exhibited strong antibacterial activity against *S. epidermidis* and *S. aureus*. In another study, Yao et al.[54] developed MNs composed of photocrosslinked methacrylated hyaluronic acid (MeHA) loaded with ZIF-8, a metal-organic framework (MOF) formed by the self-assembly of Zn^{2+} coordinated with imidazole. Their research revealed that Zn^{2+} released from ZIF-8 disrupts bacterial capsule integrity and converts oxygen molecules into free radicals, effectively killing *E. coli* and *S. aureus* to promote wound healing. Furthermore, they found that increasing the surface roughness of ZIF-8 enhanced its antibacterial activity by improving the contact area between the MOF and bacteria. Additionally, Ag^{+} has shown excellent broad-spectrum antibacterial activity against both Gram-negative and Gram-positive bacteria.

With their delicate design, MNs have enabled significant advancements in metal-based antibacterial therapy for infected wound healing. Inspired by lamprey teeth, Deng et al.[55] developed an oriented antibacterial sericin MN (OASM) system incorporating ZnO NPs to treat infected wounds. The oriented structure of MN arrays provided directional traction to promote wound closure, while the released ZnO NPs exerted antibacterial effects by generating ROS and increasing bacterial membrane permeability, thereby accelerating healing. Beyond their antibacterial capabilities, metal NP-loaded microneedles also hold promise for drug delivery and vaccination. Their ability to penetrate skin layers and deliver therapeutics directly into the dermis enhances treatment efficacy while minimizing systemic side effects compared to traditional oral or intravenous administration.

Recently, Lu et al.[56] proposed a straightforward Langmuir–Blodgett (LB) technique for fabricating antibacterial microneedles. In their study, metallic NPs (e.g., ZnO NPs, SiO_2–Ag, and ZIF-8) with varying sizes, shapes, and compositions were deposited onto poly(lactic-co-glycolic acid) (PLGA)-based MNs. Compared to the conventional dip-coating method, the LB technique offered better control over coating layers, ensuring uniform and high coverage without compromising drug-loading capacity or release kinetics. Consequently, the antibacterial MNs prepared using the LB technique exhibited superior and prolonged antibacterial performance both in vitro and in vivo, highlighting their potential for long-term transdermal drug delivery.

Despite these advancements, the development of metallic NP-loaded MNs still faces several challenges. Key issues such as limited loading capacity, uneven distribution, and complex processing parameters remain obstacles when using conventional microperfusion or coating techniques. Addressing these challenges will be essential to further advance this technology and fully realize its potential impact on global healthcare practices.

2.4.2 Chemodynamic therapy–based multimodal therapy

In the presence of H_2O_2, metal ions can catalyze the decomposition of H_2O_2 into highly toxic hydroxyl radicals (•OH) through Fenton or Fenton-like reactions, thereby killing bacteria. This process is known as chemodynamic therapy (CDT). The concept of CDT was first proposed by Shi et al.[57] in 2016 as a novel anticancer strategy. Among the commonly used CDT agents, Fe^{2+} and Fe_2O_3 are the most prominent. The antibacterial mechanism of CDT involves •OH-mediated destruction of bacterial cell membranes and critical bioactive substances (e.g., nucleic acids, proteins, and enzymes), ultimately leading to lethal damage. Additionally, •OH enhances bacterial membrane permeability, which can synergize with other treatments to improve efficacy. Unlike conventional antibiotics, CDT rarely induces bacterial resistance because it inflicts irreversible oxidative damage on essential biomolecules.

Although CDT exhibits broad-spectrum antibacterial activity, its efficacy is largely constrained by the localized H_2O_2 concentration and the inherently limited catalytic performance of CDT reagents.[58] Elevating the local temperature can accelerate Fenton or Fenton-like reactions, creating a synergistic effect with PTT to enhance antibacterial outcomes.[59] In infected wounds, the polarization of macrophages toward the M1 subtype and excessive proinflammatory cytokine release can trigger chronic inflammation, a major contributor to nonhealing wounds.[60] To simultaneously boost CDT's antibacterial efficacy and mitigate wound inflammation, Li et al.[61] developed multifunctional MM patches incorporating oxide-based NPs with dual antibacterial and immunomodulatory properties. In their design, PDA and Fe_2O_3 were first used to synthesize Fe/PDA NPs (~110 nm). GOx was then conjugated to the Fe/PDA surface via acylation, followed by HA coating to improve aqueous dispersibility and stability, yielding Fe/PDA@GOx@HA NPs (~170 nm). Meanwhile, amine-modified mesoporous silica NPs (AP-MSNs) were loaded into the microneedle tips and backing layer. The Fe/PDA@-GOx@HA NPs demonstrated dual enzymatic activity, exhibiting both GOx and POD-like catalytic properties. In the weakly acidic, high-GSH microenvironment of infected wounds, these NPs degraded, releasing GOx, Fe ions, and PDA. GOx catalyzed glucose oxidation to generate gluconic acid and H_2O_2, the latter serving as a substrate for Fe^{2+}/Fe^{3+}-mediated ·OH production to eradicate bacteria. Concurrently, PDA's high photothermal conversion efficiency enabled PTT under 808 nm laser irradiation. The combined CDT-PTT approach not only eliminated bacteria but also promoted macrophage polarization toward the antiinflammatory M2 phenotype, accelerating wound healing. Additionally, MSNs sequestered proinflammatory factors (e.g., free nucleic acids), further modulating the immune microenvironment. Remarkably, *S. aureus*-infected wounds in mice achieved complete closure within 14 days after treatment with these "all-in-one" MM patches.

2.5 Sonodynamic therapy–based antibacterial therapy

Sonodynamic therapy (SDT) is a promising treatment that utilizes nonthermal ultrasound to activate a sonosensitizer for therapeutic purposes. Although extensive research has explored SDT's mechanism, the precise process by which ultrasound activates sonosensitizers remains unclear. One widely accepted explanation involves cavitation effects, where ultrasound interacts with a liquid medium to induce the formation, growth, and collapse of microbubbles. When these bubbles collapse, they release substantial energy, generating ultrahigh temperatures and pressures at the cavitation site.[62] This energy can trigger pyrolysis-like reactions and even produce light through sonoluminescence,[63,64] which further activates the sonosensitizer to produce ROS via "hole–electron" pairs or intersystem crossing (ISC). The combined sonomechanical, sonothermal, and sonochemical effects contribute to SDT's antibacterial activity by disrupting bacterial cell walls, increasing membrane permeability, and inactivating bioactive substances.

Compared with PDT, SDT demonstrates superior efficacy in eradicating bacterial infections, particularly those deep within the skin or muscle. This enhanced performance stems from several key advantages: (1) Deeper tissue penetration: The ultrasound energy employed in SDT can reach depths exceeding 10 cm, far surpassing the limited penetration of NIR light (<3 cm) used in PDT[65]; (2) Multifaceted antibacterial mechanisms: Unlike PDT, which primarily depends on ROS with short half-lives and limited effective radius, SDT combines ROS-induced cytotoxicity with cavitation-mediated sonomechanical and sonothermal effects for more robust antibacterial activity.[66,67] Owing to these benefits, SDT has emerged as a highly promising antimicrobial therapy. Its noninvasive nature makes it appealing to both patients and clinicians, while its deep-tissue penetration enables comprehensive treatment of infections located far beneath the surface. Moreover, SDT's established track record in clinical applications further solidifies its potential as an effective antibacterial treatment.

TiO_2 is widely used inorganic sonosensitizer for SDT, but its reactive oxygen species (ROS) production efficiency is limited by the rapid recombination of electrons (e^-) and holes (h^+). To address this, researchers have recently explored modified TiO_2 NPs and alternative sonosensitizers to enhance SDT efficacy. For instance, Ouyang et al.[9] employed crystal engineering to investigate the sonocatalytic properties of TiO_2 NPs with different crystalline phases. They found that anatase-brookite TiO_2 demonstrated the highest antibacterial efficiency (99.94%) against *S. aureus* after 15 min of ultrasound irradiation. Based on this finding, they further delivered anatase-brookite TiO_2 via hyaluronic acid (HA) MNs to infected wound tissues, effectively eradicating deep-layer biofilms within 15 min of US exposure and accelerating wound healing. In another study, Liang et al.[68] developed CuO_2/TiO_2-integrated microneedles for combined sono-chemodynamic and sono-thermal antibacterial therapy (Fig. 11.4). The MNs facilitated deep dermal delivery of CuO_2/TiO_2 NPs, which, upon ultrasound activation, generated ROS and localized heat, achieving >99.9999% elimination of multidrug-resistant bacteria in vitro and in vivo within just 5 min. These results highlight the promising potential of nonantibiotic, sonosensitizer-based therapies in treating infected wounds.

MOFs have been widely used in biomedicine due to their unique structures and responsiveness to light and ultrasound. However, their sonocatalytic performance is often limited by rapid electron-hole recombination within the framework. To address this, researchers have explored structural modifications to enhance electron transfer and energy conversion efficiency. For instance, Xiang et al.[69] developed composite nanoparticles consisting of a zinc porphyrin-based MOF and ZnO (ZnTCPP@ZnO), which leveraged an interface effect to improve sonocatalytic activity and reduce the energy required for oxygen activation. These nanoparticles were then integrated into HA MNs for topical acne treatment. Under 15 min of ultrasound irradiation, the NPs generated abundant ROS, achieving a 99.73% bactericidal rate against *P. acnes* and significantly reducing acne-related inflammation. Moreover, the released Zn^{2+} ions upregulated DNA replication-related genes, further promoting fibroblast proliferation and skin repair. This engineered ultrasound-responsive MN patch demonstrated effective treatment of acne infections through ultrasound stimulation.

2.6 Starvation therapy–based antibacterial therapy

Starvation therapy (ST) has garnered significant interest in the medical field due to its potential applications in treating various diseases. ST was originally developed in the 1920s by physician Frederick Allen as a treatment for diabetes and has since expanded into other therapeutic areas.[70] In 1971, Harvard University's Professor Folkman proposed using ST to treat malignant tumors by depriving cancer cells of nutrients.[71] This concept opened new avenues in oncology and spurred further research into ST's broader therapeutic potential. One promising application of ST is in antibacterial therapy, where glucose depletion disrupts bacterial energy supply and proliferation. This approach could help combat antibiotic resistance and provide alternative treatments for bacterial infections. A key mechanism in ST involves glucose oxidase (GOx), a biocatalyst that converts glucose into gluconic acid and H_2O_2 under

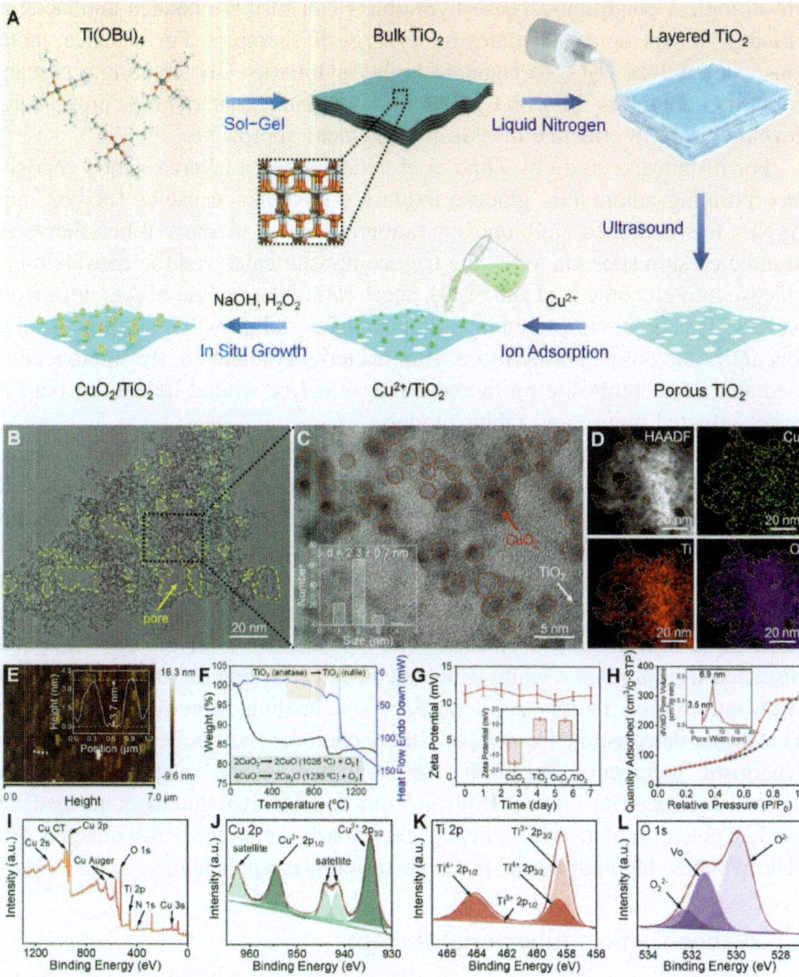

FIGURE 11.4 Synthesis and characterization of copper peroxide/titanium dioxide (CuO_2/TiO_2) heterostructures. (A) Schematic illustration of the synthetic route of CuO_2/TiO_2 nanostructures. (B, C) Transmission electron microscopy (TEM) images of CuO_2/TiO_2. Inset: The size distribution of CuO_2/TiO_2 characterized by TEM. (D) TEM mapping images of CuO_2/TiO_2: high-angle annular dark-field (HAADF) (*white*), Cu (*green*), Ti (*red*), and O (*purple*). (E) Atomic force microscopy (AFM) image of CuO_2/TiO_2. Inset: The height of CuO_2/TiO_2. (F) Thermogravimetry (TG) spectrum of CuO_2/TiO_2. (G) Zeta potential of CuO_2/TiO_2 over 7 days. Inset: Zeta potentials of CuO_2, TiO_2, and CuO_2/TiO_2 at 30 μg·mL−1. (H) Nitrogen (N_2) absorption–desorption curves and (inset) pore size distribution for CuO_2/TiO_2. (I–L) X-ray photoelectron spectroscopy (XPS) spectra of CuO_2/TiO_2: (I) survey, (J) Cu 2p, (K) Ti 2p, and (L) O 1s. *Reproduced from Ref. 68 with permission from Elsevier.*

physiological conditions. These byproducts can further enhance antibacterial efficacy by serving as substrates for synergistic therapies. For instance, metal ions can catalyze H_2O_2 to generate highly cytotoxic ·OH, creating a cascade effect that combines ST with CDT. This synergistic strategy holds promise for developing more effective disease management approaches.

For instance, a study by Zhao et al.[72] developed a layered nanocomposite incorporating apramycin, glucose oxidase (GOx) nanocapsules (nGOx), and AgNPs for combined antibiotic/starvation/metal ion therapy. When delivered to infected skin sites via MNs, the nanocomposite catalyzed the conversion of glucose into gluconic acid and H_2O_2, accelerating the release of Ag^+ ions from AgNPs. This process resulted in significant bacterial growth inhibition, even at low antibiotic doses. Furthermore, after bacterial eradication, the microneedle-mediated nanocomposite promoted rapid, scar-free wound healing in both *P. acnes*-infected mouse and rabbit models.

To address the inherent instability and rapid degradation of GOx under physiological conditions, researchers have proposed inorganic nanozymes with GOx-like activity as a promising alternative for diabetic wound treatment. For instance, Shan et al.[73] developed a catalytic microneedle patch composed of NIR–II–responsive Au–Cu_2MoS_4 (Au-CMS) nanosheets for managing infectious diabetic wounds. These nanosheets exhibited dual nanozyme activities, mimicking both GOx and catalase. They consumed local glucose to enable sustained insulin release while simultaneously catalyzing the decomposition of H_2O_2 into oxygen, thereby accelerating wound healing. Moreover, the Au-CMS nanosheets demonstrated excellent NIR-II photothermal properties, enabling a synergistic therapeutic approach that combined insulin release with photothermal antibacterial effects. Both in vitro and in vivo studies confirmed the patch's potent antibacterial efficacy, particularly against MRSA-infected diabetic wounds, highlighting its potential for clinical applications.

2.7 Nanoenzyme antibacterial therapy

The history of nanoenzyme research dates back to the late 20th century, when scientists began to focus on exploring the biocatalytic function of nanomaterials. In 2004, the term "nanoenzyme" was first proposed by Paolo Scrimin et al.[74] to describe the transphosphorylation reactivity of triazacyclononane functionalized gold NPs. In 2007, Yan et al.[75] reported the first discovery of Fe_3O_4 NPs with peroxidase-like activity, which is considered to be a pioneering study of nanoenzymes. In 2013, Wei and Wang further defined the concept of nanoenzyme as "nanomaterials with enzyme-like properties" in a review.[76] With the rapid development of nanotechnology, research on multifunctional nanozyme-based therapy has made significant progress.

Benefiting from the multiple catalytic, antibacterial, and immunomodulatory activities of nanoenzymes, nanoenzyme antibacterial therapy (NABT) has received increasing attention in treating skin infections, particularly refractory biofilm-infected or diabetic wounds.[77,78] Based on the catalytic activity, nanoenzymes are mainly divided into two types: (1) POD-like nanoenzymes, which can catalyze the decomposition of H_2O_2 into $\cdot OH$,[79] and (2) the oxidase (OXD)-like nanoenzymes, which can catalyze the decomposition of O_2 into $\cdot O_2^-$[80] (Fig. 11.5). The activity of nanoenzymes can be easily tuned and enhanced by introducing external stimuli, such as light,[81] ultrasound,[82] and metabolic regulators.[8] This enables nanoenzymes to provide multimodal antiinfection therapy by reshaping the hostile infectious, oxidative, and inflammatory microenvironment.

Biofilm infection, high glucose levels, and vascular blockages are the three major factors that hinder the healing of diabetic wounds. To remove these obstacles present in diabetic wounds, Yu et al.[7] reported a multi-enzyme cascade MN system that can hierarchically eradicate MRSA biofilms and promote angiogenesis. Specifically, α-amylase and GOx were loaded in MIL-101 NPs (MOF), followed by encapsulating in the needle of MNs for bacterial killing through vacuum casting. Methacrylic anhydride gelatin grafted with proangiogenic peptide (QHREDGS) was loaded in the base of MNs to promote tissue remodeling. After the MN patch was applied to the MRSA biofilm-infected diabetic wounds established in the mice, α-amylase was first released from the MNs to destroy the EPS surrounding the biofilm. Then, GOx consumed glucose in the wound to produce H_2O_2, which was further converted into ROS by the Fe^{3+} present in MIL-101 to eradicate bacteria encaged by the biofilm. With the decrease in tissue glucose and removal of bacteria, the inflammatory levels of wounds were decreased to promote wound healing. The QHREDGS-modified gelatin further promoted collagen deposition and

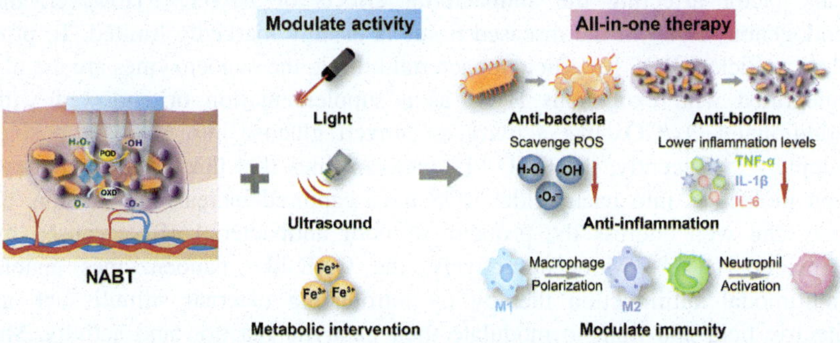

FIGURE 11.5 Schematic illustration of nanozyme-augmented antibacterial therapy (NABT) used as an "all-in-one" treatment strategy for skin bacterial infection. Based on the antibacterial mechanism, the nanoenzymes can be categorized as peroxidase-like (POD-like) and oxidase-like (OXD-like). The activity of these nanoenzymes can be modulated by external stimuli (e.g., light, ultrasound, metabolic regulators). Benefiting from the multiple antibacterial, antiinflammatory, and immunomodulatory functions, nanoenzymes enable NABT to serve as an effective "all-in-one" therapy for skin infections.

vascular remodeling for accelerated wound healing. As a result, the infectious diabetic wounds were completely healed on day 12, demonstrating the great potential of MN multi-enzyme cascade activities in managing diabetic wounds.

The healing of skin wounds usually undergoes four stages, including hemostasis, inflammation, proliferation, and remodeling. In infectious chronic wounds, the proliferation of bacteria can cause persistent inflammation, making them difficult to heal. To treat infectious chronic wounds, it would be better to eliminate bacteria and decrease the inflammation levels involved in the wounds. However, there are spatiotemporal differences in the clearance of bacteria and inflammation. To meet the demands of wound healing in different stages, Shan et al.[83] designed spatiotemporal catalytic MN patches loaded with two antagonistic nanoenzymes (termed as CMSP-MNs) to treat infected wounds. In this study, Cu_2MoS_4 (CMS) NPs with POD-like catalytic activity were loaded in the tips to generate multiple ROS for bacterial killing in the infection stage. In contrast, PDA NPs with antiinflammatory functions were loaded in the backing layers to eliminate overproduced ROS in the inflammatory phase. The CMSP-MNs displayed high bactericidal activity against MRSA and *E. coli* in vitro, eliminating 99.998% of bacteria after 1 h of treatment. In addition, the CMSP-MNs effectively scavenged ROS (e.g., H_2O_2, $O_2^{\bullet -}$, and •OH) and down-regulated the level of proinflammatory factors (e.g., TNF-α, IL-6). Benefiting from their spatiotemporal adjustable catalytic performance, the CMSP-MNs accelerated the healing of MRSA-infected diabetic wounds and pressure ulcers established in mice, which reached about 95% closure at day 12% and 100% closure at day 6, respectively.

As most of the nanoenzymes used for antibacterial therapy largely depend on their POD-like activity, the local concentration of H_2O_2 may be a significant factor affecting the antibacterial effects of NABT.[84] However, the endogenous H_2O_2 in the diseased tissue is usually markedly limited. To produce sufficient ROS for efficient bacterial killing, the nanoenzymes are mainly integrated with exogenous H_2O_2 as a supplementation or endowed with additional GOx/GOx-like activity to convert glucose into H_2O_2 as a self-supply.[78,85] Recently, the OXD-like nanoenzymes that directly convert oxygen molecules into bactericidal ROS have captured increasing attention, as they can well address the problem of poor antibacterial effect caused by deficient H_2O_2.[86] More attractively, the OXD-like nanoenzymes enable multimodal antiinfection therapy by introducing external stimuli such as electric field and light to modulate their catalytic reaction and activity. Shi et al.[87] recently developed HA MNs integrated with piezoelectric and photocatalytic nanoenzymes to provide antibacterial and antiinflammatory combination therapy for MRSA-infected wounds. Briefly, graphitic carbon nitride (C_3N_5) nanosheets were first synthesized via a thermal polymerization approach, followed by decoration with platinum–ruthenium (PtRu) nanoalloys to construct multifunctional nanoenzymes (PtRu/C_3N_5). Under the US, PtRu/

C_3N_5 showed piezoelectric-enhanced OXD-like activity, which was 3.9-fold higher than that of C_3N_5. After ultrasound for 10 min, PtRu/C_3N_5 showed a broad-spectrum antibacterial effect and killed almost 100% of the bacteria in vitro and in vivo. The PtRu nanoalloys with photocatalytic activity triggered the production of H_2 gas upon visible light irradiation to reduce inflammation levels and accelerate infectious wound healing. Benefiting from the piezoelectric and photocatalytic activities of PtRu/C_3N_5, the PtRu/C_3N_5-loaded MNs offered an "all-in-one" antibacterial and antiinflammatory therapy for infectious wounds.

In addition to the conventional treatment modalities mentioned above, iron-based nanoenzymes that can interfere the bacterial metabolism and immune cell functions have also been proposed to treat refractory bacterial infection. Zhu et al.[8] developed MN patches integrated with Fe_3O_4-based photothermal nanoenzyme to provide iron-actuated Janus ion therapy (IJIT) for refractory biofilm-induced infection. Briefly, Fe_3O_4-doped graphene oxide (FGO) nanosheets were loaded in MeHA gel to fabricate photothermal MN patches (FGO@MN) that could respond to the biofilm infection microenvironment. FGO not only catalyzes the conversion of H_2O_2 into ·OH but also sensitizes the biofilm to PTT by destroying the heat-shock proteins in bacteria. Further, FGO, with excellent photothermal performance, produced a mild photothermal treatment to trigger intracellular uptake and overload of iron, which finally caused the biofilms to undergo ferroptosis-like death. Meanwhile, the neutrophils in the biofilm infection microenvironment acquired iron as a nutrient, rejuvenating and reactivating their ability to suppress the growth of biofilms. More than 95% of biofilms were eliminated by integrating iron-sensitized PTT, iron-induced ferroptosis-like death, and iron-nourished immune reactivation. After treatment with FGO@MN for 15 days, refractory bacterial biofilm-induced infections were scavenged in biofilm-implanted mice and wound-bearing diabetic mice, suggesting that IJIT has good prospects for clinical application.

2.8 Poly(ionic liquid)-based antibacterial therapy

In recent years, ionic liquid (IL) has been exploited as a novel antibacterial agent to overcome antibiotic resistance. IL usually refers to salts that consist of definite anions and cations and have a melting point below 100 °C.[88] PIL is defined as a class of polymers made from IL monomers with highly stable physicochemical properties. Both IL and PIL have unique antibacterial mechanisms and broad-spectrum antibacterial activity. The functional cationic groups present in IL or poly(ionic liquid) (PIL), such as imidazolium, pyrrolidinium, and quaternary ammonium, form electrostatic interactions with negatively charged bacterial cell walls, and their lipophilic alkyl chains further destroy the bacterial cell membrane via hydrophobic interaction.[89,90] The combination of cationic PIL and anionic active via electrostatic interaction could be constructed as a multifunctional drug delivery system for better therapeutic outcomes. For instance, Zhang et al.[91] developed salicylic acid (SA)-loaded PIL-based MNs for enhanced treatment of acne infections. The PIL-based MNs were prepared by photo-crosslinking of an imidazolium-type IL monomer in the micromolds of MNs, followed by anion exchange with SA^-. The PIL-based MNs exerted excellent antibacterial activity against *E. coli*, *S. aureus*, and *P. acnes* and efficiently decreased the levels of proinflammatory cells and cytokines (TNF-α and IL-8) in the *P. acnes*-infected mice model, showing great potential for treating skin acne infections. In another study, Chao et al.[92] prepared four kinds of imidazole IL-based MNs and studied the effect of the carbon chain lengths of IL monomers on their antibacterial performance. It was found that the increase in substituent carbon chain lengths contributed to improved antibacterial activity. Moreover, the IL-based MNs enabled self-sterilization of the administration site, where tiny holes were created after the insertion of MNs into the skin.

In summary, have demonstrated significant success in treating skin infections. Researchers have explored novel antibacterial agents and therapies (e.g., antimicrobial peptide, PTT, PDT, MABT, CDT, SDT, ST, PIL) to effectively combat biofilm-associated skin infections. These innovative approaches enhance drug permeation into biofilms by disrupting their bioactive components. Due to their multitarget antimicrobial effects, they hold promise as potential alternatives to antibiotics, thereby mitigating the risk of multidrug-resistant bacteria. Furthermore, some promising strategies have been proposed to improve the antibacterial effect of MNs: (1) The geometry and composition of MNs critically influence their mechanical properties and drug delivery efficiency. For example, conical MNs exhibit superior mechanical strength and skin insertion capability compared to cone-cylinder, hexagonal pyramid, or rectangular pyramid designs.[93] Additionally, incorporating hyaluronidase into MNs has been shown to improve mechanical robustness and transdermal drug permeability, leading to higher drug retention in infected tissues.[94] Thus, refining MN geometry and composition offers a straightforward approach to enhancing antibacterial performance. (2) Utilizing enzyme-based (e.g., α-amylase), light-, ultrasound-, or nanomotor-driven strategies can disrupt biofilm barriers, enabling deeper drug penetration and more efficient bacterial eradication; (3) Integrating complementary bactericidal mechanisms (e.g., antibacterial, antiinflammatory, and immunomodulatory actions) can improve treatment outcomes. Advances in understanding the pathogenesis of skin infections have spurred the development of all-in-one therapies, such as NABT and IJIT, which are revolutionizing the management of chronic wound infections.

3. Conclusion and prospects

Bacterial skin infections pose a significant health burden. The growing exploration of microneedles (MNs) offers promising solutions to the challenges of subcutaneous injections and transdermal drug delivery for these infections. MNs are emerging as a versatile platform for integrating diverse antibacterial therapies. However, despite their progress in antibacterial applications, critical gaps remain in their clinical translation: (1) The long-term safety and biocompatibility of MN materials require rigorous evaluation; (2) Scalable, sterile production of MNs remains technically demanding, and high costs impede further development. For example, while PDT, PTT, CDT, and SDT introduce novel mechanisms, their specialized devices and large-scale production feasibility remain unvalidated; (3) Clinical trials regarding MNs are still limited, and standardized efficacy metrics or biocompatibility testing protocols, such as those from the FDA or EMA, have yet to be established.

To overcome these bottlenecks and accelerate translation from bench to bedside, interdisciplinary collaboration among academia, industry, and regulatory agencies is essential. Advances in materials science, fabrication techniques (e.g., 3D/4D printing), nanotechnology, and biosensors are expanding opportunities for future innovation. Although MN technology is still in its early stages, it holds great potential to redefine standards of care for bacterial skin infections. In the near future, it could enhance drug delivery efficiency, improve diagnostic accuracy, and optimize therapeutic outcomes.

References

1. Liu, W.; Gao, R.; Yang, C.; Feng, Z.; Ou-Yang, W.; Pan, X.; Huang, P.; Zhang, C.; Kong, D.; Wang, W. ECM-Mimetic Immunomodulatory Hydrogel for Methicillin-Resistant Staphylococcus aureus–infected Chronic Skin Wound Healing. *Sci. Adv.* **2022,** *8* (27), eabn7006.
2. Sartorius, B.; Gray, A. P.; Weaver, N. D.; Aguilar, G. R.; Swetschinski, L. R.; Ikuta, K. S.; Mestrovic, T.; Chung, E.; Wool, E. E.; Han, C. The Burden of Bacterial Antimicrobial Resistance in the WHO African Region in 2019: a Cross-Country Systematic Analysis. *Lancet Genet. Health* **2024,** *12* (2), e201–e216.
3. Phatale, V.; Vaiphei, K. K.; Jha, S.; Patil, D.; Agrawal, M.; Alexander, A. Overcoming Skin Barriers through Advanced Transdermal Drug Delivery Approaches. *J. Contr. Release* **2022,** *351,* 361–380.
4. Long, L.; Ji, D.; Hu, C.; Yang, L.; Tang, S.; Wang, Y. J. M. T. B. *Microneedles for in Situ Tissue Regeneration* **2023,** *19*; p 100579.
5. Wang, T.; Cornel, E. J.; Li, C.; Du, J. J. J. C. R. *Drug Delivery Approaches for Enhanced Antibiofilm Therapy* **2023,** *353*; pp 350–365.
6. Ly, K. L.; Colon-Ascanio, M.; Ou, J.; Wang, H.; Lee, S. W.; Wang, Y.; Choy, J. S.; Phillips, K. S.; Luo, X. J. B. *Dissolvable Alginate Hydrogel-Based Biofilm Microreactors for Antibiotic Susceptibility Assays* **2023,** *5*; p 100103.
7. Yu, X.; Zhao, J.; Ma, X.; Fan, D. J. C. E. J. *A Multi-Enzyme Cascade Microneedle Reaction System for Hierarchically MRSA Biofilm Elimination and Diabetic Wound Healing* **2023,** *465*; p 142933.
8. Zhu, W.; Mei, J.; Zhang, X.; Zhou, J.; Xu, D.; Su, Z.; Fang, S.; Wang, J.; Zhang, X.; Zhu, C. J. A. M. Photothermal Nanozyme-based Microneedle Patch against Refractory Bacterial Biofilm Infection via Iron-actuated Janus Ion Therapy. *Adv. Mater.* **2022,** *34* (51), 2207961.

9. Ouyang, Q.; Zeng, Y.; Yu, Y.; Tan, L.; Liu, X.; Zheng, Y.; Wu, S. J. S. Ultrasound-Responsive Microneedles Eradicate Deep-Layered Wound Biofilm Based on TiO2 Crystal Phase Engineering. *Small* **2023,** *19* (3), 2205292.
10. Chen, L.; Fang, D.; Zhang, J.; Xiao, X.; Li, N.; Li, Y.; Wan, M.; Mao, C. J. J. C.; Science, I. *Nanomotors-loaded Microneedle Patches for the Treatment of Bacterial Biofilm-Related Infections of Wound* **2023,** *647*; pp 142–151.
11. Puls, J.-S.; Brajtenbach, D.; Schneider, T.; Kubitscheck, U.; Grein, F. J. S. Inhibition of Peptidoglycan Synthesis Is Sufficient for Total Arrest of Staphylococcal Cell Division. *Sci. Adv.* **2023,** *9* (12), eade9023.
12. Hurdle, J. G.; O'neill, A. J.; Chopra, I.; Lee, R. Targeting Bacterial Membrane Function: an Underexploited Mechanism for Treating Persistent Infections. *Nat. Rev. Microbiol.* **2011,** *9* (1), 62–75.
13. Wilson, D. Ribosome-targeting Antibiotics and Mechanisms of Bacterial Resistance. *Nat. Rev. Microbiol.* **2014,** *12* (1), 35–48.
14. To, K.; Berger, M. J.; Morici, M.; Paternoga, H.; Bulatov, T.; Di Stasi, A.; Dang, T.; Mainz, A.; Raulf, K.; Crowe-McAuliffe, C. *Paenilamicins Are Context-specific Translocation Inhibitors of Protein Synthesis,* 2024; pp 1–10.
15. Jin, Q.; Xie, X.; Zhai, Y.; Zhang, H. Mechanisms of Folate Metabolism-Related Substances Affecting *Staphylococcus aureus* Infection. *Int. J. Med. Microbiol.* **2023,** *313* (2), 151577.
16. Gittard, S. D.; Ovsianikov, A.; Akar, H.; Chichkov, B.; Monteiro-Riviere, N. A.; Stafslien, S.; Chisholm, B.; Shin, C. C.; Shih, C. M.; Lin, S. J. Two Photon Polymerization-micromolding of Polyethylene Glycol-gentamicin Sulfate Microneedles. *Adv. Eng. Mater.* **2010,** *12* (4), B77–B82.
17. Keyvani, F.; GhavamiNejad, P.; Saleh, M. A.; Soltani, M.; Zhao, Y.; Sadeghzadeh, S.; Shakeri, A.; Chelle, P.; Zheng, H.; Rahman, F. Integrated Electrochemical Aptamer Biosensing and Colorimetric pH Monitoring via Hydrogel Microneedle Assays for Assessing Antibiotic Treatment. *Adv. Sci.* **2024,** *11* (41), 2309027.
18. Mir, M.; Permana, A. D.; Tekko, I. A.; McCarthy, H. O.; Ahmed, N.; Donnelly, R. P. *Microneedle Liquid Injection System Assisted Delivery of Infection Responsive Nanoparticles: A Promising Approach for Enhanced Site-specific Delivery of Carvacrol against Polymicrobial Biofilms-Infected Wounds* **2020,** *587*; p 119643.
19. Xu, J.; Danehy, R.; Cai, H.; Ao, Z.; Pu, M.; Nusawardhana, A.; Rowe-Magnus, D.; Guo, F. Microneedle Patch-Mediated Treatment of Bacterial Biofilms. *ACS Appl. Mater. Interfaces* **2019,** *11* (16), 14640–14646.
20. Ziesmer, J.; Tajpara, P.; Hempel, N. J.; Ehrström, M.; Melican, K.; Eidsmo, L.; Sotiriou, G. A. Vancomycin-loaded Microneedle Arrays against Methicillin-resistant *Staphylococcus aureus* Skin Infections. *Adv. Mater. Technol.* **2021,** *6* (7), 2001307.
21. Arshad, M. S.; Zahra, A. T.; Zafar, S.; Zaman, H.; Akhtar, A.; Ayaz, M. M.; Kucuk, I.; Maniruzzaman, M.; Chang, M.-W.; Ahmad, Z. J. P. R. *Antibiofilm Effects of Macrolide Loaded Microneedle Patches: Prospects in Healing Infected Wounds* **2021,** *38*; pp 165–177.
22. Zhang, Y.; Feng, P.; Yu, J.; Yang, J.; Zhao, J.; Wang, J.; Shen, Q.; Gu, Z. *Ros-responsive Microneedle Patch for Acne Vulgaris Treatment* **2018,** *1*; p 1800035.
23. da Silva, F. P.; Machado, M. Antimicrobial Peptides: Clinical Relevance and Therapeutic Implications. *Peptides* **2012,** *36* (2), 308–314.
24. Zhang, L.-J.; Gallo, R. Antimicrobial peptides. *Curr. Biol.* **2016,** *26* (1), R14–R19.
25. Lin, L.; Chi, J.; Yan, Y.; Luo, R.; Feng, X.; Zheng, Y.; Xian, D.; Li, X.; Quan, G.; Liu, D. Membrane-disruptive Peptides/peptidomimetics-Based Therapeutics: Promising Systems to

Combat Bacteria and Cancer in the Drug-Resistant Era. *Acta Pharmaceut. Sin. B* **2021,** *11* (9), 2609–2644.

26. Magana, M.; Pushpanathan, M.; Santos, A. L.; Leanse, L.; Fernandez, M.; Ioannidis, A.; Giulianotti, M. A.; Apidianakis, Y.; Bradfute, S.; Ferguson, A. The Value of Antimicrobial Peptides in the Age of Resistance. *Lancet Infect. Dis.* **2020,** *20* (9), e216–e230.
27. Su, Y.; Mainardi, V. L.; Wang, H.; McCarthy, A.; Zhang, Y. S.; Chen, S.; John, J. V.; Wong, S. L.; Hollins, R. R.; Wang, G. J. A. Dissolvable Microneedles Coupled with Nanofiber Dressings Eradicate Biofilms via Effectively Delivering a Database-Designed Antimicrobial Peptide. *ACS Nano* **2020,** *14* (9), 11775–11786.
28. Zhang, X.; Chen, G.; Yu, Y.; Sun, L.; Zhao, Y. J. R. Bioinspired Adhesive and Antibacterial Microneedles for Versatile Transdermal Drug Delivery. *Research* **2020**.
29. Badran, Z.; Rahman, B.; De Bonfils, P.; Nun, P.; Coeffard, V.; Verron, E. J. D. D. T. Antibacterial Nanophotosensitizers in Photodynamic Therapy: An Update. *Drug Discov. Today* **2023,** *28* (4), 103493.
30. Cieplik, F.; Deng, D.; Crielaard, W.; Buchalla, W.; Hellwig, E.; Al-Ahmad, A.; Maisch, T. J. C. Antimicrobial Photodynamic Therapy–What We Know and what We Don't. *Crit. Rev. Microbiol.* **2018,** *44* (5), 571–589.
31. Pang, X.; Li, D.; Zhu, J.; Cheng, J.; Liu, GJN-ml. *Beyond Antibiotics: Photo/sonodynamic Approaches for Bacterial Theranostics* **2020,** *12*; pp 1–23.
32. Foote, C. *Definition of Type I and Type II Photosensitized Oxidation,* **54**; Wiley Online Library, 1991; p 659.
33. Caffarel-Salvador, E.; Kearney, M.-C.; Mairs, R.; Gallo, L.; Stewart, S. A.; Brady, A. J.; Donnelly, R. Methylene Blue-Loaded Dissolving Microneedles: Potential Use in Photodynamic Antimicrobial Chemotherapy of Infected Wounds. *Pharmaceutics* **2015,** *7* (4), 397–412.
34. Wang, H.; Fu, Y.; Du, S.; Liu, P.; Ren, J.; Liu, Y.; Tao, J.; Zhang, L.; Zhu, J. Mechanically Robust Dissolving Microneedles Made of Supramolecular Photosensitizers for Effective Photodynamic Bacterial Biofilm Elimination. *ACS Appl. Mater. Interfaces* **2023,** *15* (21), 25417–25426.
35. Gong, J.-H.; Chen, L.-J.; Zhao, X.; Yan, X.-P. Persistent Production of Reactive Oxygen Species with Zn2GeO4: Cu Nanorod-Loaded Microneedles for Methicillin-Resistant *Staphylococcus aureus* Infectious Wound Healing. *ACS Appl. Mater. Interfaces* **2022,** *14* (15), 17142–17152.
36. Wen, T.; Lin, Z.; Zhao, Y.; Zhou, Y.; Niu, B.; Shi, C.; Lu, C.; Wen, X.; Zhang, M.; Quan, G.; et al. Bioresponsive Nanoarchitectonics-Integrated Microneedles for Amplified Chemo-Photodynamic Therapy against Acne Vulgaris. *ACS Appl. Mater. Interfaces* **2021,** *13* (41), 48433–48448.
37. Lei, X.-L.; Cheng, K.; Li, Y.; Zhong, Z.-T.; Hou, X.-L.; Song, L.-B.; Zhang, F.; Wang, J.-H.; Zhao, Y.-D.; Xu, Q.-R. The Eradication of Biofilm for Therapy of Bacterial Infected Chronic Wound Based on pH-Responsive Micelle of Antimicrobial Peptide Derived Biodegradable Microneedle Patch. *Chem. Eng. J.* **2023,** *462,* 142222.
38. Chen, L.; Peng, M.; Zhou, J.; Hu, X.; Piao, Y.; Li, H.; Hu, R.; Li, Y.; Shi, L.; Liu, Y. J. A. M. Supramolecular Photothermal Cascade Nano-Reactor Enables Photothermal Effect, Cascade Reaction, and In Situ Hydrogelation for Biofilm-Associated Tooth-Extraction Wound Healing. *Adv. Mater.* **2023,** *35* (31), 2301664.
39. Jiang, Y.; Huang, J.; Xu, C.; Pu, K. J. N. Activatable Polymer Nanoagonist for Second Near-Infrared Photothermal Immunotherapy of Cancer. *Nat. Commun.* **2021,** *12* (1), 742.

40. Chen, Y.; Gao, Y.; Chen, Y.; Liu, L.; Mo, A.; Peng, Q. J. J. C. R. Nanomaterials-based Photothermal Therapy and its Potentials in Antibacterial Treatment. *J. Control Release* **2020,** *328,* 251–262.
41. Hong, G.; Diao, S.; Chang, J.; Antaris, A. L.; Chen, C.; Zhang, B.; Zhao, S.; Atochin, D. N.; Huang, P. L.; Andreasson, K. I. J. N. Through-skull Fluorescence Imaging of the Brain in a New Near-Infrared Window. *Nature Photonics* **2014,** *8* (9), 723–730.
42. Wu, X.; Suo, Y.; Shi, H.; Liu, R.; Wu, F.; Wang, T.; Ma, L.; Liu, H.; Cheng, Z.-M. *Deep-tissue Photothermal Therapy Using Laser Illumination at NIR-IIa Window* **2020,** *12*; pp 1–13.
43. Lei, X.; Li, M.; Wang, C.; Cui, P.; Qiu, L.; Zhou, S.; Jiang, P.; Li, H.; Zhao, D.; Ni, X. *Degradable Microneedle Patches Loaded with Antibacterial Gelatin Nanoparticles to Treat Staphylococcal Infection-Induced Chronic Wounds* **2022,** *217*; pp 55–65.
44. Lei, X.; Li, M.; Wang, C.; Cui, P.; Qiu, L.; Zhou, S.; Jiang, P.; Li, H.; Zhao, D.; Ni, X. Degradable Microneedle Patches Loaded with Antibacterial Gelatin Nanoparticles to Treat Staphylococcal Infection-Induced Chronic Wounds. *Int. J. Biol. Macromol.* **2022,** *217,* 55–65.
45. Su, Y.; Andrabi, S. M.; Shahriar, S. S.; Wong, S. L.; Wang, G.; Xie, J. J. J. C. R. *Triggered Release of Antimicrobial Peptide from Microneedle Patches for Treatment of Wound Biofilms* **2023,** *356*; pp 131–141.
46. Feng, X.; Xian, D.; Fu, J.; Luo, R.; Wang, W.; Zheng, Y.; He, Q.; Ouyang, Z.; Fang, S.; Zhang, W. J. C. E. J. *Four-armed Host-Defense Peptidomimetics-Augmented Vanadium Carbide MXene-Based Microneedle Array for Efficient Photo-Excited Bacteria-Killing* **2023,** *456*; p 141121.
47. Yu, X.; Zhao, J.; Fan, D. J. C. E. J. *A Dissolving Microneedle Patch for Antibiotic/enzymolysis/photothermal Triple Therapy against Bacteria and Their Biofilms* **2022,** *437*; p 135475.
48. Chernousova, S.; Epple, M. Silver as Antibacterial Agent: Ion, Nanoparticle, and Metal. *Angew. Chem. Int. Ed. Engl.* **2013,** *52* (6), 1636–1653.
49. Hobman, J. L.; Crossman, L. C. J. J. Bacterial Antimicrobial Metal Ion Resistance. *J. Med. Microbiol.* **2015,** *64* (5), 471–497.
50. Tang, S.; Zheng, J. J. A. Antibacterial Activity of Silver Nanoparticles: Structural Effects. *Adv. Healthc. Mater.* **2018,** *7* (13), 1701503.
51. Liu, Y.; Lu, Y.; Xu, Z.; Ma, X.; Chen, X.; Liu, W. J. D. D. T. Repurposing of the Gold Drug Auranofin and a Review of its Derivatives as Antibacterial Therapeutics. *Drug Discov. Today* **2022,** *27* (7), 1961–1973.
52. Li, S.; Dong, S.; Xu, W.; Tu, S.; Yan, L.; Zhao, C.; Ding, J.; Chen, X. J. A. Antibacterial Hydrogels. *Adv. Sci.* **2018,** *5* (5), 1700527.
53. Gittard, S. D.; Miller, P. R.; Jin, C.; Martin, T. N.; Boehm, R. D.; Chisholm, B. J.; Stafslien, S. J.; Daniels, J. W.; Cilz, N.; Monteiro-Riviere, N. A. J. J. *Deposition of Antimicrobial Coatings on Microstereolithography-Fabricated Microneedles* **2011,** *63*; pp 59–68.
54. Yao, S.; Chi, J.; Wang, Y.; Zhao, Y.; Luo, Y.; Wang, Y. J. A. H. M. Zn-MOF Encapsulated Antibacterial and Degradable Microneedles Array for Promoting Wound Healing. *Adv. Healthc. Mater.* **2021,** *10* (12), 2100056.
55. Deng, Y.; Yang, C.; Zhu, Y.; Liu, W.; Li, H.; Wang, L.; Chen, W.; Wang, Z.; Wang, L. J. N. Lamprey-teeth-inspired Oriented Antibacterial Sericin Microneedles for Infected Wound Healing Improvement. *Nano Lett.* **2022,** *22* (7), 2702–2711.
56. Lu, Z.; Du, S.; Li, J.; Zhang, M.; Nie, H.; Zhou, X.; Li, F.; Wei, X.; Wang, J.; Liu, F. J. A. M. Langmuir–Blodgett-Mediated Formation of Antibacterial Microneedles for Long-term Transdermal. *Drug Deliv.* **2023,** *35* (38), 2303388.

57. Zhang, C.; Bu, W.; Ni, D.; Zhang, S.; Li, Q.; Yao, Z.; Zhang, J.; Yao, H.; Wang, Z.; Shi, J. Synthesis of Iron Nanometallic Glasses and Their Application in Cancer Therapy by a Localized Fenton Reaction. *Angew. Chem. Int. Ed. Engl.* **2016,** *55* (6), 2101–2106.
58. Xu, Y.; Xiao, L.; Chen, J.; Wu, Q.; Yu, W.; Zeng, W.; Shi, Y.; Lu, Y.; Liu, Y. α-Fe2O3 Based Nanotherapeutics for Near-Infrared/dihydroartemisinin Dual-Augmented Chemodynamic Antibacterial Therapy. *Acta Biomater.* **2022,** *150,* 367–379.
59. Jia, C.; Guo, Y.; Wu, F. G. J. S. Chemodynamic Therapy via Fenton and Fenton-like Nanomaterials: Strategies and Recent Advances. *Small* **2022,** *18* (6), 2103868.
60. Kharaziha, M.; Baidya, A.; Annabi, N. J. A. M. Rational Design of Immunomodulatory Hydrogels for Chronic Wound Healing. *Adv. Mater.* **2021,** *33* (39), 2100176.
61. Li, S.; Wang, X.; Yan, Z.; Wang, T.; Chen, Z.; Song, H.; Zheng, Y. J. A. S. Microneedle Patches with Antimicrobial and Immunomodulating Properties for Infected Wound Healing. *Adv. Sci.* **2023,** *10* (22), 2300576.
62. Lea-Banks, H.; O'reilly, M.; Hynynen, K. J. J. C. R. *Ultrasound-responsive Droplets for Therapy: A Review* **2019,** *293*; pp 144–154.
63. Wang, R.; Liu, Q.; Gao, A.; Tang, N.; Zhang, Q.; Zhang, A.; Cui, D. J. N. Recent Developments of Sonodynamic Therapy in Antibacterial Application. *Molecules* **2022,** *14* (36), 12999–13017.
64. Qian, X.; Zheng, Y.; Chen, Y. J. A. Micro/nanoparticle-augmented Sonodynamic Therapy (SDT): Breaking the Depth Shallow of Photoactivation. *Adv. Mater.* **2016,** *28* (37), 8097–8129.
65. Li, X.; Lee, S.; Yoon, J. J. C. S. R. Supramolecular Photosensitizers Rejuvenate Photodynamic Therapy. *Chem. Soc. Rev.* **2018,** *47* (4), 1174–1188.
66. Zhang, Y.; Zhang, X.; Yang, H.; Yu, L.; Xu, Y.; Sharma, A.; Yin, P.; Li, X.; Kim, J. S.; Sun, Y. J. C. S. R. Advanced Biotechnology-Assisted Precise Sonodynamic Therapy. *Chem. Soc. Rev.* **2021,** *50* (20), 11227–11248.
67. Guan, W.; Tan, L.; Liu, X.; Cui, Z.; Zheng, Y.; Yeung, K. W. K.; Zheng, D.; Liang, Y.; Li, Z.; Zhu, S. J. A. M. Ultrasonic Interfacial Engineering of Red Phosphorous–Metal for Eradicating MRSA Infection Effectively. *Adv. Mater.* **2021,** *33* (5), 2006047.
68. Liang, M.; Shang, L.; Yu, Y.; Jiang, Y.; Bai, Q.; Ma, J.; Yang, D.; Sui, N.; Zhu, Z. *Ultrasound Activatable Microneedles for Bilaterally Augmented Sono-Chemodynamic and Sonothermal Antibacterial Therapy* **2023,** *158*; pp 811–826.
69. Xiang, Y.; Lu, J.; Mao, C.; Zhu, Y.; Wang, C.; Wu, J.; Liu, X.; Wu, S.; Kwan, K. Y.; Cheung, K. Ultrasound-triggered Interfacial Engineering-Based Microneedle for Bacterial Infection Acne Treatment. *Sci. Adv.* **2023,** *9* (10), eadf0854.
70. Cox, C. Elizabeth Evans Hughes—Surviving Starvation Therapy for Diabetes. *Lancet* **2011,** *377* (9773), 1232–1233.
71. Folkman, J. Tumor Angiogenesis: Therapeutic Implications. *N. Engl. J. Med.* **1971,** *285* (21), 1182–1186.
72. Zhao, M.; Zhou, M.; Gao, P.; Zheng, X.; Yu, W.; Wang, Z.; Li, J.; Zhang, J. AgNPs/nGOx/Apra Nanocomposites for Synergistic Antimicrobial Therapy and Scarless Skin Recovery. *J. Mater. Chem. B* **2022,** *10* (9), 1393–1402.
73. Shan, J.; Zhang, X.; Cheng, Y.; Song, C.; Chen, G.; Gu, Z.; Zhao, Y. J. A. B. Glucose Metabolism-Inspired Catalytic Patches for NIR-II Phototherapy of Diabetic Wound Infection. *Acta. Biomater.* **2023,** *157,* 200–209.
74. Manea, F.; Houillon, F.; Pasquato, L.; Scrimin, P. Nanozymes: Gold-Nanoparticle-Based Transphosphorylation Catalysts. *Angew. Chem. Int. Ed. Engl.* **2004,** *43,* 6165–6169.

75. Gao, L.; Zhuang, J.; Nie, L.; Zhang, J.; Zhang, Y.; Gu, N.; Wang, T.; Feng, J.; Yang, D.; Perrett, S. J. N. Intrinsic Peroxidase-like Activity of Ferromagnetic Nanoparticles. *Nat. Nanotechnol.* **2007,** *2* (9), 577–583.
76. Wei, H.; Wang, E. J. C. S. R. Nanomaterials with Enzyme-like Characteristics (Nanozymes): Next-Generation Artificial Enzymes. *Chem. Soc. Rev.* **2013,** *42* (14), 6060–6093.
77. Cheng, F.; Wang, S.; Zheng, H.; Shen, H.; Zhou, L.; Yang, Z.; Li, Q.; Zhang, Q.; Zhang, H. J. S. M. Ceria Nanoenzyme-based Hydrogel with Antiglycative and Antioxidative Performance for Infected Diabetic Wound Healing. *Small Methods* **2022,** *6* (11), 2200949.
78. Wang, S.; Zheng, H.; Zhou, L.; Cheng, F.; Liu, Z.; Zhang, H.; Wang, L.; Zhang, Q. J. N. L. Nanoenzyme-reinforced Injectable Hydrogel for Healing Diabetic Wounds Infected with Multidrug Resistant Bacteria. *Nano Lett.* **2020,** *20* (7), 5149–5158.
79. Tao, Y.; Ju, E.; Ren, J.; Qu, X. J. A. M. Bifunctionalized Mesoporous Silica-Supported Gold Nanoparticles: Intrinsic Oxidase and Peroxidase Catalytic Activities for Antibacterial Applications. *Adv. Mater.* **2014,** *27* (6), 1097–1104.
80. Wang, J.; Wang, Y.; Zhang, D.; Chen, C. J. A. A. M. Intrinsic Oxidase-like Nanoenzyme Co4S3/Co (OH) 2 Hybrid Nanotubes with Broad-Spectrum Antibacterial Activity. *ACS Appl. Mater. Interfaces* **2020,** *12* (26), 29614–29624.
81. Shi, J.; Shu, R.; Shi, X.; Li, Y.; Li, J.; Deng, Y.; Yang, W. J. R. Multi-activity Cobalt ferrite/MXene Nanoenzymes for Drug-free Phototherapy in Bacterial Infection Treatment. *RSC Adv.* **2022,** *12* (18), 11090–11099.
82. Sun, D.; Pang, X.; Cheng, Y.; Ming, J.; Xiang, S.; Zhang, C.; Lv, P.; Chu, C.; Chen, X.; Liu, G. J. A. Ultrasound-switchable Nanozyme Augments Sonodynamic Therapy against Multidrug-Resistant Bacterial Infection. *ACS Nano* **2020,** *14* (2), 2063–2076.
83. Shan, J.; Zhang, X.; Wang, L.; Zhao, Y. J. S. Spatiotemporal Catalytic Nanozymes Microneedle Patches with Opposite Properties for Wound Management. *Small* **2023,** *19* (36), 2302347.
84. Chen, W. H.; González, M. V.-; Kozell, A.; Cecconello, A.; Willner, I. J. S. Cu2+-Modified Metal–Organic Framework Nanoparticles: a Peroxidase-mimicking Nanoenzyme. *Small* **2018,** *14* (5), 1703149.
85. Yan, Z.; Bing, W.; Ding, C.; Dong, K.; Ren, J.; Qu, X. J. N. AH 2 O 2-free Depot for Treating Bacterial Infection: Localized Cascade Reactions to Eradicate Biofilms In Vivo. *Nanoscale* **2018,** *10* (37), 17656–17662.
86. Singh, N.; Sherin, G.; Mugesh, G. J. A. C. Antioxidant and Prooxidant Nanozymes: From Cellular Redox Regulation to Next-Generation Therapeutics. *Angew. Chem. Int. Ed. Engl.* **2023,** *135* (33).
87. Shi, S.; Jiang, Y.; Yu, Y.; Liang, M.; Bai, Q.; Wang, L.; Yang, D.; Sui, N.; Zhu, Z. J. A. F. M. Piezo-Augmented and Photocatalytic Nanozyme Integrated Microneedles for Antibacterial and Anti-inflammatory Combination Therapy. *Adv. Funct. Mater.* **2023,** *33* (10), 2210850.
88. Zheng, L.; Li, J.; Yu, M.; Jia, W.; Duan, S.; Cao, D.; Ding, X.; Yu, B.; Zhang, X.; Xu, F. Molecular Sizes and Antibacterial Performance Relationships of Flexible Ionic Liquid Derivatives. *J. Am. Chem. Soc.* **2020,** *142* (47), 20257–20269.
89. Shi, J.; Wang, M.; Sun, Z.; Liu, Y.; Guo, J.; Mao, H.; Yan, F. J. A. Aggregation-induced Emission-Based Ionic Liquids for Bacterial Killing, Imaging, Cell Labeling, and Bacterial Detection in Blood Cells. *Acta Biomater.* **2019,** *97,* 247–259.
90. Qin, J.; Guo, J.; Xu, Q.; Zheng, Z.; Mao, H.; Yan, F. J. A. Synthesis of Pyrrolidinium-type Poly (Ionic Liquid) Membranes for Antibacterial Applications. *ACS Appl. Mater. Interfaces* **2017,** *9* (12), 10504–10511.

91. Zhang, T.; Sun, B.; Guo, J.; Wang, M.; Cui, H.; Mao, H.; Wang, B.; Yan, F. J. A. B. *Active Pharmaceutical Ingredient Poly (Ionic Liquid)-Based Microneedles for the Treatment of Skin Acne Infection* **2020,** *115*; pp 136–147.
92. Chao, X.; Zhang, C.; Li, X.; Lv, H.; Ling, G.; Zhang, P. J. B. Synthesis and Characterization of Ionic Liquid Microneedle Patches with Different Carbon Chain Lengths for Antibacterial Application. *Biomater. Sci.* **2022,** *10* (4), 1008–1017.
93. Li, Y.; Hu, X.; Dong, Z.; Chen, Y.; Zhao, W.; Wang, Y.; Zhang, L.; Chen, M.; Wu, C.; Wang, Q. *Dissolving Microneedle Arrays with Optimized Needle Geometry for Transcutaneous Immunization* **2020,** *151*; p 105361.
94. Huang, Y.; Peng, T.; Hu, W.; Gao, X.; Chen, Y.; Zhang, Q.; Wu, C.; Pan, X. J. J. C. R. *Fully Armed Photodynamic Therapy with Spear and Shear for Topical Deep Hypertrophic Scar Treatment* **2022,** *343*; pp 408–419.

Chapter 12

Microneedle technology for cosmetic applications

Yanping Fu and Guilan Quan
State Key Laboratory of Bioactive Molecules and Druggability Assessment/International Cooperative Laboratory of Traditional Chinese Medicine Modernization and Innovative Drug Development of Ministry of Education (MOE) of China/College of Pharmacy, Jinan University, Guangzhou, China

1. Pathological scars

Pathological (Fig. 12.1) scars are defined as abnormal skin tissues that result from an aberrant wound healing process following a variety of injuries including surgeries, burns, scalds, and severe trauma.[1] In general, scarring occurs as the result of an injury. During healing, the body may choose to generate fibrotic tissue expeditiously rather than attempting to reconstruct the original tissue structure. This represents an intense healing response that determines the type of scar, whether it is a keloid or hypertrophic one. In contrast to hypertrophic scars, which remain confined to the boundaries of the wound, keloid scars expand beyond those boundaries.[2,3] Pathological scars are characterized by a raised, inflamed, and red surface, and frequently result in discomfort, pruritus, and skin tightening. In comparison to healthy skin, pathological scars exhibit a reduction in the number of ancillary structures, including hair follicles, skin glands, and dermal papillae. Such structural alterations can also impede mobility and prevent the body from effectively cooling itself because of the absence of sweat glands. The resulting physical, psychological, and cosmetic issues are numerous and significant.[4] Therefore, the prevention and treatment of scars are of the utmost importance.

1.1 Wound healing and scar formation

Pathological scarring is an aberrant consequence of the wound healing process, emerging from malfunctions that occur during the various stages of healing.[5] The process of wound healing is composed of four consecutive overlapping

Microneedle Technology. https://doi.org/10.1016/B978-0-443-45544-5.00005-1

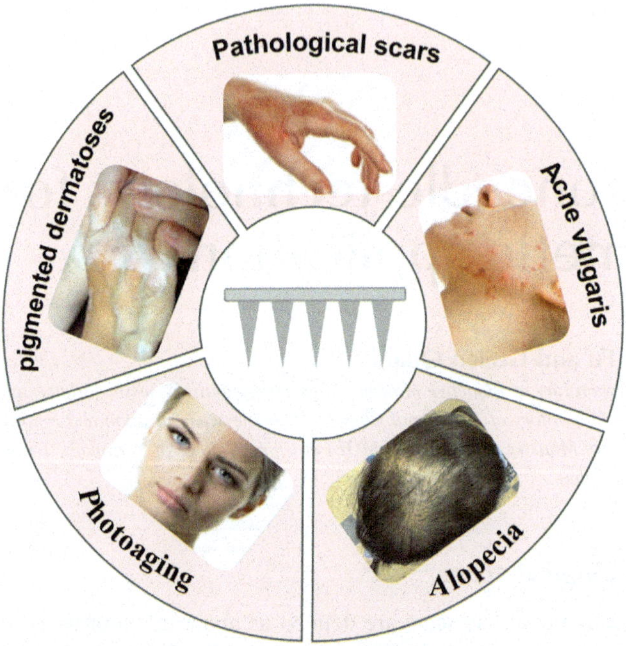

FIGURE 12.1 Microneedle technology for cosmetic applications.

phases that commence in response to tissue injury: hemostasis, inflammation, proliferation, and remodeling (Fig. 12.2).

In the event of an injury, the need for hemostasis is immediate and urgent. This process is initiated by vasoconstriction, followed by the aggregation of platelets and their direct interaction with degranulation. Subsequently, this initiates the clotting cascade, resulting in the formation of a blood clot composed of fibronectin cross-linked with fibrin. The clotting action serves two purposes: first, it curtails excessive blood loss; and second, it functions as a provisional scaffold. This scaffold establishes a chemotactic gradient that facilitates repair. Moreover, it forms a provisional eschar that acts as a barrier, minimizing the invasion of microorganisms.[7,8]

The second stage of wound healing, inflammation, is a protective response that occurs from vascular damage. In this phase, immune cells begin to migrate toward the wound area. The activation and recruitment of these immune cells are triggered and rely on signals produced by damage- and pathogen-associated molecular patterns, along with several crucial chemokines. Fibrocytes, a distinct subpopulation of leukocytes, undergo proliferation at the wound site after injury and transform into fibroblast-like cells. These fibrocytes have been linked to the development of hypertrophic and keloid scars.[9,10]

The third phase of wound healing, proliferation, occurs approximately 2–10 days following the initial injury. During this period, new tissue forms

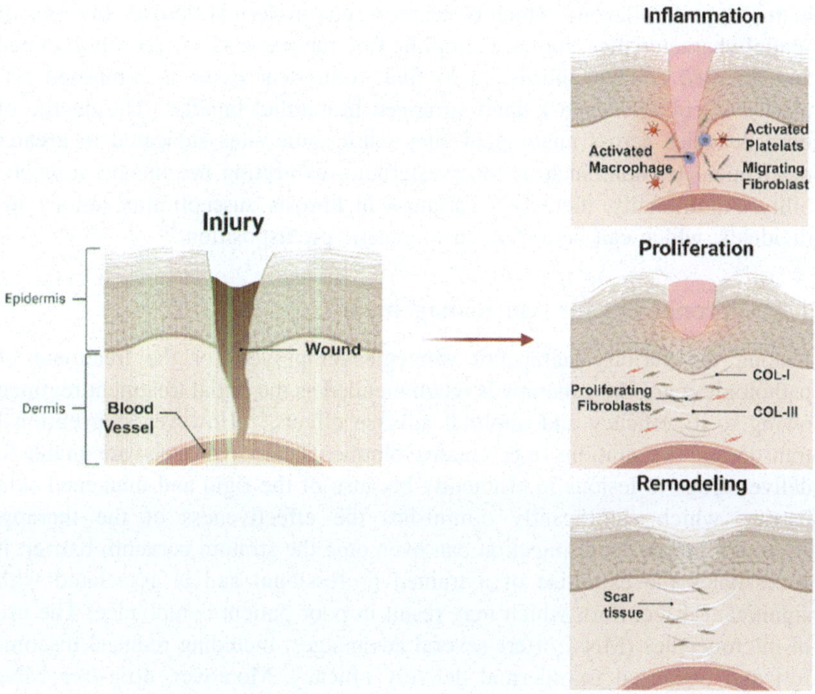

FIGURE 12.2 Process of wound healing from injury to scar.[6]

within the wound through a series of processes including reepithelialization, the generation of extracellular matrix (ECM), the development of granulation tissue, and neovascularization. This phase is characterized by the proliferation and migration of a variety of cellular types. In the initial stages of the process, keratinocytes migrate across the damaged region. Subsequently, the formation of new blood vessels becomes evident. Capillary sprouts associated with macrophages and fibroblasts replace the granulation tissue in lieu of the fibrin matrix. Subsequently, collagen, a principal component of the ECM, is produced and deposited by myofibroblasts and fibroblasts. During excessive collagen production, the formation of scars is inevitable. It has been observed that maintaining equilibrium between the deposition and degradation of ECM components is of paramount importance, as any disruption to this process may result in the formation of pathological scars.[11,12]

The remodeling phase is distinguished by the active reorganization of the ECM, which results in an increase in the total number of cross-links among collagen fibers, facilitating the maturation of scars. This phase represents the inevitable conclusion of the healing process for cutaneous wounds. Subsequently, injured tissue is replaced by connective tissue rich in collagen type I (during the early repair stages, immature type III collagen is replaced by mature type I collagen), which typically results in dermal fibrosis. In terms of spatial characteristics, this is exemplified by the presence of densely bundled and oriented collagen fibrils. In its final form, scar tissue is composed primarily of collagen type I fibers arranged in parallel bundles. The degree of fibrosis varies across anatomical sites, with some sites subjected to greater mechanical tension, such as the presternum, exhibiting heightened susceptibility. Additionally, there is a variation in fibrosis susceptibility among individuals, which can be attributed to genetic predisposition.[13]

1.2 Microneedles for scar management

Among the various therapeutic strategies employed for the treatment of pathologic scars, drug therapy is recommended as the initial treatment regimen owing to its efficacy and minimal adverse effects.[14] However, conventional transdermal preparations (e.g., creams, ointments, and hydrogels) are unable to deliver drugs to lesions in efficiently because of the rigid and thickened skin barrier, which significantly diminishes the effectiveness of the therapy. Although intralesional injection can overcome the stratum corneum barrier, it necessitates the expertise of a trained professional and is associated with significant discomfort, which may result in poor patient compliance. The use of microneedles (MNs) offers several advantages, including reduced discomfort and enhanced transdermal delivery efficacy. Moreover, drug-free MNs enable effective mechanotherapy for antiscarring by reducing fibroblast-generated mechanical stress and attenuating integrin-focal adhesion kinase signaling.[15] The therapeutic efficacy of MNs can be readily modified by changing their density and size.[16] These findings suggest that MNs may represent a promising approach to managing scars.

Treatment of hypertrophic scars. The use of MNs for the delivering assorted therapeutic agents (e.g., chemicals, photosensitizers, and nucleic acid drugs) has been explored by numerous researchers as a prospective avenue for the efficacious treatment of hypertrophic scars. However, two significant obstacles impede the efficacy of MNs delivery. (1) The rigid stratum corneum represents the initial barrier against drug permeation, necessitating that MNs possess sufficient mechanical strength to penetrate the stiff stratum corneum.[4] (2) Furthermore, the dense ECM presents an additional barrier to the diffusion of drugs to their intended cells.

To surmount these two physical impediments, Lin et al.[4] constructed mechanically enhanced MNs containing triamcinolone (TA) for the treatment of hypertrophic scars (Fig. 12.3A). Hydroxypropyl-β-cyclodextrin was added to sodium hyaluronic acid (HA), the needle material. The hydrogen interaction between hydroxypropyl-β-cyclodextrin and HA restricted the mobility of the molecular chains and subsequently increased the mechanical strength of the MNs. Hydroxypropyl-β-cyclodextrin also increased the drug loading of TA in MNs by increasing the solubility of TA. It was found that the scar elevation

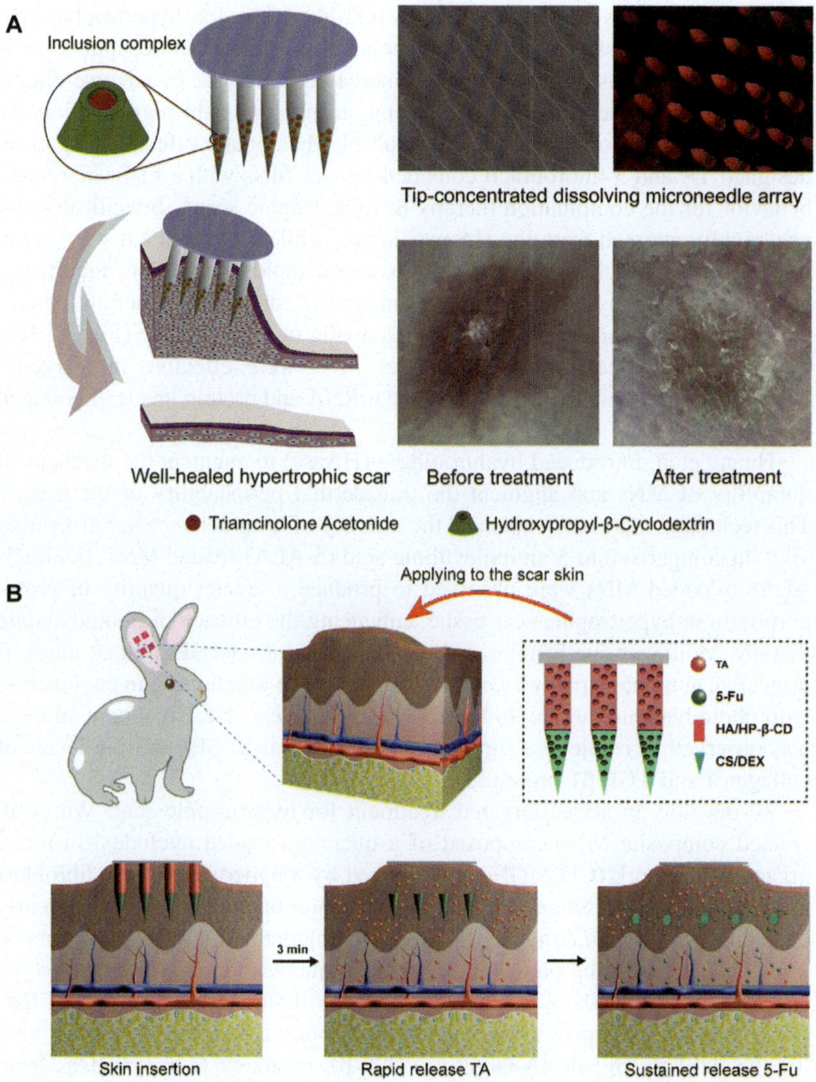

FIGURE 12.3 (A) Schematic illustration of mechanically enhanced microneedles (MNs) containing triamcinolone (TA) for the treatment of hypertrophic scars.[4] (B) Schematic illustration of bilayer MNs with biphasic release behavior for the combination therapy of hypertrophic scar.[17] *HA*, hyaluronic acid.

index (SEI) decreased to normal, together with the downregulation of mRNA expressions of collagen I and transforming growth factor-β1 (TGF-β1) following the administration of TA-MNs. In addition, physical stimulation mediated by drug-free dissolving MNs (DMNs) promoted the regression of scars.

Although TA is the recommended first-line drug for hypertrophic scar therapy, a high dosage is typically necessary to achieve an acceptable antiscarring effect. This may result in an increased incidence of adverse effects associated with glucocorticoids, including atrophy and skin pigmentation. To reduce the dosage of TA and obtain multiple therapeutic effects, Yang et al. designed TA and 5-fluorouracil coloaded bilayer MNs with a biphasic release behavior for the combination therapy of hypertrophic scar.[17] Specifically, TA was rapidly released from the HA needle tail, while 5-fluorouracil was slowly released from the chitosan needle tip to avoid rapid elimination, facilitating combination therapy. The bilayer MN-mediated combination therapy exhibited superior antiscarring efficacy in comparison to monotherapy (Fig. 12.3B). The findings indicated that the bilayer MNs were effective in markedly reducing the SEI and downregulating the mRNA and protein levels of collagen I and TGF-β1.

Huang et al. introduced hyaluronidase (HAase) to augment the mechanical durability of MNs and augment the transdermal permeability of the drug.[18] This technique markedly enhanced the efficacy of drug delivery in vitro and in vivo. In comparison to 5-aminolevulinic acid (5-ALA)-loaded MNs, HAase/5-ALA-co-loaded MNs were observed to produce a greater quantity of protoporphyrin in hypertrophic scar tissue, enhancing the efficacy of photodynamic therapy. Moreover, metformin-loaded MNs were introduced to block the self-protective autophagy pathway, with the objective of functioning in conjunction with photodynamic therapy to treat hypertrophic scar. This treatment strategy was observed to result in a significant reduction in the SEI and the levels of collagen I and TGF-β1 proteins.

To develop an active targeted treatment for hypertrophic scar, Wu et al. created composite MNs composed of a quercetin-loaded cyclodextrin metal organic framework (CD-MOF) camouflaged by a hypertrophic scar fibroblast membrane.[19] The MN mediated delivery of biomimetic CD-MOF efficiently decreased the SEI and collagen I/III levels by regulating the Wnt/β-catenin and JAK2/STAT3 signaling pathways. This technique was more effective than the use of nonfunctionalized CD-MOF with or without the assistance of MNs. Wang et al.[20] developed an upconversion nanoparticle-embedded MN system for the delivery of siRNA-targeting TGF-β1 receptors to be used as gene therapy for hypertrophic scar. The porous silica shell was employed to load siRNA, facilitating enhanced cellular internalization. The core, composed of upconverted nanoparticle, enabled the monitoring of gene delivery and gene expression via imaging by upconversion luminescence. Once internalized, the released siRNA silenced the target mRNA and inhibited fibroblast hyperproliferation and collagen overproduction.

Treatment of keloids. To assess the efficacy and safety of TA-loaded MN arrays for the treatment of keloids, Tan et al.[21] conducted a two-phase clinical trial that lasted for 8 weeks. Two keloids per patient were treated for 4 weeks with MNs applied daily, followed by a subsequent 4-week period of no intervention. In comparison to the control group, the group treated with MNs demonstrated a notable reduction in keloid volume, which was observed to be dose-dependent. Park et al.[22] developed stainless-steel solid MNs coated with 5-fluorouracil–loaded carboxymethyl chitosan nanoparticles to treat keloid scars. Nanoparticles released from the MNs were observed to diffuse into the skin, inhibiting the proliferation of human keloid fibroblasts.

Treatment of acne scars. In contrast to pathological scars, acne scars are characterized by the presence of dermal depressions and the destruction of collagen, which occurs as a consequence of the inflammatory process associated with acne recovery.[23] It is estimated that approximately 95% of individuals with acne will develop a series of facial scarring, which has been demonstrated to have a significant impact on psychological well-being.[24] A number of therapeutic regimens have been investigated for the treatment of acne scars, including surgical techniques (e.g., subcision and punch graft), ablative and nonablative laser treatment, and the use of dermal fillers.[23] In a study, minimally invasive microneedling with a dermaroller was employed as an alternative strategy to treat atrophic and hypertrophic acne scars in a clinical setting. This approach has been demonstrated to reduce the risk of depigmentation, scarring, and infections with prolonged healing.[23] As a collagen induction therapy, repetitive skin puncture using microneedling can also facilitate the release of various growth factors for neovascularization and neocollagenation, promoting the leveling of atrophic scars. The combination of microneedling with therapeutic agents such as platelet-rich plasma, [25] 5-ALA,[26] and glycolic acid peel,[27] has been demonstrated to be a more effective treatment for scars, making it a favorable alternative to invasive techniques.

2. Acne vulgaris

Acne vulgaris is one of the most prevalent dermatological conditions, affecting approximately 650 million individuals globally.[28] Acne vulgaris is regarded as a chronic ailment owing to its extended duration, the characteristic pattern of reoccurrence and regression, and the diverse manifestations it presents, such as sudden and intense flare-ups or a gradual and insidious commencement. Severe forms of acne vulgaris are not only a source of significant discomfort, they can also result in deformities and scarring. For some patients, the condition can have a profound impact on self-esteem and mental well-being.[29]

2.1 Pathophysiology of acne vulgaris

Acne vulgaris is a multifactorial dermatological disorder that primarily affects the pilosebaceous unit. It involves complex interactions among sebaceous glands, keratinocytes, and the skin microbiota[30] (Fig. 12.4). It is commonly observed during the pubertal period, owing to hormonal fluctuations that result

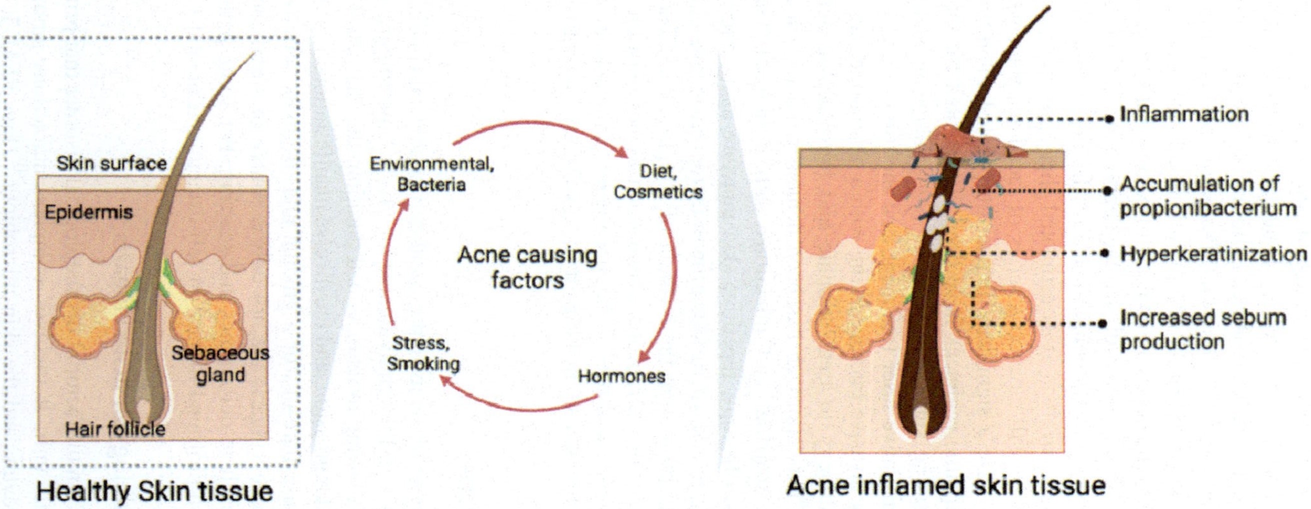

FIGURE 12.4 Factors that contribute to the formation and development of acne vulgaris.[32]

in an increase in sebum production, a critical factor in the development of acne. Sebum is a complex mixture of triglycerides, wax esters, squalene, and fatty acids, secreted by sebocytes in the sebaceous glands. In the absence of underlying pathology, sebum plays a beneficial role in maintaining the integrity of the skin barrier. However, excessive sebum production, particularly under the influence of androgens, can contribute to the formation of microcomedones, the precursors to acne lesions. The initial step in the pathogenesis of acne is follicular hyperkeratinization, which refers to the abnormal shedding of keratinocytes, leading to the obstruction of hair follicles.[31]

This blockage traps sebum within the follicle, creating an environment conducive to the growth of *Propionibacterium acnes (P. acnes)*, a bacterium that naturally inhabits the skin. The accumulation of sebum and the overgrowth of *P. acnes* induce an inflammatory response characterized by the recruitment of immune cells, including neutrophils and T lymphocytes, to the site of infection. Proinflammatory cytokines, including IL-1β, IL-8, and TNF-α, are upregulated in response to this inflammatory process, exacerbating the condition. *P. acnes* has a crucial role in acne inflammation by interacting with the immune system. Specific strains of *P. acnes*, particularly type III, have been linked to more severe inflammatory responses.[33] These strains have been observed to upregulate the expression of proinflammatory molecules, including matrix metalloproteinases, which have been shown to contribute to tissue damage and scarring. Additionally, *P. acnes* have been demonstrated to induce the activation of the NLRP3 inflammasome in sebocytes, leading to the release of IL-1β and further intensifying the inflammatory cascade.[29,34,35]

Moreover, genetic factors have been identified as a contributing factor in the development of acne. A number of genetic polymorphisms have been identified, including those related to insulin-like growth factor 1 and the *PPARG* gene, which may influence sebum production and the inflammatory response. Furthermore, multiple susceptibility loci have been identified in diverse populations, including Han Chinese and European descent groups, which suggests a genetic predisposition to acne.[36] Environmental factors, including diet, stress, and lifestyle, also influence the pathogenesis of acne. High-glycemic diets and increased androgen levels can exacerbate sebum production, whereas stress-induced neuropeptides such as substance P may stimulate sebaceous gland proliferation and activity. These factors, in combination with genetic susceptibility and microbial dysbiosis, create a highly conducive environment for the development of acne vulgaris.[30]

2.2 Microneedles for acne vulgaris treatment

Previous studies indicated that oral retinoic acid has the potential to reduce sebum secretion effectively and exert a beneficial impact on the treatment of acne. Nevertheless, the clinical application of retinoic acid is constrained by its teratogenic adverse effects.[31,37] Fractional radiofrequency MNs (FRMs) are being utilized as minimally invasive devices to treat acne and rejuvenation of the skin. The device has the capacity to induce long-term dermal remodeling in the reticular dermis, including neoelasticization and neocollagen formation,[38–40] while avoiding the adverse effects associated with retinoic acid. In a study conducted by Lee et al., the therapeutic efficacy of the FRM device was evaluated in Asian patients with acne vulgaris, and objective measurements were proposed to explore its impact on sebum secretion.[41] The study included 20 patients with acne vulgaris in South Korea, who received a single FRM treatment. The results demonstrated that FRM had an evident inhibitory effect on sebum secretion in patients with acne vulgaris.

To enhance transdermal drug penetration and treatment efficacy, Zhang et al. developed a drug-loaded reactive oxygen species (ROS)-responsive MN patch utilizing a polyvinyl alcohol matrix (Fig. 12.5A). This approach enables the targeted release of drugs in a controlled manner while minimizing potential adverse effects.[42] Compared with the commonly used antiacne cream, the MN system could significantly promote transdermal drug penetration, improving efficacious treatment for dermal lesions. In addition, HA and diatomite could form a patch with high physical adsorption capacity to absorb pus and dead cell fragments, which was beneficial for skin healing. An in vivo study in a mouse acne model showed that this system could also effectively reduce skin swelling and inhibit bacterial growth. In another similar study, Zhang and colleagues developed an active pharmaceutical ingredient, a poly(ionic liquid)-based MN patch containing salicylic acid for *P. acnes* infection treatment.[43] Studies found that MNs with high mechanical strength could generate microchannels in the skin, effectively injecting salicylic acid into the skin to exert antibacterial effects. In addition, in a mouse antiinfection model, this system effectively eliminated mouse ear swelling and inhibited the growth of *P. acnes* compared with free drug. Simultaneously, this system effectively reduced the number of inflammatory cells to suppress the secretion of inflammatory factors. Wen et al.[44] designed an MN patch integrated with bioresponsive-inspired nanoarchitectonics for the eradication of *P. acnes*-induced acne vulgaris through combined chemo-photodynamic therapy. Specifically, the photosensitizer indocyanine green (ICG) was loaded into the zeolitic imidazolate framework-8 (ZIF-8) to improve photostability, which would be triggered by 808-nm laser irradiation to generate cytotoxic ROS to result in oxidative damage and disturbed metabolic activities of *P. acnes*. The fabricated ZIF-8-ICG@MNs had outstanding synergistic antiacne efficiency both in vitro and in vivo (Fig. 12.5B).

3. Alopecia

The prevalence of alopecia has increased considerably in society; approximately 50% of males and 15%–30% of females experience hair loss.[45] Alopecia is a common autoimmune disorder with a multitude of potential

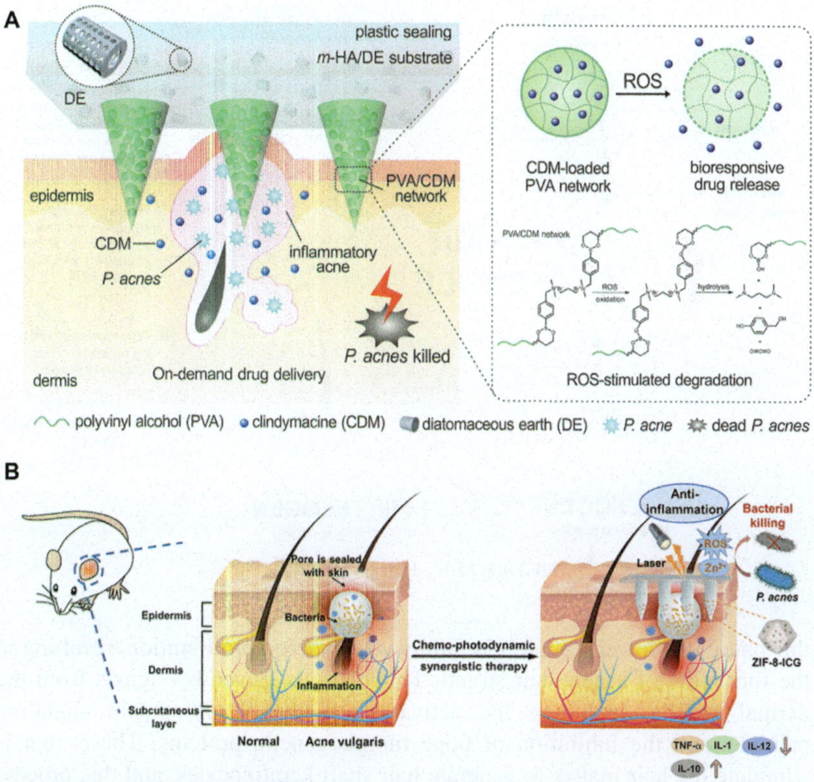

FIGURE 12.5 (A) Schematic illustration of reactive oxygen species (ROS)-responsive microneedle (MN) patch for the treatment of acne vulgaris.[42] (B) Schematic illustration of multifunctional MNs for the treatment of *Propionibacterium acnes*-induced acne vulgaris through combined chemo-photodynamic therapy.[44] *ICG*, indocyanine green; *ZIF-8*, zeolitic imidazolate framework-8.

etiologies, including medication, aging, underlying disease, and psychological stress. It has a profound impact on an individual's physical appearance, social interactions, and emotional well-being.

3.1 Hair growth cycle

The cyclical rounds of anagen (proliferation and growth phase), catagen (apoptosis and regression phase), telogen (resting phase), and exogen (shedding phase) that human hair follicles undergo play a crucial role in the regeneration and shedding of hair[46] (Fig. 12.6). This cycle is regulated by hair follicle stem cells, signaling molecules, and interactions between epithelial and mesenchymal cells.[47]

Anagen phase. The anagen phase represents the active growth stage of the hair follicle, characterized by rapid cell division and differentiation. During

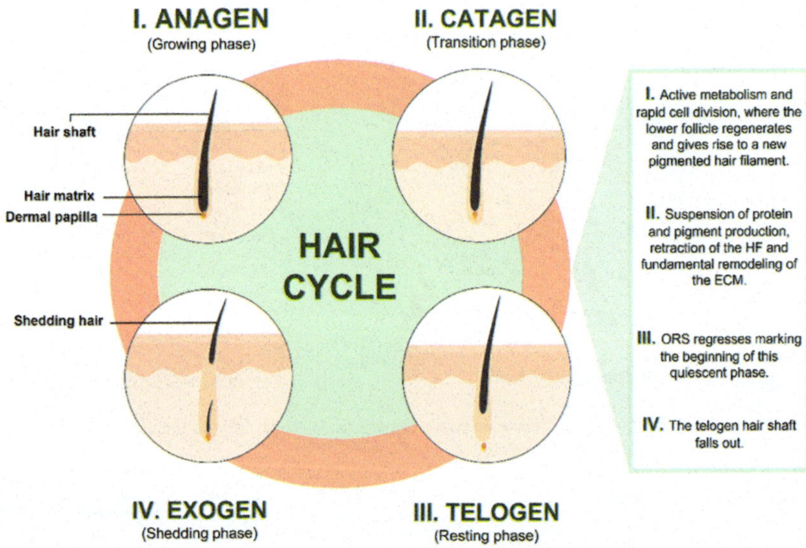

FIGURE 12.6 Hair growth cycle.[46]

this phase, the lower follicle's keratinocytes undergo proliferation, resulting in the formation of a new hair strand. This phase is driven by signals from the dermal papilla, including the activation of the Wnt/β-catenin signaling pathway and the inhibition of bone morphogenetic proteins. These signals stimulate the hair matrix to generate hair shaft keratinocytes, and this process also induces the proliferation of bulge stem cells through sonic hedgehog signaling. As the hair follicle undergoes regeneration, cells from the hair germ at the base of the follicle facilitate downward hair growth, contributing to the continued growth phase. It is estimated that approximately 85% of the scalp's hair is in the anagen phase at any given time, with this phase lasting between 2 and 7 years.[48]

Catagen phase. Subsequent to anagen, the hair follicle commences the catagen phase, which is characterized by the cessation of hair production and the advent of follicular regression. During the catagen phase, the hair follicle undergoes a reduction in size and the lower portion of the follicle undergoes apoptosis, while the upper portion becomes attached to the dermal papilla. This phase is initiated by the upregulation of fibroblast growth factor (FGF)-5, which acts as a key regulator of catagen. This stage lasts for approximately 2 to 4 weeks and represents a critical time for follicle remodeling, setting the stage for the next anagen cycle.[48,49]

Telogen and exogen phase. The telogen phase represents the resting period of the hair follicle cycle. During this phase, the follicle is in a state of dormancy, with the cessation of hair production. However, the follicle itself remains intact beneath the skin surface. The outer root sheath undergoes regression, and minimal DNA and RNA are produced by the follicle cells.[50] The dermal papilla remains situated beneath the bulge, and interactions between mesenchymal and epithelial cells maintain the follicle in a quiescent state. Approximately 10%–15% of the scalp's hair is in the telogen phase at any given time, with this phase lasting approximately 3 months. Subsequently, external or internal stimuli prompt the reactivation of the follicle, initiating the regeneration of the hair follicle and the return to the anagen phase.[51] The exogen phase represents a recent addition to the understanding of the hair cycle. It involves the shedding of the old hair shaft at the end of the telogen phase, which occurs in a regulated, active manner rather than a passive process. Following exogen, the follicle begins a new cycle marked by the reactivation of stem cells in the bulge area and the regeneration of the follicle, leading into a new anagen phase.[52,53]

3.2 Microneedles for alopecia treatment

Androgenetic alopecia (AGA) and alopecia areata (AA) represent the two principal categories of alopecia. Additionally, AGA is often considered a form of male pattern baldness, accounting for over 90% of cases of alopecia. AGA is distinguished by a horseshoe pattern, in which hair loss typically occurs in the temporal and vertex regions, while the occipital region exhibits sparse hair.[54] AA presents as glossy, well-demarcated, and round spots of hair loss on the scalp that do not atrophy to a specific area.[45] Topical drug therapy represents the primary treatment method for stimulating hair regrowth. Numerous researchers have investigated methods to enhance the efficiency of drug delivery to hair follicles. With the exception of transdermal drug delivery, MNs may serve as an adjunctive strategy for facilitating hair regrowth in alopecia, as evidenced by the observation that genes associated with hair growth, including Wnt3a, β-catenin, vascular endothelial growth factor (VEGF), and Wnt10, exhibit a notable increase following repeated MN stimulation. Many studies have reported that the use of MNs in the treatment of AGA and AA is an effective approach.[55]

Treatment of AGA. AGA is a chronic progressive disease with hair follicle niche disorders. The use of specific type II 5α reductase inhibitors such as finasteride (FNS) to halt the conversion of testosterone into dihydrotestosterone was proposed as the primary treatment for AGA. However, the oral administration of FNS tablets has been associated with significant adverse effects, including decreased libido, erectile dysfunction, and ejaculation disorders.[56,57] The intradermal delivery of FNS via MNs and the topical treatment of AGA can enhance therapeutic outcomes while circumventing the potential for systemic adverse effects. Kim et al.[56] developed FNS powder-carrying MNs with a topical diffusion enhancer to enhance the dissolution and release of FNS in the skin. This system facilitated the sustained release of FNS for 3 days and resulted in a greater quantity and density of regenerated hair than was observed with the topical FNS gel.

It is widely accepted that the two primary causes of AGA are excessive ROS production and insufficient vascularization surrounding hair follicles. Accordingly, Yuan et al.[58] synthesized CeO_2 nanoenzyme-loaded MNs to scavenge ROS in the perifollicular region and trigger neovascularization through mechanical stimulation (Fig. 12.7A). The technique demonstrated

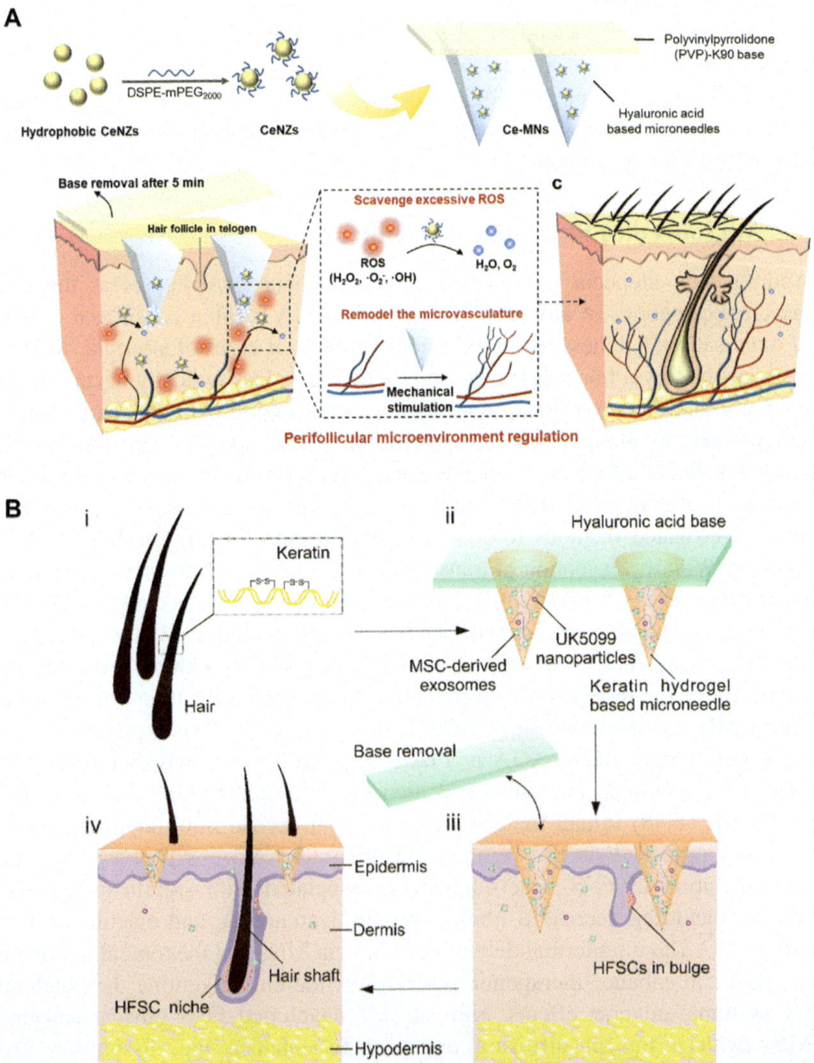

FIGURE 12.7 (A) Schematic illustration of androgenetic alopecia therapy through a CeO_2 nanozyme-integrated microneedle (MN) patch.[58] (B) Schematic illustration of hair loss therapy through a detachable MN patch system.[59] *ROS*, reactive oxygen species.

efficacy in accelerating hair regrowth in mice with AGA, exhibiting superior therapeutic outcomes compared to topical minoxidil at a reduced frequency of administration.

In the presence of specific stimuli, hair follicle stem cells can facilitate hair growth by promoting the transition from the telogen to the anagen phase. Prior research indicated that elevating lactic acid accumulation in cells to stimulate hair follicle stem cells with the small-molecule drug UK5099[60] or chitosan lactate[61] represents a promising avenue for efficacious AGA therapy. Yang et al.[59] developed hair-derived keratin MNs for the codelivery of UK5099 and mesenchymal stem cell-derived exosomes, which serve as activators of hair follicle stem cells (Fig. 12.7B). This technique facilitated the transition of the hair follicle cycle and promoted hair regrowth in a shaved mouse model within a 6-day period.

To enhance the targeted delivery of drugs to hair follicles for the treatment of AGA, Cao et al.[57] developed a novel approach utilizing bioinspired nanostructured lipid carriers (NLCs) loaded within dissolving MN arrays. This strategy demonstrated high therapeutic efficacy in AGA treatment. The NLCs were synthesized from intrinsic glycerides and squalene to target the delivery of FNS to the hair follicles, where it would preferentially fuse and interact with sebum. The bioinspired MNs demonstrated superior efficacy in promoting hair regrowth compared to FNS-NLC or commercial FNS. This was evidenced by a significant elevation in the gene expression of β-catenin, insulin-like growth factor-1, and VEGF, accompanied by a reduction in the expression of SRD5A2, TGF-β1, and Dickkopf-related protein.

Treatment of AA. AA is an autoimmune disorder characterized by scarless hair loss; it affects approximately 2% of the general population.[62] AA is typically accompanied by an inflammatory response at the hair follicles, with initial onset before the age of 30. The etiology of AA is complex and may involve a combination of genetic and environmental factors.[63] The clinical manifestations of AA include alopecia with monofocal or multifocal patches on the scalp and, in some cases, diffuse alopecia affecting the entire body.[64,65] The available treatments for AA include immunomodulators (corticosteroids, cyclosporine, and methotrexate), minoxidil, and contact immunotherapy, which may be administered topically or systemically. Minoxidil serves as a primary treatment for AA, demonstrating significant efficacy in promoting hair growth. Nonetheless, its oral and topical applications are associated with challenges such as low skin permeability, limited bioavailability, and the potential for adverse local reactions such as scalp irritation, burning sensations, and hair loss.[66] Afika et al.[67] formulated minoxidil into three-layer DMNs (TDMN) to increase drug penetration and avoid irritation. The TDMN consisted of three layers: a special drug-filled tip, a second layer of a drug-free fast-dissolving base, and a third layer of base plate. The experimental results showed that the drug penetration rate of TDMN was four times higher than that of commercially available creams, and drug retention in the skin was higher in the TDMN group than in the commercially available cream preparation group. In addition, the TDMN was nonirritating and nontoxic according to the HET-CAM test and hemolysis test.

Current treatments are effective for patients with only mild disease and have a high recurrence rate. Giorgio et al.[63] combined 5-ALA–mediated photodynamic therapy and MNs to treat moderate to severe AA and confirmed the efficacy of this method by conducting a clinical study involving 41 patients. The researchers observed that the MNs enhanced the penetration of 5-ALA into the skin and augmented its immunosuppressive activity, achieving the optimal clinical outcome. Hair regrowth was observed in 16 of 17 patients.

Studies revealed a reduction in Tregs within AA lesions, demonstrating the critical role of Tregs in the progression of AA. It was shown that impairment of skin-specific Tregs alone is sufficient to induce alopecia and hair follicle inflammation. Younis et al.[68] constructed hydrogel-based MN composed of HA to deliver CCL22 and IL-2 for Treg recruitment and expansion, respectively. In an immune-mediated mouse model of AA treated with hydrogel-based MNs, localized increases in Treg numbers led to sustained hair regrowth and the attenuation of inflammatory pathways. Furthermore, hydrogel-based MNs induced Treg amplification in human skin without peripheral immunosuppression, as demonstrated in a humanized skin graft mouse model.

4. Photoaging

Photoaging, is a consequence of prolonged exposure to ultraviolet (UV) radiation and is one of the most prevalent forms of skin aging. It is characterized by the presence of coarse wrinkles, a roughened texture, skin sagging, and a diminished barrier function. Furthermore, photoaging has been linked to a range of dermatological conditions, including lentigo maligna, actinic keratosis, and various forms of skin cancer.[69] Therefore, the continuous pursuit of decelerating the aging process renders skin antiaging an essential objective for which there is a need to develop effective treatments.

4.1 Ultraviolet-induced photoaging

Environmental concerns regarding global warming and ozone layer depletion have resulted in awareness about an increase in UV radiation, which undoubtedly represents a substantial and imminent threat to skin health. UV radiation is classified into three distinct categories: UVA (320–400 nm), UVB (280–320 nm), and UVC (200–280 nm), based on their respective wavelength intervals. Of these, UVB has been identified as the primary factor responsible for skin photoaging and the elevation of skin cancer risk. The capacity of UVB to penetrate the epidermis and reach the upper dermis results in considerable photochemical damage to cells, which disrupts the collagen structure, promotes the generation of free radicals, and triggers inflammatory responses.[70] One of the most significant consequences of UV-induced damage is the excessive production of ROS, including hydrogen peroxide and superoxide anion, which plays a pivotal role in accelerating photoaging. UV irradiation simultaneously reduces the skin's natural antioxidant defenses. This imbalance results in oxidative stress, which leads to structural and functional alterations in cellular proteins and genes. These changes compromise the ECM, particularly collagen, and activate key signaling pathways involving molecules such as Raf, MEKK1, and protein tyrosine phosphatase. Collectively, these processes contribute to inflammation, dermal matrix remodeling, and visible signs of skin aging.[71] Therefore, antioxidants and the promotion of epidermal repair are two of the most effective treatments for skin photoaging.

4.2 Microneedles for photoaging therapy

Treatments for photoaging include topical agents, cosmeceuticals, and clinical interventions. Commonly utilized pharmacological interventions encompass antioxidant, cytokines, and stem cells. These compounds target key aspects of photoaging by promoting collagen synthesis, scavenging free radicals, and enhancing skin renewal. Despite the benefits of these treatments, limitations are often encountered owing to differences in drug properties, including molecular size, lipophilicity, and acidity, which can impede effective skin penetration. To address these limitations, physical skin penetration enhancement technologies have been developed, offering innovative means to deliver therapeutic agents more effectively. However, many of these technologies encounter obstacles in the transition from the clinical setting to commercial adoption, with only a few ultimately reaching the market. Among them, fractional laser-assisted drug delivery has demonstrated considerable potential, although it necessitates specialized equipment and expertise, which constrains its widespread accessibility.[72] These constraints in current strategies underscore the urgent need for more effective and adaptable strategies to combat photoaging, particularly those that enhance drug delivery while maintaining safety and efficacy.

MN technology has emerged as a promising approach to addressing the challenges of photoaging therapy. In comparison to conventional drug delivery methods, MNs have the capacity to circumvent the stratum corneum, the primary impediment to effective transdermal drug absorption. This enables the precise and efficient delivery of active compounds directly to target sites within the dermis. Furthermore, MNs are less invasive than other physical enhancement methods, such as fractional lasers, which makes them more accessible and potentially less prone to complications. In addition to enhancing drug delivery, MNs themselves can stimulate the skin's intrinsic repair processes. The mechanical microstimulation that occurs as a result of inserting MNs initiates a cascade of biological responses, triggering wound healing. This process entails the recruitment of growth factors and inflammatory cells to the site of injury, which leads to the synthesis and remodeling of collagen. This is a key protein in maintaining the structural integrity and elasticity of the skin.[73] Consequently, MNs not only facilitate the penetration of antiaging agents but also promote collagen production and tissue regeneration, offering a dual approach to treating photoaging.

A study conducted by Kim et al. [74] demonstrated the efficacy of dissolving MN patches loaded with retinyl retinoate and ascorbic acid . The results indicated a notable improvement in skin roughness and a reduction in wrinkles. MNs facilitate the efficient and controlled delivery of these active ingredients to the dermis, where they can exert the most significant impact on collagen synthesis and skin rejuvenation. In addition to antioxidants, other active ingredients, including HA and growth factors, are frequently employed in the treatment of photoaging. For example, HA is of great importance in maintaining the skin's hydration and elasticity, which are essential for reversing the visible effects of photoaging. However, its poor permeability as a macromolecule has severely limited the therapeutic effect. Tang et al.[70] developed an HA-based MN prepared by the functional modification of HA with sulfonate-rich or phosphatidylcholine-rich polymer (HA/PX) (Fig. 12.8). The results indicated that the transdermal delivery of HA/PX via MNs could significantly mitigate wrinkles and skin damage in photoaged nude mice and treat skin photoaging by enhancing epidermal thickness, promoting collagen deposition, and reducing oxidative stress. Similarly, growth factors such FGF can stimulate collagen production and accelerate the skin's natural regeneration processes. Yang et al.[75] created an MN patch loaded with FGF-2 and FGF-21. FGF-2/FGF-21 MNs exhibited significant improvements in UV-induced acute skin inflammation and reduced mouse skin wrinkles.

Biotherapy has shown great potential for antiaging and rejuvenation. Extracellular vesicles (EVs) are important carriers secreted by a variety of cells and mediate intracellular communication. Stem cell-derived EVs have been reported to be rich in peptides, cytokines, microRNAs and lncRNAs, molecules that can regulate a variety of biological processes in target cells such as collagen synthesis, immunomodulation, cellular senescence, and DNA repair. Cao et al.[76] investigated the antiphotoaging effects of extracellular vesicles of adipose-derived stem cells (ADSC-EVs) assisted by MNs. The results indicated that MN treatment alone could improve the epidermal structure and function of photoaging skin, and a combination with ADSC-EVs would accelerate the restoration of inflammation caused by MN and improve the content of collagen.

These findings collectively highlight the dual efficacy of MNs in photoaging therapy. They demonstrate that MNs not only enhance the delivery of active compounds but also stimulate the skin's intrinsic healing processes, such as collagen synthesis and tissue regeneration. This offers a safe, effective, and versatile approach to improving the skin's health and appearance.

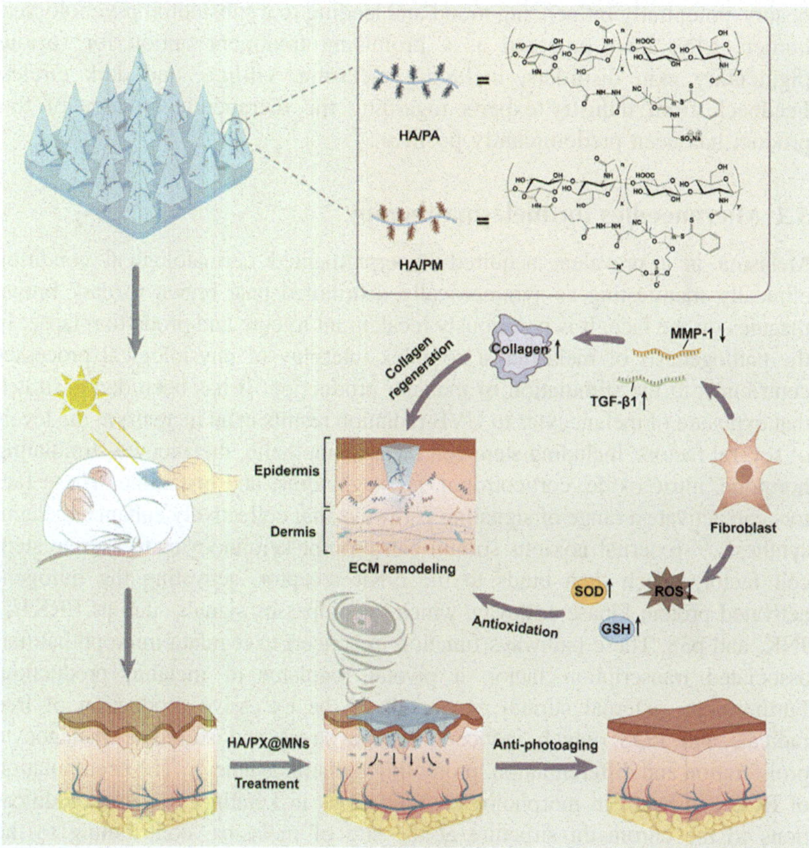

FIGURE 12.8 Schematic illustration of ultraviolet B-induced photoaged skin protection via hyaluronic acid (HA)/PA and HA/PM loaded microneedles. *ECM*, extracellular matrix; *TGF-β1*, transforming growth factor-β1; *ROS*, reactive oxygen species.

5. Pigmented dermatoses

The two primary forms of melanin, eumelanin and pheomelanin, are responsible for diverse colorations observed in the human skin. Eumelanin is responsible for the production of melanin, which results in darker skin tones, while pheomelanin is responsible for the production of lighter skin tones.[77] The synthesis and distribution of melanin are pivotal in determining the pigmentation of the skin. Disruption of this process can result in the development of pigmentary disorders. Pigmented dermatosis is a series of diseases caused by a decrease or increase in pigmentation. The pathogenesis of these diseases may be related to an abnormality in melanin synthesis, transportation, and metabolism.[78] While the majority of pigmented skin diseases do not pose a significant health risk, they frequently affect an individual's perception of beauty, potentially influencing mood and leading to a substantial psychological burden. MNs have emerged as a promising treatment option for various pigmentary skin disorders, including melasma, vitiligo, and dark circles. Feedback from industry experts regarding the therapeutic efficacy of this product has been predominantly positive.

5.1 Microneedles for melasma therapy

Melasma is a prevalent acquired hyperpigmented dermatological condition clinically manifesting as symmetrically distributed pale brown to dark brown macules on the face. It is notoriously recalcitrant to cure and prone to relapse. In the pathogenesis of melasma, a complex interplay of physiological processes contributes to the stimulation of melanin production. It has been demonstrated that exposure of melanocytes to UVB radiation results in an increase in the levels of several factors, including stem cell factor, endothelin, melanocyte-stimulating hormone, nitric oxide, corticotropin, prostaglandins, and histamine. These factors can activate a range of signaling pathways that collectively enhance melanin synthesis.[79] External noxious stimuli can prompt keratinocytes to secrete stem cell factor, which then binds to the c-Kit receptor, activating the mitogen-activated protein kinase pathway, which encompasses signals such as ERK1/2, JNK, and p38. These pathways function in concert to regulate microphthalmia-associated transcription factor, a pivotal regulator of melanin production. Furthermore, external stimuli can result in the excessive production of free radicals and ROS, which induce oxidative stress and stimulate melanocyte proliferation and differentiation, promoting melanin synthesis. The accumulation of ROS can result in morphological alterations in keratinocytes and modifications to the chromatin structure at the site of melasma, contributing to the degradation of dermal connective tissue and the acceleration of pigmentation. Furthermore, VEGF has been identified as a factor in the formation of hyperpigmented spots by stimulating blood vessel growth, which can result in the leakage of red blood cells and the deposition of hemosiderin in affected area.[80]

MNs have demonstrated efficacy in the treatment of melasma by stimulating fibroblast proliferation and remodeling dermal collagen, which facilitates the repair of damage to the dermis and basement membrane. This process results in a reduction of pigmentation-stimulating signals released from the dermis and an acceleration of melanin metabolism in the epidermis, which ultimately restores the normal structure of hyperpigmented skin. A study conducted by Cassiano et al.[81] demonstrated that the use of MNs resulted in a 9% reduction in the Melasma Area and Severity Index compared to the use of sunscreen. This was accompanied by a significant decrease in visible melanin density and the rapid reconstruction of collagen in the basement membrane and dermis. Additionally, MNs have been shown to assist in maintaining equilibrium between oxidative and antioxidant systems. The mechanical energy generated during MN application is converted into heat, dilating local capillaries, improving skin microcirculation, and alleviating hypoxia. This reduces the accumulation of ROS and modulates VEGF levels.[82] MNs also influence the skin's surface microbiota by enhancing skin immunity, decreasing pigmentation, and reducing the colonization of pigmentation-associated microbes.

Additionally, MNs facilitate the formation of microchannels in the skin, enhancing the penetration of topical pharmaceutical agents. Tranexamic acid (TA) has been demonstrated to function as an effective tyrosinase inhibitor, and thus it may be a valuable addition to the treatment of melasma. However, long-term oral administration of TA may result in adverse reactions in the gastrointestinal tract and an increased risk of thrombosis. In a randomized, half-face controlled trial of 30 patients, Xu et al.[83] observed that the improvement rate of the topical TA combined with MNs increased by 25% compared with the control group. To enhance the efficacy of transdermal delivery of TA, it was incorporated into HA-based MNs, which provided the requisite amount of drug to treat melasma. The results demonstrated that MNs significantly reduced the thickness of the epidermis in melasma mice, decreased melanin production, and decreased the expression of dopachrome tautomerase.[84] In addition, Xing et al.[85] conducted a study that examined the efficacy of a combination of TA and licorice extract in the treatment of melasma. Specifically, licorice extract was employed as an additive to TA MNs because of its good whitening and anti-inflammatory effects. In a pharmacodynamics investigation, TA-dissolving MNs obviously reduced melanin deposition in the skin of melasma guinea pigs after eight consecutive administrations. In particular, the combination of TA and licorice extract dissolving MNs worked better than TA alone. Therefore, the combined administration of drugs via the introduction of MNs represents an efficacious approach to treating melasma.

5.2 Microneedles for vitiligo therapy

Vitiligo is a pigmentation disorder characterized by the loss of melanocytes in the epidermis, which results in the formation of asymptomatic white patches on the skin and mucous membranes. The prevalence of vitiligo is estimated to affect approximately 0.5%–1.0% of the global population, with a gradual increase in incidence over time. The condition has a complex etiology involving genetic predisposition, autoimmune responses, oxidative stress, and microvascular abnormalities, all of which contribute to the destruction or impaired function of melanocytes.[86] Antiinflammatory therapeutics, including JAK-STAT inhibitor (tofacitinib) and a calcineurin inhibitor (tacrolimus), have been extensively applied in the clinical treatment of vitiligo. Nonetheless, delivering these drugs effectively to the dermal–epidermal junction remains challenging. While systemic administration carries the risk of adverse side effects, topical formulations such as creams and ointments are hindered by poor skin permeability, ultimately limiting their therapeutic effectiveness.

Developments indicate that MN technology may offer a promising avenue for repigmentation in patients with treatment-resistant vitiligo.

Advancements in understanding the mechanisms of vitiligo have highlighted the critical role of autoreactive cytotoxic $CD8^+$ T cells in targeting melanocytes, driving disease progression through the local production of IFN-γ. This cytokine induces neighboring fibroblasts to secrete chemokines, creating a positive feedback loop that recruits additional T cells to the skin.[87] To achieve successful repigmentation in vitiligo, two therapeutic objectives must be met: suppressing autoimmunity and promoting the regeneration of melanocytes from their stem cell niche within hair follicles. Targeted immunotherapy, including the use of immunosuppressants, is designed to inhibit cytokine signaling, while α-melanocyte-stimulating hormone (α-MSH) analogs and phototherapy are utilized to stimulate melanocyte proliferation and migration. Calcineurin inhibitors such as tacrolimus and pimecrolimus have been employed in the treatment of vitiligo by inhibiting calcineurin in T cells, suppressing the production and release of proinflammatory cytokines. In addition, these inhibitors facilitate melanocyte migration while exerting minimal effects on melanocyte proliferation. Studies have highlighted the advantages of combining MNs with calcineurin inhibitors. Ebrahim et al.[88] reported a significantly higher repigmentation rate of 66.6% in patients treated with MNs and tacrolimus compared to 33.3% in those receiving topical tacrolimus alone. Furthermore, Ebrahim et al.[89] demonstrated the enhanced efficacy and tolerability of MNs combined with tacrolimus compared to tacrolimus ointment alone, with the combination therapy inducing an upregulation of c-kit (a cell surface receptor crucial for melanocyte survival and proliferation) expression in the vitiliginous epidermis.

α-MSH, a neuroendocrine hormone, plays a crucial role in stimulating melanin production and shielding melanocytes from oxidative stress.[90] Notably, reduced expression levels of α-MSH have been identified in both the serum and skin lesions of individuals with vitiligo.[91] Chong et al. demonstrated that MNs composed of silk fibroin methacrylate and α-MSH significantly enhanced melanin synthesis, promoted the transport of melanin granules to surrounding cells, and ultimately improved epidermal repigmentation. In another study, hydrogel MNs made from dextran methacrylate and a cyclodextrin-adamantane host-guest assembly polymer were employed to encapsulate tofacitinib and α-MSH. This strategy notably improved repigmentation in both the skin and hair of an animal model of vitiligo.[92]

5.3 Microneedles for infraorbital dark circle therapy

Infraorbital dark circles are defined as the appearance of bilateral, semicircular areas of hyperpigmentation beneath the eyes. This condition is a common manifestation of periorbital melanosis and is frequently addressed in dermatological consultations. In addition to its esthetic implications, infraorbital dark circles can have a profound impact on an individual's quality of life.[93] Infraorbital dark circles are classified as pigmented, vascular, structural, and mixed types based on their primary pathogenic cause. The possible etiologic factors of dark circles include dermal melanin deposition, the superficial positioning of vasculature, and shadowing caused by skin laxity, among others. Given this multifactorial etiology, effective management often necessitates a multimodal therapeutic approach incorporating options such as topical depigmenting agents (phenylethyl resorcinol), chemical peels (trichloroacetic acid), laser therapy, and surgical interventions.

The barrier effect of the stratum corneum significantly limits the absorption of topical drugs, making it challenging to achieve desired therapeutic outcomes. To address this issue, MNs are employed as an adjunct to enhance the delivery of local therapeutic agents in the treatment of dark circles. In a randomized, controlled trial of 20 patients, Yu et al.[94] evaluated the efficacy and safety of MN-assisted phenylethyl resorcinol for the treatment of infraorbital dark circles. The results demonstrated that the mean value of melanin index in the MN-assisted phenylethyl resorcinol group was significantly lower than the baseline. In another study, 13 female patients with pigmented dark circles were treated with 10% trichloroacetic acid via MNs, and 92.3% of patients experienced a notable improvement.[95]

References

1. Peng, T.; Chen, Y.; Hu, W.; Huang, Y.; Zhang, M.; Lu, C.; Pan, X.; Wu, C. Microneedles for Enhanced Topical Treatment of Skin Disorders: Applications, Challenges, and Prospects. *Engineering* **2023,** *30,* 170–189.
2. Konieczny, P.; Naik, S. Healing without Scarring. *Science* **2021,** *372* (6540), 346–347.
3. Limandjaja, G. C.; Niessen, F. B.; Scheper, R. J.; Gibbs, S. Hypertrophic Scars and Keloids: Overview of the Evidence and Practical Guide for Differentiating between These Abnormal Scars. *Exp. Dermatol.* **2021,** *30* (1), 146–161.
4. Lin, S.; Quan, G.; Hou, A.; Yang, P.; Peng, T.; Gu, Y.; Qin, W.; Liu, R.; Ma, X.; Pan, X.; et al. Strategy for Hypertrophic Scar Therapy: Improved Delivery of Triamcinolone Acetonide Using Mechanically Robust Tip-Concentrated Dissolving Microneedle Array. *J. Contr. Release* **2019,** *306,* 69–82.
5. Peña, O. A.; Martin, P. Cellular and Molecular Mechanisms of Skin Wound Healing. *Nat. Rev. Mol. Cell Biol.* **2024,** *25* (8), 599–616.
6. Li, X.; Wang, X.; Shen, T.; Xiong, J.; Ma, Q.; Guo, G.; Zhu, F. Advances in Photodynamic Therapy of Pathologic Scar. *Photodiagnosis Photodyn. Ther.* **2024,** *46,* 104040.
7. Greaves, N. S.; Ashcroft, K. J.; Baguneid, M.; Bayat, A. Current Understanding of Molecular and Cellular Mechanisms in Fibroplasia and Angiogenesis during Acute Wound Healing. *J. Dermatol. Sci.* **2013,** *72* (3), 206–217.
8. Hodgkinson, T.; Bayat, A. Dermal Substitute-Assisted Healing: Enhancing Stem Cell Therapy with Novel Biomaterial Design. *Arch. Dermatol. Res.* **2011,** *303* (5), 301–315.
9. Kidzeru, E. B.; Lebeko, M.; Sharma, J. R.; Nkengazong, L.; Adeola, H. A.; Ndlovu, H.; Nonhlanhla, P. K.; Bayat, A. Immune Cells and Associated Molecular Markers in Dermal Fibrosis with Focus on Raised Cutaneous Scars. *Exp. Dermatol.* **2023,** *32* (5), 570–587.

10. Iqbal, S. A.; Sidgwick, G. P.; Bayat, A. Identification of Fibrocytes from Mesenchymal Stem Cells in Keloid Tissue: a Potential Source of Abnormal Fibroblasts in Keloid Scarring. *Arch. Dermatol. Res.* **2012,** *304* (8), 665–671.
11. Gurtner, G. C.; Werner, S.; Barrandon, Y.; Longaker, M. T. Wound Repair and Regeneration. *Nature* **2008,** *453* (7193), 314–321.
12. Cabral, J.; Ryan, A. E.; Griffin, M. D.; Ritter, T. Extracellular Vesicles as Modulators of Wound Healing. *Adv. Drug Deliv. Rev.* **2018,** *129,* 394–406.
13. Jeschke, M. G.; Wood, F. M.; Middelkoop, E.; Bayat, A.; Teot, L.; Ogawa, R.; Gauglitz, G. G.; Scars *Nat. Rev. Dis. Primers* **2023,** *9* (1), 64.
14. Hietanen, K. E.; Järvinen, T. A.; Huhtala, H.; Tolonen, T. T.; Kuokkanen, H. O.; Kaartinen, I. S. Treatment of Keloid Scars with Intralesional Triamcinolone and 5-fluorouracil Injections - a Randomized Controlled Trial. *J. Plast. Reconstr. Aesthetic Surg.* **2019,** *72* (1), 4–11.
15. Zhang, Q.; Shi, L.; He, H.; Liu, X.; Huang, Y.; Xu, D.; Yao, M.; Zhang, N.; Guo, Y.; Lu, Y.; et al. Down-Regulating Scar Formation by Microneedles Directly via a Mechanical Communication Pathway. *ACS Nano* **2022,** *16* (7), 10163–10178.
16. Yeo, D. C.; Balmayor, E. R.; Schantz, J. T.; Xu, C. Microneedle Physical Contact as a Therapeutic for Abnormal Scars. *Eur. J. Med. Res.* **2017,** *22* (1), 28.
17. Yang, B.; Dong, Y.; Shen, Y.; Hou, A.; Quan, G.; Pan, X.; Wu, C. Bilayer Dissolving Microneedle Array Containing 5-fluorouracil and Triamcinolone with Biphasic Release Profile for Hypertrophic Scar Therapy. *Bioact. Mater.* **2021,** *6* (8), 2400–2411.
18. Huang, Y.; Peng, T.; Hu, W.; Gao, X.; Chen, Y.; Zhang, Q.; Wu, C.; Pan, X. Fully Armed Photodynamic Therapy with Spear and Shear for Topical Deep Hypertrophic Scar Treatment. *J. Contr. Release* **2022,** *343,* 408–419.
19. Wu, T.; Hou, X.; Li, J.; Ruan, H.; Pei, L.; Guo, T.; Wang, Z.; Ci, T.; Ruan, S.; He, Y.; et al. Microneedle-Mediated Biomimetic Cyclodextrin Metal Organic Frameworks for Active Targeting and Treatment of Hypertrophic Scars. *ACS Nano* **2021,** *15* (12), 20087–20104.
20. Wang, M.; Han, Y.; Yu, X.; Liang, L.; Chang, H.; Yeo, D. C.; Wiraja, C.; Wee, M. L.; Liu, L.; Liu, X.; et al. Upconversion Nanoparticle Powered Microneedle Patches for Transdermal Delivery of siRNA. *Adv. Healthcare Mater.* **2020,** *9* (2), e1900635.
21. Tan, C. W. X.; Tan, W. D.; Srivastava, R.; Yow, A. P.; Wong, D. W. K.; Tey, H. L. Dissolving Triamcinolone-Embedded Microneedles for the Treatment of Keloids: A Single-Blinded Intra-individual Controlled Clinical Trial. *Dermatol. Ther.* **2019,** *9* (3), 601–611.
22. Park, J.; Kim, Y. C. Topical Delivery of 5-Fluorouracil-Loaded Carboxymethyl Chitosan Nanoparticles Using Microneedles for Keloid Treatment. *Drug Deliv Transl Res* **2021,** *11* (1), 205–213.
23. Sitohang, I. B. S.; Sirait, S. A. P.; Suryanegara, J. Microneedling in the Treatment of Atrophic Scars: A Systematic Review of Randomised Controlled Trials. *Int. Wound J.* **2021,** *18* (5), 577–585.
24. Layton, A. M.; Henderson, C. A.; Cunliffe, W. J. A Clinical Evaluation of Acne Scarring and its Incidence. *Clin. Exp. Dermatol.* **1994,** *19* (4), 303–308.
25. Gupta, M.; Barman, K. D.; Sarkar, R. A Comparative Study of Microneedling Alone versus along with Platelet-Rich Plasma in Acne Scars. *J. Cutan. Aesthetic Surg.* **2021,** *14* (1), 64–71.
26. Yan, D.; Zhao, H.; Li, C.; Xia, A.; Zhang, J.; Zhang, S.; Yun, Q.; Li, X.; Huang, F.; Tian, Y. A Clinical Study of Carbon Dioxide Lattice Laser–Assisted or Microneedle-Assisted 5-aminolevulinic Acid–Based Photodynamic Therapy for the Treatment of Hypertrophic Acne Scars. *Photodermatol. Photoimmunol. Photomed.* **2022,** *38* (1), 53–59.

27. Sharad, J. Combination of Microneedling and Glycolic Acid Peels for the Treatment of Acne Scars in Dark Skin. *J. Cosmet. Dermatol.* **2011,** *10* (4), 317–323.
28. White, G. M. Recent Findings in the Epidemiologic Evidence, Classification, and Subtypes of Acne Vulgaris. *J. Am. Acad. Dermatol.* **1998,** *39* (2), S34–S37.
29. Halvorsen, J. A.; Stern, R. S.; Dalgard, F.; Thoresen, M.; Bjertness, E.; Lien, L. Suicidal Ideation, Mental Health Problems, and Social Impairment Are Increased in Adolescents with Acne: A Population-Based Study. *J. Invest. Dermatol.* **2011,** *131* (2), 363–370.
30. Tuchayi, S. M.; Makrantonaki, E.; Ganceviciene, R.; Dessinioti, C.; Feldman, S. R.; Zouboulis, C. C. Acne Vulgaris. *Nat. Rev. Dis. Primers* **2015,** *1* (1), 15029.
31. Choi, C. W.; Choi, J. W.; Park, K. C.; Youn, S. W. Facial Sebum Affects the Development of Acne, Especially the Distribution of Inflammatory Acne. *J. Eur. Acad. Dermatol. Venereol.* **2013,** *27* (3), 301–306.
32. Vasam, M.; Korutla, S.; Bohara, R. A. Acne Vulgaris: A Review of the Pathophysiology, Treatment, and Recent Nanotechnology Based Advances. *Biochem. Biophys. Rep.* **2023,** *36,* 101578.
33. Nagy, I.; Pivarcsi, A.; Kis, K.; Koreck, A.; Bodai, L.; McDowell, A.; Seltmann, H.; Patrick, S.; Zouboulis, C. C.; Kemény, L. Propionibacterium Acnes and Lipopolysaccharide Induce the Expression of Antimicrobial Peptides and Proinflammatory Cytokines/chemokines in Human Sebocytes. *Microb. Infect.* **2006,** *8* (8), 2195–2205.
34. Zouboulis, C. C. Acne as a Chronic Systemic Disease. *Clin. Dermatol.* **2014,** *32* (3), 389–396.
35. Das, S.; Reynolds, R. V. Recent Advances in Acne Pathogenesis: Implications for Therapy. *Am. J. Clin. Dermatol.* **2014,** *15* (6), 479–488.
36. Evans, D. M.; Kirk, K. M.; Nyholt, D. R.; Novac, C.; Martin, N. G. Teenage Acne Is Influenced by Genetic Factors. *Br. J. Dermatol.* **2005,** *152* (3), 579–581.
37. Kobayashi, T.; Tamada, S. Selective Electrothermolysis of the Sebaceous Glands: Treatment of Facial Seborrhea. *Dermatol. Surg.* **2007,** *33* (2), 169–177.
38. Seo, K. Y.; Yoon, M. S.; Kim, D. H.; Lee, H. J. Skin Rejuvenation by Microneedle Fractional Radiofrequency Treatment in Asian Skin; Clinical and Histological Analysis. *Laser Surg. Med.* **2012,** *44* (8), 631–636.
39. Cho, S. I.; Chung, B. Y.; Choi, M. G.; Baek, J. H.; Cho, H. J.; Park, C. W.; Lee, C. H.; Kim, H. O. Evaluation of the Clinical Efficacy of Fractional Radiofrequency Microneedle Treatment in Acne Scars and Large Facial Pores. *Dermatol. Surg.* **2012,** *38* (7), 1017–1024.
40. Hantash, B. M.; Ubeid, A. A.; Chang, H.; Kafi, R.; Renton, B. Bipolar Fractional Radiofrequency Treatment Induces Neoelastogenesis and Neocollagenesis. *Laser Surg. Med.* **2009,** *41* (1), 1–9.
41. Lee, K. R.; Lee, E. G.; Lee, H. J.; Yoon, M. S. Assessment of Treatment Efficacy and Sebosuppressive Effect of Fractional Radiofrequency Microneedle on Acne Vulgaris. *Laser Surg. Med.* **2013,** *45* (10), 639–647.
42. Zhang, Y.; Feng, P.; Yu, J.; Yang, J.; Zhao, J.; Wang, J.; Shen, Q.; Gu, Z. ROS-responsive Microneedle Patch for Acne Vulgaris Treatment. *Adv. Ther.* **2018,** *1* (3), 1800035.
43. Zhang, T.; Sun, B.; Guo, J.; Wang, M.; Cui, H.; Mao, H.; Wang, B.; Yan, F. Active Pharmaceutical Ingredient Poly(ionic Liquid)-Based Microneedles for the Treatment of Skin Acne Infection. *Acta Biomater.* **2020,** *115,* 136–147.
44. Wen, T.; Lin, Z.; Zhao, Y.; Zhou, Y.; Niu, B.; Shi, C.; Lu, C.; Wen, X.; Zhang, M.; Quan, G.; et al. Bioresponsive Nanoarchitectonics-Integrated Microneedles for Amplified Chemo-Photodynamic Therapy against Acne Vulgaris. *ACS Appl. Mater. Interfaces* **2021,** *13* (41), 48433–48448.

45. Yang, G.; Chen, G.; Gu, Z. Transdermal Drug Delivery for Hair Regrowth. *Mol. Pharm.* **2021,** *18* (2), 483–490.
46. Correia, M.; Lopes, J.; Lopes, D.; Melero, A.; Makvandi, P.; Veiga, F.; Coelho, J. F. J.; Fonseca, A. C.; Paiva-Santos, A. C. Nanotechnology-based Techniques for Hair Follicle Regeneration. *Biomaterials* **2023,** *302,* 122348.
47. Rompolas, P.; Deschene, E. R.; Zito, G.; Gonzalez, D. G.; Saotome, I.; Haberman, A. M.; Greco, V. Live Imaging of Stem Cell and Progeny Behaviour in Physiological Hair-Follicle Regeneration. *Nature* **2012,** *487* (7408), 496–499.
48. Suen, W. J.; Li, S. T.; Yang, L. T. Hes1 Regulates Anagen Initiation and Hair Follicle Regeneration through Modulation of Hedgehog Signaling. *Stem Cell.* **2020,** *38* (2), 301–314.
49. Shirokova, V.; Biggs, L. C.; Jussila, M.; Ohyama, T.; Groves, A. K.; Mikkola, M. L. Foxi3 Deficiency Compromises Hair Follicle Stem Cell Specification and Activation. *Stem Cell.* **2016,** *34* (7), 1896–1908.
50. Paus, R.; Cotsarelis, G. The Biology of Hair Follicles. *N. Engl. J. Med.* **1999,** *341* (7), 491–497.
51. Chen, C. C.; Murray, P. J.; Jiang, T. X.; Plikus, M. V.; Chang, Y. T.; Lee, O. K.; Widelitz, R. B.; Chuong, C. M. Regenerative Hair Waves in Aging Mice and Extra-follicular Modulators Follistatin, Dkk1, and Sfrp4. *J. Invest. Dermatol.* **2014,** *134* (8), 2086–2096.
52. Abreu, C. M.; Marques, A. P. Recreation of a Hair Follicle Regenerative Microenvironment: Successes and Pitfalls. *Bioeng. Transl. Med.* **2022,** *7* (1), e10235.
53. Higgins, C. A.; Westgate, G. E.; Jahoda, C. A. From Telogen to Exogen: Mechanisms Underlying Formation and Subsequent Loss of the Hair Club Fiber. *J. Invest. Dermatol.* **2009,** *129* (9), 2100–2108.
54. Lolli, F.; Pallotti, F.; Rossi, A.; Fortuna, M. C.; Caro, G.; Lenzi, A.; Sansone, A.; Lombardo, F. Androgenetic Alopecia: a Review. *Endocrine* **2017,** *57* (1), 9–17.
55. Iriarte, C.; Awosika, O.; Rengifo-Pardo, M.; Ehrlich, A. Review of Applications of Microneedling in Dermatology. *Clin. Cosmet. Invest. Dermatol.* **2017,** *10,* 289–298.
56. Kim, S.; Eum, J.; Yang, H.; Jung, H. Transdermal Finasteride Delivery via Powder-Carrying Microneedles with a Diffusion Enhancer to Treat Androgenetic Alopecia. *J. Contr. Release* **2019,** *316,* 1–11.
57. Cao, S.; Wang, Y.; Wang, M.; Yang, X.; Tang, Y.; Pang, M.; Wang, W.; Chen, L.; Wu, C.; Xu, Y. Microneedles Mediated Bioinspired Lipid Nanocarriers for Targeted Treatment of Alopecia. *J. Contr. Release* **2021,** *329,* 1–15.
58. Yuan, A.; Xia, F.; Bian, Q.; Wu, H.; Gu, Y.; Wang, T.; Wang, R.; Huang, L.; Huang, Q.; Rao, Y.; et al. Ceria Nanozyme-Integrated Microneedles Reshape the Perifollicular Microenvironment for Androgenetic Alopecia Treatment. *ACS Nano* **2021,** *15* (8), 13759–13769.
59. Yang, G.; Chen, Q.; Wen, D.; Chen, Z.; Wang, J.; Chen, G.; Wang, Z.; Zhang, X.; Zhang, Y.; Hu, Q.; et al. A Therapeutic Microneedle Patch Made from Hair-Derived Keratin for Promoting Hair Regrowth. *ACS Nano* **2019,** *13* (4), 4354–4360.
60. Flores, A.; Schell, J.; Krall, A. S.; Jelinek, D.; Miranda, M.; Grigorian, M.; Braas, D.; White, A. C.; Zhou, J. L.; Graham, N. A.; et al. Lactate Dehydrogenase Activity Drives Hair Follicle Stem Cell Activation. *Nat. Cell Biol.* **2017,** *19* (9), 1017–1026.
61. Shi, Y.; Zhao, J.; Li, H.; Yu, M.; Zhang, W.; Qin, D.; Qiu, K.; Chen, X.; Kong, M. A Drug-Free, Hair Follicle Cycling Regulatable, Separable, Antibacterial Microneedle Patch for Hair Regeneration Therapy. *Adv. Healthcare Mater.* **2022,** *11* (19), 2200908.
62. Strazzulla, L. C.; Wang, E. H. C.; Avila, L.; Lo Sicco, K.; Brinster, N.; Christiano, A. M.; Shapiro, J. Alopecia Areata: Disease Characteristics, Clinical Evaluation, and New Perspectives on Pathogenesis. *J. Am. Acad. Dermatol.* **2018,** *78* (1), 1–12.

63. Giorgio, C. M.; Babino, G.; Caccavale, S.; Russo, T.; Rosa, AB De; Alfano, R.; Fulgione, E.; Argenziano, G. Combination of Photodynamic Therapy with 5-aminolaevulinic Acid and Microneedling in the Treatment of Alopecia Areata Resistant to Conventional Therapies: Our Experience with 41 Patients. *Clin. Exp. Dermatol.* **2020,** *45* (3), 323–326.
64. Sterkens, A.; Lambert, J.; Bervoets, A. Alopecia Areata: a Review on Diagnosis, Immunological Etiopathogenesis and Treatment Options. *Clin. Exp. Med.* **2021,** *21* (2), 215–230.
65. Zhou, C.; Li, X.; Wang, C.; Zhang, J. Alopecia Areata: an Update on Etiopathogenesis, Diagnosis, and Management. *Clin. Rev. Allergy Immunol.* **2021,** *61* (3), 403–423.
66. Strazzulla, L. C.; Wang, E. H. C.; Avila, L.; Lo Sicco, K.; Brinster, N.; Christiano, A. M.; Shapiro, J. Alopecia Areata: An Appraisal of New Treatment Approaches and Overview of Current Therapies. *J. Am. Acad. Dermatol.* **2018,** *78* (1), 15–24.
67. Afika, N.; Saniy, A. F.; Dharma, A. A. F.; Ko, C. K.; Kamran, R.; Permana, A. D. Trilayer Dissolving Microneedle for Transdermal Delivery of Minoxidil: a Proof-Of-Concept Study. *J. Biomater. Sci. Polym. Ed.* **2024,** *35* (11), 1750–1770.
68. Younis, N.; Puigmal, N.; Kurdi, A. E.; Badaoui, A.; Zhang, D.; Morales-Garay, C.; Saad, A.; Cruz, D.; Rahy, N. A.; Daccache, A.; et al. Microneedle-Mediated Delivery of Immunomodulators Restores Immune Privilege in Hair Follicles and Reverses Immune-Mediated Alopecia. *Adv. Mater.* **2024,** *36* (31), 2312088.
69. Panich, U.; Sittithumcharee, G.; Rathviboon, N.; Jirawatnotai, S. Ultraviolet Radiation-Induced Skin Aging: The Role of DNA Damage and Oxidative Stress in Epidermal Stem Cell Damage Mediated Skin Aging. *Stem Cell. Int.* **2016**, (1), 7370642.
70. Tang, Z.; Liu, Z.; Zhang, Y.; Luo, S.; Xu, Y.; Ren, L. Functional Hyaluronic Acid Microneedles for Skin Photoaging Based on Collagen Induction and Oxidative Stress Regulation Strategies. *Int. J. Biol. Macromol.* **2024,** *277,* 134080.
71. Rittié, L.; Fisher, G. J. UV-light-induced Signal Cascades and Skin Aging. *Ageing Res. Rev.* **2002,** *1* (4), 705–720.
72. Yamada, M.; Prow, T. W. Physical Drug Delivery Enhancement for Aged Skin, UV Damaged Skin and Skin Cancer: Translation and Commercialization. *Adv. Drug Deliv. Rev.* **2020,** *153,* 2–17.
73. Nair, P. A.; Tandel, J.; Microneedling, D. S. A Means of Collagen Induction Therapy. *J. Dermatol. Dermatol. Surg.* **2021,** *25* (2), 49–53.
74. Kim, M.; Yang, H.; Kim, H.; Jung, H.; Jung, H. Novel Cosmetic Patches for Wrinkle Improvement: Retinyl Retinoate- and Ascorbic Acid-Loaded Dissolving Microneedles. *Int. J. Cosmet. Sci.* **2014,** *36* (3), 207–212.
75. Yang, G.; Hu, S.; Jiang, H.; Cheng, K. Peelable Microneedle Patches Deliver Fibroblast Growth Factors to Repair Skin Photoaging Damage. *Nanotheranostics* **2023,** *7* (4), 380–392.
76. Cao, Z.; Jin, S.; Wang, P.; He, Q.; Yang, Y.; Gao, Z.; Wang, X. Microneedle Based Adipose Derived Stem Cells-Derived Extracellular Vesicles Therapy Ameliorates UV-Induced Photoaging in SKH-1 Mice. *J. Biomed. Mater. Res.* **2021,** *109* (10), 1849–1857.
77. Lin, J. Y.; Fisher, D. E. Melanocyte Biology and Skin Pigmentation. *Nature* **2007,** *445* (7130), 843–850.
78. Wan, L.; Ai, Z.; Chen, J.; Jiang, Q.; Chen, H.; Li, Q.; Lu, Y.; Chen, L. Detection Algorithm for Pigmented Skin Disease Based on Classifier-Level and Feature-Level Fusion. *Front. Public Health* **2022,** *10,* 1034772.
79. Passeron, T.; Picardo, M. Melasma, a Photoaging Disorder. *Pigm. Cell Melanoma Res.* **2018,** *31* (4), 461–465.
80. Espósito, A. C. C.; Cassiano, D. P.; Bagatin, E.; Miot, H. A. Regarding the Alterations in Oxidative Stress Status Induced by Melasma Treatments. *Arch. Dermatol. Res.* **2021,** *313* (8), 705–706.

81. Cassiano, D. P.; Espósito, A. C. C.; Hassun, K. M.; Lima, E. V. A.; Bagatin, E.; Miot, H. A. Early Clinical and Histological Changes Induced by Microneedling in Facial Melasma: A Pilot Study. *Indian J. Dermatol. Venereol. Leprol.* **2019,** *85* (6), 638–641.
82. Amani, H.; Shahbazi, M. A.; D'Amico, C.; Fontana, F.; Abbaszadeh, S.; Santos, H. A. Microneedles for Painless Transdermal Immunotherapeutic Applications. *J. Contr. Release* **2021,** *330,* 185–217.
83. Xu, Y.; Ma, R.; Juliandri, J.; Wang, X.; Xu, B.; Wang, D.; Lu, Y.; Zhou, B.; Luo, D. Efficacy of Functional Microarray of Microneedles Combined with Topical Tranexamic Acid for Melasma: A Randomized, Self-Controlled, Split-Face Study. *Medicine* **2017,** *96* (19), e6897.
84. Zhang, C.; Song, X.; Ma, C.; Wang, M.; Deng, H.; Yuan, M.; Zhu, W.; Shen, L.; Li, Y.; Shi, L. Hyaluronic Acid Dissolving Microneedle Patch Loaded with Tranexamic Acid for Melasma Treatment. *Int. J. Biol. Macromol.* **2024,** *270,* 132255.
85. Xing, M.; Wang, X.; Zhao, L.; Zhou, Z.; Liu, H.; Wang, B.; Cheng, A.; Zhang, S.; Gao, Y. Novel Dissolving Microneedles Preparation for Synergistic Melasma Therapy: Combined Effects of Tranexamic Acid and Licorice Extract. *Int. J. Pharm.* **2021,** *600,* 120406.
86. Ezzedine, K.; Eleftheriadou, V.; Whitton, M.; van Geel, N. Vitiligo. *Lancet* **2015,** *386* (9988), 74–84.
87. Khaitan, B. K.; Sindhuja, T. Autoimmunity in Vitiligo: Therapeutic Implications and Opportunities. *Autoimmun. Rev.* **2022,** *21* (1), 102932.
88. Ebrahim, H. M.; Albalate, W. Efficacy of Microneedling Combined with Tacrolimus versus Either One Alone for Vitiligo Treatment. *J. Cosmet. Dermatol.* **2020,** *19* (4), 855–862.
89. Ebrahim, H. M.; Elkot, R.; Albalate, W. Combined Microneedling with Tacrolimus vs Tacrolimus Monotherapy for Vitiligo Treatment. *J. Dermatol. Treat.* **2021,** *32* (8), 999–1004.
90. Pillaiyar, T.; Manickam, M.; Jung, S. H. Recent Development of Signaling Pathways Inhibitors of Melanogenesis. *Cell. Signal.* **2017,** *40,* 99–115.
91. Abdalla, M.; Nayaf, M.; Hussein, S. Z. Correlation between Serum α-MSH and Vitamin D Levels in Vitiligo Patients. *Int. J. Dermatol.* **2020,** *23,* 163–167.
92. Liang, J.; Yu, Y.; Li, C.; Li, Q.; Chen, P.; Li, W.; Liu, W.; Li, Z.; Liu, Y.; Zhang, S.; et al. Tofacitinib Combined with Melanocyte Protector α-MSH to Treat Vitiligo through Dextran Based Hydrogel Microneedles. *Carbohydr. Polym.* **2023,** *305,* 120549.
93. Sarkar, R.; Ranjan, R.; Garg, S.; Garg, V. K.; Sonthalia, S.; Bansal, S. Periorbital Hyperpigmentation: A Comprehensive Review. *J. Clin. Aesthet. Dermatol.* **2016,** *9* (1), 49–55.
94. Yu, C.; Gu, X.; Chen, H.; Li, Y. Clinical Efficacy and Safety of Nano-Microneedle-Assisted Phenylethyl Resorcinol for the Treatment of Infraorbital Dark Circles. *J. Cosmet. Dermatol.* **2021,** *20* (3), 884–889.
95. Kontochristopoulos, G.; Kouris, A.; Platsidaki, E.; Markantoni, V.; Gerodimou, M.; Antoniou, C. Combination of Microneedling and 10% Trichloroacetic Acid Peels in the Management of Infraorbital Dark Circles. *J. Cosmet. Laser Ther.* **2016,** *18* (5), 289–292.

Chapter 13

Microneedle technology for biological monitoring and diagnosis*

Ying Chen and Guilan Quan
State Key Laboratory of Bioactive Molecules and Druggability Assessment/International Cooperative Laboratory of Traditional Chinese Medicine Modernization and Innovative Drug Development of Ministry of Education (MOE) of China/College of Pharmacy, Jinan University, Guangzhou, China

1. Introduction

The advent of scientific and technological advancement has led to the extensive utilisation of needle blood collection technology in the domain of biological monitoring and diagnosis. Nevertheless, the prevalence of this technology has also resulted in a number of problematic occurrences. For example, the repeated use of unsterilized needles has the potential to facilitate the spread of pathogens, which may result in significant health risks.[1] In the event that the wound resulting from the needling procedure is not treated promptly, the potential exists for infection and trauma to occur, which would ultimately compromise patient safety.[2] Furthermore, a subset of individuals experiences a condition known as "needle phobia," which can manifest as dizziness, panic attacks, nausea, pallor, and generalized perspiration in the event of needling. In extreme cases, syncope and other phenomena may occur.[3] Furthermore, patients who are subjected to prolonged monitoring may exhibit reduced compliance due to the invasive nature of the procedure, which may potentially compromise the reliability of bioindicator testing.[4]

Therefore, researchers have shifted their focus from blood to other interstitial fluids (ISF). The skin is the largest organ in the human body and contains

* This chapter is adapted from the Chinese paper entitled '[Research Progress in Biological Monitoring and Diagnosis Based on Microneedles]' published in Progress in Pharmaceutical Sciences, [2024], Vol.[48], No.[4], pp.[289-300], DOI: [10.20053/j.issn1001-5094.2024.04.006].

Microneedle Technology. https://doi.org/10.1016/B978-0-443-45544-5.00011-7

a multitude of health-related markers.[5,6] Many of the biomarkers present in the fluid within the skin are highly similar to those found in the blood, which serves as a vital conduit for transmitting physiological signals.[7] Consequently, the sampling and testing of skin biofluids by point-of-care testing (POCT) has increasingly captured the attention of researchers.[8] The extraction of intradermal fluid is typically conducted in a minimally invasive manner, and MN technology offers distinctive advantages in the domain of transdermal biomonitoring and diagnosis.

MNs were initially fabricated by Henry et al. in 1998 through microfabrication technology, comprising an array of needles with lengths of 100–1000 μm attached to a base.[9] MNs are capable of directionally penetrating the skin stratum corneum barrier, forming hundreds of micron-sized channels within the skin. This enables the markedly enhanced transdermal drug delivery efficacy.[10] Due to its advantages of convenience, noninvasiveness, and low pain, MNs have been widely applied in the treatment of malignant tumors,[11] hypertrophic scar,[12] diabetes mellitus,[13] alopecia,[14] and other diseases.

In recent years, MNs have made significant contributions to the field of transdermal drug delivery, and MN-based biosensing research has also garnered increasing attention in the scientific community.[15–18] MNs are capable of penetrating the skin and forming microchannels through which biomarker molecules in the blood or ISF can be sampled.[19–22] In comparison to conventional blood sampling techniques, MN-based monitoring and diagnosis systems have the potential to overcome numerous limitations and represent a highly promising approach for painless and minimally invasive testing. This chapter will review the research progress of MN-based monitoring and diagnosis, with a focus on the classification of MN-based monitoring and diagnosis strategies, the types of samples that can be extracted from MNs, and the applications of MNs in monitoring and diagnosis. Additionally, it will provide an outlook on the future development and applications of MNs.

2. Classification of microneedle-based monitoring and diagnosis strategies

MN-based monitoring and diagnosis systems can be classified into two categories:

In vivo sampling of analytes and follow-up in vitro testing, and in vivo capture and detection of analytes.[19–22]

2.1 In vivo sampling and in vitro detection of analytes in biological samples

MNs can extract biofluids (e.g., ISF or blood) from the skin based on negative pressure or capillary force. The extracted biofluids are then subjected to subsequent in vitro assays. These MNs typically lack electronic sensors for the detection of biofluids, and are used exclusively for in vivo sampling (Fig. 13.1).

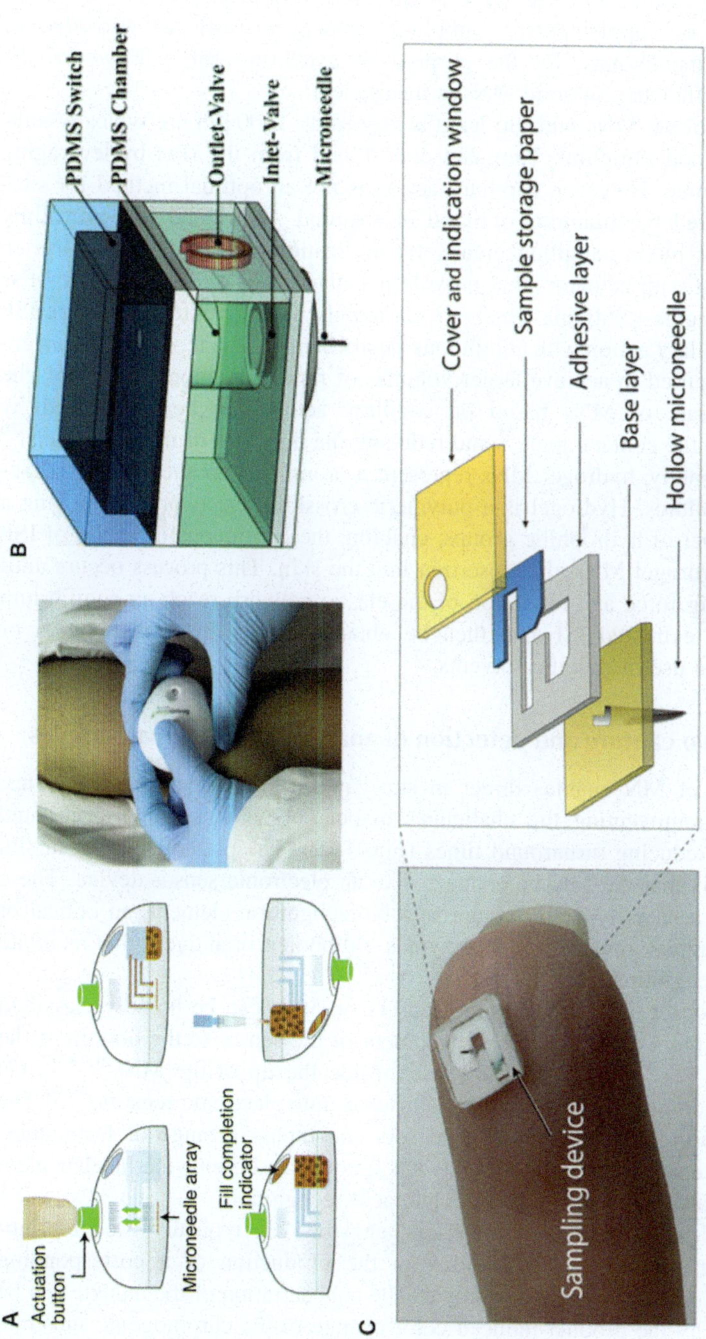

FIGURE 13.1 In vivo sampling and subsequent in vitro detection of analytes in biological samples (A Schematic diagram of collecting capillary blood samples using a microneedle(MN)-based device[28]; B: Schematic diagram of elastic self-healing actuator integrated with MNs for blood extraction[24]; C: Schematic diagram of painless sampling and subsequent processing of interstitial fluid (ISF) using an MN system.[29]) *(A) Adapted with permission from Ref.[28], Copyright 2018 Springer Nature Publisher. (B) Adapted with permission from Ref.[24], Copyright 2012 John Wiley and Sons Publisher. (C) Adapted with permission from Ref.[29], Copyright 2023 John Wiley and Sons Publisher.*

The application of solid MNs to the skin is typically preceded by the utilization of supplementary sampling techniques, such as attractive or absorbent membranes, for the purpose of extracting ISF.[23] However, the sampling efficiency of solid MNs is insufficient.

Long hollow MNs with tip lengths exceeding 1500 μm are typically suitable for blood sampling. They can draw blood from the skin by leveraging capillary force. However, capillary action is not an optimal method for situations where large quantities of blood are required. Accordingly, the prevailing method for blood sampling entails the utilization of long hollow MNs in conjunction with actuators that provide negative pressure.[24] For hollow MNs with tip lengths <900 μm, which are commonly used for extracting tissue ISF due to capillary action with continuous microchannels inside, several hours are usually required to achieve larger volumes of fluid extraction. In general, the sampling rate of MNs based on capillary action can be accelerated by decreasing the contact angle[25] and reducing the diameter of microchannels.[26]

Additionally, hydrogel MNs represent a promising avenue for the extraction of biofluids. Hydrogel is a polymeric crosslinked network comprising a high number of hydrophilic groups, enabling the continuous diffusion of ISF into the hydrogel MNs after insertion into the skin. This process occurs until the swelling force and retraction of the elastic network reach an equilibrium state. The extracted ISF can then be obtained either by centrifugation or through the use of specific solvents.[27]

2.2 In vivo capture and detection of analytes in biological samples

This type of MN enables direct in situ capture and detection of analytes, thereby circumventing the challenges associated with sample transfer and markedly reducing turnaround times (Fig. 13.2). The prerequisite for in situ detection is that the MN be connected to an electronic sensor device. These electronic sensors typically comprise a biorecognition element, an optical or electrochemical transducer, and enzymes, antibodies, or antigens that facilitate analyte recognition and capture.

The modification of electronic sensors on MNs can be broadly classified into three main categories: (1) Attachment of the sensor to the bottom of the MNs[30–32]; (2) Mounting of the sensor inside the tip of the MNs[21,33,34]; (3) Modification of MNs to transform themselves into electronic sensors.[35–37] The working principles of electronic sensors encompass a range of techniques, including colorimetric assay, sandwich immunoassay, enzyme-labeled electrochemical immunoassay, and nucleic acid recognition.

In MN-based colorimetric sensors, enzymes are typically used for biorecognition and catalysis, resulting in the production of a corresponding product in the presence of an analyte. The concentration of the analyte can be estimated by the product-induced color change of the chromogenic substrate on a paper strip.[38] In MN-based immunosensors, the primary antibody and

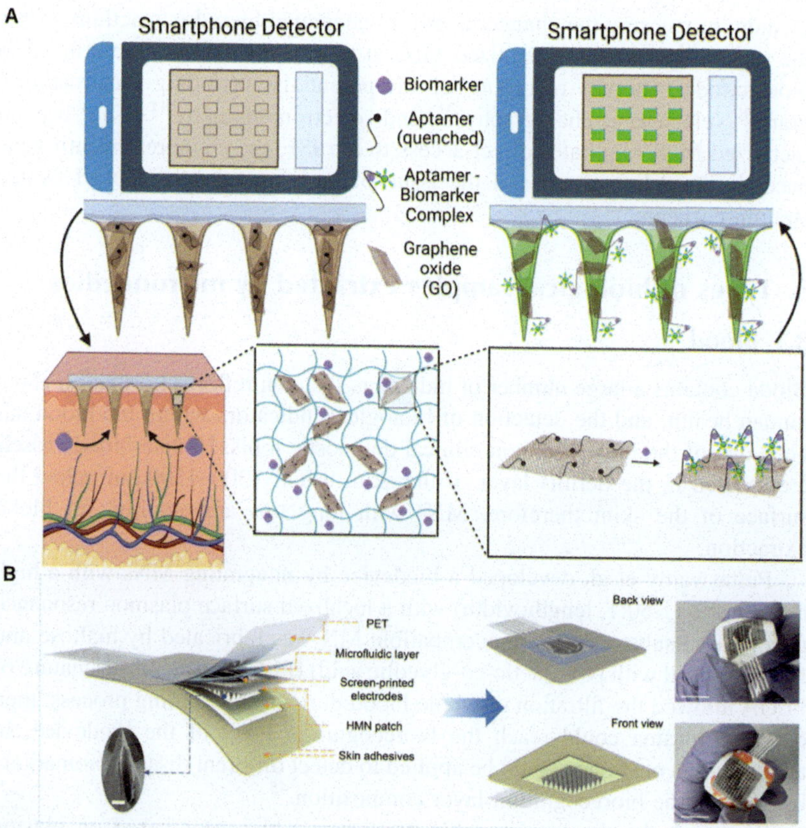

FIGURE 13.2 Schematic diagram of in situ capture and detection of analytes in biological samples (A: A smartphone-based microneedle (MN) system for in situ detection of small molecules and proteins[43]; B: Wearable touch-controlled device integrating hollow MNs for on-demand extraction of interstitial fluid (ISF) and electrochemical sensing.[44]) *(A) Adapted with permission from Ref.[43], Copyright 2023 John Wiley and Sons Publisher. (B) Adapted with permission from Ref.[44], Copyright 2023 John Wiley and Sons Publisher.*

antigen are usually attached to the surface of MNs to capture proteins in vivo. The secondary antibody is often modified with a fluorescent moiety, enabling its use in absorbance or enzyme-labeled colorimetric measurements to estimate the protein content in vivo. The typical MN-based immunosensor is the enzyme-linked immunosorbent assay.

However, the aforementioned MN-based sensors are typically disposable, necessitating their combination with electrochemical electrodes or the modification of MNs into electrochemical electrodes when continuous detection is required. Electrochemical electrodes are primarily composed of sensing materials and conductive electrodes. MN-based electrochemical sensors are predominantly used to analyze analyte concentrations by employing a redox medium to measure the change in current generated by redox reactions within the sensor.[39] Moreover, metalized MNs affixed to the skin can be utilized as bioelectrodes for the recording of bio-potentials, including electrocardiogram,[40] electroencephalography,[20] and electromyography.[41] The use of metalized MNs facilitates direct access to the ISF, thereby circumventing the necessity for electrolytes and the issue of high impedance induced by the stratum corneum.[42]

3. Types of biological samples extracted by microneedles

3.1 Blood

Blood contains a large number of indicators that can reflect information about human health, and the detection of biological indicators within the blood has been one of the most common clinical diagnostic tools. Human blood vessels are located in the dermis layer of the skin, usually 300–1500 μm below the surface of the skin; therefore, MNs with long tips are required for blood extraction.

Puttaswamy et al. developed a biodevice by integrating MNs with a high aspect ratio (>60:1, length:width) with a localized surface plasmon resonance paper-based substrate. The biocompatible MN was fabricated by maltose and further coated with poly(lactic-co-glycolic acid) (PLGA). The porous nature of PLGA allowed the filtration of whole blood during the sampling process, then only the plasma could reach the biorecognition layer of the biodevice. In addition, this biodevice could be applied to detect different disease biomarkers by altering the biorecognition layer composition.[45]

Wu et al. developed a portable paper-based biosensor based on plasma fluorescence-labeled MNs for quantitative detection of anemia biomarkers in whole blood, with a foldable paper origami-based device capable of achieving sequential analytical steps, including rapid serum isolation, target enrichment, and multiplexed data readout. The employment of plasma fluorescence-labeled MN arrays enabled efficient sampling of trace targets from ng/mL to pg/mL, which improved the detection sensitivity by approximately 7.6-fold as compared with the flat substrate-based assay. The biosensor was capable of simultaneous quantification of four common anemia biomarkers with unique advantages such as a small sample size (150 μL), short detection time (20 min), low cost ($2), good stability, and user-friendliness, which is expected to rapidly and accurately diagnose anemia in practical applications.[46]

3.2 ISF

As blood vessels are usually located 300–1500 μm below the skin surface, blood sampling requires long needles, which can be painful and frightening for the patient. Recently, ISF has attracted much attention, which is mainly located in the dermis.[47] ISF is derived from plasma and is maintained through fluid dynamics and osmolarity; thus, its composition is similar or related to that of plasma. The components within the ISF that are commonly used to reflect human health include glucose, cholesterol, cortisol, lactate, lipids, sodium, potassium, chloride, and some biomarkers.[48]

Zheng et al. developed a stretchable MN-based field-effect transistor biosensor with fast response ability for minimally invasive real-time detection of sodium in ISF. This sensor had high sensitivity, low detection limit, excellent biocompatibility, and good mechanical stability in the human body. Moreover, this system integrated a wireless data transmitter with an Internet of Things cloud, thus enabling real-time monitoring and long-term analysis.[49]

However, extracting ISF is not an easy task. The variability of the skin in different locations usually exists, including the thickness of the dermis,[50–52] the density of the glands and hair follicles,[53,54] and the mechanical properties of skin 50. In clinic, ISF is commonly extracted for blood glucose monitoring,[55] drug bioavailability evaluation,[56] and tumor microenvironment analysis.[57] During MN invasion, the area sampled may trigger a local inflammatory response, which can alter the composition of ISF. Thus, the use of MNs to extract a reasonable volume of ISF for quantitative analysis remains a challenge.

3.3 Suction effusion fluid

Suction effusion fluid (SEF) is the fluid contained in a blister formed by negative pressure attraction, which occurs as a result of separation between the epidermis and dermis due to mechanical stimulation or friction. SEF is composed primarily of extracellular fluid, proteins, cytokines, cellular debris, and other solutes. The composition of SEF is essentially similar to that of blood and ISF, but there are some biomolecules that are unique in SEF.[58,59] Nowadays, SEF is mostly used in the diagnosis of diseases such as psoriasis[60] and atopic dermatitis.[61]

3.4 Gingival crevicular fluid

Gingival crevicular fluid (GCF) is the fluid present between the gingival epithelium at the edge of the gums and the roots of the teeth, produced by the gingival epithelial cells. In healthy individuals, the amount of GCF is minimal. Conversely, the secretion of GCF increases when inflammation or periodontal disease occurs around the teeth.[62] It mainly consists of plasma exudate components, cellular debris, bacteria, inflammatory mediators, and other solutes.[63] Therefore, monitoring of GCF can determine the progression of periodontal disease and plays an important role in the assessment of associated diseases.

There are no studies using MNs to extract GCF, which can be explained by the following reasons. Firstly, locating and controlling the depth of GCF is challenging. GCF is located in the confined space between the gingival epithelium at the gingiva's margin and the tooth roots, with variable depths and locations. When employing MNs for extraction, it is imperative to accurately ascertain and regulate the depth of the needle tip, and failure to do so can result in damage to the gums or discomfort. Secondly, the secretion amount of GCF is relatively small, and the solute concentration is low at the micro to smaller level, making it difficult to collect enough samples quickly by MNs.

4. Microneedle-mediated biomonitoring and diagnosis applications

4.1 Microneedle-mediated biomonitoring applications

4.1.1 Glucose monitoring

Traditional blood glucose monitoring relies on blood glucose test strips and blood glucose meters, and usually requires the pricking of fingers, which is obviously painful during operation. Therefore, the use of MN-based sensors for glucose monitoring can reduce the pain to a certain extent, thereby improving patient compliance. Moreover, MN-based sensors can improve the precision of glucose monitoring, which is beneficial to better control blood glucose levels. Currently, there are various types of MN-based devices for glucose monitoring, which are mainly classified into two categories: colorimetric MN glucose sensors and electrochemical MN glucose sensors.[64]

The colorimetric MN glucose sensor was first reported by Mukerjee et al. in 2004, who designed silicon MNs for extracting ISF from the skin. Each needle tip in the MN array had a hole at the top, and a waiting period of 20–30 minutes was required for the ISF to fill up the tips after insertion into the skin. Then, the concentration of blood glucose was indicated by color changes.[65] More recently, studies have integrated extraction and analysis to increase measurement precision. For example, Zeng et al. developed a glucose-responsive colloidal crystal microneedle (GCC-MN) patch for minimally invasive, painless, and visually recognizable glucose colorimetric monitoring. The GCC-MN consists of a polymer core and a glucose-responsive colloidal crystal shell, which converts glucose concentration into a color change distinguishable to the naked eye in less than 5 minutes.[66]

In electrochemical MN glucose sensors, the concentration of glucose is converted into an electrical signal via MN-based electrodes. Yin et al. developed an MN patch for blood glucose monitoring in which glucose oxidase and carbon-nanotube-modified hollow microneedles were employed as electrochemical sensors, while integrated circuits were utilized for signal processing and transmission. Real-time blood glucose levels were displayed on a smartphone via Bluetooth. The MN design significantly improved the detection sensitivity, allowing continuous real-time measurement of blood glucose with micromolar-level sensitivity in live animals, while extending the lifetime to more than 14 days[67] (Fig. 13.3).

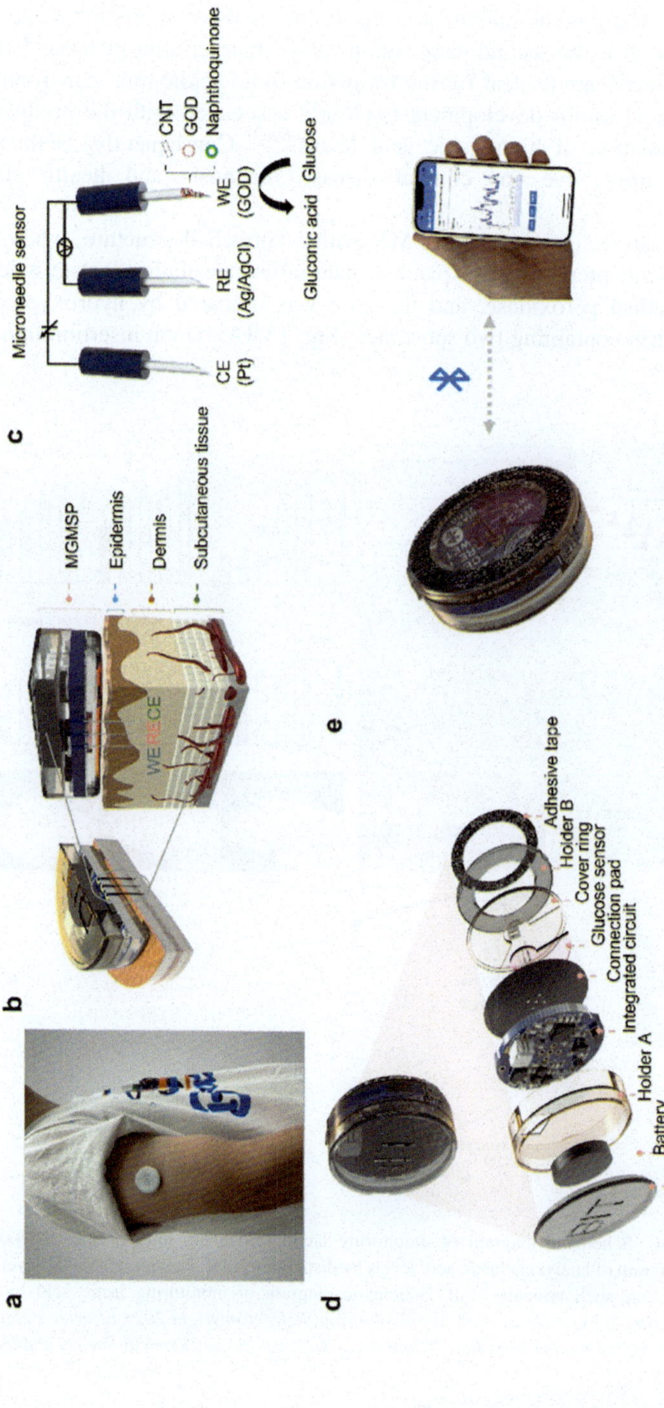

FIGURE 13.3 Schematic diagram of blood glucose monitoring using an electrochemical sensor based on microneedles.[67] *Adapted with permission from Ref.*[67] *Copyright 2024 Elsevier Publisher.*

4.1.2 Lactic acid monitoring

Lactic acid is a pivotal analytical target in the fields of sports medicine and clinical care. It is the second most significant biomarker after glucose.[68] High lactic acid levels are critical for the formation of an acidic microenvironment in tumors, and tumor development is closely associated with the production and accumulation of high lactic acid levels.[69,70] Consequently, lactic acid testing is imperative for clinical disease diagnosis and health status monitoring.

Yang et al. designed a bilayer MN with a core-shell structure, where the shell layer was prepared by expanded material encapsulating lactate oxidase and horseradish peroxidase, and the core was prepared by hydroxypropyl-β-cyclodextrin containing two substrates (Fig. 13.4A). Upon insertion into the

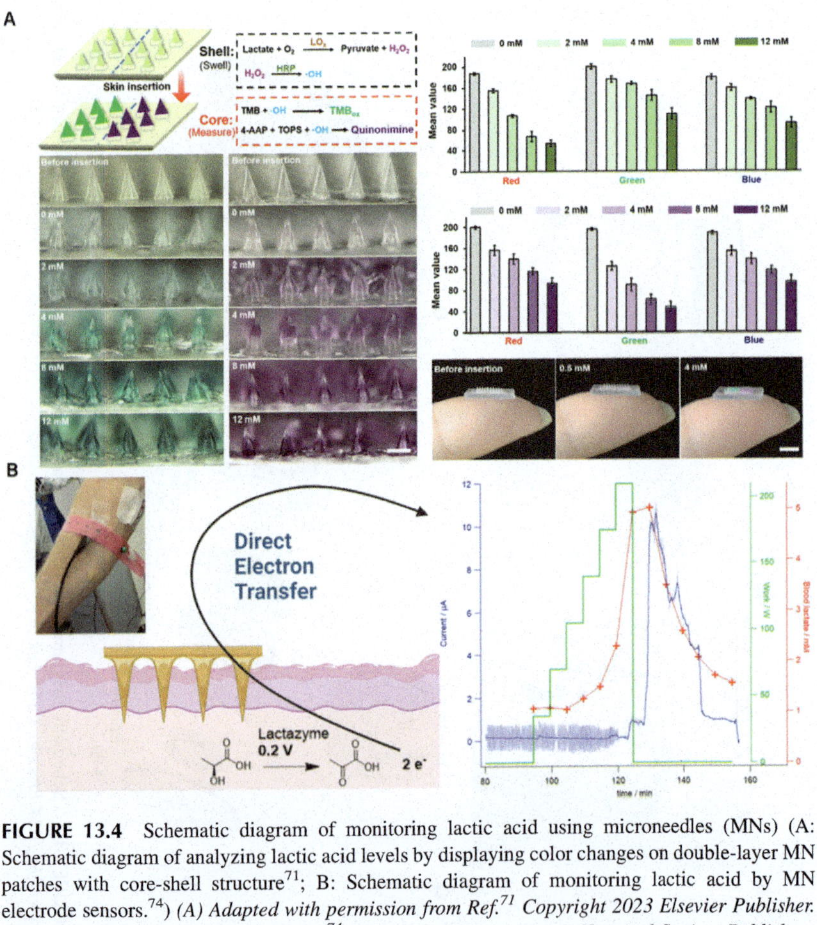

FIGURE 13.4 Schematic diagram of monitoring lactic acid using microneedles (MNs) (A: Schematic diagram of analyzing lactic acid levels by displaying color changes on double-layer MN patches with core-shell structure[71]; B: Schematic diagram of monitoring lactic acid by MN electrode sensors.[74]) *(A) Adapted with permission from Ref.[71] Copyright 2023 Elsevier Publisher. (B) Adapted with permission from Ref.[74] Copyright 2023 American Chemical Society Publisher.*

skin, the shell layer rapidly absorbed sufficient tissue fluid and converted lactic acid to hydroxyl radicals in the presence of the enzymes within the shell. Then the produced hydroxyl radicals reacted with the substrate in the core to display two visible color changes. This core-shell MN design allows rapid analysis of lactic acid levels in a short period of time by extracting tissue fluids from the skin with a high degree of detection accuracy.[71]

However, there are many oxidizable molecules present in ISF, such as ascorbic acid and acetaminophen, which can be oxidized at 0.7 V or below to interfere with lactic acid monitoring. Therefore, it is necessary to develop sensors that operate at lower voltages to minimize the off-target currents generated by such molecules and thus improve the overall device specificity and signal-to-noise ratio.[72,73] Freeman et al. investigated an MN electrode sensor by immobilizing the lactase on the surface of the MN electrode, and the lactate detection was realized by the direct electron transfer mechanism.[74] The sensor demonstrated high sensitivity and selectivity, enabling continuous lactic acid monitoring and effectively mitigating the impact of interfering substances on measurement outcomes (Fig. 13.4B).

4.1.3 Therapeutic drug concentration monitoring (TDM)

TDM is the process of quantitatively analyzing the concentration of drugs or their metabolites in biological samples using modern testing methods. The objective of TDM is to ascertain the safety range of blood drug concentrations and to formulate the optimal drug administration scheme (dose, route, delivery interval). This process culminates in the realization of individualized drug therapy that is characterized by optimal safety, effectiveness, and economic efficiency.[75] Conventional TDM methods generally necessitate blood draws, which can lead to diminished patient compliance. To address this issue, Ranamukhaarachchi et al. proposed the use of hollow MNs to extract a very small volume (<1 nL) of ISF to measure drug concentration. As shown in Fig. 13.5A, the inner cavity of the MNs is functionalized to be used as a microreactor during the extraction process to capture and bind the target drug candidate without the need for sample transfer. An optofluidic device was integrated with MN to rapidly quantify highly sensitive drug analytes. The MNs also were employed to monitor vancomycin blood concentrations, with vancomycin detection volume ranging from 50–100 μL and a detection limit of 1.35 μM, demonstrating its great potential for reducing healthcare cost and patient suffering.[76]

Apart from antibiotics, Levodopa is likewise one of the drugs that need to be monitored precisely.[77] The precise regulation of levodopa dosage is imperative, as improper dosing can result in adverse effects, including anxiety and depression when the dosage is insufficient, and movement disorders and psychiatric disorders when the dosage is excessive. As shown in Fig. 13.5B, Goud et al. designed wearable MN electrochemical detection electrodes; this

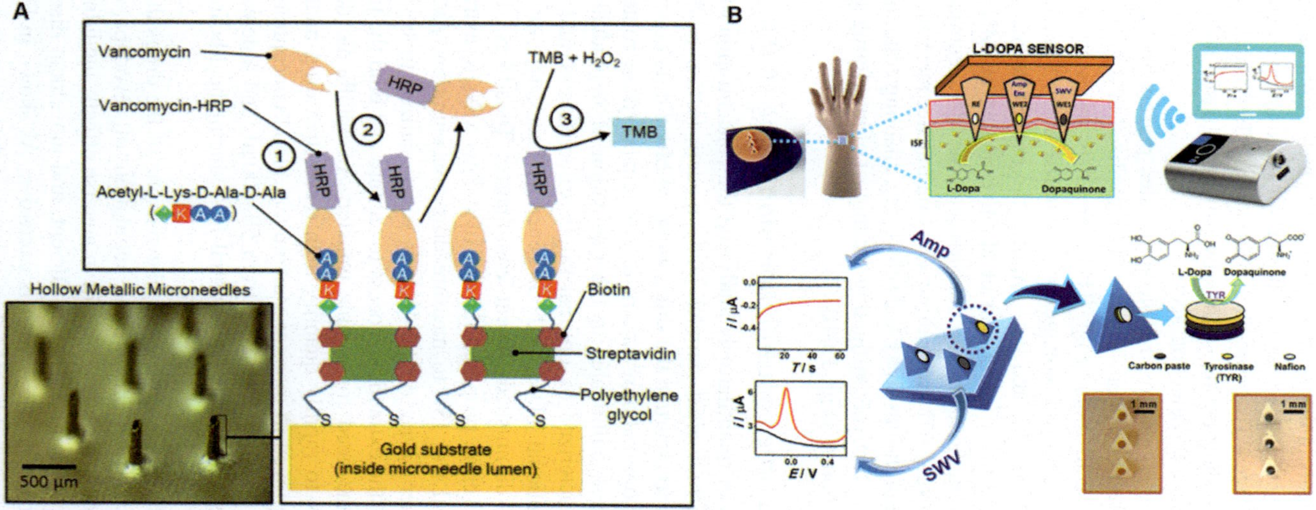

FIGURE 13.5 Schematic diagram of monitoring the concentration of therapeutic drugs using microneedles (MNs) (A: Schematic diagram of quantifying drug levels by integrating MNs with optofluidic devices[76]; B: Schematic diagram of detecting levodopa using a wearable MN containing an electrochemical detection electrode.[78]) *(A) Adapted with permission from Ref.[76] Copyright 2016 Springer Nature Publisher. Adapted with permission from Ref.[78] Copyright 2019 American Chemical Society Publisher.*

system integrated dual working electrodes to detect levodopa independently by square wave voltammetry and chronoamperometry, respectively. And the two different modes of detection sensing significantly improved the detection accuracy.[78]

4.1.4 Cell-free DNA monitoring

Currently, most MN-based biomonitoring systems have focused on small molecules or electrolytes, whereas the monitoring of macromolecular markers is also extraordinarily important.[79–81] In view of this, Yang et al. designed a hydrogel MN patch (Fig. 13.6A), which could rapidly capture Epstein–Barr virus cell-free DNA (EBV cfDNA) from ISF, with a maximum capture efficiency of 93.6% within 15 min. Subsequently, offline quantification of EBV cfDNA was achieved in vitro by electrochemical recombinase polymerase amplification wearable flexible microfluidics.[82] In another study, Yang et al. developed an instant online wearable MN patch for in vivo extraction and long-term monitoring of cfDNA (Fig. 13.6B), which was capable of real-time monitoring of Epstein–Barr virus, sepsis, and renal transplant cfDNA with 60% fetal bovine serum immunity and demonstrated stable sensitivity in vivo for up to 10 days. This wearable patch holds promise for long-term in vivo monitoring of cfDNA and for early disease screening and prognosis.[83]

4.1.5 Ion concentration monitoring

Ion concentration in blood is an important reference indicator for many diseases. For example, the concentration of Na^+ and Ca^{2+} is closely related to hypernatremia and hypokalaemia. Therefore, real-time monitoring of ion concentration in blood is particularly important for understanding one's health.[84] Huang et al. developed an integrated MN sensing system, in which lamellar MNs fabricated by laser micromachining were assembled to address the limitations of fabricating three-dimensional MN structures with multiple electrode channels. The ion-sensing MN system was sensitive enough to detect real-time changes in sodium, potassium, and calcium ion concentrations with good linear response, detection selectivity, and sustained stability. Finally, the integration of these MN electrodes into a wearable monitoring system could serve as an important tool for real-time monitoring of human health and potentially provide informative feedback for diseases related to physiological ion changes.[85]

4.1.6 Physiological pH monitoring

Monitoring of physiological pH can provide physicochemical information about the body's regulatory systems and is of great clinical importance. Slight changes in physiological pH can have an impact on cell, tissue, and organ function, and thus can be a good indicator for early disease detection.[86,87] In this regard, Odinotski et al. proposed the first conductive hydrogel MN

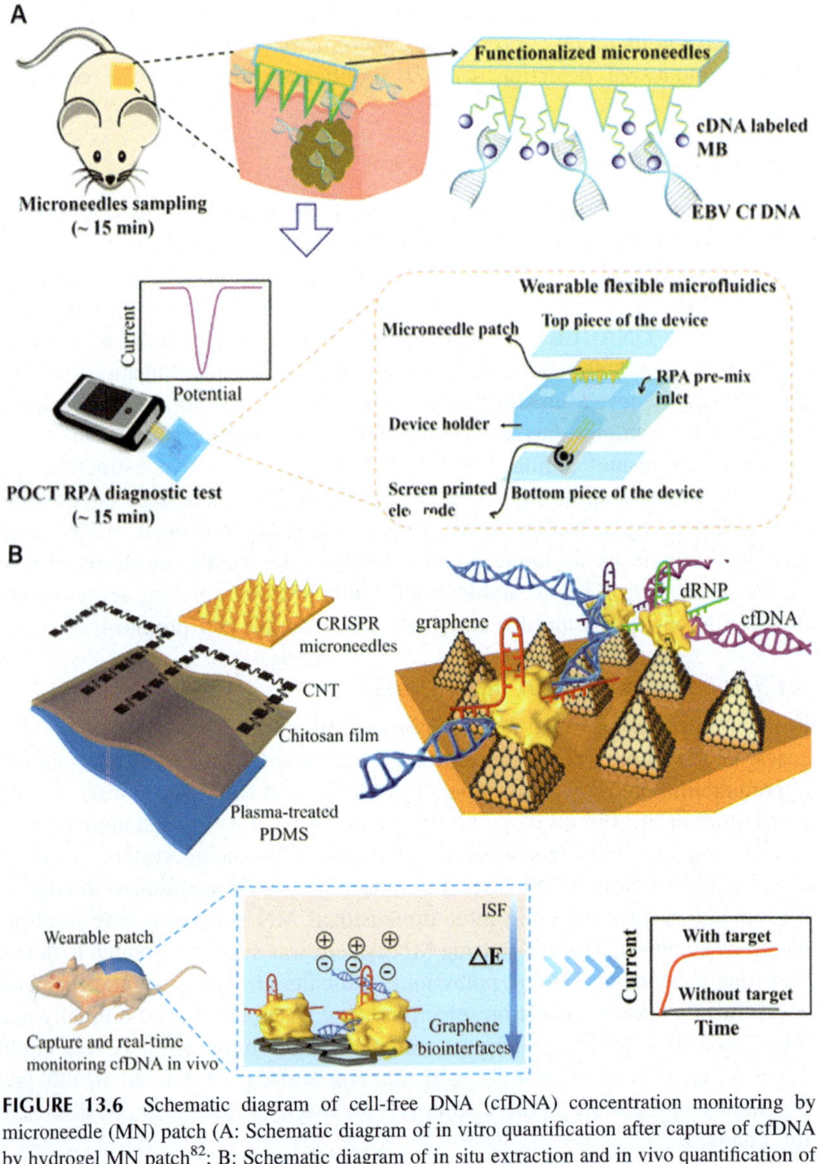

FIGURE 13.6 Schematic diagram of cell-free DNA (cfDNA) concentration monitoring by microneedle (MN) patch (A: Schematic diagram of in vitro quantification after capture of cfDNA by hydrogel MN patch[82]; B: Schematic diagram of in situ extraction and in vivo quantification of cfDNA by real-time online wearing MN patch.[83]) *(A) Adapted with permission from Ref.[82] Copyright 2019 American Chemical Society Publisher. Adapted with permission from Ref.[83] Copyright 2022 Springer Nature Publisher.*

electrode, which was capable of detecting pH without the need for post-processing. This array was fabricated using dopamine-modified hyaluronic acid hydrogel embedded with poly(3,4-ethylenedioxythiophene) polystyrene sulfonic acid to increase conductivity. The hydrogel MN enabled in vivo measurement with 93% accuracy compared to conventional pH probe meters. This hydrogel MN technology bridged the gap between traditional metal electrochemical biosensors and direct ISF extraction, which provides a platform for the development of polymer-based wearable sensors capable of detection on MNs.[88]

4.1.7 Lymphedema monitoring

Lymphedema is a localized tissue edema mainly caused by an abnormal increase of fluid in the tissue interstitial space due to obstruction of lymphatic fluid return in certain parts of the body. The main manifestations are deposition of lipids, liposclerosis, and abnormal proliferation of subcutaneous fibrous connective tissue, which in later stages can lead to thickening and roughness of the skin[89,90]. Early detection of lymphedema is crucial as the disease can have a major impact on quality of life. As the cost and complexity of most imaging strategies still have significant drawbacks, a lymphatic imaging technique using near-infrared tracers has been developed in clinical practice.[91] On this basis, Babity et al. developed polymeric MN patches loaded with cyanine 7.5-conjugated polyethylene glycol conjugates, taking advantage of the minimally invasive nature of MNs, as well as the good solubility and lymphatic selectivity of polyethylene glycolated Near Infrared tracers. This patch was capable of delivering the polyethylene glycolated cyanine 7.5 tracer to the skin, resulting in a potentially fluorescent medical tattoo for lymphedema monitoring. Further integrating this system with portable detection technology demonstrated the potential for simple, minimally invasive, and home-based monitoring of lymphatic drainage.[92]

4.1.8 Reactive oxygen species concentration monitoring

Oxidative stress induced by reactive oxygen species (ROS) is closely related to numerous diseases in the human body,[93] and therefore monitoring ROS can directly or indirectly reflect health status or disease progression. Among numerous diseases, atopic dermatitis is associated with elevated levels of oxidative stress and lipid peroxidation markers as well as decreased levels of systemic antioxidants, suggesting that ROS may drive the disease process to some extent.[94–96] Babity et al. developed a detection system based on polyacetylhydantoin polymeric MNs. As shown in Fig. 13.7A and B, the active sensors in combination with MNs could form fluorescent tattoos for ratiometric detection and imaging of inflammatory skin under monitored (pathological) physiological conditions.[97] In addition to atopic dermatitis, there is a close relationship between acute peritonitis and ROS. Acute peritonitis, mostly caused by chemical irritation and bacterial infection, is an acute inflammatory response in the abdomen. Hydrogen peroxide, a reactive oxygen molecule, has been demonstrated to be a by-product and specific biomarker of acute peritonitis.[98] Mei et al. developed a labeled, wearable MN patch based on surface-

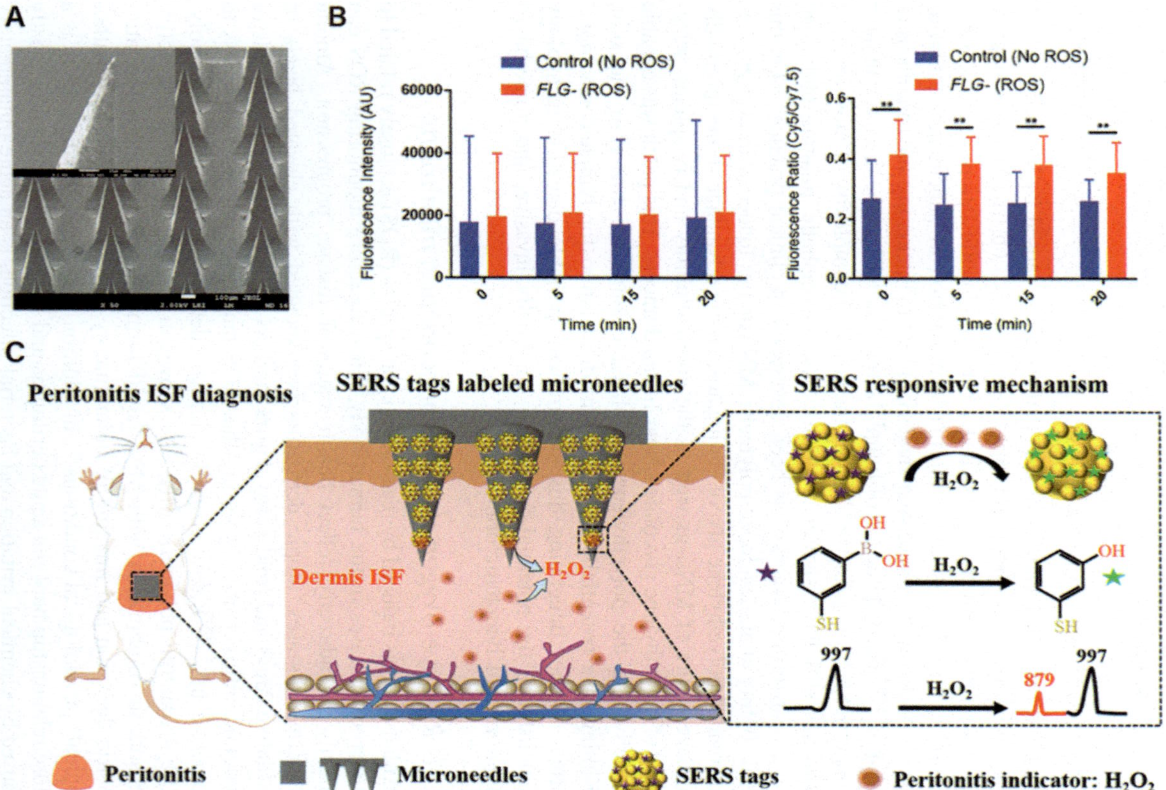

FIGURE 13.7 Schematic diagram of monitoring reactive oxygen species (ROS) concentration using microneedles MNs: (A–B: Fluorescence monitoring of physiological ROS concentration based on polyacetylcyanamide polymer MNs[97]; C: Schematic diagram of qualitative and quantitative hydrogen peroxide (H_2O_2) analysis using wearable MN patches and biosensors.[99]) *(A and B) Adapted with permission from Ref.[97] Copyright 2022 John Wiley and Sons Publisher. (C) Adapted*

enhanced Raman scattering (Fig. 13.7C), which could be used as a biologic sensor to puncture the epidermis of mouse abdominal skin in response to hydrogen peroxide in the dermal ISF. Depending on the extent of the altered surface-enhanced Raman scattering signal, it is feasible to qualitatively and quantitatively analyze hydrogen peroxide to diagnose the development of peritonitis.[99]

4.2 Microneedle-based biodiagnosis applications

4.2.1 Melanoma

Cutaneous melanoma is the most aggressive form of skin cancer, characterized by its high propensity for recurrence, metastasis, and mortality.[100] Early diagnosis and surgical excision of melanoma can improve patient survival, but the diagnosis of melanoma remains challenging due to the lack of reliable biomarkers for the disease. Tyrosinase (TYR) is a critical polyphenol oxidase in the synthesis of melanin. Its expression and accumulation in skin cells have been demonstrated to be intimately linked with melanoma progression. Consequently, TYR is progressively being regarded as a potential biomarker for melanoma diagnosis.[101,102]

Huang et al. developed wearable MNs containing trimetallic Au@Ag-Pt nanoparticles for the purpose of in situ detection of TYR. This system has the potential to facilitate the screening process for potential melanoma, thereby assisting general practitioners and inexperienced dermatologists in differentiating between melanomas and suspicious lesions, consequently reducing the number of unnecessary biopsies. The design principle entails the modification of dopamine on MN to incorporate a functionalized catechol group, which reacts with 4-mercaptophenylboronic acid (4-MPBA)-modified Au@Ag-Pt nanoenzymes (M/Au@Ag-Pt) to construct a stable MN/M/Au@Ag-Pt platform. The aggregation of M/Au@Ag-Pt on MN in this process can produce a strong surface-enhanced Raman scattering signal, while the remaining M/Au@Ag-Pt in solution can oxidize colorless 3,3‘,5,5’-tetramethylbenzidine (TMB) to blue oxTMB to produce a colorimetric signal. When MN was inserted into the skin and stimulated by TYR, the catechol group was specifically converted to benzoquinone oxide under its catalysis, which blocked the interaction between M/Au@Ag-Pt and MN, triggering the SERS signal to be “off” and the colorimetric signal to be “on,” to achieve dual-mode detection of TYR.[103]

4.2.2 Breast cancer

Breast cancer is the most prevalent malignancy affecting women’s health on a global scale. The monitoring of biomarkers is imperative for the early diagnosis and prognosis of the disease.[104] In the diagnosis of breast cancer, epidermal growth factor receptor 2 is a key biomarker.[105] Dervisevic et al.

developed a high-density gold-plated silicon MN array for epidermal growth factor receptor 2 detection. This sensing platform can be used as both a biomarker extraction platform and an electrochemical sensor. It is capable of selective immunocapture of epidermal growth factor receptor 2 and has a linear detection range of 10–250 ng/mL, with a detection limit of 4.8 ng/mL. This level is notably lower than the expected biomarker levels in breast cancer patients, thereby opening up new avenues for the development of high-performance wearable care devices.[106]

4.2.3 Infectious diseases

Infectious diseases are a major cause of morbidity and mortality in developing countries. Early diagnosis is paramount in preventing the deterioration of certain infectious diseases and reducing morbidity and mortality.[107] The emergence of a novel coronavirus at the end of 2019, designated as the severe acute respiratory syndrome coronavirus 2, caused the occurrence and spread of novel coronavirus infections. Bao et al. developed a novel coronavirus infection diagnostic device by integrating porous MNs with immunochromatographic biosensors. The paper-based immunoassay structure in the device was designed with both vertical and lateral flow to achieve continuous sampling from the MN arrays to the immunoassay biosensor. The paper-based immunoassay structure in the device is designed with vertical and lateral flow, enabling continuous sampling from the MN array to the immunoassay biosensor. The results of the immunoassay are visible to the naked eye within 3 min. The diagnostic device can be used as an effective complementary method to other diagnostic tests and has great potential for rapid screening of various infectious diseases.[108]

5. Conclusions and perspectives

In recent years, MN-based biomonitoring and diagnosis technologies have achieved significant breakthroughs in several domains. Firstly, MNs enable painless or low-pain collection of biological samples, such as blood and tissue fluids, providing a more comfortable and convenient way for clinical diagnosis. Furthermore, the integration of MNs with sensor technology facilitates the real-time and continuous monitoring of a broad spectrum of physiological indices, thereby providing substantial data support for the implementation of personalized healthcare strategies.

Nonetheless, there are still certain challenges in promoting research and application in this field, such as the stability of MNs and sensors, long-term reliability, and data processing algorithms that need to be addressed. Furthermore, there is an imperative to enhance collaboration with medical practitioners and healthcare institutions to facilitate the translation and implementation of MN-based biomonitoring and diagnostic findings. As technology advances and improves, microneedle-based biomonitoring and diagnostic technology is poised to enhance people's health management and disease treatment.

References

1. Sagoe-Moses, C.; Pearson, R. D.; Perry, J.; Jagger, J. Risks to Health Care Workers in Developing Countries - Reply. *N. Engl. J. Med.* **2001,** *345,* 1916.
2. Ayala, E. S.; Meuret, A. E.; Ritz, T. Treatments for Blood-Injury-Injection Phobia: A Critical Review of Current Evidence. *J. Psychiatr. Res.* **2009,** *43,* 1235–1242.
3. Liu, G. S.; Kong, Y. F.; Wang, Y. S.; Luo, Y. H.; Fan, X. D.; Xie, X.; Yang, B. R.; Wu, M. X. Microneedles for Transdermal Diagnostics: Recent Advances and New Horizons. *Biomaterials* **2020,** *232,* 16.
4. Wang, X. H.; Wang, Z. F.; Xiao, M.; Li, Z. H.; Zhu, Z. G. Advances in Biomedical Systems Based on Microneedles: Design, Fabrication, and Application. *Biomater. Sci.* **2023,** *34.*
5. Choi, J.; Ghaffari, R.; Baker, L. B.; Rogers, J. A. Skin-interfaced Systems for Sweat Collection and Analytics. *Sci. Adv.* **2018,** *4,* 9.
6. He, X. C.; Xu, T. L.; Gu, Z.; Gao, W.; Xu, L. P.; Pan, T. R.; Zhang, X. J. Flexible and Superwettable Bands as a Platform toward Sweat Sampling and Sensing. *Anal. Chem.* **2019,** *91,* 4296–4300.
7. Heikenfeld, J.; Jajack, A.; Feldman, B.; Granger, S. W.; Gaitonde, S.; Begtrup, G.; Katchman, B. A. Accessing Analytes in Biofluids for Peripheral Biochemical Monitoring. *Nat. Biotechnol.* **2019,** *37,* 407–419.
8. Lee, H.; Hong, Y. J.; Baik, S.; Hyeon, T.; Kim, D. H. Enzyme-based Glucose Sensor: From Invasive to Wearable Device. *Adv. Healthcare Mater.* **2018,** *7,* 14.
9. Henry, S.; McAllister, D. V.; Allen, M. G.; Prausnitz, M. R. Microfabricated Microneedles: a Novel Approach to Transdermal Drug Delivery. *J. Pharmacol. Sci.* **1998,** *87,* 922–925.
10. Xue, P.; Zhang, L.; Xu, Z. G.; Yan, J. J.; Gu, Z.; Kang, Y. J. Blood Sampling Using Microneedles as a Minimally Invasive Platform for Biomedical Diagnostics. *Appl. Mater. Today* **2018,** *13,* 144–157.
11. Lan, X. M.; She, J. C.; Lin, D. A.; Xu, Y.; Li, X.; Yang, W. F.; Lui, V. W. Y.; Jin, L. J.; Xie, X.; Su, Y. X. Microneedle-mediated Delivery of Lipid-Coated Cisplatin Nanoparticles for Efficient and Safe Cancer Therapy. *ACS Appl. Mater. Interfaces* **2018,** *10,* 33060–33069.
12. Zhang, X. X.; Chen, G. P.; Sun, L. Y.; Ye, F. F.; Shen, X.; Zhao, Y. J. Claw-inspired Microneedle Patches with Liquid Metal Encapsulation for Accelerating Incisional Wound Healing. *Chem. Eng. J.* **2021,** *406,* 8.
13. Wu, M. X.; Zhang, Y. J.; Huang, H.; Li, J. W.; Liu, H. Y.; Guo, Z. Y.; Xue, L. J.; Liu, S.; Lei, Y. F. Assisted 3D Printing of Microneedle Patches for Minimally Invasive Glucose Control in Diabetes. *Mater. Sci. Eng. C Mater.* **2020,** *117,* 10.
14. Pensato, R.; Padula, S La. A Comprehensive Review of Microneedling as a Potential Treatment Option for Androgenetic Alopecia. *Aesthetic Plast. Surg.* **2023,** *47,* 237–238.
15. Samper, I. C.; Gowers, S. A. N.; Booth, M. A.; Wang, C.; Watts, T.; Phairatana, T.; Vallant, N.; Sandhu, B.; Papalois, V.; Boutelle, M. G. Portable Microfluidic Biosensing System for Real-Time Analysis of Microdialysate in Transplant Kidneys. *Anal. Chem.* **2019,** *91,* 14631–14638.
16. Kim, K. B.; Choi, H.; Jung, H. J.; Oh, Y. J.; Cho, C. H.; Min, J. H.; Yoon, S.; Kim, J.; Cho, S. J.; Cha, H. J. Mussel-inspired Enzyme Immobilization and Dual Real-Time

Compensation Algorithms for Durable and Accurate Continuous Glucose Monitoring. *Biosens. Bioelectron.* **2019,** *143,* 7.

17. Xie, X.; Doloff, J. C.; Yesilyurt, V.; Sadraei, A.; McGarrigle, J. J.; Commis, M.; Veiseh, O.; Farah, S.; Isa, D.; Ghanis, S.; et al. Reduction of Measurement Noise in a Continuous Glucose Monitor by Coating the Sensor with a Zwitterionic Polymer. *Nat. Biomed. Eng.* **2018,** *2,* 894–906.
18. Dhanjai, S. M. M.; Lu, W. H. Modified Stainless Steel Microneedle Electrode for Polyphenolics Detection. *Anal. Bioanal. Chem.* **2020,** *412,* 7063–7072.
19. Sharma, S.; El-Laboudi, A.; Reddy, M.; Jugnee, N.; Sivasubramaniyam, S.; El Sharkawy, M.; Georgiou, P.; Johnston, D.; Oliver, N.; Cass, A. E. G. A Pilot Study in Humans of Microneedle Sensor Arrays for Continuous Glucose Monitoring. *Anal. Methods* **2018,** *10,* 2088–2095.
20. Wang, R. X.; Jiang, X. M.; Wang, W.; Li, Z. H. A Microneedle Electrode Array on Flexible Substrate for Long-Term EEG Monitoring. *Sens. Actuator B Chem.* **2017,** *244,* 750–758.
21. Mohan, A. M. V.; Windmiller, J. R.; Mishra, R. K.; Wang, J. Continuous Minimally-Invasive Alcohol Monitoring Using Microneedle Sensor Arrays. *Biosens. Bioelectron.* **2017,** *91,* 574–579.
22. Valdés-Ramírez, G.; Li, Y. C.; Kim, J.; Jia, W. Z.; Bandodkar, A. J.; Nuñez-Flores, R.; Miller, P. R.; Wu, S. Y.; Narayan, R.; Windmiller, J. R.; Polsky, R.; Wang, J. Microneedle-based Self-Powered Glucose Sensor. *Electrochem. Commun.* **2014,** *47,* 58–62.
23. Tariq, N.; Ashraf, M. W.; Tayyaba, S. A Review on Solid Microneedles for Biomedical Applications. *J. Pharm. Innov.* **2022,** *17,* 1464–1483.
24. Li, C. G.; Lee, K.; Lee, C. Y.; Dangol, M.; Jung, H. A Minimally Invasive Blood-Extraction System: Elastic Self-Recovery Actuator Integrated with an Ultrahigh- Aspect-Ratio Microneedle. *Adv. Mater.* **2012,** *24,* 4583–4586.
25. Chakraborty, S.; Tsuchiya, K. Development and Fluidic Simulation of Microneedles for Painless Pathological Interfacing with Living Systems. *J. Appl. Phys.* **2008,** *103,* 9.
26. Strambini, L. M.; Longo, A.; Scarano, S.; Prescimone, T.; Palchetti, I.; Minunni, M.; Giannessi, D.; Barillaro, G. Self-powered Microneedle-Based Biosensors for Pain-free High-Accuracy Measurement of Glycaemia in Interstitial Fluid. *Biosens. Bioelectron.* **2015,** *66,* 162–168.
27. Ahmed, E. M. Hydrogel: Preparation, Characterization, and Applications: A Review. *J. Adv. Res.* **2015,** *6,* 105–121.
28. Blicharz, T. M.; Gong, P.; Bunner, B. M.; Chu, L. L.; Leonard, K. M.; Wakefield, J. A.; Williams, R. E.; Dadgar, M.; Tagliabue, C. A.; El Khaja, R.; et al. Microneedle-based Device for the One-step Painless Collection of Capillary Blood Samples. *Nat. Biomed. Eng.* **2018,** *2,* 151–157.
29. Ribet, F.; Bendes, A.; Fredolini, C.; Dobielewski, M.; Böttcher, M.; Beck, O.; Schwenk, J. M.; Stemme, G.; Roxhed, N. Microneedle Patch for Painless Intradermal Collection of Interstitial Fluid Enabling Multianalyte Measurement of Small Molecules, SARS-CoV-2 Antibodies, and Protein Profiling. *Adv. Healthcare Mater.* **2023,** *12,* 11.
30. Chua, B.; Desai, S. P.; Tierney, M. J.; Tamada, J. A.; Jina, A. N. Effect of Microneedles Shape on Skin Penetration and Minimally Invasive Continuous Glucose Monitoring *In Vivo*. *Sens. Actuator A Phys.* **2013,** *203,* 373–381.
31. Nagamine, K.; Kubota, J.; Kai, H.; Ono, Y.; Nishizawa, M. An Array of Porous Microneedles for Transdermal Monitoring of Intercellular Swelling. *Biomed. Microdevices* **2017,** *19,* 6.

32. Miller, P. R.; Skoog, S. A.; Edwards, T. L.; Lopez, D. M.; Wheeler, D. R.; Arango, D. C.; Xiao, X. Y.; Brozik, S. M.; Wang, J.; Polsky, R.; et al. Multiplexed Microneedle-Based Biosensor Array for Characterization of Metabolic Acidosis. *Talanta* **2012,** *88,* 739–742.
33. Windmiller, J. R.; Zhou, N. D.; Chuang, M. C.; Valdés-Ramírez, G.; Santhosh, P.; Miller, P. R.; Narayan, R.; Wang, J. Microneedle Array-Based Carbon Paste Amperometric Sensors and Biosensors. *Analyst* **2011,** *136,* 1846–1851.
34. Ribet, F.; Stemme, G.; Roxhed, N. Real-time Intradermal Continuous Glucose Monitoring Using a Minimally Invasive Microneedle-Based System. *Biomed. Microdevices* **2018,** *20,* 10.
35. Corrie, S. R.; Fernando, G. J. P.; Crichton, M. L.; Brunck, M. E. G.; Anderson, C. D.; Kendall, M. A. F. Surface-modified Microprojection Arrays for Intradermal Biomarker Capture, with Low Non-specific Protein Binding. *Lab Chip* **2010,** *10,* 2655–2658.
36. Keum, D. H.; Jung, H. S.; Wang, T.; Shin, M. H.; Kim, Y. E.; Kim, K. H.; Ahn, G. O.; Hahn, S. K. Microneedle Biosensor for Real-Time Electrical Detection of Nitric Oxide for *In Situ* Cancer Diagnosis during Endomicroscopy. *Adv. Healthcare Mater.* **2015,** *4,* 1153–1158.
37. Gowers, S. A. N.; Freeman, D. M. E.; Rawson, T. M.; Rogers, M. L.; Wilson, R. C.; Holmes, A. H.; Cass, A. E.; Hare, D. O.'. Development of a minimally invasive microneedle-based sensor for continuous monitoring of β-Lactam antibiotic concentrations in vivo. *ACS Sens.* **2019,** *4* (1072-+).
38. Blake, D. A.; McLean, N. V. A Colorimetric Assay for the Measurement of D-Glucose Consumption by Cultured Cells. *Anal. Biochem.* **1989,** *177,* 156–160.
39. Pappa, A. M.; Parlak, O.; Scheiblin, G.; Mailley, P.; Salleo, A.; Owens, R. M. Organic Electronics for Point-Of-Care Metabolite Monitoring. *Trends Biotechnol.* **2018,** *36,* 45–59.
40. Pei, W. H.; Zhang, H.; Wang, Y. J.; Guo, X. H.; Xing, X.; Huang, Y.; Xie, Y. X.; Yang, X. W.; Chen, H. D. Skin-potential Variation Insensitive Dry Electrodes for ECG Recording. *IEEE Trans. Biomed. Eng.* **2017,** *64,* 463–470.
41. Ren, L.; Xu, S. J.; Gao, J.; Lin, Z.; Chen, Z. P.; Liu, B.; Liang, L.; Jiang, L. L. Fabrication of Flexible Microneedle Array Electrodes for Wearable Bio-Signal Recording. *Sensors* **2018,** *18,* 11.
42. Griss, P.; Enoksson, P.; Tolvanen-Laakso, H. K.; Meriläinen, P.; Ollmar, S.; Stemme, G. Micromachined Electrodes for Biopotential Measurements. *J. Microelectromech. Syst.* **2001,** *10,* 10–16.
43. Keyvani, F.; Zheng, H. J.; Kaysir, M. R.; Mantaila, D. F.; Nejad, P. G.; Rahman, F. A.; Quadrilatero, J.; Ban, D. Y.; Poudineh, M. A Hydrogel Microneedle Assay Combined with Nucleic Acid Probes for On-Site Detection of Small Molecules and Proteins. *Angew. Chem. Int. Edit.* **2023,** *62,* 9.
44. Abbasiasl, T.; Mirlou, F.; Mirzajani, H.; Bathaei, M. J.; Istif, E.; Shomalizadeh, N.; Cebecioglu, R. E.; Özkahraman, E. E.; Yener, U. C.; Beker, L. A Wearable Touch-Activated Device Integrated with Hollow Microneedles for Continuous Sampling and Sensing of Dermal Interstitial Fluid. *Adv. Mater.* **2023,** *17.*
45. Puttaswamy, S. V.; Lubarsky, G. V.; Kelsey, C.; Zhang, X. S.; Finlay, D.; McLaughlin, J. A.; Bhalla, N. Nanophotonic-carbohydrate Lab-On-A-Microneedle for Rapid Detection of Human Cystatin C in Finger-Prick Blood. *ACS Nano* **2020,** *14,* 11939–11949.
46. Wu, J.; Liu, Y.; Peng, L. J.; Liu, Q. S. J.; Wang, D. W.; Jing, X.; Hu, Y. Z.; Lin, J.; Fu, H. Y.; Ji, X. M.; et al. A Plasmonic Fluor-Lightened Microneedle Array Enables Ultrasensitive Multitarget Whole Blood Diagnosis of Anemia in a Paper Origami-Based Device. *Small* **2023,** *19,* 14.

47. Samant, P. P.; Niedzwiecki, M. M.; Raviele, N.; Tran, V.; Mena-Lapaix, J.; Walker, D. I.; Felner, E. I.; Jones, D. P.; Miller, G. W.; Prausnitz, M. R. Sampling Interstitial Fluid from Human Skin Using a Microneedle Patch. *Sci. Transl. Med.* **2020,** *12,* 15.
48. Gervais, L.; de Rooij, N.; Delamarche, E. Microfluidic Chips for Point-Of-Care Immunodiagnostics. *Adv. Mater.* **2011,** *23,* H151–H176.
49. Zheng, Y. B.; Omar, R.; Zhang, R. J.; Tang, N.; Khatib, M.; Xu, Q.; Milyutin, Y.; Saliba, W.; Broza, Y. Y.; Wu, W. W.; et al. A Wearable Microneedle-Based Extended Gate Transistor for Real-Time Detection of Sodium in Interstitial Fluids. *Adv. Mater.* **2022,** *34,* 11.
50. Diridollou, S.; Black, D.; Lagarde, J. M.; Gall, Y.; Berson, M.; Vabre, V.; Patat, F.; Vaillant, L. Sex- and Site-dependent Variations in the Thickness and Mechanical Properties of Human Skin In Vivo. *Int. J. Cosmetic* **2000,** *22,* 421–435.
51. Sandby-Moller, J.; Poulsen, T.; Wulf, H. C. Epidermal Thickness at Different Body Sites: Relationship to Age, Gender, Pigmentation, Blood Content, Skin Type and Smoking Habits. *Acta Derm. Venereol.* **2003,** *83,* 410–413.
52. Tsukahara, K.; Takema, Y.; Moriwaki, S.; Fujimura, T.; Imokawa, G. Dermal Fluid Translocation Is an Important Determinant of the Diurnal Variation in Human Skin Thickness. *Br. J. Dermatol.* **2001,** *145,* 590–596.
53. Taylor, N. A.; Machado-Moreira, C. A. Regional Variations in Transepidermal Water Loss, Eccrine Sweat Gland Density, Sweat Secretion Rates and Electrolyte Composition in Resting and Exercising Humans. *Extreme Physiol. Med.* **2013,** *2,* 4.
54. Otberg, N.; Richter, H.; Schaefer, H.; Blume-Peytavi, U.; Sterry, W.; Lademann, J. Variations of Hair Follicle Size and Distribution in Different Body Sites. *J. Invest. Dermatol.* **2004,** *122,* 14–19.
55. GhavamiNejad, P.; GhavamiNejad, A.; Zheng, H. J.; Dhingra, K.; Samarikhalaj, M.; Poudineh, M. A Conductive Hydrogel Microneedle-Based Assay Integrating PEDOT:PSS and Ag-Pt Nanoparticles for Real-Time, Enzyme-Less, and Electrochemical Sensing of Glucose. *Adv. Healthcare Mater.* **2023,** *12,* 11.
56. Bodenlenz, M.; Tiffner, K. I.; Raml, R.; Augustin, T.; Dragatin, C.; Birngruber, T.; Schimek, D.; Schwagerle, G.; Pieber, T. R.; Raney, S. G.; Kanfer, I.; Sinner, F. Open Flow Microperfusion as a Dermal Pharmacokinetic Approach to Evaluate Topical Bioequivalence. *Clin. Pharmacokinet.* **2017,** *56* (91), 99.
57. Wiig, H.; Tenstad, O.; Iversen, P. O.; Kalluri, R.; Bjerkvig, R. Interstitial Fluid: the Overlooked Component of the Tumor Microenvironment? *Fibrogenesis Tissue Repair* **2010,** *3,* 12.
58. Sjöbom, U.; Christenson, K.; Hellström, A.; Nilsson, A. K. Inflammatory Markers in Suction Blister Fluid: A Comparative Study between Interstitial Fluid and Plasma. *Front. Immunol.* **2020,** *11,* 9.
59. Niedzwiecki, M. M.; Samant, P.; Walker, D. I.; Tran, V.; Jones, D. P.; Prausnitz, M. R.; Miller, G. W. Human Suction Blister Fluid Composition Determined Using High-Resolution Metabolomics. *Anal. Chem.* **2018,** *90,* 3786–3792.
60. Schepko, M.; Stumpf, K. C.; Tumala, S.; Peters, E. M.; Kupfer, J. P.; Schut, C. Study Protocol: Neuro-Inflammatory Parameters as Mediators of the Relationship between Social Anxiety and Itch Intensity: A Cross-Sectional, Controlled Laboratory Study in Patients with Psoriasis and Healthy Skin Controls. *PLoS One* **2023,** *18,* 15.
61. Rindler, K.; Krausgruber, T.; Thaler, F. M.; Alkon, N.; Bangert, C.; Kurz, H.; Fortelny, N.; Rojahn, T. B.; Jonak, C.; Griss, J.; et al. Spontaneously Resolved Atopic Dermatitis Shows Melanocyte and Immune Cell Activation Distinct from Healthy Control Skin. *Front. Immunol.* **2021,** *12,* 14.

62. Lin, S. J.; Chen, Y. L.; Kuo, M. Y. B.; Li, C. L.; Lu, H. K. Measurement of Gp130 Cytokines - Oncostatin M and IL-6 in Gingival Crevicular Fluid of Patients with Chronic Periodontitis. *Cytokine* **2005,** *30,* 160–167.
63. Kido, J.; Bando, M.; Hiroshima, Y.; Iwasaka, H.; Yamada, K.; Ohgami, N.; Nambu, T.; Kataoka, M.; Yamamoto, T.; Shinohara, Y.; et al. Analysis of Proteins in Human Gingival Crevicular Fluid by Mass Spectrometry. *J. Periodontal. Res.* **2012,** *47,* 488–499.
64. Ju, J.; Li, L.; Regmi, S.; Zhang, X. Y.; Tang, S. X. Microneedle-Based Glucose Sensor Platform: From *Vitro* to Wearable Point-Of-Care Testing Systems. *Biosensors (Basel)* **2022,** *12,* 18.
65. Mukerjee, E.; Collins, S. D.; Isseroff, R. R.; Smith, R. L. Microneedle Array for Transdermal Biological Fluid Extraction and In Situ Analysis. *Sens. Actuator A Phys.* **2004,** *114,* 267–275.
66. Zeng, Y.; Wang, J. Q.; Wang, Z. J.; Chen, G. J.; Yu, J. C.; Li, S.; Li, Q. W.; Li, H. J.; Wen, D.; Gu, Z. Z.; et al. Colloidal Crystal Microneedle Patch for Glucose Monitoring. *Nano Today* **2020,** *35* (7).
67. Yin, S. J.; Yu, Z. Q.; Song, N. N.; Guo, Z. J.; Li, W.; Ma, J. H.; Wang, X. Y.; Liu, J. Y.; Liang, M. M. A Long Lifetime and Highly Sensitive Wearable Microneedle Sensor for the Continuous Real-Time Monitoring of Glucose in Interstitial Fluid. *Biosens. Bioelectron.* **2024,** *244,* 8.
68. Rassaei, L.; Olthuis, W.; Tsujimura, S.; Sudhölter, E. J. R.; van den Berg, A. Lactate Biosensors: Current Status and Outlook. *Anal. Bioanal. Chem.* **2014,** *406,* 123–137.
69. Pereira-Nunes, A.; Afonso, J.; Granja, S.; Baltazar, F. Lactate and Lactate Transporters as Key Players in the Maintenance of the Warburg Effect. In *Tumor Microenvironment: The Main Driver of Metabolic Adaptation;* Serpa, J., Ed.; Springer International Publishing Ag: Cham, 2020; pp 51–74.
70. Pundir, C. S.; Narwal, V.; Batra, B. Determination of Lactic Acid with Special Emphasis on Biosensing Methods: A Review. *Biosens. Bioelectron.* **2016,** *86,* 777–790.
71. Yang, H. Y.; Jiang, X.; Zeng, Y. N.; Zhang, W.; Yuan, Q. Q.; Yin, M. R.; Wu, G. S.; Li, W. A Swellable Bilateral Microneedle Patch with Core-Shell Structure for Rapid Lactate Analysis and Early Melanoma Diagnosis. *Chem. Eng. J.* **2023,** *455,* 8.
72. Toghill, K. E.; Compton, R. G. Electrochemical Non-enzymatic Glucose Sensors: A Perspective and an Evaluation. *Int. J. Electrochem. Sci.* **2010,** *5,* 1246–1301.
73. Samant, P. P.; Prausnitz, M. R. Mechanisms of Sampling Interstitial Fluid from Skin Using a Microneedle Patch. *Proc. Natl. Acad. Sci. U. S. A.* **2018,** *115,* 4583–4588.
74. Freeman, D. M. E.; Ming, D. K.; Wilson, R.; Herzog, P. L.; Schulz, C.; Felice, A. K. G.; Chen, Y. C.; Hare, D. O.'; Holmes, A. H.; Cass, A. E. G. Continuous Measurement of Lactate Concentration in Human Subjects through Direct Electron Transfer from Enzymes to Microneedle Electrodes. *ACS Sens.* **2023,** *9.*
75. Schulz, M.; Schmoldt, A.; Andresen-Streichert, H.; Iwersen-Bergmann, S. Revisited: Therapeutic and Toxic Blood Concentrations of More Than 1100 Drugs and Other Xenobiotics. *Crit. Care* **2020,** *24,* 4.
76. Ranamukhaarachchi, S. A.; Padeste, C.; Dübner, M.; Häfeli, U. O.; Stoeber, B.; Cadarso, V. J. Integrated Hollow Microneedle-Optofluidic Biosensor for Therapeutic Drug Monitoring in Sub-nanoliter Volumes. *Sci. Rep.* **2016,** *6,* 10.
77. Lin, L.; Lian, H. T.; Sun, X. Y.; Yu, Y. M.; Liu, B. An L-Dopa Electrochemical Sensor Based on a Graphene Doped Molecularly Imprinted Chitosan Film. *Anal. Methods* **2015,** *7,* 1387–1394.

78. Goud, K. Y.; Moonla, C.; Mishra, R. K.; Yu, C. M.; Narayan, R.; Litvan, I.; Wang, J. Wearable Electrochemical Microneedle Sensor for Continuous Monitoring of Levodopa: Toward Parkinson Management. *ACS Sens.* **2019,** *4,* 2196–2204.
79. Yang, Y. R.; Song, Y.; Bo, X. J.; Min, J. H.; Pak, O. S.; Zhu, L. L.; Wang, M. Q.; Tu, J. B.; Kogan, A.; Zhang, H. X.; et al. A Laser-Engraved Wearable Sensor for Sensitive Detection of Uric Acid and Tyrosine in Sweat. *Nat. Biotechnol.* **2020,** *38,* 217.
80. Lee, G. H.; Moon, H.; Kim, H.; Lee, G. H.; Kwon, W.; Yoo, S.; Myung, D.; Yun, S. H.; Bao, Z.; Hahn, S. K. Multifunctional Materials for Implantable and Wearable Photonic Healthcare Devices. *Nat. Rev. Mater.* **2020,** *5,* 149–165.
81. Shrivastava, S.; Trung, T. Q.; Lee, N. E. Recent Progress, Challenges, and Prospects of Fully Integrated Mobile and Wearable Point-Of-Care Testing Systems for Self-Testing. *Chem. Soc. Rev.* **2020,** *49,* 1812–1866.
82. Yang, B.; Fang, X. E.; Kong, J. L. *In Situ* Sampling and Monitoring Cell-free DNA of the Epstein-Barr Virus from Dermal Interstitial Fluid Using Wearable Microneedle Patches. *ACS Appl. Mater. Interfaces* **2019,** *11,* 38448–38458.
83. Yang, B.; Kong, J. L.; Fang, X. E. Programmable CRISPR-Cas9 Microneedle Patch for Long-Term Capture and Real-Time Monitoring of Universal Cell-free DNA. *Nat. Commun.* **2022,** *13,* 16.
84. Teymourian, H.; Tehrani, F.; Mahato, K.; Wang, J. Lab under the Skin: Microneedle Based Wearable Devices. *Adv. Healthcare Mater.* **2021,** *10,* 19.
85. Huang, X.; Zheng, S.; Liang, B.; He, M.; Wu, F.; Yang, J.; Chen, H.; Xie, X. 3D-assembled Microneedle Ion Sensor-Based Wearable System for the Transdermal Monitoring of Physiological Ion Fluctuations. *Microsys. Nanoeng.* **2023,** *9.*
86. Ghoneim, M. T.; Nguyen, A.; Dereje, N.; Huang, J.; Moore, G. C.; Murzynowski, P. J.; Dagdeviren, C. Recent Progress in Electrochemical pH-Sensing Materials and Configurations for Biomedical Applications. *Chem. Rev.* **2019,** *119,* 5248–5297.
87. García-Guzmán, J. J.; Pérez-Ràfols, C.; Cuartero, M.; Crespo, G. A. Toward *In Vivo* Transdermal pH Sensing with a Validated Microneedle Membrane Electrode. *ACS Sens.* **2021,** *6,* 1129–1137.
88. Odinotski, S.; Dhingra, K.; GhavamiNejad, A.; Zheng, H. J.; GhavamiNejad, P.; Gaouda, H.; Mohammadrezaei, D.; Poudineh, M. A Conductive Hydrogel-Based Microneedle Platform for Real-Time pH Measurement in Live Animals. *Small* **2022,** *18,* 10.
89. Grada, A. A.; Phillips, T. J. Lymphedema Pathophysiology and Clinical Manifestations. *J. Am. Acad. Dermatol.* **2017,** *77,* 1009–1020.
90. Rockson, S. G.; Rivera, K. K. Estimating the Population Burden of Lymphedema. In *Lymphatic Continuum Revisited;* Rockson, S. G., Ed.; Wiley-Blackwell: Malden, 2008; pp 147–154.
91. Rasmussen, J. C.; Sevick-Muraca, E. M. IEEE, Clinical Translation and Discovery with Near-Infrared Fluorescence Lymphatic Imaging. In *Conference on Lasers and Electro-Optics (CLEO), Ieee, San Jose, CA;* 2014.
92. Babity, S.; Polomska, A. K.; Couture, F.; Bonmarin, M.; Fehr, D.; Detmar, M.; Brambilla, D. Rational Design of a Fluorescent Microneedle Tattoo for Minimally Invasive Monitoring of Lymphatic Function. *J. Contr. Release* **2020,** *327,* 350–359.
93. Young, L. B. S. Reactive Oxygen Species in Tumor Necrosis Factor-α-Activated Primary Human Keratinocytes: Implications for Psoriasis and Inflammatory Skin Disease (Vol 128, Pg 2606, 2008). *J. Invest. Dermatol.* **2009,** *129,* 1838.

94. Tsukahara, H.; Shibata, R.; Ohshima, Y.; Todoroki, Y.; Sato, S.; Ohta, N.; Hiraoka, M.; Yoshida, A.; Nishima, S.; Mayumi, M. Oxidative Stress and Altered Antioxidant Defenses in Children with Acute Exacerbation of Atopic Dermatitis. *Life Sci.* **2003,** *72,* 2509–2516.
95. Chang, H. Y.; Suh, D. I.; Yang, S. I.; Kang, M. J.; Lee, S. Y.; Lee, E.; Choi, I. A.; Lee, K. S.; Shin, Y. J.; Shin, Y. H.; et al. Prenatal Maternal Distress Affects Atopic Dermatitis in Offspring Mediated by Oxidative Stress. *J. Allergy Clin. Immunol.* **2016,** *138,* 468.
96. Patelarou, E.; Giourgouli, G.; Lykeridou, A.; Vrioni, E.; Fotos, N.; Siamaga, E.; Vivilaki, V.; Brokalaki, H. Association between Biomarker-Quantified Antioxidant Status during Pregnancy and Infancy and Allergic Disease during Early Childhood: A Systematic Review. *Nutr. Rev.* **2011,** *69,* 627–641.
97. Babity, S.; Couture, F.; Campos, E. V. R.; Hedtrich, S.; Hagen, R.; Fehr, D.; Bonmarin, M.; Brambilla, D. A Naked Eye-Invisible Ratiometric Fluorescent Microneedle Tattoo for Real-Time Monitoring of Inflammatory Skin Conditions. *Adv. Healthcare Mater.* **2022,** *11,* 15.
98. Li, Q. Q.; Ge, X. G.; Ye, J. M.; Li, Z.; Su, L. C.; Wu, Y.; Yang, H. H.; Song, J. B. Dual Ratiometric SERS and Photoacoustic Core-Satellite Nanoprobe for Quantitatively Visualizing Hydrogen Peroxide in Inflammation and Cancer. *Angew. Chem. Int. Ed.* **2021,** *60,* 7323–7332.
99. Mei, R. C.; Wang, Y. Q.; Zhao, X. Z.; Shi, S.; Wang, X. Y.; Zhou, N.; Shen, D. Z.; Kang, Q.; Chen, L. X. Skin Interstitial Fluid-Based SERS Tags Labeled Microneedles for Tracking of Peritonitis Progression and Treatment Effect. *ACS Sens.* **2023,** *9.*
100. Schadendorf, D.; van Akkooi, A. C. J.; Berking, C.; Griewank, K. G.; Gutzmer, R.; Hauschild, A.; Stang, A.; Roesch, A.; Ugurel, S. Melanoma. *Lancet* **2018,** *392,* 971–984.
101. Boyle, J. L.; Haupt, H. M.; Stern, J. B.; Mulhaupt, H. A. B. Tyrosinase Expression in Malignant Melanoma, Desmoplastic Melanoma, and Peripheral Nerve Tumors - an Immunohistochemical Study. *Arch. Pathol. Lab Med.* **2002,** *126,* 816–822.
102. Zhou, J.; Shi, W.; Li, L. H.; Gong, Q. Y.; Wu, X. F.; Li, X. H.; Ma, H. M. Detection of Misdistribution of Tyrosinase from Melanosomes to Lysosomes and its Upregulation under Psoralen/ultraviolet a with a Melanosome-Targeting Tyrosinase Fluorescent Probe. *Anal. Chem.* **2016,** *88,* 4557–4564.
103. Huang, X. Q.; Chen, L. Z.; Sha, T. J.; Lin, Y. J.; Zeng, R. M.; Xu, J.; Chen, S. Z.; Cai, H. H.; Zhang, J. L.; Zhou, H. B.; et al. *In Situ* Tyrosinase Monitoring by Wearable Microneedle Patch toward Clinical Melanoma Screening. *ACS Nano* **2023,** *17,* 20073–20086.
104. Akram, M.; Iqbal, M.; Daniyal, M.; Khan, A. U. Awareness and Current Knowledge of Breast Cancer. *Biol. Res.* **2017,** *50,* 23.
105. Loibl, S.; Poortmans, P.; Morrow, M.; Denkert, C.; Curigliano, G. Breast Cancer. *Lancet* **2021,** *397,* 1750–1769.
106. Dervisevic, M.; Alba, M.; Adams, T. E.; Prieto-Simon, B.; Voelcker, N. H. Electrochemical Immunosensor for Breast Cancer Biomarker Detection Using High-Density Silicon Microneedle Array. *Biosens. Bioelectron.* **2021,** *192,* 9.
107. Allen, C.; Biryukov, S.; Giussani, G.; Gugnani, H.; Hay, S. I.; Millear, A.; Truelsen, T.; Cortinovis, M. GBD 2015 DALYs and HALE Collaborators Global, Regional, and National Disability-Adjusted Life-Years (DALYs) for 315 Diseases and Injuries and Healthy Life Expectancy (HALE), 1990-2015: a Systematic Analysis for the Global Burden of Disease Study 2015 (Vol 388, Pg 1603, 2016). *Lancet* **2017,** *389,* E1.
108. Bao, L. L.; Park, J.; Qin, B. Y.; Kim, B. Anti-SARS-CoV-2 IgM/IgG Antibodies Detection Using a Patch Sensor Containing Porous Microneedles and a Paper-Based Immunoassay. *Sci. Rep.* **2022,** *12,* 16.

Index

'*Note:* Page numbers followed by "f" indicate figures and "t" indicate tables.'

C

D

E

F

R

S

T

Printed and bound by CPI Group (UK) Ltd, Croydon, CR0 4YY
14/09/2026
02216409-0003